Synthesis Gas

Scrivener Publishing
100 Cummings Center, Suite 541J
Beverly, MA 01915-6106

Publishers at Scrivener
Martin Scrivener (martin@scrivenerpublishing.com)
Phillip Carmical (pcarmical@scrivenerpublishing.com)

Synthesis Gas

Production and Properties

James G. Speight

WILEY

This edition first published 2020 by John Wiley & Sons, Inc., 111 River Street, Hoboken, NJ 07030, USA and Scrivener Publishing LLC, 100 Cummings Center, Suite 541J, Beverly, MA 01915, USA
© 2020 Scrivener Publishing LLC
For more information about Scrivener publications please visit www.scrivenerpublishing.com.

All rights reserved. No part of this publication may be reproduced, stored in a retrieval system, or transmitted, in any form or by any means, electronic, mechanical, photocopying, recording, or otherwise, except as permitted by law. Advice on how to obtain permission to reuse material from this title is available at http://www.wiley.com/go/permissions.

Wiley Global Headquarters
111 River Street, Hoboken, NJ 07030, USA

For details of our global editorial offices, customer services, and more information about Wiley prod-ucts visit us at www.wiley.com.

Limit of Liability/Disclaimer of Warranty
While the publisher and authors have used their best efforts in preparing this work, they make no representations or warranties with respect to the accuracy or completeness of the contents of this work and specifically disclaim all warranties, including without limitation any implied warranties of merchant-ability or fitness for a particular purpose. No warranty may be created or extended by sales representatives, written sales materials, or promotional statements for this work. The fact that an organization, website, or product is referred to in this work as a citation and/or potential source of further information does not mean that the publisher and authors endorse the information or services the organization, website, or product may provide or recommendations it may make. This work is sold with the understanding that the publisher is not engaged in rendering professional services. The advice and strategies contained herein may not be suitable for your situation. You should consult with a specialist where appropriate. Neither the publisher nor authors shall be liable for any loss of profit or any other commercial damages, including but not limited to special, incidental, consequential, or other damages. Further, readers should be aware that websites listed in this work may have changed or disappeared between when this work was written and when it is read.

Library of Congress Cataloging-in-Publication Data

ISBN 978-1-119-70772-1

Cover image: Oil Refinery - Fotogigi85, Coal Image - Photographer | Dreamstime.com
Cover design by Kris Hackerott

Set in size of 11pt and Minion Pro by Manila Typesetting Company, Makati, Philippines

Contents

Preface	xiii
Part 1: Production	**1**

1 Energy Sources and Energy Supply — 3
- 1.1 Introduction — 3
- 1.2 Typical Energy Sources — 8
  - 1.2.1 Natural Gas and Natural Gas Hydrates — 9
  - 1.2.2 The Crude Oil Family — 10
  - 1.2.3 Extra Heavy Crude Oil and Tar Sand Bitumen — 12
- 1.3 Other Energy Sources — 15
  - 1.3.1 Coal — 16
  - 1.3.2 Oil Shale — 19
  - 1.3.3 Biomass — 21
  - 1.3.4 Solid Waste — 25
- 1.4 Energy Supply — 28
  - 1.4.1 Economic Factors — 28
  - 1.4.2 Geopolitical Factors — 29
  - 1.4.3 Physical Factors — 29
  - 1.4.4 Technological Factors — 30
- 1.5 Energy Independence — 31
- References — 36

2 Production of Synthesis Gas — 41
- 2.1 Introduction — 41
- 2.2 Synthesis Gas Generation — 44
- 2.3 Feedstocks — 46
  - 2.3.1 Natural Gas — 46
  - 2.3.2 Crude Oil Resid, Heavy Crude Oil, Extra Heavy Crude Oil, and Tar Sand Bitumen — 47
  - 2.3.3 Refinery Coke — 50
  - 2.3.4 Coal — 50
  - 2.3.5 Biomass — 52
  - 2.3.6 Solid Waste — 56
  - 2.3.7 Black Liquor — 59
  - 2.3.8 Mixed Feedstocks — 61

		2.3.8.1 Biomass and Coal	62
		2.3.8.2 Biomass and Municipal Solid Waste	62
2.4	Influence of Feedstock Quality		63
2.5	Gasification Processes		65
	2.5.1	Feedstock Pretreatment	66
	2.5.2	Feedstock Devolatilization	67
	2.5.3	Char Gasification	68
	2.5.4	General Chemistry	68
	2.5.5	Stage-by-Stage Chemistry	72
		2.5.5.1 Primary Gasification	72
		2.5.5.2 Secondary Gasification	74
		2.5.5.3 Water Gas Shift Reaction	76
		2.5.5.4 Carbon Dioxide Gasification	77
		2.5.5.5 Hydrogasification	78
		2.5.5.6 Methanation	79
		2.5.5.7 Catalytic Gasification	80
	2.5.6	Physical Effects	80
2.6	Products		82
	2.6.1	Gaseous Products	83
		2.6.1.1 Low Btu Gas	84
		2.6.1.2 Medium Btu Gas	85
		2.6.1.3 High Btu Gas	86
		2.6.1.4 Synthesis Gas	86
	2.6.2	Liquid Products	87
	2.6.3	Tar	88
	References		89
3	**Gasifier Types and Gasification Chemistry**		**95**
3.1	Introduction		95
3.2	Gasifier Types		96
	3.2.1	Fixed-Bed Gasifier	102
	3.2.2	Fluid-Bed Gasifier	105
	3.2.3	Entrained-Bed Gasifier	108
	3.2.4	Molten Salt Gasifier	109
	3.2.5	Plasma Gasifier	111
	3.2.6	Other Types	113
	3.2.7	Gasifier Selection	113
3.3	General Chemistry		115
	3.3.1	Devolatilization	118
	3.3.2	Products	118
3.4	Process Options		119
	3.4.1	Effects of Process Parameters	120
	3.4.2	Effect of Heat Release	121
	3.4.3	Other Effects	121
	References		122

4 Gasification of Coal — 125
- 4.1 Introduction — 125
- 4.2 Coal Types and Properties — 128
- 4.3 Gas Products — 130
  - 4.3.1 Coal Devolatilization — 131
  - 4.3.2 Char Gasification — 131
  - 4.3.3 Gasification Chemistry — 132
  - 4.3.4 Other Process Options — 133
    - 4.3.4.1 Hydrogasification — 133
    - 4.3.4.2 Catalytic Gasification — 134
    - 4.3.4.3 Plasma Gasification — 134
  - 4.3.5 Process Optimization — 135
- 4.4 Product Quality — 136
  - 4.4.1 Low Btu Gas — 136
  - 4.4.2 Medium Btu Gas — 138
  - 4.4.3 High Btu Gas — 138
  - 4.4.4 Methane — 139
  - 4.4.5 Hydrogen — 139
  - 4.4.6 Other Gases — 140
- 4.5 Chemicals Production — 140
  - 4.5.1 Coal Tar Chemicals — 140
  - 4.5.2 Fischer-Tropsch Chemicals — 143
    - 4.5.2.1 Fischer-Tropsch Catalysts — 143
    - 4.5.2.2 Product Distribution — 144
- 4.6 Advantages and Limitations — 145
- References — 145

5 Gasification of Heavy Feedstocks — 149
- 5.1 Introduction — 149
- 5.2 Heavy Feedstocks — 152
  - 5.2.1 Crude Oil Residua — 153
  - 5.2.2 Heavy Crude Oil — 155
  - 5.2.3 Extra Heavy Crude Oil — 155
  - 5.2.4 Tar Sand Bitumen — 155
  - 5.2.5 Other Feedstocks — 156
    - 5.2.5.1 Crude Oil Coke — 157
    - 5.2.5.2 Solvent Deasphalter Bottoms — 158
- 5.3 Synthesis Gas Production — 159
  - 5.3.1 Partial Oxidation Technology — 160
    - 5.3.1.1 Shell Gasification Process — 162
    - 5.3.1.2 Texaco Process — 162
    - 5.3.1.3 Phillips Process — 163
  - 5.3.2 Catalytic Partial Oxidation — 163
- 5.4 Products — 164
  - 5.4.1 Gas Purification and Quality — 165
  - 5.4.2 Process Optimization — 166

	5.5	Advantages and Limitations	166
		5.5.1 Other Uses of Residua	167
		5.5.2 Gasification in the Future Refinery	167
		References	169
6	**Gasification of Biomass**	**173**
	6.1	Introduction	173
	6.2	Gasification Chemistry	177
		6.2.1 General Aspects	178
		6.2.2 Reactions	181
		6.2.2.1 Water Gas Shift Reaction	184
		6.2.2.2 Carbon Dioxide Gasification	185
		6.2.2.3 Hydrogasification	186
		6.2.2.4 Methanation	186
	6.3	Gasification Processes	187
		6.3.1 Gasifiers	188
		6.3.2 Fischer-Tropsch Synthesis	192
		6.3.3 Feedstocks	193
		6.3.3.1 Biomass	193
		6.3.3.2 Gasification of Biomass with Coal	194
		6.3.3.3 Gasification of Biomass with Other Feedstocks	198
	6.4	Gas Production and Products	199
		6.4.1 Gas Production	199
		6.4.2 Gaseous Products	201
		6.4.2.1 Synthesis Gas	201
		6.4.2.2 Low-Btu Gas	203
		6.4.2.3 Medium-Btu Gas	203
		6.4.2.4 High-Btu Gas	204
		6.4.3 Liquid Products	205
		6.4.4 Solid Products	205
	6.5	The Future	206
		References	210
7	**Gasification of Waste**	**217**
	7.1	Introduction	217
	7.2	Waste Types	219
		7.2.1 Solid Waste	220
		7.2.2 Municipal Solid Waste	221
		7.2.3 Industrial Solid Waste	221
		7.2.4 Bio-Solids	222
		7.2.5 Biomedical Waste	223
		7.2.6 Sewage Sludge	223
	7.3	Feedstock Properties	224
	7.4	Fuel Production	224
		7.4.1 Preprocessing	225
		7.4.2 Process Design	227

	7.5	Process Products	228
		7.5.1 Synthesis Gas	228
		7.5.2 Carbon Dioxide	228
		7.5.3 Tar	229
		7.5.4 Particulate Matter	231
		7.5.5 Halogens/Acid Gases	231
		7.5.6 Heavy Metals	232
		7.5.7 Alkalis	233
		7.5.8 Slag	233
	7.6	Advantages and Limitations	234
		References	235
8	**Reforming Processes**		**239**
	8.1	Introduction	239
	8.2	Processes Requiring Hydrogen	242
		8.2.1 Hydrotreating	243
		8.2.2 Hydrocracking	244
	8.3	Feedstocks	245
	8.4	Process Chemistry	246
	8.5	Commercial Processes	248
		8.5.1 Autothermal Reforming	249
		8.5.2 Combined Reforming	249
		8.5.3 Dry Reforming	250
		8.5.4 Steam-Methane Reforming	251
		8.5.5 Steam-Naphtha Reforming	253
	8.6	Catalysts	254
		8.6.1 Reforming Catalysts	254
		8.6.2 Shift Conversion Catalysts	256
		8.6.3 Methanation Catalysts	256
	8.7	Hydrogen Purification	257
		8.7.1 Wet Scrubbing	257
		8.7.2 Pressure-Swing Adsorption Units	257
		8.7.3 Membrane Systems	258
		8.7.4 Cryogenic Separation	258
	8.8	Hydrogen Management	259
		References	260
9	**Gas Conditioning and Cleaning**		**263**
	9.1	Introduction	263
	9.2	Gas Streams	265
	9.3	Synthesis Gas Cleaning	270
		9.3.1 Composition	270
		9.3.2 Process Types	272
	9.4	Water Removal	274
		9.4.1 Absorption	275
		9.4.2 Adsorption	276
		9.4.3 Cryogenics	278

9.5	Acid Gas Removal	278
	9.5.1 Adsorption	279
	9.5.2 Absorption	280
	9.5.3 Chemisorption	281
	9.5.4 Other Processes	285
9.6	Removal of Condensable Hydrocarbons	289
	9.6.1 Extraction	291
	9.6.2 Absorption	292
	9.6.3 Fractionation	292
	9.6.4 Enrichment	293
9.7	Tar Removal	294
	9.7.1 Physical Methods	294
	9.7.2 Thermal Methods	296
9.8	Other Contaminant Removal	296
	9.8.1 Nitrogen Removal	296
	9.8.2 Ammonia Removal	298
	9.8.3 Particulate Matter Removal	298
	9.8.4 Siloxane Removal	298
	9.8.5 Alkali Metal Salt Removal	299
	9.8.6 Biological Methods	299
	9.8.6.1 Biofiltration	300
	9.8.6.2 Bioscrubbing	302
	9.8.6.3 Bio-Oxidation	303
9.9	Tail Gas Cleaning	303
	9.9.1 Claus Process	304
	9.9.2 SCOT Process	305
	References	306

Part 2: Fuels and Chemicals from Synthesis Gas — 311

10 The Fischer-Tropsch Process — 313

10.1	Introduction	313
10.2	History and Development of the Process	317
10.3	Synthesis Gas	320
10.4	Production of Synthesis Gas	323
	10.4.1 Feedstocks	323
	10.4.2 Product Distribution	326
10.5	Process Parameters	327
10.6	Reactors and Catalysts	330
	10.6.1 Reactors	330
	10.6.2 Catalysts	332
10.7	Products and Product Quality	336
	10.7.1 Products	336
	10.7.2 Product Quality	337

	10.8	Fischer-Tropsch Chemistry		339
		10.8.1 Chemical Principles		340
		10.8.2 Refining Fischer-Tropsch Products		344
		References		346

11 Synthesis Gas in the Refinery — 349

- 11.1 Introduction — 349
- 11.2 Processes and Feedstocks — 350
  - 11.2.1 Gasification of Residua — 353
  - 11.2.2 Gasification of Residua with Coal — 354
  - 11.2.3 Gasification of Residua with Biomass — 354
  - 11.2.4 Gasification of Residua with Waste — 356
- 11.3 Synthetic Fuel Production — 358
  - 11.3.1 Fischer-Tropsch Synthesis — 359
  - 11.3.2 Fischer-Tropsch Liquids — 360
  - 11.3.3 Upgrading Fischer-Tropsch Liquids — 362
    - 11.3.3.1 Gasoline Production — 363
    - 11.3.3.2 Diesel Production — 365
- 11.4 Sabatier-Senderens Process — 366
  - 11.4.1 Methanol Production — 367
  - 11.4.2 Dimethyl Ether Production — 368
- 11.5 The Future — 369
- References — 373

12 Hydrogen Production — 377

- 12.1 Introduction — 377
- 12.2 Processes — 381
  - 12.2.1 Feedstocks — 382
  - 12.2.2 Commercial Processes — 383
    - 12.2.2.1 Hydrocarbon Gasification — 384
    - 12.2.2.2 Hypro Process — 385
    - 12.2.2.3 Hydrogen from Pyrolysis Processes — 386
    - 12.2.2.4 Hydrogen from Refinery Gas — 387
    - 12.2.2.5 Other Options — 387
  - 12.2.3 Process Chemistry — 388
- 12.3 Hydrogen Purification — 390
  - 12.3.1 Wet Scrubbing — 391
  - 12.3.2 Pressure-Swing Adsorption — 391
  - 12.3.3 Membrane Systems — 392
  - 12.3.4 Cryogenic Separation — 393
- 12.4 Hydrogen Management — 394
- References — 395

13 Chemicals from Synthesis Gas — 399

- 13.1 Introduction — 399
- 13.2 Historical Aspects and Overview — 410
- 13.3 The Petrochemical Industry — 412

13.4	Petrochemicals		417
	13.4.1 Primary Petrochemicals		417
	13.4.2 Products and End Use		418
	13.4.3 Production of Petrochemicals		419
	13.4.4 Gaseous Fuels and Chemicals		425
		13.4.4.1 Ammonia	425
		13.4.4.2 Hydrogen	427
		13.4.4.3 Synthetic Natural Gas	427
	13.4.5 Liquid Fuels and Chemicals		428
		13.4.5.1 Fischer-Tropsch Liquids	428
		13.4.5.2 Methanol	428
		13.4.5.3 Dimethyl Ether	429
		13.4.5.4 Methanol-to-Gasoline and Olefins	429
		13.4.5.5 Other Processes	429
13.5	The Future		430
	References		437

14 Technology Integration — 439

14.1	Introduction	439
14.2	Applications and Products	440
	14.2.1 Chemicals and Fertilizers	441
	14.2.2 Substitute Natural Gas	441
	14.2.3 Hydrogen for Crude Oil Refining	442
	14.2.4 Transportation Fuels	443
	14.2.5 Transportation Fuels from Tar Sand Bitumen	445
	14.2.6 Power Generation	445
	14.2.7 Waste-to-Energy Gasification	446
	14.2.8 Biomass Gasification	447
14.3	Environmental Benefits	449
	14.3.1 Carbon Dioxide	450
	14.3.2 Air Emissions	450
	14.3.3 Solids Generation	450
	14.3.4 Water Use	450
14.4	A Process for Now and the Future	451
	14.4.1 The Process	451
	14.4.2 Refinery of the Future	453
	14.4.3 Economic Aspects	454
	14.4.4 Market Outlook	455
14.5	Conclusions	455
	References	457

Coversion Factors — 459

Glossary — 463

About the Author — 489

Index — 491

Preface

The projections for the continued use of fossil fuels indicate that there will be at least another five decades of fossil fuel use (especially coal and petroleum) before biomass and other forms of alternate energy take hold. Furthermore, estimations that the era of fossil fuels (petroleum, coal, and natural gas) will be almost over when the cumulative production of the fossil resources reaches 85% of their initial total reserves may or may not have some merit. In fact, the relative scarcity (compared to a few decades ago) of petroleum was real but it seems that the remaining reserves make it likely that there will be an adequate supply of energy for several decades. The environmental issues are very real and require serious and continuous attention.

In preparation for the depletion of fossil fuel resources, gasification can be proposed as a viable and reliable alternative solution for energy recovery from a variety of feedstocks. Gasification processes can accept a variety of feedstocks but the reactor must be selected on the basis of the feedstock properties and behavior in the process, especially when coal, biomass, and various wastes are considered as gasification feedstocks. The focus will be on the production of synthesis gas as an intermediate in the production of the necessary fuels and chemicals.

On the other hand, the gasification process still faces some technical and economic problems, mainly related to the highly heterogeneous nature of unconventional feedstocks such as biomass and municipal solid wastes and the relatively limited number of gasification plants worldwide based on this technology that have continuous operating experience under commercial conditions.

Synthesis gas (syngas) is a fuel gas mixture consisting predominantly of carbon monoxide and hydrogen and is typically a product of a gasification. The gasification process is applicable to many carbonaceous feedstocks including natural gas, petroleum resids, coal, biomass, by reaction of the feedstock with steam (steam reforming), carbon dioxide (dry reforming) or oxygen (partial oxidation). Synthesis gas is a crucial intermediate resource for production of hydrogen, ammonia, methanol, and synthetic hydrocarbon fuels.

It is the purpose of this book to present an overview of the issues related to the production and use of synthesis gas and to present to the reader the means by which the continually evolving synthesis gas technology will play a role in future production of fuels and chemicals.

Dr. James G. Speight
Laramie, Wyoming, USA

Part 1
PRODUCTION

1
Energy Sources and Energy Supply

1.1 Introduction

The major sources of energy have been, and continue to be, the various fossil fuels of which the major component of the group are (i) natural gas, (ii) crude oil, and (iii) coal with tar sand bitumen and oil shale available in considerable quantities for use on an as-needed basis. However, the Earth contains a finite supply of fossil fuels – although there are questions about the real amounts of these fossil fuels remaining. The best current estimates for the longevity of each fossil fuel is estimated from the reserves/production ratio (BP, 2019) which gives an indication (in years) of how long each fossil fuel will last at the current rates of production. Thus, estimates vary from at least 50 years of crude oil at current rates of consumption to 300 years of coal at current rates of consumption with natural gas varying between the two extremes. In addition, the amounts of natural gas and crude oil located in tight sandstone formations and in shale formations has added a recent but exciting twist to the amount of these fossil fuels remaining. Peak energy theory proponents are inclined to discount the tight formations and shale formation as a mere aberration (or a hiccup) in the depletion of these resources while opponents of the peak energy theory take the opposite view and consider tight formations and shale formations as prolonging the longevity of natural gas and crude oil by a substantial time period. In addition, some areas of the Earth are still relatively unexplored or have been poorly analyzed and (using crude oil as the example) knowledge of in-ground resources increases dramatically as an oil reservoir is exploited.

Energy sources have been used since the beginning of recorded history and the fossil fuel resources will continue to be recognized as major sources of energy for at least the foreseeable future (Crane *et al.*, 2010; World Energy Council, 2008; Gudmestad *et al.*, 2010; Speight, 2011a, 2011b, Khoshnaw, 2013; Speight, 2014a; BP, 2019). Fossil fuels are those fuels, namely natural gas, crude oil (including heavy crude oil), extra heavy crude oil, tar sand bitumen, coal, and oil shale produced by the decay of plant remains over geological time represent an unrealized potential, with liquid fuels from crude oil being only a fraction of those that could ultimately be produced from heavy oil and tar sand bitumen (Speight, 1990, 1997, 2011a; 2013d, 2013e, 2014).

In fact, at the present time, the majority of the energy consumed worldwide is produced from the fossil fuels (crude oil: approximately 38 to 40%, coal: approximately 31 to 35%, natural gas: approximately 20 to 25%) with the remainder of the energy requirements to come from nuclear and hydroelectric sources. As a result, fossil fuels (in varying amounts depending upon the source of information) are projected to be the major sources of energy for the next 50 years (Crane *et al.*, 2010; World Energy Council, 2008; Gudmestad *et al.*, 2010; Speight, 2011a, 2011b; Khoshnaw, 2013; BP, 2014; Speight, 2014; BP, 2019).

Fuels from fossil fuels (especially the crude oil-based fuels) are well-established products that have served industry and domestic consumers for more than 100 years and for the foreseeable future various fuels will still be largely based on hydrocarbon fuels derived from crude oil. Although the theory of peak oil is questionable, there is no doubt that crude oil, once considered inexhaustible, is being depleted at a measurable rate. The supposition by peak oil proponents is that supplies of crude oil are approaching a precipice in which fuels that are currently available may, within a foreseeable short time frame, be no longer available. While such a scenario is considered to be unlikely (Speight and Islam, 2016), the need to consider alternate technologies to produce liquid fuels that could mitigate the forthcoming effects of the shortage of transportation fuels is necessary and cannot be ignored.

The best current estimates for the longevity of each fossil fuel is estimated from the reserves/production ratio (BP, 2019) which gives an indication (in years) of how long each fossil fuel will last at the current rates of production.

Alternate fuels produced from a source other than crude oil are making some headway into the fuel demand. For example, diesel from plant sources (biodiesel) is similar in performance to diesel from crude oil and has the added advantage of a higher cetane rating than crude oil-derived diesel. However, the production of liquid fuels from sources other than crude oil has a checkered history. The on-again-off-again efforts that are the result of the inability of the political decision-makers to formulate meaningful policies has caused the production of non-conventional fuels to move slowly, if at all (Yergin, 1991; Bower, 2009; Wihbey, 2009; Speight, 2011a, 2011b, Yergin, 2011; Speight, 2014a).

This is due in no small part to the price fluctuations of crude oil and the common fuel products (i.e., gasoline and diesel fuel) and the lack of planning and associated foresight by various levels of government. It must be realized that for decades the price of crude oil produced in the crude oil-exporting nations has always been maintained at a level that was sufficiently low to discourage the establishment of a domestic synthetic fuels industry in many of the crude oil-consuming countries (Speight, 2011a). However, in spite of additional supplies of crude oil and natural gas coming from tight formations and shale formation, the time will come when the lack of preparedness for the production of non-conventional fuels may set many a national government on its heels. It is not a matter of "if the lack of preparedness come to fruition" but "when will the lack of preparedness come to fruition?"

In the near term, the ability of conventional fuel sources and technologies to support the global demand for energy will depend on how efficiently the energy sector can match available energy resources (Figure 1.1) with the end user and how efficiently and cost effectively the energy can be delivered. These factors are directly related to the continuing evolution of a truly global energy market. In the long term, a sustainable energy future cannot be created by treating energy as an independent topic (Zatzman, 2012). Rather, the role of the energy and the interrelationship of the energy market with other markets and the various aspects of market infrastructure demand further attention and consideration. Greater energy efficiency will depend on the developing the ability of the world market to integrate energy resources within a common structure (Gudmestad et al., 2010; Speight, 2011b; Khoshnaw, 2013).

World petro-politics are now in place (Bentley, 2002; Speight, 2011a) for the establishment of a synthetic fuels (including a biofuels) industry and, without being unduly dismissive of such efforts, it is up to various levels of government not only to promote the establishment of such an industry but to lead the way recognizing that it is not only a

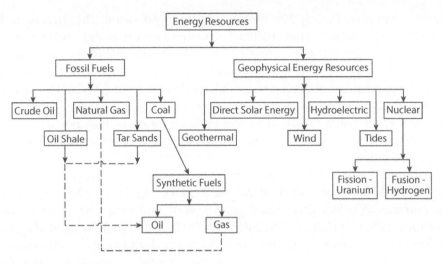

Figure 1.1 Types of energy resources.

matter of supply and demand but of the available and variable technology. Unfortunately, although there may be sufficient crude oil remaining to maintain the Crude Oil Age (or the Petroleum Age, that is, the age in which the developed countries of the world operate) for another 50 years (Speight, 2011a, 2011b), the time to prepare is *now*. The world is not yet on the precipice of energy deficiency (as many alarmists claim) but it is necessary that the politicians in the various levels of (national) governments of oil-consuming nations look beyond the next election with an eye to the future. It should also be the focus of the proponents of biofuels production and use to ensure that sufficient feedstocks are available to successfully operate a biofuels refinery thereby contributing alternate fuels to the gradual (but not drastic) reduction of crude oil-based fuels (Speight, 2008; Giampietro and Mayumi, 2009; Speight, 2011a, 2011b). However, it is time for procrastination to cease, since delay will not help in getting beyond the depletion of crude oil and natural gas resources, and various levels of government must start being serious in terms of looking to the future for other sources of energy to supplement and even replace the current source of hydrocarbon fuels.

In addition, and in keeping with the preferential use of lighter crude oil as well as maturation effect in the reservoir, crude oil available currently to the refinery is somewhat different in composition and properties from those available approximately 50 years ago (Parkash, 2003; Gary *et al.*, 2007; Speight, 2008; Siefried and Witzel, 2010; Speight, 2011a, 2014a, 2015b; Hsu and Robinson, 2017; Speight, 2017). The current crude oils are somewhat heavier insofar as they have higher proportions of non-volatile (asphaltic) constituents. In fact, by the standards of yesteryear, many of the crude oils currently in use would have been classified as heavy feedstocks, bearing in mind that they may not approach the definitions that should be used based on the method of recovery. Changes in feedstock character, such as this tendency to more viscous (heavier) crude oils, require adjustments to refinery operations to handle these heavier crude oils to reduce the amount of coke formed during processing and to balance the overall product slate (Speight, 2011a, 2014a).

As the 21st century matures, there will continue to be an increased demand for energy to support the needs of commerce industry and residential uses – in fact, as the 2040 to 2049 decade approaches, commercial and residential energy demand is expected to rise considerably – by approximately 30% over current energy demand. This increase is due, in part, to developing countries, where national economies are expanding and the move away from rural living to city living is increasing. In addition, the fuel of the rural population (biomass) is giving way to the fuel of the cities (transportation fuels, electric power) as the lifestyles of the populations of developing countries changes from agrarian to metropolitan. Furthermore, the increased population of the cities requires more effective public transportation systems as the rising middle class seeks private means of transportation (automobiles). As a result, fossil fuels will continue to be the predominant source of energy for at least the next 50 years.

However, there are several variables that can impact energy demand from fossil fuels. For example, coal (as a source of electrical energy) faces significant challenges from governmental policies to reduce greenhouse gas emissions, and fuels from crude oil can also face similar legislation (Speight, 2013a, 2013b, 2014a) in addition to the types of application and use, location and regional resources, cost of energy, cleanness and environmental factors, safety of generation and utilization, and socioeconomic factors, as well as global and regional politics (Speight, 2011a). More particularly, the recovery of natural gas and crude oil from tight sandstone and shale formations face challenges related to hydraulic fracturing. Briefly, hydraulic fracturing is an extractive method used by crude oil and natural gas companies to open pathways in tight (low-permeability) geologic formations so that the oil or gas trapped within can be recovered at a higher flow rate. When used in combination with horizontal drilling, hydraulic fracturing has allowed industry to access natural gas reserves previously considered uneconomical, particularly in shale formations. Although, hydraulic fracturing creates access to more natural gas supplies, but the process requires the use of large quantities of water and fracturing fluids, which are injected underground at high volumes and pressure. Oil and gas service companies design fracturing fluids to create fractures and transport sand or other granular substances to prop open the fractures. The composition of these fluids varies by formation, ranging from a simple mixture of water and sand to more complex mixtures with a multitude of chemical additives. Hydraulic fracturing has opened access to vast domestic reserves of natural gas that could provide an important stepping stone to a clean energy future. Yet questions related to the safety of hydraulic fracturing persist and the technology has been the subject of both enthusiasm and increasing environmental and health concerns in recent years, especially in relation to the possibility (some would say *reality*) of contaminated drinking water because of the chemicals used in the process and the disturbance of the geological formations (Speight, 2015a).

The danger revealed by the peak energy theory is that the world is approaching an energy precipice in which (apparently) crude oil that is available one year will not be available the next year. On the other hand, the peak energy opponents take a more realistic view in that the depletion of fossil fuels will occur gradually and, and with the current trends in considering other sources of energy, the concept of the energy precipice is not logical (Speight and Islam, 2016).

The most unrealistic variable in the peak energy scenario arises from the misuse of data that supposedly indicate that the world is approaching the energy precipice in which fossil fuel will no longer be available for use as energy sources – the date of the energy precipice is not only wildly speculative but, in many cases, totally unrealistic. Fossil fuel energy sources will undoubtedly reach a depletion point in the future when these energy sources are no longer available – but not at the moment or even in the present century. At the same time, new gas-fired generating units use highly efficient technologies and are supported by abundant gas supplies. As a result, gas is increasingly viewed as the most economical fossil fuel choice for electricity generation for the United States. Finally, a word on reserve estimation. There are a number of different methods by which crude oil and natural gas reserves can be calculated. These methods can be grouped into three general categories: (i) volumetric methods, (ii) materials balance method, and (iii) the decline curve method or production performance method.

The methods designated as *volumetric methods* represent attempts to determine the amount of oil-in-place by using the size of the reservoir as well as the physical properties of the reservoir rock(s) and the reservoir fluids. In the calculation process, a recovery factor is assumed, using data (and assumptions) from other crude oil and natural gas fields with similar characteristics to the field under evaluation. Based on these assumptions, the estimated amount of crude oil or natural gas in-place is multiplied by the recovery factor (derived from the other (similar) fields to arrive at an estimate of the reserves in-place. Current recovery factors for oil fields around the world typically range between 10 and 60% v/v of the crude oil and natural gas in-place while some recovery factors are in excess of 80% v/v of the crude oil and natural gas in place. The wide variance is due largely to the diversity of fluid and reservoir characteristics for different deposits. The method is most useful early in the life of the reservoir, before significant production has occurred. However, site specificity, which arise because of the differences in reservoir character (for example reservoir mineralogy, porosity, permeability) and the character of the reservoir fluids must also be given serious consideration, otherwise the estimation of the reserves in-place may be in error (by an order of magnitude above or below the real amount in-place. Such is the difficulty of estimating the reserves.

In addition, the materials *balance method* for a crude oil field or natural gas field uses an equation (or derivation thereof) that relates (in the case of a crude oil reservoir with associated natural gas) the volume of crude oil, water and gas that has been produced from a reservoir, and the change in reservoir pressure, to calculate the remaining crude oil or natural gas. The calculation uses the assumption that as fluids from the reservoir are produced, there will be a change in the reservoir pressure that depends on the remaining volume of oil and gas. The method requires extensive pressure-volume-temperature analysis as well as an accurate pressure history of the field. If the pressure history of the field is not available, the calculation requires some production to occur (or to have occurred) (typically 5% to 10% v/v of ultimate recovery), unless reliable pressure history can be used from a field with similar reservoir rock characteristics as well as the characteristics of the reservoir fluids.

The *decline curve method* (also known as the *production performance method*) uses known production data to fit a decline curve and estimate future oil production – the three most common forms of decline curves are *exponential*, *hyperbolic*, and *harmonic*. The decline curve analysis is a long-established tool for developing future outlook for crude oil and/or natural gas production from an individual well or from an entire oilfield. Depletion has a fundamental role in the extraction of finite resources and is one of the driving mechanisms for oil flows within a reservoir and the depletion rate can be connected to decline curves. Consequently, depletion analysis is a useful tool for analysis and forecasting crude oil and natural gas production.

In the calculation, it is assumed that the production will decline on a reasonably smooth curve, and so allowances must be made for wells shut in and production restrictions.

The curve can be expressed mathematically or plotted on a graph to estimate future production. It has the advantage of (implicitly) including all reservoir characteristics. However, the method requires a sufficient well or reservoir history to establish a statistically significant trend, ideally when production is not curtailed by regulatory or other artificial conditions.

Generally, the initial estimates of the size of newly discovered oil fields are usually too low. As crude oil and/or natural gas production continues, successive estimates of the ultimate recovery of fields tend to increase. The term *reserve growth* refers to the typical increases in estimated ultimate recovery that occur as oil fields are developed and produced. Reserve growth has now become an important part of estimating total potential reserves of an individual province or country. As the worldwide crude oil reserves continue to decline, there is the need for the reserve estimates to be more precise through application of the reserve-growth concept. In fact, the concept could be applied even to the so-called undiscovered resources with some qualifications as to the inherent risk (Speight, 2011a, 2014a).

1.2 Typical Energy Sources

The widespread use of fossil fuels has been one of the most important stimuli of economic growth and has allowed the consumption of energy at a greater rate than it is being replaced and presents an unprecedented risk management problem (Yergin, 1991; Hirsch, 2005; Hirsch *et al.*, 2005; Yergin, 2011). A peak in the production of crude oil will happen, but whether it will occur slowly or abruptly is not certain – given appropriate warnings, the latter is likely to be the case. The adoption of alternate technologies to supplant the deficit in oil production will require a substantial time period on the order of at least 10 to 20 years.

Global energy consumption is increasing and is expected to rise by 41% over the period to 2035 – compared to a 52% rise over the last 20 years and 30% rise over the last decade. Of the growth in demand, 95% is expected to come from the emerging economies, while energy use in the advanced economies of North America, Europe and Asia as a group is expected to grow only very slowly – and begin to decline in the later years of the forecast period (BP, 2019). The data for reserve estimates indicate that there are sufficient reserves to cover this trend at least to and even beyond 2035. Crude oil and its associate remain the leading fuel and source of chemicals (Speight, 2014a, 2019a).

For many decades, coal has been the primary feedstock for gasification units but due to recent concerns about the use of fossil fuels and the resulting environmental pollutants, irrespective of the various gas cleaning processes and gasification plant environmental cleanup efforts, there is a move to feedstocks other than coal for gasification processes (Speight, 2013a, 2013b, 2014b). But more pertinent to the present text, the gasification process can also use carbonaceous feedstocks which would otherwise have been discarded and unused, such as waste biomass and other similar biodegradable wastes. Various feedstocks such as biomass, crude oil resids, and other carbonaceous wastes can be used to their fullest potential. In fact, the refining industry has seen fit to use crude oil resid gasification as a source of hydrogen for the past several decades (Speight, 2014a).

Gasification processes can accept a variety of feedstocks but the reactor must be selected on the basis of feedstock properties and behavior in the process. The advantage of the gasification process when a carbonaceous feedstock (a feedstock containing carbon) or hydrocarbonaceous feedstock (a feedstock containing carbon and hydrogen) is employed is that the product of focus – synthesis gas – is potentially more useful as an energy source and results in an overall cleaner process. The production of synthesis gas is a more efficient production of an energy source than, say, the direct combustion of the original feedstock because synthesis gas can be (i) combusted at higher temperatures, (ii) used in fuel cells, (iii) used to produce methanol, (iv) used as a source of hydrogen, and (v) particularly because the synthesis gas can be converted via the Fischer-Tropsch process into a range of synthesis liquid fuels suitable for use gasoline engines, for diesel engines, or for wax production.

Therefore, a brief comment about each of the potential energy sources is presented below.

1.2.1 Natural Gas and Natural Gas Hydrates

It is rare that crude oil and also heavy crude oil do not occur without an accompanying cover of gas (Speight, 2014a, 2019b). It is therefore important, when describing reserves of crude oil, to also acknowledge the occurrence, properties, and character of the natural gas. In recent years, natural gas has gained popularity among a variety of industrial sectors. Natural gas burns cleaner than coal or crude oil, thus providing environmental benefits. Natural gas is distributed mainly via pipeline, and some in a liquid phase (LNG) transported across oceans by tanker.

Assuming that the current level of natural gas consumption for the world is maintained, the reserve would be enough to last for another 64 years. However, in this estimation of natural gas longevity, factors such as the increase in annual consumption, the discovery of new reservoirs, and advances in discovery/recovery technology, and utilization of natural gas hydrates are not included. As a result of discoveries of gas in tight shale formations – which has offset more than the annual consumption – the world reserves of natural gas have been in a generally upward trend, due to discoveries of major natural gas fields.

Natural gas liquids (NGLs) – which are the higher-boiling constituents of natural gas separated from natural gas at a gas processing plant, and include ethane, propane, butane, and pentanes – have taken on a new prominence as shale gas production has increased and prices have fallen (Ratner and Tiemann, 2014). As a result, most producers are accepting the challenges with the opportunism and have shifted production to tight formations, such as the Bakken formation in North Dakota and Montana, to capitalize on the occurrence of natural gas liquids in shale gas development (Speight, 2013f; Sandrea, 2014; Speight, 2015a).

Methane hydrates (also often referred to as methane clathrates) is a resource in which a large amount of methane is trapped within a crystal structure of water, forming a solid similar to ice (Chapter 1) (Collett, 2009). Methane hydrates exist as methane (the chief constituent of natural gas) trapped in a cage-like lattice of ice which, if either warmed or depressurized (with suitable caution), revert back to water and natural gas. When brought to the surface of the Earth, one cubic meter of gas hydrate releases 164 cubic meters of natural gas.

Gas hydrates occur in two discrete geological situations: (i) marine shelf sediments and (ii) on-shore Polar Regions beneath permafrost (Kvenvolden 1993; Kvenvolden and Lorenson, 2000). These two Hydrates occur in these two types of settings because these are the settings where the pressure-temperature conditions are within the hydrate stability field (Lerche and Bagirov, 1998). Gas hydrates can be detected seismically as well as by well logs (Goldberg and Saito, 1998; Hornbach *et al.*, 2003).

When drilling in crude oil-bearing and gas-bearing formations submerged in deep water, the reservoir gas may flow into the well bore and form gas hydrates owing to the low temperatures and high pressures found during deep water drilling. The gas hydrates may then flow upward with drilling mud or other discharged fluids. When the hydrates rise, the pressure in the annulus decreases and the hydrates dissociate into gas and water. The rapid gas expansion ejects fluid from the well, reducing the pressure further, which leads to more hydrate dissociation and further fluid ejection.

1.2.2 The Crude Oil Family

Crude oil and the equivalent term *petroleum*, cover a wide assortment of materials consisting of mixtures of hydrocarbon derivatives and other compounds containing variable amounts of sulfur, nitrogen, and oxygen, which may vary widely in volatility, specific gravity, and viscosity. Metal-containing constituents, notably those compounds that contain vanadium and nickel, usually occur in the more viscous crude oils in amounts up to several thousand parts per million and can have serious consequences during processing of these feedstocks. Because crude oil is a mixture of widely varying constituents and proportions, its physical properties also vary widely and the color from colorless to black. The crude oil family consists of various types of crude oil: (i) conventional crude oil, (ii) crude oil from tight formations, (iii) opportunity crude oils, (iv) high acid crude oil, (v) foamy oil, (vi) eavy crude oil.

The total amount of crude oil is indeed finite, and, therefore, production will one day reach a peak and then begin to decline. This is common sense, as explained in the resource depletion theory which, in this case, assumes that reserves of crude oil will not be replenished (i.e., that abiogenic replenishment is negligible) and future world oil production must inevitably reach a peak and then decline as these reserves are exhausted (Hubbert, 1956, 1962). There is no doubt that crude oil and natural gas are being consumed at a steady rate but whether or not the Hubbert peak oil theory will affect the consumption of crude oil is another issue. It is a theory that is based on reserve estimates and reserve consumption. No one will disagree that hydrocarbon resources (in the form of crude oil and natural gas) are finite resources and will run out at some future point in time but the proponents of an energy precipice must recognize that this will not be the case, at least not for now (Speight and Islam, 2016). The issue is the timing of this event – whether it is tomorrow, next week, next month, next year, or in 50 or more years remains to be seen. Current evidence (Speight, 2011a, 2011c; BP, 2019) favors a lifetime of more than 50 years for the current reserves of crude oil and natural gas, perhaps longer if hydraulic fracturing continues to play a dominant role in crude oil and natural gas production (Speight, 2015a). Thus, controversy surrounds the theory – not so much from the theory itself which is quite realistic but from the way in which the theory is used by varying collections of alarmists – since predictions for the time of the global peak is dependent on the past production and discovery data used in the calculation.

To date, crude oil production on a worldwide basis has come almost exclusively from what are considered to be conventional crude oil reservoirs from which crude oil can be produced using tried-and-true recovery technologies compared with non-conventional sources that require more complex or more expensive technologies to extract – examples of such resources are tar sand bitumen, liquids from coal, liquids from biomass, and liquids from oil shale (Lee, 1990; Scouten, 1990; Lee, 1991; Speight, 2008, 2011b, 2012, 2013a, 2014b, 2016).

Generally, *crude oil from tight formations* (sometimes referred to as unconventional tight oil resources) are found at considerable depths in sedimentary rock formations that are characterized by very low permeability. While some of the tight oil plays produce oil directly from shales, tight oil resources are also produced from low-permeability siltstone formations, sandstone formations, and carbonate formations that occur in close association with a shale source rock.

Oil from tight shale formation is characterized by a low content of high-boiling (resid) constituents, low-sulfur content, and a significant molecular weight distribution of the paraffinic wax content (Speight, 2014a, 2015b). Finally, the properties of crude oils from tight formations are highly variable. Density and other properties can show wide variation, even within the same field. The Bakken crude is light and sweet with an API of 42° and a sulfur content of 0.19% w/w. Similarly, Eagle Ford is a light sweet feed, with a sulfur content of approximately 0.1% w/w and with published API gravity between 40° API and 62° API.

There is also the need for a refinery to be configured to accommodate *opportunity crude oils* and/or *high acid crude oils* which, for many purposes are often included with heavy feedstocks (Speight, 2014a, 2014b; Yeung, 2014). *Opportunity crude oils* are either new crude oils with unknown or poorly understood properties relating to processing issues or are existing crude oils with well-known properties and processing concerns (Ohmes, 2014). Opportunity crude oils are often, but not always, heavy crude oils but in either case are more difficult to process due to high levels of solids (and other contaminants) produced with the oil, high levels of acidity, and high viscosity. These crude oils may also be incompatible with other oils in the refinery feedstock blend and cause excessive equipment fouling when processed either in a blend or separately (Speight, 2015b). There is also the need for a refinery to be configured to accommodate *opportunity crude oils* and/or *high acid crude oils* which, for many purposes are often included with heavy feedstocks.

Opportunity crude oils, while offering initial pricing advantages, may have composition problems which can cause severe problems at the refinery, harming infrastructure, yield, and profitability. Before refining, there is the need for comprehensive evaluations of opportunity crudes, giving the potential buyer and seller the needed data to make informed decisions regarding fair pricing and the suitability of a particular opportunity crude oil for a refinery. This will assist the refiner to manage the ever-changing crude oil quality input to a refinery – including quality and quantity requirements and situations, crude oil variations, contractual specifications, and risks associated with such opportunity crudes.

High-acid crude oils are crude oils that contain considerable proportions of naphthenic acids which, as commonly used in the crude oil industry, refers collectively to all of the organic acids present in the crude oil (Shalaby, 2005; Speight, 2014b). In many instances, the high-acid crude oils are actually the heavier crude oils (Speight, 2014a, 2014b). The total acid matrix is therefore complex and it is unlikely that a simple titration, such as the traditional methods for measurement of the total acid number, can give meaningful results to use in predictions of problems. An alternative way of defining the relative organic acid fraction of crude oils is therefore a real need in the oil industry, both upstream and downstream.

High acid crude oils cause corrosion in the refinery – corrosion is predominant at temperatures in excess of 180°C (355°F) (Kane and Cayard, 2002; Ghoshal and Sainik, 2013; Speight, 2014c) – and occurs particularly in the atmospheric distillation unit (the first point of entry of the high-acid crude oil) and also in the vacuum distillation units. In addition, overhead corrosion is caused by the mineral salts, magnesium, calcium and sodium chloride which are hydrolyzed to produce volatile hydrochloric acid, causing a highly corrosive condition in the overhead exchangers. Therefore, these salts present a significant contamination in opportunity crude oils. Other contaminants in opportunity crude oils which are shown to accelerate the hydrolysis reactions are inorganic clay minerals and organic acids.

Foamy oil is oil-continuous foam that contains dispersed gas bubbles produced at the well head from heavy crude oil reservoirs under solution gas drive (Maini, 1999; Sheng *et al.*, 1999; Maini, 2001). The nature of the gas dispersions in oil distinguishes foamy oil behavior from conventional heavy crude oil. The gas that comes out of solution in the reservoir does not coalesce into large gas bubbles nor into a continuous flowing gas phase. Instead it remains as small bubbles entrained in the crude oil, keeping the effective oil viscosity low while providing expansive energy that helps drive the oil toward the producing. Foamy oil accounts for unusually high production in heavy crude oil reservoirs under solution-gas drive.

During primary production of heavy crude oil from solution gas drive reservoirs, the oil is pushed into the production wells by energy supplied by the dissolved gas. As fluid is withdrawn from the production wells, the pressure in the reservoir declines and the gas that was dissolved in the oil at high pressure starts to come out of solution (*foamy oil*). As pressure declines further with continued removal of fluids from the production wells, more gas is released from solution and the gas already released expands in volume. The expanding gas, which at this point is in the form of isolated bubbles, pushes the oil out of the pores and provides energy for the flow of oil into the production well. This process is very efficient until the isolated gas bubbles link up and the gas itself starts flowing into the production well. Once the gas flow starts, the oil has to compete with the gas for available flow energy. Thus, in some heavy crude oil reservoirs, due to the properties of the oil and the sand and also due to the production methods, the released gas forms foam with the oil and remains subdivided in the form of dispersed bubbles much longer.

Heavy crude oil is a *type* of crude oil that is different from conventional crude oil insofar as it is much more difficult to recover from the subsurface reservoir. Heavy crude oil, particularly heavy crude oil formed by biodegradation of organic deposits, is found in shallow reservoirs, formed by unconsolidated sands. This characteristic, which causes difficulties during well drilling and completion operations, may become a production advantage due to higher permeability. In simple terms, heavy crude oil is a type of crude oil which is very viscous and does not flow easily. The common characteristic properties (relative to conventional crude oil) are high specific gravity, low hydrogen to carbon ratios, high carbon residues, and high contents of asphaltenes, heavy metal, sulfur and nitrogen. Specialized refining processes are required to produce more useful fractions, such as: naphtha, kerosene, and gas oil.

1.2.3 Extra Heavy Crude Oil and Tar Sand Bitumen

In addition to conventional crude oil and heavy crude oil, there remains an even more viscous material that offers some relief to the potential shortfalls in supply (Meyer and De Witt, 1990; Meyer and Attanasi, 2003; Speight, 2014; BP, 2019). This is the *bitumen* known as extra heavy crude oil and the bitumen found in *tar sand* (*oil sand*) deposits (Table 1.1).

Table 1.1 Simplified differentiation between conventional crude oil, tight oil, heavy crude oil, extra heavy crude oil, and tar sand bitumen*.

Conventional Crude Oil
Mobile in the reservoir; API gravity: >25°
High-permeability reservoir
Primary recovery
Secondary recovery
Tight Oil
Similar properties to the properties of conventional crude oil; API gravity: >25°
Immobile in the reservoir
Low-permeability reservoir
Horizontal drilling into reservoir
Fracturing (typically multi-fracturing) to release fluids/gases
Heavy Crude Oil
More viscous than conventional crude oil; API gravity: 10-20°
Mobile in the reservoir
High-permeability reservoir
Secondary recovery
Tertiary recovery (enhanced oil recovery – EOR; e.g. steam stimulation)
Extra Heavy Crude Oil
Similar properties to the properties of tar sand bitumen; API gravity: <10°
Mobile in the reservoir
High-permeability reservoir
Secondary recovery
Tertiary recovery (enhanced oil recovery – EOR; e.g. steam stimulation)
Tar Sand Bitumen
Immobile in the deposit; API gravity: <10°
High-permeability reservoir
Mining (often preceded by explosive fracturing)
Steam assisted gravity draining (SAGD)

(*Continued*)

Table 1.1 Simplified differentiation between conventional crude oil, tight oil, heavy crude oil, extra heavy crude oil, and tar sand bitumen*. (*Continued*)

Solvent methods (VAPEX)
Extreme heating methods
Innovative methods**

*This list is not intended for use as a means of classification.
**Innovative methods excludes tertiary recovery methods and methods such as steam assisted gravity drainage (SAGD) and vapor assisted extraction (VAPEX) methods but does include variants or hybrids thereof.

However, many of these reserves are only available with some difficulty and optional refinery scenarios will be necessary for conversion of these materials to liquid products (Speight, 2000, 2014a) because of the substantial differences in character between conventional crude oil and tar sand bitumen (Speight, 2014a).

Extra heavy crude oil is a nondescript term (related to viscosity) of little scientific meaning that is usually applied to tar sand bitumen, which is generally capable of free flow under reservoir conditions (Table 1.1). Thus, general difference is that extra heavy crude oil, which may have properties similar to tar sand bitumen in the laboratory but, unlike tar sand bitumen in the deposit, has some degree of mobility in the reservoir or deposit (Table 1.1) (Delbianco and Montanari, 2009; Speight, 2014a). Extra heavy crude oils can flow at reservoir temperature and can be produced economically, without additional viscosity-reduction techniques, through variants of conventional processes such as long horizontal wells, or multilaterals. This is the case, for instance, in the Orinoco basin (Venezuela) or in offshore reservoirs of the coast of Brazil but, once outside of the influence of the high reservoir temperature, these oils are too viscous at surface to be transported through conventional pipelines and require heated pipelines for transportation. Alternatively, the oil must be partially upgraded or fully upgraded or diluted with a light hydrocarbon (such as aromatic naphtha) to create a mix that is suitable for transportation (Speight, 2014a, 2017).

Tar sand (referred to as *oil sand* in Canada) deposits are found in various countries throughout the world, but in vast quantities in Alberta and Venezuela. There have been many attempts to define tar sand deposits and the bitumen contained therein. In order to define conventional crude oil, heavy crude oil, and bitumen, the use of a single physical parameter such as viscosity is not sufficient. Other properties such as API gravity, elemental analysis, composition, and, most of all, the properties of the bulk deposit must also be included in any definition of these materials. Only then will it be possible to classify crude oil and its derivatives.

In fact, the most appropriate and workable definition of *tar sand* is found in the writings of the United States government (US Congress, 1976), which is not subject to any variation in chemical or physical properties that can vary depending upon the method of property determination and the accuracy of that method (Speight, 2014), viz.:

> *Tar sands are the several rock types that contain an extremely viscous hydrocarbon which is not recoverable in its natural state by conventional oil well production methods including currently used enhanced recovery techniques. The hydrocarbon-bearing rocks are variously known as bitumen-rocks oil, impregnated rocks, oil sands, and rock asphalt.*

This definition speaks to the character of the bitumen through the method of recovery (Speight, 2014, 2016). Thus, the bitumen found in tar sand deposits is an extremely viscous material that is *immobile under reservoir conditions* and cannot be recovered through a well by the application of secondary or enhanced recovery techniques.

By inference and by omission, conventional crude oil and heavy crude oil are also included in this definition. Extra heavy oil can also be accommodated under this definition because the oil approximates the properties and behavior of tar sand bitumen at ambient conditions but is mobile because the reservoir temperature is higher than the pour point of the oil (Table 1.1). Crude oil is the material that can be recovered by conventional oil well production methods whereas heavy crude oil is the material that can be recovered by enhanced recovery methods. Tar sand currently recovered by a mining process followed by separation of the bitumen by the hot water process. The bitumen is then used to produce hydrocarbon derivatives by a conversion process.

1.3 Other Energy Sources

All fossil fuels are non-renewable, and as such they will get eventually depleted. Since they are based on finite resources and their distributions are heavily localized in certain areas of the world, they will become expensive. Further, energy generation from fossil fuels requires combustion, thus damaging the environment with pollutants and greenhouse gas emission (Speight and Lee, 2000). In order to sustain the future of the world with clean environment and non-depletive energy, renewable energy is a right choice. Renewable energy sources include solar energy, wind energy, geothermal energy, biomass, and hydrogen. Most renewable energy except geothermal energy comes directly or indirectly from sun. Benefits of renewable energy are numerous and they include: (i) environmental cleanness without pollutant emission, (ii) non-depletive nature, (iii) availability throughout the world, (iv) no cause for global warming, (v) waste reduction, (vi) stabilization of energy cost, and (vii) creation of jobs.

Alternate fuels produced from sources other than crude oil are making some headway into the fuel demand. For example, diesel from plant sources (biodiesel) is similar in performance to diesel from crude oil and has the added advantage of a higher cetane rating than crude oil-derived diesel. However, the production of liquid fuels from sources other than crude oil has a checkered history. The on-again-off-again efforts that are the result of the inability of the political decision-makers to formulate meaningful policies has caused the production of non-conventional fuels to move slowly, if at all (Yergin, 1991; Bower, 2009; Wihbey, 2009; Speight, 2011a, 2011b, Yergin, 2011; Speight, 2014a).

Non-fossil fuels are alternative sources of energy that do not rely on continued consumption of the limited supplies of crude oil, coal, and natural gas. Examples of the non-fossil fuel energy sources include: (i) biomass, wind, solar, geothermal, tidal, nuclear, and hydrogen sources (Nersesian, 2007; Speight, 2008, 2011c). Such resources are considered to be extremely important to the future of energy generation because they are renewable energy sources that could be exploited continuously and not suffer depletion. In addition, energy production using non-fossil-based sources is claimed to generate much less pollution than the fossil fuel energy sources. This is considered crucial by many governments who are looking for ways to reduce the amount of pollution produced by their countries.

The advantages of fossil fuel resources are often considered to include the know-how and ease of production, many opponents of fossil fuel use cite the adverse effects on the environment (Speight, 2008, 2013a, 2013b, 2014a) and consider non-fossil fuels as a much better way to generate energy.

While there are now methods of burning gas and similar products very efficiently, as clean fossil fuels, a certain amount of pollution is still generated. Accordingly, various initiatives now exist, especially in Western countries, to encourage corporations and energy companies to invest in methods of producing energy from renewable (non-fossil fuel).

1.3.1 Coal

Coal of various types (Table 1.2) is an organic sedimentary rock that is formed from the accumulation and preservation of plant materials, usually in a swamp environment (Speight, 2013a, 2013b). Coal is a combustible rock and along with crude oil and natural gas it is one of the three most important fossil fuels, such as for the generation of electricity and provides approximately 40% of electricity production on a worldwide basis. In many countries these data are much higher: Poland relies on coal for approximately 94% of its electricity; South Africa relies on coal for approximately 92% of its electricity; China relies on coal for approximately 77% of its electricity; and Australia relies on coal for approximately 76% of its electricity.

Table 1.2 Types of coal.

Coal type	Description
Lignite	Also referred to as brown coal.
	The lowest rank of coal.
	Used almost exclusively as fuel for steam-electric power generation.
	Jet is a compact form of lignite that is sometimes polished and has been used as an ornamental stone since the Iron Age.
Sub-bituminous coal	The properties range from those of lignite to those of bituminous coal.
	Primarily as fuel for steam-electric power generation.
Bituminous coal	A dense coal, usually black, sometimes dark brown, often with well-defined bands of brittle and dull material,
	Used primarily as fuel in steam-electric power generation.
	Also used for heat and power applications in manufacturing and to produce coke.
Anthracite	The highest rank coal.
	A hard, glossy, black coal.
	Used primarily for residential and commercial space heating.

Total recoverable reserves of coal around the world are estimated at 891,531 million tons – the United States has sufficient coal reserves to last (at current rates of consumption) in excess of 250 years (BP, 2019). Even though coal deposits are distributed widely throughout the world, deposits in three countries account for approximately 57% of the world recoverable coal reserves, viz., United States (27%), Russian Federation (18%), and China (13%), and another six countries account for 30% of the total reserves: Australia (9%), India (7%), Ukraine (4%), Kazakhstan (4%), South Africa (3%), and Japan (3%). Coal is also very unequally and unevenly distributed in the world, just as other fossil fuels such as crude oil and natural gas.

Coal has been studied extensively for conversion into gaseous and liquid fuels as well as hydrocarbon feedstocks. Largely thanks to its relative abundance and stable fuel price on the market, coal has been a focal target for synthetic conversion into other forms of fuels, i.e., synfuels. Research and development work on coal conversion has seen peaks (highs) and valleys (lows) due to external factors including the comparative fossil fuel market as well as the international energy outlook of the era. Coal can be gasified, liquefied, pyrolyzed, and co-processed with other fuels including oil, biomass, scrap tires, and municipal solid wastes (Speight, 2008, 2011b, 2011c, 2013a, 2014b). Secondary conversion of coal-derived gas and liquids can generate a wide array of petrochemical products as well as alternative fuels.

For the past two centuries, coal played this important role – providing coal gas for lighting and heating and then electricity generation with the accompanying importance of coal as an essential fuel for steel and cement production, as well as a variety of other industrial activities. On a worldwide basis, in excess of 4 billion tons (4.0×10^9 tons) of coal is consumed by a variety of sectors – including power generation (steam coal and/or lignite), iron and steel production (coking coal), cement manufacturing, and as a solid fuel or a source of liquid fuels (Speight, 2013a, 2103b). In fact, coal remains an important source of energy in many countries, and is used to provide approximately 40% of electricity worldwide, but this does not give the true picture of the use of coal for electricity production. During recent times the coal industry has been pressured into consideration of the environmental aspects of coal use and has responded with a variety of on-stream coal-cleaning and gas-cleaning technologies (Speight 2013a).

Coal is the largest single source of fuel for the generation of electricity worldwide, as well as the largest source of carbon dioxide emissions, which have been implicated as the primary cause of global climate change, although the debate still rages as to the actual cause (or causes) of climate change. Coal is found as successive layers, or seams, sandwiched between strata of sandstone and shale and extracted from the ground by coal mining – either underground coal seams (underground mining) or by open-pit mining (surface mining).

Coal remains in adequate supply and at current rates of recovery and consumption, the world global coal reserves have been variously estimated to have a reserves/production ratio of at least 155 years. However, as with all estimates of resource longevity, coal longevity is subject to the assumed rate of consumption remaining at the current rate of consumption and, moreover, to technological developments that dictate the rate at which the coal can be mined. But most importantly, coal is a fossil fuel and an *unclean* energy source that will only add to global warming. In fact, the next time electricity is advertised as a clean energy source, just consider the means by which the majority of electricity is produced – almost

50% of the electricity generated in the United States derives from coal (EIA, 2007; Speight, 2013a).

Coal occurs in different forms or *types* (Speight, 2013). Variations in the nature of the source material and local or regional variations in the coalification processes cause the vegetal matter to evolve differently. Various classification systems thus exist to define the different types of coal. Using the ASTM system of classification (ASTM D388), the coal precursors are transformed over time (as geological processes increase their effect over time).

Chemically, coal is a hydrogen-deficient hydrocarbon with an atomic hydrogen-to-carbon ratio near 0.8, as compared to crude oil hydrocarbon derivatives, which have an atomic hydrogen-to-carbon ratio approximately equal to 2, and methane (CH_4) that has an atomic carbon-to-hydrogen ratio equal to 4. For this reason, any process used to convert coal to alternative fuels must add hydrogen or redistribute the hydrogen in the original coal to generate hydrogen-rich products and coke (Speight, 2013).

The chemical composition of the coal is defined in terms of its proximate and ultimate (elemental) analyses (Speight, 2013). The parameters of proximate analysis are moisture, volatile matter, ash, and fixed carbon. Elemental analysis (ultimate analysis) encompasses the quantitative determination of carbon, hydrogen, nitrogen, sulfur, and oxygen within the coal. Additionally, specific physical and mechanical properties of coal and particular carbonization properties are also determined.

Carbon monoxide and hydrogen are produced by the gasification of coal in which a mixture of gases is produced. In addition to carbon monoxide and hydrogen, methane and other hydrocarbon derivatives are also produced depending on conditions. Gasification may be accomplished either *in situ* or in processing plants. *In situ* gasification is accomplished by controlled, incomplete burning of a coal bed underground while adding air and steam. The gases are withdrawn and may be burned to produce heat, generate electricity or are utilized as synthesis gas in indirect liquefaction as well as for the production of chemicals.

Producing diesel and other fuels from coal can be performed through the conversion of coal to synthesis gas, a combination of carbon monoxide, hydrogen, carbon dioxide, and methane. Synthesis gas is subsequently reacted through Fischer-Tropsch Synthesis processes to produce hydrocarbon derivatives that can be refined into liquid fuels. By increasing the quantity of high-quality fuels from coal (while reducing costs), research into this process could help mitigating the dependence on ever-increasingly expensive and depleting stocks of crude oil.

While coal is an abundant natural resource, its combustion or gasification produces both toxic pollutants and greenhouse gases. By developing adsorbents to capture the pollutants (mercury, sulfur, arsenic, and other harmful gases), scientists are striving not only to reduce the quantity of emitted gases but also to maximize the thermal efficiency of the cleanup.

Gasification thus offers one of the cleanest and versatile ways to convert the energy contained in coal into electricity, hydrogen, and other sources of power. Turning coal into synthesis gas isn't a new concept; in fact the basic technology dates back to pre-World War II. In fact, a gasification unit can process virtually all the residua and wastes that are produced in refineries leading to enhanced yields of high-value products (and hence their competitiveness in the market) by deeper upgrading of their crude oil.

1.3.2 Oil Shale

Just like the term *oil sand* (tar sand in the United States), the term *oil shale* is a misnomer since the mineral does not contain oil nor is it always *shale*. The organic material is chiefly *kerogen* and the *shale* is usually a relatively hard rock, called marl. *Oil shale* is a complex and intimate mixture of organic and inorganic materials that vary widely in composition and properties. In general terms, oil shale is a fine-grained sedimentary rock that is rich inorganic matter and yields oil when heated. Some oil shale is genuine shale but others have been misclassified and are actually siltstones, impure limestone, or even impure coal. Oil shale does not contain oil and only produces oil when it is heated to approximately 500°C (approximately 930°F), when some of the organic material is transformed into a distillate similar to crude oil (Lee, 1990; Scouten, 1990; Lee, 1991; Speight, 2008, 2012).

Thus, when properly processed, kerogen can be converted into a substance somewhat similar to crude oil which is often better than the lowest grade of oil produced from conventional oil reservoirs but of lower quality than conventional light oil. *Shale oil* (*retort oil*) is the liquid oil condensed from the effluent in oil shale retorting and typically contains appreciable amounts of water and solids, as well as having an irrepressible tendency to form sediments. Oil shale is an inorganic, non-porous sedimentary marlstone rock containing various amounts of solid organic material (known as *kerogen*) that yields hydrocarbon derivatives, along with non-hydrocarbon derivatives, and a variety of solid products, when subjected to pyrolysis (a treatment that consists of heating the rock at high temperature) (Lee, 1990; Scouten, 1990; Lee, 1991; Speight, 2008, 2012).

Oil production potential from oil shale is measured by a laboratory pyrolysis method called Fischer Assay (Speight, 1994, 2008, 2012) and is reported in barrels per ton (42 US gallons per barrel, approximately 35 Imperial gallons per barrel). Rich oil shale zones can yield more than 40 US gallons per ton, while most shale zones produce 10 to 25 US gallons per ton. Yields of shale oil in excess of 25 US gallons per ton are generally viewed as the most economically attractive, and hence, the most favorable for initial development. Thus, oil shale has, though, a definite potential for meeting energy demand in an environmentally acceptable manner (Lee, 1990; Scouten, 1990; Lee, 1991; Bartis *et al.*, 2005; Andrews, 2006; Speight, 2008; 2012).

The oil shale deposits in the western United States contain approximately 15% w/w organic material (kerogen) (Lee, 1990; Scouten, 1990; Lee, 1991; Speight, 2012). By heating oil shale to high temperatures (>500°C, >930°F), the kerogen is decomposed and converted to a volatile liquid product. However, shale oil is sufficiently different from crude oil and refining processing shale oil presents some unusual problems but, nevertheless, shale oil can be refined into a variety of liquid fuels, gases, and high-value products for the petrochemical industry.

The United States has vast known oil shale resources that could translate into as much as 2.2 trillion barrels of known oil-*in-place*. Oil shale deposits in the United States are concentrated mainly in the Green River Formation in the states of Colorado, Wyoming and Utah, which account for nearly three-quarters of this potential (Lee, 1990; Scouten, 1990; Lee, 1991; Bartis *et al.*, 2005; Andrews, 2006; Speight, 2008; 2012). Because of the abundance and geographic concentration of the known resource, oil shale has been recognized as a potentially valuable United States energy resource since as early as 1859, the same year

Colonel Drake completed his first oil well in Titusville, Pennsylvania (Chapter 1). Early products derived from shale oil included kerosene and lamp oil, paraffin, fuel oil, lubricating oil and grease, naphtha, illuminating gas, and ammonium sulfate fertilizer.

Since the beginning of the 20th century, when the United States Navy converted its ships from coal to fuel oil, and the economy of the United States was transformed by gasoline-fueled automobiles and diesel-fueled trucks and trains, concerns have been raised related to assuring adequate supplies of liquid fuels at affordable prices to meet the growing needs of the nation and its consumers. Thus, it is not surprising that the abundant resources of oil shale in the United States were given consideration as a major source for these fuels. In fact, the Mineral Leasing Act of 1920 made crude oil and oil shale resources on federal lands available for development under the terms of federal mineral leases. This enthusiasm for oil shale resources was mitigated to a large extent by the discoveries of more economically producible and easy-to-refine liquid crude oil in commercial quantities, which caused the interest in oil shale to decline markedly.

However, the interest in oil shale resumed after World War II, when military fuel demand and domestic fuel rationing and rising fuel prices made the economic and strategic importance of the oil shale resource more apparent. After the war, the booming postwar economy drove demand for fuels ever higher, starting with the commencement of the development, in 1946, of the Anvil Point, Colorado, oil shale demonstration project by the United States Bureau of Mines. Significant investments were made by commercial companies to define and develop the US oil shale resource and to develop commercially viable technologies and processes to mine, produce, retort, and upgrade oil shale into viable refinery feedstocks and by-products. Once again, however, major crude oil discoveries in the lower-48 United States, off-shore, and in Alaska, as well as other parts of the world, reduced the foreseeable need for shale oil and interest and associated activities again diminished.

By 1970, oil discoveries were slowing, demand was rising, and crude oil imports into the United States, largely from the Middle Eastern oil-producing nations, were rising to meet demand. Global oil prices, while still relatively low, were also rising, reflecting the changing market conditions. Ongoing oil shale research and testing projects were reenergized and new projects were envisioned by numerous energy companies seeking alternative fuel feedstocks (Speight, 2008, 2011c). These efforts were significantly amplified by the impact of the 1973 Arab oil embargo which demonstrated the vulnerability of the oil-consuming nations, particularly the United States, to oil import supply disruptions, and were underscored by a new supply disruption associated with the 1979 Iranian Revolution.

By 1982, however, technology advances and new discoveries of offshore oil resources in the North Sea and other bodies of water provided new sources for oil imports into the United States. Thus, despite significant investments by energy companies, numerous variations and advances in mining, restoration, retorting, and in-situ processes, the costs of oil shale production relative to current and foreseeable oil prices, made continuation of most commercial efforts impractical.

Despite the huge resources, oil shale is an underutilized energy resource. In fact, one of the issues that arise when dealing with fuels from oil shale is the start-stop-start episodic nature of the various projects. The projects have varied in time and economic investment and viability. The reasons comprise competition from cheaper energy sources, heavy front-end investments and, of late, an unfavorable environmental record. Oil shale has, though, a definite potential for meeting energy demand in an environmentally acceptable manner (Bartis *et al.*, 2005; Andrews, 2006; Speight, 2020).

1.3.3 Biomass

Biomass is a renewable resource that has received considerable attention due to environmental considerations and the increasing demands of energy worldwide (Brown, 2003; NREL 2003; Wright *et al.*, 2006; Tsai *et al.*, 2007; Speight, 2008; Langeveld *et al.*, 2010; Speight, 2011c; Lee and Shah, 2013; Hornung, 2014). Biomass is produced by a photosynthetic process (photosynthesis) which involves chemical reactions occurring on the Earth between sunlight and green plants within the plants in the form of chemical energy. In the process, solar energy is absorbed by green plants and some microorganisms to synthesize organic compounds from low-energy carbon dioxide (CO_2) and water (H_2O). For example,

$$6CO_2 + 6H_2 \rightarrow 6CH_2O + 3O_2$$
$$\text{Biomass}$$

The general formula of the organic material produced during photosynthesis process is $(CH_2O)_n$ which is mainly carbohydrate material. Some of the simple carbohydrates involved in this process are the simple carbohydrates glucose ($C_6H_{12}O_6$) and sucrose ($C_{12}H_{22}O_{11}$) and constitute biomass, which is a renewable energy source due to its natural and repeated occurrence in the environment in the presence of sunlight. The amount of biomass that can be grown certainly depends on the availability of sunlight to drive the conversion of carbon dioxide and water into carbohydrates. In addition to limitations of sunlight, there is a limit placed by the availability of appropriate land, temperature, climate and nutrients, namely nitrogen, phosphorus and trace minerals in the soil.

Biomass is clean for it has negligible content of sulfur, nitrogen and ash, which give lower emissions of sulfur dioxide, nitrogen oxides and soot than conventional fossil fuels. Biomass resources are many and varied, including (i) forest and mill residues, (ii) agricultural crops and waste, (iii) wood and wood waste, (iv) animal waste/s, (v) livestock operation residues, (vi) aquatic plants, (vii) fast-growing trees and plants, (viii) municipal waste, and (ix) and industrial waste. The role of wood and forestry residues in terms of energy production is as old as fire itself and in many societies wood is still the major source of energy. In general, biomass can include anything that is not a fossil fuel that is based on bio-organic materials other than natural gas, crude oil, heavy crude oil, extra heavy crude oil and tar sand bitumen (Lucia *et al.*, 2006; Speight, 2008, 2011c; Lee and Shah, 2013).

There are many types of biomass resources that can be used and replaced without irreversibly depleting reserves and the use of biomass will continue to grow in importance as replacements for fossil fuel sources and as feedstocks for a range of products (Narayan, 2007; Speight, 2008, 2011c; Lee and Shah, 2013). Some biomass materials also have particular unique and beneficial properties which can be exploited in a range of products including pharmaceuticals and certain lubricants. In this context, the increased use of biofuels should be viewed as *one of a range of possible measures* for achieving self-sufficiency in energy, rather than a panacea to completely replace the fossil fuels (Crocker and Crofcheck, 2006; Worldwatch Institute, 2006; Freeman, 2007; Nersesian, 2007).

Worldwide biofuels production is still small at approximately 70 billion (70 x 10^9 tons) oil equivalent. With the volatility of crude oil prices of crude oil, the relative competitiveness of renewable and alternative fuels is drastically improving. Further, technological advances in the alternative renewable energy areas as well as public awareness backed by strong governmental supports and incentives, make the outlook of the alternative and renewable energy very promising (Energy Security Leadership Council, 2013). There are seven countries that can be considered the leaders in biofuels (particularly bioethanol) production. The leaders are (as a percentage of the total production): United States (43.5%), Brazil (24%), France (3%), Germany (4%), Argentina (3%), China (3%), and Indonesia (2.5%). But this still represents minor amounts of the total energy consumption in these countries.

Ethanol (bio-ethanol) from corn has been used in an accelerated pace as gasoline blending fuel as well as a new brand of fuel *E85*, which contains 85% ethanol and 15% gasoline. The majority of the gas stations in the United States supply E85 fuels regularly and many automakers are offering multiple lines of automobiles that can be operated on either conventional gasoline or E85. This not only contributes to cleaner burning fuels but also supplements the amount of gasoline sold.

One extremely important aspect of biomass use as a process feedstock is the preparation of the biomass (also referred to as biomass cleaning or biomass pretreatment), the removal of any contaminants that could have an adverse effect of the process and on the yields and quality of the products. Thus, feedstock preparation is, essentially, the pretreatment of the biomass feedstock to assist in the efficiency of the conversion process. In fact, pretreatment of biomass is considered one of the most important steps in the overall processing in a biomass-to-biofuel program and can occur using acidic or alkaline reagents (Table 1.3) as well as using a variety of physical methods (Table 1.4) and the method of choice depends very much upon the process needs. With the strong advancement in developing lignocellulose biomass-based refinery and algal biomass-based biorefinery, the major focus has been on developing pretreatment methods and technologies that are technically and economically feasible (Pandey *et al.*, 2015).

Table 1.3 Acidic and alkaline methods for biomass treatment.

Method	Conditions	Outcome
Acid based methods	Low pH using an acid (H_2SO_4, H_3PO_4)	Hydrolysis of the hemicellulose to monomer sugars Minimizes the need for hemicellulases
Neutral conditions		Steam pretreatment and hydrothermolysis Solubilizes most of the hydrocarbons by conversion to acetic acid Does not usually result in total conversion to monomer sugars Requires hemicellulases acting on soluble oligomers
Alkaline methods		Leaves a part of the hydrocarbon in the solid fraction Requires hemicellulases acting hydrocarbons

Table 1.4 Summation of the methods for the pretreatment of biomass feedstocks.

Physical methods	Miscellaneous methods
Milling:	Explosion:*
-Ball milling	- Steam, NH_3, CO_2, SO_2, Acids, Alkali
-Two-roll milling	- NaOH, NH_3, $(NH_4)_2SO_3$
-Hammer milling	Acid:
Irradiation:	- Sulfuric, hydrochloric, and phosphoric acids
-Gamma-ray irradiation	Gas:
-Electron-beam irradiation	- Chlorine dioxide, nitrous oxide, sulfur dioxide
-Microwave irradiation	Oxidation:
Other methods:	- Hydrogen peroxide
- Hydrothermal	- Wet oxidation
- High pressure steaming	- Ozone
- Extrusion	Solvent extraction of lignin:
- Pyrolysis	- Ethanol-water extraction
	- Benzene-water extraction
	- Butanol-water extraction
	Organic solvents Ionic liquids

*The feedstock material is subjected to the action of steam and high-pressure carbon dioxide before being discharged through a nozzle.

Typically, the fundamental steps in the pretreatment of biomass involve processes such as (i) washing/separation of inorganic matter such as stones/pebbles, (ii) size reduction which involves grinding, milling, and crushing, and (iii) separation of soluble matter (Table 1.4). Also, the pretreatment process that is selected is, depending upon the character of the biomass and the process, likely to be different for different raw materials and desired products.

One aspect of feedstock preparation in the light of the processes in which the biomass is to be used and converted, is the concept of torrefaction which is used as a pretreatment step for biomass conversion techniques, such as gasification and cofiring. The thermal treatment not only destructs the fibrous structure and tenacity of biomass, but is also known to increase the calorific value of the biomass (Prins *et al.*, 2006a, 2006b, 2006c). Typically, torrefaction commences when the temperature reaches 200°C (390°F) and ends when the process is again cooled from the specific temperature to 200°C (390°F). During the process the biomass partly devolatilizes, leading to a decrease in mass, but the initial energy content of the torrefied biomass is biomass which makes it more attractive for, for example, transportation (to a conversion site). After torrefaction at, say, temperatures up to 300°C (570°F) the grindability of raw biomass shows an improvement in grindability.

A wide range of biomass feedstocks can be used in pyrolysis processes. The pyrolysis process is very dependent on the moisture content of the feedstock, which should be around 10%. At higher moisture contents, high levels of water are produced and at lower levels there is a risk that the process only produces dust instead of oil. High-moisture waste streams, such as sludge and processing wastes, require drying before subjecting to pyrolysis. Thus, the efficiency and nature of the pyrolysis process is dependent on the particle size of feedstocks. Most of the pyrolysis technologies can only process small particles to a maximum of 2 mm keeping in view the need for rapid heat transfer through the particle. The demand for small particle size means that the feedstock has to be size-reduced before being used for pyrolysis.

Moisture in the biomass is another consideration for feedstock preparation because moisture in the feedstock will simply vaporize during the process and then recondense with the bio-oil product which has an adverse impact on the resulting quality of the bio-oil. It should also be noted that water is formed as part of the thermochemical reactions occurring during pyrolysis. Thus, if dry biomass is subjected to the thermal requirements for fast pyrolysis the resulting bio-oil will still contain water (as much as 12 to 15% w/w). This water is process-originated water that is the result of the dehydration of carbohydrate derivatives in the feedstocks as well as the result of reactions occurring between the hydrogen and oxygen at the high temperature (500°C, 930°F) of the process environment.

The moisture in the feedstock acts as heat sink and competes directly with the heat available for pyrolysis. Ideally it would be desirable to have little or no moisture in the feedstock but practical considerations make this unrealistic. Moisture levels on the order of 5 to 10% w/w are generally considered acceptable for the pyrolysis process technologies currently in use. As with the particle size, the moisture levels in the feedstock biomass are a trade-off between the cost of drying and the heating value penalty paid by having moisture in the feedstock. If the moisture content in biomass feedstock is too high, the bio-oil may be produced with high moisture content which eventually reduces its calorific value. Therefore biomass should undergo a pretreatment (drying) process to reduce the water content before pyrolysis is carried out (Dobele *et al.*, 2007). In contrast, high temperature during the drying process could be a critical issue for the possibility of producing thermal-oxidative reactions, causing a cross-linked condensed system of the components and higher thermal stability of the biomass complex.

To achieve high yields of the products (gases, liquids, and solids), it is also necessary to prepare the solid biomass feedstock in such a manner that it can facilitate the required heat transfer rates in the pyrolysis process. There are three primary heat transfer mechanisms available to engineers in designing reaction vessels: (i) convection, (ii) conduction, and (iii) radiation. To adequately exploit one or more of these heat transfer mechanisms as applied to biomass pyrolysis, it is necessary to have a relatively small particle for introduction to the reaction vessel. This ensures a high surface area per unit volume of particle and, as a result of the small particle size the whole particle achieves the desired temperature in a very short residence time. Another reason for the conversion of the feedstock to small particles is the physical transition of biomass as it undergoes pyrolysis when char develops at the surface of the particle. The char can act as an insulator that impedes the transfer of heat into the center of the particle and therefore runs counter to the requirements needed for pyrolysis. The smaller the particle the less of an affect this has on heat transfer (Bridgwater *et al.*, 2001).

1.3.4 Solid Waste

Energy generation utilizing biomass and municipal solid wastes (MSW) are also promising in regions where landfill space is very limited. Technological advances in the fields have made this option efficient and environmentally safe, possibility even supplementing refinery feedstocks as sources of energy through the installation of gasification units (Speight, 2008, 2011a, 2011b, 2011c, 2013b).

Waste may be municipal solid waste (MSW) which had minimal presorting, or refuse-derived fuel (RDF) with significant pretreatment, usually mechanical screening and shredding. Other more specific waste sources (excluding hazardous waste) and possibly including crude oil coke may provide niche opportunities for co-utilization (Bridgwater, 2003; Arena, 2012; Basu, 2013; Speight, 2013, 2014b). The traditional waste-to-energy plant, based on mass-burn combustion on an inclined grate, has a low public acceptability despite the very low emissions achieved over the last decade with modern flue gas clean-up equipment. This has led to difficulty in obtaining planning permissions to construct needed new waste to energy plants. After much debate, various governments have allowed options for advanced waste conversion technologies (gasification, pyrolysis and anaerobic digestion), but will only give credit to the proportion of electricity generated from non-fossil waste.

Use of waste materials as co-gasification feedstocks may attract significant disposal credits (Ricketts *et al.*, 2002). Cleaner biomass materials are renewable fuels and may attract premium prices for the electricity generated. Availability of sufficient fuel locally for an economic plant size is often a major issue, as is the reliability of the fuel supply. Use of more-predictably available coal alongside these fuels overcomes some of these difficulties and risks. Coal could be regarded as the base feedstock which keeps the plant running when the fuels producing the better revenue streams are not available in sufficient quantities.

Wood fuels are fuels derived from natural forests, natural woodlands and forestry plantations, namely fuelwood and charcoal from these sources. These fuels include sawdust and other residues from forestry and wood processing activities. Over 50% of all wood used in the world is fuelwood. Most of the fuelwood is used in developing countries. In developing countries wood makes up about 80% of all wood used.

Size of the wood waste resource depends upon how much wood is harvested for lumber, pulp and paper. Finally, fuelwood can be grown in plantations like a crop. Fast-growing species such as poplar, willow or eucalyptus can be harvested every few years. With short-rotation poplar coppices grown in three 7-year rotations, it is now possible to obtain 10 to 13 tons of dry matter per hectare annually on soil of average or good quality. Waste wood from the forest products industry such as bark, sawdust, board ends, etc., are widely used for energy production. This industry, in many cases, is now a net exporter of electricity generated by the combustion of wastes.

Overall, wood wastes of all types make excellent biomass fuels and can be used in a wide variety of biomass technologies. Combustion of woody fuels to generate steam or electricity is a proven technology and is the most common biomass-to-energy process. Different types of woody fuels can typically be mixed together as a common fuel, although differing moisture content and chemical makeup can affect the overall conversion rate or efficiency of a biomass project.

There are at least six subgroups of woody fuels: (i) forestry residues which include in-forest woody debris and slash from logging and forest management activities, (ii) mill residues which include byproducts such as sawdust, hog fuel, and wood chips from lumber mills, plywood manufacturing, and other wood processing facilities, (iii) agricultural residues which includes byproducts of agricultural activities including crop wastes, waste from vineyard and orchard pruning, and rejected agricultural products, (iv) urban wood and yard wastes which includes residential organics collected by municipal programs or recycling centers and construction wood wastes, (v) dedicated biomass crops which includes trees, corn, oilseed rape, and other crops grown as dedicated feedstocks for a biomass project, and (vi) chemical recovery fuels – sometimes known as black liquor) – which includes woody residues recovered out of the chemicals used to separate fiber for the pulp and paper industry. Mill residues are a much more economically attractive fuel than forestry residues, since the in-forest collection and chipping are already included as part of the commercial mill operations. Biomass facilities collocated with and integral to the mill operation have the advantage of eliminating transportation altogether and thus truly achieve a no-cost fuel.

Softwood residues are generally in high demand as feedstocks for paper production, but hardwood timber residues have less demand and fewer competing uses. In the past, as much as 50% of the tree was left on site at the time of harvest. Whole tree harvest systems for pulp chips recover a much larger fraction of the wood. Wood harvests for timber production often generate residues which may be left on the site or recovered for pulp production. Economics of wood recovery depend greatly on accessibility and local demand. Underutilized wood species include Southern red oak, poplar, and various small-diameter hardwood species. Unharvested dead and diseased trees can comprise a major resource in some regions. When such timber has accumulated in abundance, it comprises a fire hazard and must be removed. Such low-grade wood generally has little value and is often removed by prescribed burns in order to reduce the risk of wildfires.

Agricultural residues are basically biomass materials that are byproducts of agriculture. This includes materials such as cotton stalks, wheat and rice straw, coconut shells, maize and jowar cobs, jute sticks, and rice husks. Many developing countries have a wide variety of agricultural residues in ample quantities. Large quantities of agricultural plant residues are produced annually worldwide and are vastly underutilized. The most common agricultural residue is the rice husk, which makes up approximately 25% w/w of the rice.

Corn stalks and wheat straws are the two agricultural residues produced in the largest quantities. However, many other residues such as potato and beet waste may be prevalent in some regions. In addition to quantity it is necessary to consider density and water content (which may restrict the feasibility of transportation) and seasonality which may restrict the ability of the conversion plant to operate on a year-round basis. Facilities designed to use seasonal crops will need adequate storage space and should also be flexible enough to accommodate alternative feedstocks such as wood residues or other wastes in order to operate year-around. Some agricultural residues need to be left in the field in order to increase tilth (the state of aggregation of soil and its condition for supporting plant growth and to reduce erosion) but some residues such as corncobs can be removed and converted without much difficultly.

Agricultural residues can provide a substantial amount of biomass fuel. Similar to the way mill residues provide a significant portion of the overall biomass consumption in areas that are copiously forested, agricultural residues from sugar cane harvesting and processing provide a significant portion of the total biomass consumption in other parts of the world. One significant issue with agricultural residues is the seasonal variation of the supply. Large residue volumes follow harvests, but residues throughout the rest of the year are minimal. Biomass facilities that depend significantly on agricultural residues must either be able to adjust output to follow the seasonal variation, or have the capacity to stockpile a significant amount of fuel.

Dry animal manure, which is typically defined as having a moisture content less than 30% w/w, is produced by feedlots and livestock corrals, where the manure is collected and removed only once or twice a year. Manure that is scraped or flushed on a more frequent schedule can also be separated, stacked, and allowed to dry. Dry manure can be composted or can fuel a biomass-to-energy combustion project. Animal manure does have value to farmers as fertilizer, and a biomass-to-energy project would need to compete for the manure. However, the total volume of manure produced in many livestock operations exceeds the amount of fertilizer required for the farmlands and, in some areas/countries, nutrient management plans are beginning to limit the over-fertilization of farmlands. Therefore, although there are competitive uses for the manure and low-cost disposal options at this time, manure disposal is going to become more costly over time, and the demand for alternative disposal options, including biomass-to-energy, will only increase.

Biomass technologies present attractive options for mitigating many of the environmental challenges of manure wastes. The most common biomass technologies for animal manures are combustion, anaerobic digestion, and composting. Moisture content of the manure and the amount of contaminants, such as bedding, determine which technology is most appropriate.

Urban wastes include municipal solid waste that is generated by household and commercial activities and liquid waste or sewage. Most municipal solid waste is currently disposed of in landfill sites. However, the disposal of this waste is a growing problem worldwide. Much of the waste could be used for energy production though incineration and processes. Japan currently incinerates more than 80% of the available municipal solid waste. It is also possible to use the methane produced in landfill sites for energy production.

Urban wood and yard wastes are similar in nature to agricultural residues in many regards. A biomass facility will rarely need to purchase urban wood and yard wastes, and most likely can charge a tipping fee to accept the fuel. Many landfills are already sorting waste material by isolating wood waste. This waste could be diverted to a biomass project, and although the volume currently accepted at the landfills would not be enough on its own to fuel a biomass project, it could be an important supplemental fuel and could provide more value to the community in which the landfill resides through a biomass project than it currently does as daily landfill cover.

Municipal solid wastes are produced and collected each year with the vast majority being disposed of in open fields. The biomass resource in municipal solid waste comprises the putrescible materials, paper and plastic and averages 80% of the total municipal solid waste collected. Municipal solid waste can be converted into energy by direct combustion, or by natural anaerobic digestion in the engineered landfill. At the landfill sites the gas produced by the natural decomposition of municipal solid waste (approximately 50% methane and 50% carbon dioxide) is collected from the stored material and scrubbed and cleaned before feeding into internal combustion engines or gas turbines to generate heat and power. The organic fraction of municipal solid waste can be anaerobically stabilized in a high-rate digester to obtain biogas for electricity or steam generation.

1.4 Energy Supply

Energy supply (energy flow) is the delivery of fuels or transformed fuels to point of consumption and the term potentially encompasses the extraction, transmission, generation distribution, and storage of fuels. This supply of energy can be disrupted by several factors, including imposition of higher energy prices due to action by OPEC (in the case of crude oil) or other cartel, war, political disputes, economic disputes, or physical damage to the energy infrastructure due to terrorism. The security of energy supply is a major concern of national security and energy independence.

1.4.1 Economic Factors

Crude oil economics is, as might be expected if the peak oil theory is true, dominated by the so-called occurrence of the remaining reserves that are to be depleted in the very near future. While there may be varied volatility in the price of crude oil, when there is a belief that the price in the future will be higher than the price for immediate delivery there will be a flurry of economic activity in the form of the purchase of short-term contracts. This, in turn, leads to a so-called equilibrium price level that is invariable higher than the previous price level which can influence producers to defer production activities in anticipation of a rise in price (Fleming, 2000). In actual fact, the price circle is indeed a vicious circle insofar as one arc of the price circle affects the other arcs (of the price circle) leading to what might be called economic anarchy. This anarchy might easily be laid at the feet of the peak oil theorists who base their theory on faulty or uncertain data and speculation. Moreover, the economic principles, which explain how a market economy works, tend to break down when applied to natural resources such as oil. In fact, there are two ways in which the principles of market economics do not apply to crude oil: (i) the current price of oil has virtually no influence on the rate at which it is discovered and (ii) the rules of supply and demand do not always hold, and (iii) a rise in the price of crude oil does not always lead to an increase in production.

While current high oil prices may encourage development and adoption of alternatives to oil, if high oil prices are not sustained, efforts to develop and adopt alternatives may fall by the wayside. The high oil prices and fears of running out of oil in the 1970s and early 1980s encouraged investments in alternative energy sources (including synthetic fuels made from coal and oil shale) but when oil prices decreased, investments in these alternatives became uneconomic. In fact, the development of renewable energy systems needs to be supported by decisive, well-coordinated action by governments, in sustained multi-decade programs. Many oil-consuming nations are moving to alternate fuel development rather than be faced with a *destabilizing energy gap*.

1.4.2 Geopolitical Factors

The true picture of oil supply may never be known. Difficult as it is because of a variety of factors, reporting data on oil production or oil reserves is a political act (Laherrère, 2001). The United States Securities and Exchange Commission (SEC) obliges the oil companies listed on the US stock market to report only *proved reserves* and to omit *probable reserves* that are reported in the rest of the world. This practice of reporting only proved reserves can lead to a strong reserve growth since 90% of the annual reserves oil addition come from revisions of old fields, showing that the assessment of the fields was poorly reported. In fact, reporting of production is not much better and may give a false impression of oil abundance (Simmons, 2000) – technical data do exist and must be included in any peak theory model.

1.4.3 Physical Factors

Crude oil reserves (Speight, 2011a, 2014a) are the estimated quantities of oil that are claimed to be recoverable under existing operating and economic conditions. However, because of reservoir characteristics and the limitations of current recovery technologies only a fraction of this oil can be brought to the surface; it is this producible fraction that is considered to be *reserves*. Crude oil recovery varies greatly from oil field to oil field based on the character of the field and the operating history as well as in response to changes in technology and economics.

According to current estimates, more than three-quarters of the oil reserves of the world are located in OPEC countries. The bulk of OPEC oil reserves is located in the Middle East, with Saudi Arabia, Iran and Iraq contributing 41.8% to the OPEC total. OPEC member countries have made significant contributions to their reserves in recent years by adopting best practices in the industry. As a result, OPEC proven reserves currently stand at 1214.2 billion barrels (1214.2 x 10^9 bbls) which represented 71.9% of the total crude oil reserves (BP, 2019).

There has been surprise at the OPEC estimates of proven reserves (Campbell and Laherrère, 1998) since OPEC estimates increased sharply in the 1980s, corresponding to a change in quota rules instituted by OPEC that linked a member production quota by a member country in part to its remaining proven reserves. Indeed, companies that are not subject to the federal securities laws in the United States and their related liability standards, include companies wholly owned by various OPEC member countries where the majority of reserves are located. In addition, many OPEC countries' reported reserves remained relatively unchanged during the 1990s, even as they continued high levels of oil production. For example, estimates of reserves in Kuwait were unchanged from 1991 to 2002, even though the country produced more than 8 billion barrels (8 x 10^9 bbls) of crude oil over that period and did not report any new oil discoveries. The potential disbelief in the data reported by OPEC is problematic with respect to predicting the timing of a peak in oil production because OPEC holds most of the current estimated proven oil reserves of the world.

The United States Geological Survey provides oil resources estimates, which are different from proved reserves estimates. Oil resources estimates are significantly higher because they estimate the total oil resource base of the world, rather than just what is now proven to be economically producible. Estimates of the resource by the United States Geological

Survey base include past production and current reserves as well as the potential for future increases in current conventional oil reserves (often referred to as *reserves growth*) and the amount of estimated conventional oil that has the potential to be added to these reserves. Estimates of reserves growth and those resources that have the potential to be added to oil reserves are important in determining when oil production may peak.

Further contributing to the uncertainty of the timing of a peak is the lack of a comprehensive assessment of oil from nonconventional sources. For example, estimates of crude oil longevity have only recently started to include oil from non-conventional sources (BP, 2019) and oil from these sources was not included in early peak oil theories. Yet oil from non-conventional sources exists in substantial amounts, which could greatly delay the onset of a peak in production. However, challenges facing this production (Speight, 2008, 2011a, 2014a, 2016) indicate that the amount of nonconventional oil that will eventually be produced is, like the peak oil theory, highly speculative. However, despite this *apparent* uncertainty, development and production of oil (synthetic crude oil) from the Alberta tar sands and Venezuelan extra-heavy crude oil production are under way now and the refining technologies are being adapted to produce liquid fuels from these sources (Speight, 2008, 2013b, 2013c, 2013d, 2013e, 2014a).

1.4.4 Technological Factors

In region after region, there are reports of (i) aging and depleted fields, (ii) poor quality – heavy oil, (iii) the need for enhanced recovery methods, and (iv) new areas turning out to be dry well, leading to the claim that peak oil has arrived. For example, for whatever reason, the fields in Alaska, the former Soviet Union, Mexico, Venezuela, and Norway (North Sea) are all claimed to be past their peak. It is grudgingly admitted by the peak oil theorists that there is (or there may be) a (remote or even unlikely) possibility of new finds of oil fields off the coast of West Africa, but their development is still years away, and these new finds will not be on a scale capable of making a difference. It is also further claimed that the only producers with an oil resource which may be capable of keeping oil flowing into the world market at a roughly constant level are the Middle East OPEC five – Saudi Arabia, Iran, Iraq, Kuwait and the United Arab Emirates (Fleming, 2000). However, because of much speculation on the part of the peak oil theorists, there is some difficulty when it comes to projecting the timing of a peak in oil production because (i) technological advances, (ii) increased efficiency in recovery methods and hence reduced or stable recovery economics, and (iii) environmental challenges make it unclear how much additional oil can ultimately be recovered from proven reserves or from hard-to-reach locations and from non-conventional sources.

Worldwide, industry analysts report that deepwater (depths of 1,000 to 5,000 feet) and ultra-deepwater (5,000 to 10,000 feet) drilling efforts are concentrated offshore in Africa, Latin America, and North America, and capital expenditures for these efforts are expected to grow through at least the 2020s decade. In the United States, deepwater and ultra-deepwater drilling, primarily in the Gulf of Mexico, could increase the production of crude oil but at deepwater depths, penetrating the earth and efficiently operating drilling equipment is difficult because of the extreme pressure and temperature (Speight, 2015b).

Ultimately, however, the consequences of a peak and permanent decline in oil production could be even more prolonged and severe than those of past oil supply shocks. Even then the decline rate is the subject of speculation but like death and taxes, the decline rate is happening! The most important variable is the amount of oil left in the reservoirs but, even then, this is subject to debate and error leaving the decline rate for fields in production difficult to assess (Eagles, 2006; Gerdes, 2007; Jackson, 2006, 2007). At best, generalities can be calculated. For example, for current fields in production a low decline rate would be followed by a more moderate decline rate which would result in peak oil in the near future.

1.5 Energy Independence

Energy independence has been a non-partisan political issue in the United States since the first Arab oil embargo in 1973. Since that time, the speeches of various presidents and the Congress of the United States have continued to call for an end to the dependence on foreign oil by the United States. Nevertheless, the United States has grown more dependent on foreign oil with no end in sight. For example, in 1970 the United States imported approximately one-third of the daily oil requirement. Currently, the amount of imported oil is two-thirds of the daily requirement! The congressional rhetoric of energy independence continues but meaningful suggestions of how to address this issue remain few and far between. The economy of the United States feeds on oil and the country consumes far more oil than it can produce.

Generally, the concept of energy independence for the United States runs contrary to the trend of the internationalization of trade. The US government continues to reduce trade barriers through policies such as the North American Free Trade Agreement (NAFTA), the elimination of tariffs, as well as other free trade agreements. As a result, the percentage of the US economy that comes from international trade is steadily rising.

Increased world trade is beneficial both economically and politically insofar as it is supposed to help establish amicable relations between countries. Through mutually beneficial exchange, a great deal of this increased interrelationship will, in theory, establish opportunities for personal ties that make war (or other forms of military action) less likely. However, there are also contrary cases where countries acquire the means to be more destructive, if they so choose, through expanded economic opportunities. This type of argument has been used with regard to Iran.

Economic interdependence also makes the domestic economy more susceptible to disruptions in distant and unstable regions of the globe, such as the Middle East, South America and Africa. In fact, in many countries with proven reserves, oil production could be shut down by wars, strikes, and other political events, thus reducing the flow of oil to the world market. If these events occurred repeatedly, or in many different locations, they could constrain exploration and production, resulting in a peak despite the existence of proven oil reserves. Using a measure of political risk that assesses the likelihood that events such as civil wars, coups, and labor strikes will occur in a magnitude sufficient to reduce the gross domestic product (GDP) growth rate of a country over the next five years, four countries (Iran, Iraq, Nigeria, and Venezuela) possess proven oil reserves greater than 10 billion barrels (high reserves) and which countries contain almost one-third of worldwide oil reserves, face high levels of political risk. In fact, countries with medium or high levels of political risk contain 63% of proven worldwide oil reserves.

For example, in past years, disputes leading to withdrawal of labor (strikes) by workers in Venezuela have caused reductions in the crude oil (approximately 1,500,000 barrels per day) imported from that country. Similarly, conflicts in Nigeria between ethnic groups can disrupt the amount of crude oil (approximately 570,000 barrels per day) imported from that country. Thus, in a short time and by events out of its control, the United States can suffer an oil shortage of imported oil (to the tune of 2,000,000 barrels per day) that leaves a large gap in the required 18,000,000 barrels per day currently refined in the United States.

Furthermore, the oil industry itself has been a story of vast swings between periods of overproduction, when low prices and profits led oil producers to devise ways to restrict output and raise prices, and periods when oil supplies appeared to be on the brink of exhaustion, stimulating a global search for new supplies. This cycle may now be approaching an end. It appears that world oil supplies may truly be reaching their natural limits. With proven world oil reserves anticipated to last less than 40 years, the age of oil that began near Titusville may be coming to an end. In the years to come, the search for new sources of oil will be transformed into a quest for entirely new sources of energy.

Political and investment risk factors continue to affect future oil exploration and production and, ultimately, the timing of peak oil production. These factors include changing political conditions and investment climates in many countries that have large proven oil reserves. These factors are important in affecting future oil exploration and production.

Even in the United States, political considerations may affect the rate of exploration and production. For example, restrictions imposed to protect environmental assets mean that some oil may not be produced. The Minerals Management Service of the United States Department of the Interior estimates that approximately 76 billion barrels (76 x 10^9 bbls) of oil lie in undiscovered fields offshore in the outer continental shelf of the United States (which is necessary for a measure of energy security). Nevertheless, Congress enacted moratoriums on drilling and exploration in this area to protect coastlines from unintended oil spills. In addition, policies on federal land use need to take into account multiple uses of the land including environmental protection. Environmental restrictions may affect a peak in oil production by barring oil exploration and production in environmentally sensitive areas.

The government must adopt policies that ensure our energy independence. The US Congress is no longer believable when the members of the Congress lay the blame on foreign governments or events for an impending crisis. The Congress needs to look north and the positive role played by the government of Canada in the early 1960s when the decision was made to encourage development of the Alberta tar sands. Synthetic crude oil production is now in excess of 1,000,000 million barrels per day – less than 6% of the daily liquid fuels requirement in the United States but a much higher percentage of Canadian daily liquid fuels requirement. In the United States, the issue to be faced is not so much oil reserves but oil policies.

The economics of crude oil inventories provides the key to unlocking this mystery. The net cost of carrying inventories is equal to the interest rate, plus the cost of physical storage, minus the convenience yield. The convenience yield is driven by the precautionary demand for the storage. When the convenience yield is zero, a market is in *full carry*, future prices exceed spot prices and inventories are abundant. Alternatively, when the precautionary demand for oil is high, spot prices are strong and exceed future prices, and inventories are unusually low. The strategic petroleum reserve should be used to give the country a measure of energy independence and to thwart the efforts of the cartel to control crude oil – a world commodity.

A measure of dependency on crude oil can be viewed by various importing countries (Alhajji and Williams, 2003). In the United States, increasing crude oil imports is considered a threat to national security but there is also the line of thinking that the level of imports has no significant impact on energy security, or even national security. However, the issue becomes a problem when import vulnerability increases as crude oil imports rise which occurs when oil-consuming countries increase the share of crude oil imports from politically unstable areas of the world.

More generally, there are four measures of crude oil dependence: (i) crude oil imports as a percentage of total crude oil consumption, (ii) the number of days total crude oil stocks cover crude oil imports, (iii) the number of days total stocks cover consumption, and (iv) the percentage of crude oil in total energy consumption (Alhajji and Williams, 2003).

The dependency on foreign oil in the United States has increased steadily since 1986 and has reached record highs in the past two decades – the degree of import dependence as percentage of consumption increased from about 50% in the early-to-mid 1980s to 60% in the early 1990s (because of higher economic growth and lower oil prices on the demand side and declining US production on the supply side). Currently, 65-70% of the daily oil and oil products is imported into the United States. There have been some minor fluctuations but the changes in the amount of oil imported into the United States are usually related to changes in the US economy and the US oil production.

The United States is the only oil-importing country with significant production which is also a net importer. It is also unique in that among the net importers it has the lowest dependence in terms of net imports as a percent of consumption but it also has the highest absolute level of imports. The geo-political and economic interests and commitments raise the level of concern by US policy makers about dependence on imported oil.

In addition, commercial oil stocks in the United States have been at their lowest level in three decades. Total crude oil inventories, which include commercial and stocks in the Strategic Petroleum Reserve (SPR) are relatively low, in terms of daily coverage. Current commercial inventories are near the level at which spot shortages can occur. The past decade has seen scenarios in which the decline in commercial stocks is greater than the increase in the Strategic Petroleum Reserve, and the capacity of the Strategic Petroleum Reserve and commercial stocks to deal with a crisis is less than before the refilling program began (Williams and Alhajji, 2003). Moreover, the premature release of crude oil from the Strategic Petroleum Reserve can jeopardize national security in case of continued political problems in the oil-producing countries and weakens the ability of the United States to respond to real shortages.

Although some of the oil-importing countries have made progress in reducing their dependence on oil, the dependence of the United States on crude oil has increased in recent years from 38% of total energy consumption in 1995 to approximately 40% at the current time. This indicates two possible areas of concern regarding the extent to which crude oil influences energy security: (i) the increase in the crude oil share of energy use, and (ii) the inability or unwillingness of the United States to reduce dependence on imported oil.

It might be argued that the degree of dependence has no impact on energy security as long as foreign oil is imported form secure sources. However, if the degree of dependence on non-secure sources increases, energy security would be in jeopardy. In this case, vulnerability would increase and economic and national security of individual oil-importing countries would be compromised.

The percentage of imports from the top five suppliers can be used as a measure of the supply vulnerability to an interruption by one or more key suppliers. This is an important measure of the vulnerability of the United States to supply disruption because it shows the high level of import concentration by importing from few suppliers. The top five suppliers to oil-importing countries are Saudi Arabia, the Commonwealth of Independent States (former Soviet republics), Norway, Venezuela, and Mexico. The United States has a unique arrangement as well as location with Canada and Mexico but the political stability of Saudi Arabia is always open to discussion and question. In addition, problems in Venezuela underline the United States vulnerability to interruption from a major supplier. Venezuela has had severe problems and the potential for an interruption in oil supply always exist. Iraq crude has been off the market for several years (oil supply from Iraq is only just starting again) and conflict in Nigeria has significantly influenced oil output.

Another important measure of vulnerability is the share of world crude coming from the Gulf region. The Gulf region has been viewed historically as a politically unstable area. Incidents in the region led to the three energy crisis in 1973, 1979, and the two Gulf Wars. While a smaller share of imported oil from the Gulf producers means lesser vulnerability, it may increase vulnerability by having to rely on other sources, such as Venezuela. Whichever measure is used to assess energy dependence, the United States remains susceptible to an energy crisis because of the high dependence on imported oil. The United States has not made a significant reduction in dependence on imports since the mid-1980s and continues to import almost 70% of the total crude oil consumption.

The crude oil share in the total energy supply also reflects the dependence on crude oil by the United States. In recent years, however, this share has increased in the United States. Because of this, the United States is highly vulnerable to oil supply disruptions. Indeed, the possibility of energy crisis in the foreseeable future is greater than in previous years. Furthermore, the use of the Strategic Petroleum Reserve or government-controlled stocks to lessen the impact of an energy crisis is subject to debate. In fact, the premature release of oil stocks from the Strategic Petroleum Reserve may exacerbate an energy crisis as it depletes the stocks while shortages still exist since it can lead to stabilized (or even lower) prices and increased consumption (Alhajji and Williams, 2003).

Dependency and vulnerability to oil imports in the United States and, for that matter, in other oil-importing countries, can be reduced by diversification of suppliers and by energy diversification. In addition, diversification of suppliers has the potential to lower the relative impact of supply disruption on most countries. The political instability that swings back and forth in countries such as Venezuela, Nigeria, and Iraq emphasizes the need for diversification of suppliers and so removing the reliance on a small number of oil-producing countries.

The projections for the continued use of fossil fuels indicate that there will be at least another five decades of fossil fuel use (especially natural gas, crude oil, and coal) before biomass and other forms of alternate energy take hold. Furthermore, estimations that the era of fossil fuels (natural gas, crude oil, and coal) will be almost over when the cumulative production of the fossil resources reaches 85% of their initial total reserves may or may not have some merit. In fact, the relative scarcity (compared to a few decades ago) of crude oil was real but it seems that the remaining reserves make it likely that there will be an adequate supply of energy for several decades. The environmental issues are very real and require serious and continuous attention.

Synthesis gas (synthesis gas) fuel gas mixture consisting predominantly of carbon monoxide and hydrogen and is typically a product of a gasification. The gasification process is applicable to many carbonaceous feedstocks including natural gas, crude oil resids, coal, biomass, by reaction of the feedstock with steam (steam reforming), carbon dioxide (dry reforming) or oxygen (partial oxidation). Synthesis gas is a crucial intermediate resource for production of hydrogen, ammonia, methanol, a variety of chemicals, as well as synthetic hydrocarbon fuels.

Thus, as the reserves of natural gas, crude oil, and other forms of conventional energy are depleted, there will be the need to seek other sources, some of which are outlined in the previous sections.

Energy production such as electricity production or combined electricity and heat production remain the most likely area for the application of gasification or co-gasification. The lowest investment cost per unit of electricity generated is the use of the gas in an existing large power station. This has been achieved in several large utility boilers, often with the gas fired alongside the main fuel. This option allows a comparatively small thermal output of gas to be used with the same efficiency as the main fuel in the boiler as a large, efficient steam turbine can be used. It is anticipated that addition of gas from a biomass or wood gasifier into the natural gas feed to a gas turbine will be technically possible but there will be concerns as to the balance of commercial risks to a large power plant and the benefits of using the gas from the gasifier.

Furthermore, the disposal of municipal and industrial waste has become an important problem because the traditional means of disposal, landfill, are much less environmentally acceptable than previously. Much stricter regulation of these disposal methods will make the economics of waste processing for resource recovery much more favorable. One method of processing waste streams is to convert the energy value of the combustible waste into a fuel. One type of fuel attainable from waste is a low heating value gas, usually 100 to 150 Btu/scf, which can be used to generate process steam or to generate electricity. Co-processing such waste with coal is also an option (Speight, 2008, 2013, 2014b).

Co-gasification technology varies, being usually site specific and high feedstock dependent. At the largest scale, the plant may include the well-proven fixed-bed and entrained-flow gasification processes. At smaller scales, emphasis is placed on technologies which appear closest to commercial operation. Pyrolysis and other advanced thermal conversion processes are included where power generation is practical using the on-site feedstock produced. However, the needs to be addressed are (i) core fuel handling and gasification/pyrolysis technologies, (ii) fuel gas clean-up, and (iii) conversion of fuel gas to electric power (Ricketts *et al.*, 2002).

Waste may be municipal solid waste (MSW) which had minimal presorting, or refuse-derived fuel (RDF) with significant pretreatment, usually mechanical screening and shredding. Other more specific waste sources (excluding hazardous waste) and possibly including crude oil coke, may provide niche opportunities for co-utilization. The traditional waste-to-energy plant, based on mass-burn combustion on an inclined grate, has a low public acceptability despite the very low emissions achieved over the last decade with modern flue gas clean-up equipment. This has led to difficulty in obtaining planning permissions to construct needed new waste-to-energy plants. After much debate, various governments have allowed options for advanced waste conversion technologies (gasification, pyrolysis and anaerobic digestion), but will only give credit to the proportion of electricity generated from non-fossil waste.

Co-utilization of waste and biomass with coal may provide economies of scale that help achieve the above-identified policy objectives at an affordable cost. In some countries, governments propose co-gasification processes as being *well suited for community-sized developments*, suggesting that waste should be dealt with in smaller plants serving towns and cities, rather than moved to large, central plants (satisfying the so-called *proximity principle*).

In fact, neither biomass nor wastes are currently produced, or naturally gathered at sites in sufficient quantities to fuel a modern large and efficient power plant. Disruption, transport issues, fuel use, and public opinion all act against gathering hundreds of megawatts (MWe) at a single location. Biomass or waste-fired power plants are therefore inherently limited in size and hence in efficiency (labor costs per unit electricity produced) and in other economies of scale. The production rates of municipal refuse follow reasonably predictable patterns over time periods of a few years. Recent experience with the very limited current *biomass for energy* harvesting has shown unpredictable variations in harvesting capability with long periods of zero production over large areas during wet weather.

The potential unreliability of biomass, longer-term changes in refuse and the size limitation of a power plant using only waste and/or biomass can be overcome combining biomass, refuse and coal. It also allows benefit from a premium electricity price for electricity from biomass and the gate fee associated with waste. If the power plant is gasification-based, rather than direct combustion, further benefits may be available. These include a premium price for the electricity from waste, the range of technologies available for the gas to electricity part of the process, gas cleaning prior to the main combustion stage instead of after combustion and public image, which is currently generally better for gasification as compared to combustion. These considerations lead to current studies of co-gasification of wastes/biomass with coal (Speight, 2008).

For large-scale power generation (>50 MWe), the gasification field is dominated by plant based on the pressurized, oxygen-blown, entrained-flow or fixed-bed gasification of fossil fuels. Entrained gasifier operational experience to date has largely been with well-controlled fuel feedstocks with short-term trial work at low co-gasification ratios and with easily-handled fuels.

Use of waste materials as co-gasification feedstocks may attract significant disposal credits. Cleaner biomass materials are renewable fuels and may attract premium prices for the electricity generated. Availability of sufficient fuel locally for an economic plant size is often a major issue, as is the reliability of the fuel supply. Use of more-predictably available coal alongside these fuels overcomes some of these difficulties and risks. Coal could be regarded as the *flywheel* which keeps the plant running when the fuels producing the better revenue streams are not available in sufficient quantities.

References

Bentley, R.W. 2002. Global Oil and Gas Depletion: An Overview. *Energy Policy*, 30: 189-205.
Bower, T. 2009. *Oil: Money, Politics, and Power in the 21st Century*. Grand Central Publishing, Hachette Book Group, Inc., New York.

BP. 2019. BP Statistical Review of World Energy. BP PLC, London, United Kingdom. https://www.bp.com/content/dam/bp/business-sites/en/global/corporate/pdfs/energy-economics/statistical-review/bp-stats-review-2019-full-report.pdf

Bridgwater, A.V.; Czernik, S.; Piskorz, J. 2001. An Overview of Fast Pyrolysis. Prog. *Thermochem. Biomass Convers.*, 2: 977-997.

Campbell, C.J. and Laherrère, J.H. 1998. The End of Cheap Oil. *Scientific American*. 278: 78-83.

Collett, T.S. 2009. Natural Gas Hydrates: A Review. In: *Natural Gas Hydrates – Energy Resource Potential and Associated Geologic Hazards*. Collett, T.S. Johnson, A.H. Knapp, C.C., and Ray Boswell; R. (Editors). AAPG Memoir 89, p. 146-219. American Association of Petroleum Geologists, Tulsa, Oklahoma.

Crane, H.D., Kinderman, E.M., and Malhotra, R. 2010. *A Cubic Mile of Oil: Realties and Options for Averting the Looming Global Energy Crises*. Oxford University Press, Oxford, United Kingdom.

Crocker, M., and Crofcheck, C. 2006. Reducing national dependence on imported oil. *Energeia* Vol. 17, No. 6. Center for Applied Energy Research, University of Kentucky, Lexington, Kentucky.

Delbianco, A., and Montanari, R. 2009. *Encyclopedia of Hydrocarbons*, Volume III *New Developments: Energy, Transport, Sustainability*. Eni S.p.A., Rome, Italy.

Dobele, G., Urbanovich, I., Volpert, A., Kampars, V., and Samulis, E. 2007. Fast pyrolysis – Effect of wood drying on the yield and properties of bio-oil. *Bioresources*, 2: 699-706.

Eagles, L. 2006. Medium Term Oil Market Report. OECD/International Transport Forum Roundtable. International Energy Agency, Paris, France.

Energy Security Leadership Council. 2013. A National Strategy for Energy Security: Harnessing American Resources and Innovation. Energy Security Leadership Council, Washington, DC.

Fleming, D. 2000. After Oil. *Prospect Magazine*, November. Issue No. 57: pp 12-13. http://www.prospect-magazine.co.uk.

Gary, J.G., Handwerk, G.E., and Kaiser, M.J. 2007. *Petroleum Refining: Technology and Economics*, 5th Edition. CRC Press, Taylor & Francis Group, Boca Raton, Florida.

Gerdes, J. 2007. Modest Non-OPEC Supply Growth Underpins $60+ Oil Price. SunTrust Robinson Humphrey, Atlanta, Georgia.

Ghoshal, S., and Sainik, V. 2013. Monitor and Minimize Corrosion in High-TAN Crude Processing. *Hydrocarbon Processing*, 92(3): 35-38.

Giampietro, M., and Mayumi, K. 2009. *The Biofuel Delusion: The Fallacy of Large-Scale Agro-biofuel Production*. TJ International, Padstow, Cornwall, United Kingdom; also published by Earthscan, London, United Kingdom.

Goldberg, D., and Saito, S. 1998. Detection of Gas Hydrates Using Downhole Logs. In: *Hydrates - Relevance to World Margin Stability and Climate Change Gas*. J.-P. Henriet and J. Mienert (Editors). The Geological Society, London, United Kingdom. Page 129-132.

Gudmestad, O.T., Zolotukhin, and Jarlsby, E.T. 2010. *Petroleum Resources with Emphasis on Offshore Fields*. WIT Press, Billerica, Massachusetts.

Hirsch, R.L. 2005. The Inevitable Peaking of World Oil Production. *The Atlantic Council of the United States Bulletin*, XVI(3): 1-9.

Hirsch, R.L., Bezdek, R., and Wendling, R. 2005. Peaking Of World Oil Production: Impacts, Mitigation, and Risk Management. February. http://www.netl.doe.gov/publications/ others/pdf/Oil_Peaking_NETL.pdf

Hornbach, M.J., Holbrook, W.S., Gorman, A.R., Hackwith, K.L., Lizarralde, D., and Pecher, I. 2003. Direct Seismic Detection of Methane Hydrate on the Blake Ridge. *Geophysics*, 68: 92-100.

Hornung, A. 2014. *Transformation of Biomass: Theory to Practice*. John Wiley & Sons Inc., Chichester, West Sussex, United Kingdom.

Hsu, C.S., and Robinson, P.R. (Editors). 2017. *Handbook of Petroleum Technology*. Springer International Publishing AG, Cham, Switzerland.

Hubbert, M.K. 1956. *Nuclear Energy and the Fossil Fuels - Drilling and Production Practice*. American Petroleum Institute, Washington, DC.

Hubbert, M.K. 1962. Energy Resources. Report to the Committee on Natural Resources, National Academy of Sciences, Washington, DC.

Jackson, P.M. 2006. Why the Peak Oil Theory Falls Down: Myths, Legends, and the Future of Oil Resources. Cambridge Energy Research Associates, Cambridge, Massachusetts. November 14.

Jackson, P.M. 2007. Finding the Critical Numbers: What Are the Real Decline Rates for Global Oil Production? Cambridge Energy Research Associates, Cambridge, Massachusetts.

Kane, R.D., and Cayard, M.S. 2002. A Comprehensive Study on Naphthenic Acid Corrosion. Corrosion 2002. NACE International, Houston, Texas.

Khoshnaw, F.M. (Editor). 2013. *Crude oil and Mineral Resources*, WIT Press, Billerica, Massachusetts.

Kvenvolden, K.A. 1993. Gas Hydrates as a Potential Energy Resource – A Review of Their Methane Content. In: *The Future of Energy Gases*. D.G. Howell (Editor). Professional Paper No. 1570. United States Geological Survey, Washington, DC. Page 555-561.

Kvenvolden, K.A., and Lorenson, T.D. 2000. The Global Occurrence of Natural Gas Hydrate. In: *Natural Gas Hydrates: Occurrence, Distribution, and Dynamics*. C.K. Paull, and W.P. Dillon (Editors). AGU Monograph, American Geophysical Union, Washington, DC. Page 55.

Laherrère, J.H. 2001. Proceedings. EMF/IEA/IEW Meeting IIASA, Laxenburg, Austria. Plenary Session I: Resources, June 19.

Langeveld, H., Sanders, J., and Meeusen, M. 2010 (Editors). *The Biobased Economy: Biofuels, Materials, and Chemicals in the Post-oil Era*. TJ International, Padstow, Cornwall, United Kingdom; also published by Earthscan, London, United Kingdom.

Lee, S. 1990. *Oil Shale Technology*. CRC Press, Taylor & Francis Group, Boca Raton, Florida.

Lee, S. 1991. *Oil Shale Technology*. CRC Press, Taylor & Francis Group, Boca Raton, Florida.

Lee, S., and Shah, Y.T. 2013. *Biofuels and Bioenergy: Processes and Technologies*. CRC Press, Taylor & Francis Group, Boca Raton Florida.

Lerche, I., and Bagirov, E. 1998. Guide to Gas Hydrate Stability in Various Geological Settings. *Marine and Petroleum Geology*, 15: 427-438.

Luciani, G. 2013. *Security of Oil Supplies: Issues and Remarks*. Claeys & Casteels, Deventer, Netherlands.

Maini, B.B. 1999. Foamy Oil Flow in Primary Production of Heavy Oil under Solution Gas Drive. Paper No. SPE 56541. *Proceedings. Annual Technical Conference and Exhibition, Houston, Texas USA, October 3-6*. Society of Petroleum Engineers, Richardson, Texas.

Maini, B.B. 2001. Foamy Oil Flow. Paper No. SPE 68885. SPE J. Pet. Tech., Distinguished Authors Series. Page 54 – 64, October. Society of Petroleum Engineers, Richardson, Texas.

Meyer, R.F., and De Witt, W. Jr. 1990. Definition and World Resources of Natural Bitumens. Bulletin No. 1944. US Geological Survey, Reston, Virginia.

Meyer, R.F., and Attanasi, E.D. 2003. Heavy Oil and Natural Bitumen - Strategic Petroleum Resources. Fact Sheet 70-03. United States Geological Survey, Washington, DC. August 2003. http://pubs.usgs.gov/fs/fs070-03/fs070-03.html.

Nersesian, R.L. 2007. *Energy for the 21st Century: A Comprehensive Guide to Conventional and Alternative Fuel Sources*. M.E. Sharpe Inc., Armonk, New York.

Ohmes, R. 2014. Characterizing and Tracking Contaminants in Opportunity Crudes. Digital Refining. http://www.digitalrefining.com/article/1000893,Characterising_and_tracking_contaminants_in_opportunity_crudes_.html#.VJhFjV4AA; accessed November 1, 2014.

Pandey, A., Negi, S., Binod, P., and Larroche, C. (Editors) 2015. *Pretreatment of Biomass: Processes and Technologies*. Elsevier BV, Amsterdam, Netherlands.

Parkash, S. 2003. *Refining Processes Handbook*. Gulf Professional Publishing, Elsevier, Amsterdam, Netherlands.

Prins, M.J., Ptasinski, K.J., and Janssen, F.J.J.G. 2006a. More Efficient Biomass Gasification via Torrefaction. *Energy*, 31(15): 3458-3470.

Prins, M.J., Ptasinski, K.J., and Janssen, F.J.J.G. 2006b. Torrefaction of Wood: Part 1. Weight Loss Kinetics. *Journal of Analytical and Applied Pyrolysis*, 77(1): 28-34.

Prins, M.J., Ptasinski, K.J., and Janssen, F.J.J.G. 2006c. Torrefaction of Wood: Part 2. Analysis of Products. *Journal of Analytical and Applied Pyrolysis*, 77(1): 35-40.

Ratner, M., and Tiemann, M. 2014. *An Overview of Unconventional Oil and Natural Gas: Resources and Federal Actions*. CRS Report. Congressional Research Service, Washington, DC. January 23.

Sandrea, I. 2014. *US Shale Gas and Tight Oil: Industry Performance: Challenges and Opportunities*. Oxford Institute for Energy Studies, University of Oxford, Oxford, United Kingdom. http://www.oxfordenergy.org/wpcms/wp-content/uploads/2014/03/US-shale-gas-and-tight-oil-industry-performance-challenges-and-opportunities.pdf; accessed October 14, 2014.

Scouten, C.S. 1990. Oil Shale. In: *Fuel Science and Technology Handbook*. J.G. Speight (Editor). Marcel Dekker, New York. Chapters 25-31. Page 795-1053.

Shalaby, H.M. 2005. Refining of Kuwait's Heavy Crude Oil: Materials Challenges. *Proceedings. Workshop on Corrosion and Protection of Metals*. Arab School for Science and Technology. December 3-7, Kuwait.

Sheng, J.J., Maini, B.B., Hayes, R.E., and Tortike, W.S. 1999. Critical Review of Foamy Oil Flow. *Trans. Porous Media*, 35: 157-187.

Siefried, D., and Witzel, A.W. 2010. *Renewable Energy – The Facts*. London, United Kingdom. Formerly published by Energieagentur Regio Freiburg, Freiburg, Germany.

Simmons M.R. 2000. Fighting Rising Demand and Rising Decline Curves: Can the challenge be met? SPE Asia Pacific Oil and Gas Conference, Yokohama, April 25. http://www.simmons-cointl.com/web/downloads/spe.pdf

Speight, J.G. (Editor). 1990. *Fuel Science and Technology Handbook*, Marcel Dekker, New York.

Speight, J.G. 1997. Tar Sands. *Kirk-Othmer Encyclopedia of Chemical Technology* 4th Edition. 23: 717.

Speight. J.G. 2000. *The Desulfurization of Heavy Oils and Residua* 2nd Edition. Marcel Dekker Inc., New York.

Speight, J.G. and Lee, S. 2000. *Environmental Technology Handbook*, Taylor & Francis, Philadelphia, Pennsylvania.

Speight, J.G. 2008. *Synthetic Fuels Handbook: Properties, Processes, and Performance*. McGraw-Hill, New York.

Speight, J.G. 2011a. *An Introduction to Petroleum Technology, Economics, and Politics*. Scrivener Publishing, Salem, Massachusetts.

Speight, J.G. (Editor). 2011b. *Biofuels Handbook*. Royal Society of Chemistry, London, United Kingdom.

Speight, J.G. 2011c. *The Refinery of the Future*. Gulf Professional Publishing, Elsevier, Oxford, United Kingdom.

Speight, J.G. 2012. *Shale Oil Production Processes*. Gulf Professional Publishing, Elsevier, Oxford, United Kingdom.

Speight, J.G. 2013a. *The Chemistry and Technology of Coal* 3rd Edition. CRC Press, Taylor & Francis Group, Boca Raton, Florida.

Speight, J.G. 2013b. *Coal-Fired Power Generation Handbook*. Scrivener Publishing, Salem Massachusetts.

Speight, J.G. 2013c. *Heavy and Extra Heavy Oil Upgrading Technologies*. Gulf Professional Publishing, Elsevier, Oxford, United Kingdom.

Speight, J.G. 2013d. *Oil Sand Production Processes*. Gulf Professional Publishing, Elsevier, Oxford, United Kingdom.

Speight, J.G. 2013e. *Heavy Oil Production Processes*. Gulf Professional Publishing, Elsevier, Oxford, United Kingdom.

Speight, 2013f. *Shale Gas Production Processes*. Gulf Professional Publishing, Elsevier, Oxford, United Kingdom.

Speight, J.G. 2014a. *The Chemistry and Technology of Petroleum* 5th Edition. CRC Press, Taylor & Francis Group, Boca Raton, Florida.

Speight, J.G. 2014b. *The Gasification of Unconventional Feedstocks*. Gulf Professional Publishing, Elsevier, Oxford, United Kingdom.

Speight, J.G. 2014c. *High Acid Crudes*. Gulf Professional Publishing, Elsevier, Oxford, United Kingdom.

Speight, J.G. 2015a. *Handbook of Hydraulic Fracturing*. John Wiley & Sons Inc., Hoboken, New Jersey.

Speight, J.G. 2015b. *Offshore Oil and Gas*. Gulf Professional Publishing, Elsevier, Oxford, United Kingdom.

Speight, J.G. 2016. *Introduction to Enhanced Recovery Methods for Heavy Oil and Tar Sands* 2nd Edition. Gulf Professional Publishing Company, Elsevier, Oxford, United Kingdom.

Speight, J.G., and Islam, M.R. 2016. *Peak Energy – Myth or Reality*. Scrivener Publishing, Beverly, Massachusetts.

Speight, J.G. 2017. *Handbook of Petroleum Refining*. CRC Press, Taylor & Francis Group, Boca Raton, Florida.

Speight, J.G. 2019a. *Handbook of Petrochemical Processes*. CRC Press, Taylor & Francis Group, Boca Raton, Florida.

Speight, J.G. 2019b. *Natural Gas: A Basic Handbook* 2nd Edition. Gulf Publishing Company, Elsevier, Cambridge, Massachusetts.

Speight, J.G. 2020. *Shale Gas and Shale Oil Production Processes*, Elsevier Inc., Cambridge, Massachusetts.

US Congress. 1976. Public Law FEA-76-4. United States Library of Congress, Washington, DC.

Wihbey, P.M. 2009. *The Rise of the New Oil Order: The Facts behind the New Massive Supplies, the Exciting New Technologies, and the Emerging Powers of the XXIst Century*. Academy & Finance SA, Geneva, Switzerland.

World Energy Council. 2008. World Energy Outlook: Global Trends to 2030. International Energy Agency, Paris, France.

Wright, L., Boundy, R, Perlack, R., Davis, S., and Saulsbury. B. 2006. *Biomass Energy Data Book: Edition 1*. Office of Planning, Budget and Analysis, Energy Efficiency and Renewable Energy, United States Department of Energy. Contract No. DE-AC05-00OR22725. Oak Ridge National Laboratory, Oak Ridge, Tennessee.

Yergin, Y. 1991. *The Prize: The Epic Quest for Oil, Money and Power*. Free Press, New York.

Yergin, D. 2011. *The Quest: Energy Security and the Making of the Modern World*. Penguin Press HC, New York.

Yeung, T.W. 2014. Evaluating Opportunity Crude Processing. Digital Refining. http://www.digital refining.com/article/1000644; accessed October 25, 2014.

Zatzman, G.M. 2012. *Sustainable Resource Development*. Scrivener Publishing LLC, Beverly, Massachusetts.

2
Production of Synthesis Gas

2.1 Introduction

Synthesis gas (also frequently referred to by the abbreviated name *syngas*) is a fuel gas consisting primarily of carbon monoxide and hydrogen with, on occasion depending upon the feedstock and the production process, a smaller amount of carbon dioxide. Synthesis gas has been, for decades, a product of coal gasification and the main application has been, and continues to be, the generation of electricity.

In addition to coal as the (formerly) primary feedstock, synthesis gas can be produced from many sources, including natural gas, coal, biomass, or virtually any hydrocarbon feedstock, by reaction with steam (the steam reforming process), with carbon dioxide (the dry reforming process), or with oxygen (the partial oxidation process). Synthesis gas has become a crucial intermediate resource for production of hydrogen, ammonia, methanol, and synthetic hydrocarbon fuels. It is also used as an intermediate in the production of (i) synthetic crude oil, (ii) hydrocarbon fuels and lubricants by way of the Fischer-Tropsch process, and (iii) the methanol-to-gasoline process. In each case, the production methods include (i) steam refoming of natural gas or liquid hydrocarbon derivatives to produce hydrogen and (ii) the gasification of carbonaceous feedstocks such as coal (Chapter 4), crude oil resid (Chapter 5), biomass (Chapter 6), and waste (Chapter 7).

The gasification process is a process that converts organic (carbonaceous) feedstocks into carbon monoxide, carbon dioxide and hydrogen by reacting the feedstock at high temperatures (>700°C, 1290°F), without combustion, with a controlled amount of oxygen and/or steam (Lee *et al.*, 2007; Speight, 2008, 2013). The resulting gas mixture (*synthesis gas*, also commonly referred to as *syngas*) – which is a mixture of carbon monoxide, CO, and hydrogen H_2 – is itself a fuel as well as a source of a wide variety of chemicals (Table 2.1).

The power derived from carbonaceous feedstocks and gasification followed by the combustion of the product gas(es) is considered to be a source of renewable energy if the derived gaseous products are generated from a source (e.g., biomass) other than a fossil fuel (Speight, 2008). The gasification of a carbonaceous feedstock or a derivative (i.e., char produced from the carbonaceous material) is the conversion of a carbonaceous feedstock, such as coal, by any one of a variety of processes to produce gaseous products that are combustible or can be used for the production of a range of chemicals (Figure 2.1).

The advantage of gasification is that the use of synthesis gas is potentially more efficient as compared to direct combustion of the original fuel because it can be (i) combusted at higher temperatures, (ii) used in fuel cells, (iii) used to produce methanol and hydrogen, (iv) converted via the Fischer-Tropsch process into a range of synthesis liquid fuels suitable for use in gasoline engines or diesel engines. The gasification process can also utilize

Table 2.1 Example of chemicals produced from synthesis gas.

Synthesis gas	Fuel gas			
	Town gas			
	Hydrogen	Ammonia		
		Urea		
	Fischer-Tropsch liquids	Synthetic natural gas		
		Naphtha		
		Kerosene		
		Waxes		
	Methanol	Dimethyl ether	Ethylene	Polyolefins
			Propylene	Polyolefins
		Acetic acid	Methyl acetate	
			Acetate esters	
			Polyvinyl acetate	
			Acetic anhydride	

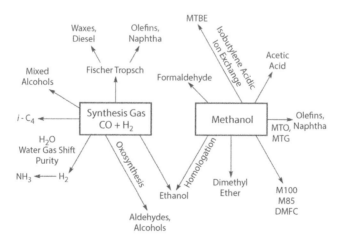

Figure 2.1 Potential products from synthesis gas produced from gasification of carbonaceous feedstocks.

carbonaceous feedstocks which would otherwise have been disposed of (e.g., biodegradable waste). In addition, the high-temperature process causes corrosive ash elements including metal chlorides and potassium salts which allow clean gas production from otherwise problematic fuels.

Coal has been the primary feedstock for gasification units for many decades. However, there is a move to carbonaceous feedstocks other than coal for gasification processes with the concern on the issue of environmental pollutants and the potential shortage for coal in some areas (except in the United States) as well as concerns related to the effect of increased coal use on the environment. Nevertheless, coal use still prevails and will continue to do so for at least several decades into the future, if not well into the next century (Speight, 2013).

In fact, gasification plants are cleaner with respect to standard pulverized coal combustion facilities, producing fewer sulfur and nitrogen byproducts, which contribute to smog and acid rain. For this reason, gasification appeals as a way to utilize relatively inexpensive and expansive coal reserves, while reducing the environmental impact. Indeed, the mounting interest in coal gasification technology reflects a convergence of two changes in the electricity generation marketplace: (i) the maturity of gasification technology, and (ii) the extremely low emissions from integrated gasification combined cycle (IGCC) plants, especially air emissions, and the potential for lower cost control of greenhouse gases than other coal-based systems. Fluctuations in the costs associated with natural gas-based power, which is viewed as a major competitor to coal-based power, can also play a role. Furthermore, gasification permits the utilization of a range of carbonaceous feedstocks (such as crude oil resids, coal, biomass, and carbonaceous domestic and industrial wastes) to their fullest potential. Thus, power developers would be well advised to consider gasification as a means of converting a carbonaceous feedstock to gas.

Liquid fuels, including gasoline, diesel, naphtha and jet fuel, are usually processed via refining of crude oil (Speight, 2014a, 2017). Due to the direct distillation, crude oil is the most suited raw material for liquid fuel production. However, with fluctuating and rising prices of crude oil, coal-to-liquids (CTL) and biomass-to-liquids (BTL) processes are currently starting to be considered as alternative routes used for liquid fuels production. Both feedstocks are converted to synthesis gas which is subsequently converted into a mixture of liquid products by Fischer-Tropsch (FT) processes. The liquid fuel obtained after FT synthesis is eventually upgraded using known crude oil refinery technologies to produce gasoline, naphtha, diesel fuel and jet fuel (Dry, 1976; Chadeesingh, 2011; Speight, 2014a, 2017). Gasification processes can accept a variety of feedstocks but the reactor must be selected on the basis of feedstock properties and behavior in the process. The future depends very much on the effect of gasification processes on the surrounding environment. It is these environmental effects and issues that will direct the success of gasification.

Clean Coal Technologies (CCTs) are a new generation of advanced coal utilization processes that are designed to enhance both the efficiency and the environmental acceptability of coal extraction, preparation and use (Speight, 2013). These technologies reduce emissions, reduce waste, and increase the amount of energy gained from coal. The goal of the program was to foster development of the most promising clean coal technologies such as improved methods of cleaning coal, fluidized bed combustion, integrated gasification combined cycle, furnace sorbent injection, and advanced flue-gas desulfurization.

In fact, there is the distinct possibility that within the foreseeable future the gasification process will increase in popularity in crude oil refineries – some refineries may even be known as gasification refineries (Speight, 2011b). A gasification refinery would have, as the center piece, gasification technology as is the case with the Sasol refinery in South Africa (Couvaras, 1997). The refinery would produce synthesis gas (from the carbonaceous feedstock) from which liquid fuels would be manufactured using the Fischer-Tropsch synthesis technology.

In fact, gasification to produce synthesis gas can proceed from any carbonaceous material, including biomass. Inorganic components of the feedstock, such as metals and minerals, are trapped in an inert and environmentally safe form as char, which may have use as a fertilizer. Biomass gasification is therefore one of the most technically and economically convincing energy possibilities for a potentially carbon neutral economy.

The manufacture of gas mixtures of carbon monoxide and hydrogen has been an important part of chemical technology for approximately a century. Originally, such mixtures were obtained by the reaction of steam with incandescent coke and were known as *water gas*. Eventually, steam reforming processes, in which steam is reacted with natural gas (methane) or crude oil naphtha over a nickel catalyst, found wide application for the production of synthesis gas.

A modified version of steam reforming known as autothermal reforming, which is a combination of partial oxidation near the reactor inlet with conventional steam reforming further along the reactor, improves the overall reactor efficiency and increases the flexibility of the process. Partial oxidation processes using oxygen instead of steam also found wide application for synthesis gas manufacture, with the special feature that they could utilize low-value feedstocks such as heavy crude oil residues. In recent years, catalytic partial oxidation employing very short reaction times (milliseconds) at high temperatures (850 to 1000°C) is providing still another approach to synthesis gas manufacture (Hickman and Schmidt, 1993).

In a gasifier, the carbonaceous material undergoes several different processes: (i) pyrolysis of carbonaceous fuels, (ii) combustion, and (iii) gasification of the remaining char. The process is very dependent on the properties of the carbonaceous material and determines the structure and composition of the char, which will then undergo gasification reactions.

As crude oil supplies decrease, the desirability of producing gas from other carbonaceous feedstocks will increase, especially in those areas where natural gas is in short supply. It is also anticipated that costs of natural gas will increase, allowing gasification of other carbonaceous feedstocks to compete as an economically viable process. Research in progress on a laboratory and pilot-plant scale should lead to the invention of new process technology by the end of the century, thus accelerating the industrial use of gasification processes.

The conversion of the gaseous products of gasification processes to synthesis gas, a mixture of hydrogen (H_2) and carbon monoxide (CO), in a ratio appropriate to the application, needs additional steps, after purification. The product gases – carbon monoxide, carbon dioxide, hydrogen, methane, and nitrogen – can be used as fuels or as raw materials for chemical or fertilizer manufacture.

2.2 Synthesis Gas Generation

The synthesis gas generation process is a non-catalytic process for producing synthesis gas (principally hydrogen and carbon monoxide) for the ultimate production of high-purity hydrogen from gaseous or liquid hydrocarbon derivatives. In this process, a controlled mixture of preheated feedstock and oxygen is fed to the top of the generator where carbon monoxide and hydrogen emerge as the products.

Soot, produced in this part of the operation, is removed in a water scrubber from the product gas stream and is then extracted from the resulting carbon-water slurry with naphtha and transferred to a fuel oil fraction. The oil-soot mixture is burned in a boiler or recycled to the generator to extinction to eliminate carbon production as part of the process. The soot-free synthesis gas is then charged to a shift converter where the carbon monoxide reacts with steam to form additional hydrogen and carbon dioxide at the stoichiometric rate of 1 mole of hydrogen for every mole of carbon monoxide charged to the converter.

The reactor temperatures vary from 1095 to 1490°C (2000 to 2700°F), while pressures can vary from approximately atmospheric pressure to approximately 2000 psi (13,790 kPa). The process has the capability of producing high-purity hydrogen although the extent of the purification procedure depends upon the use to which the hydrogen is to be put. For example, carbon dioxide can be removed by scrubbing with various alkaline reagents, while carbon monoxide can be removed by washing with liquid nitrogen or, if nitrogen is undesirable in the product, the carbon monoxide should be removed by washing with copper-amine solutions. This particular process (a partial oxidation process) has also been applied to a whole range of liquid feedstocks for hydrogen production. There is now serious consideration being given to hydrogen production by the partial oxidation of solid feedstocks such as crude oil coke (from both delayed and fluid-bed reactors), lignite, and coal, as well as crude oil residua.

The chemistry of the process, using naphthalene as an example, may be simply represented as the selective removal of carbon from the hydrocarbon feedstock and further conversion of a portion of this carbon to hydrogen:

$$C_{10}H_8 + 5O_2 \rightarrow 10CO + 4H_2$$

$$10CO + 10H_2O \rightarrow 10CO_2 + 10H_2$$

Although these reactions may be represented very simply using equations of this type, the reactions can be complex and result in carbon deposition on parts of the equipment thereby requiring careful inspection of the reactor.

The gasification process is the culmination of five processes: (i) feedstock preparation, (ii) gasification, (iii) steam generation, (iv) air separation, and (v) synthesis gas cleaning, each of which will be described in detail in the subsequent chapters. Briefly, and at this point, a brief mention of the processes is warranted.

The feedstock preparation section will, more than likely, include unit operations such as (i) size reduction, (ii) screening, and (iii) slurrying, the extent of each segment is feedstock dependent and also dependent on the properties of the raw material feed and on the requirements of the gasification technology that is selected.

The gasifier is a reactor for the partial oxidation of the carbon-based raw material to produce raw synthesis gas. This is a high temperature conversion process, and in the gasification zone of the gasifier the temperature is in the range 900 to 1600°C (1650 to 2910°F). The conversion of the raw material feed to synthesis gas is endothermic. Part of the feedstock is completely oxidized to carbon dioxide and water to provide the energy for syngas production. The carbon monoxide, hydrogen, carbon dioxide, and water equilibrate at reaction conditions so that the composition of these compounds in the raw synthesis gas product is governed by the water gas shift equilibrium at outlet conditions. All elements that can be volatilized at the gasification temperature will end up in the raw synthesis gas. For example, if the raw material contained sulfur or nitrogen, the raw synthesis gas will also contain hydrogen sulfide (H2S) and ammonia (NH3). Non-volatilized carbon matter and mineral matter are rejected from the gasification process either as soot, ash, or slag. Another important aspect of gasification is heat recovery, which determines the thermal efficiency of the gasifier.

In the steam generation section, the synthesis gas composition can be manipulated by the ratio of steam to the feedstock that is co-fed to the gasifier. The steam is a source of hydrogen and is essential for feed materials with a low effective hydrogen-to-carbon ratio.

Gasification is a partial oxidation and pure oxygen is usually the most efficient oxidant. It is possible to use air as oxidant, but air should not be considered as an oxidant in a gasification process where the energy is supplied by direct combustion of part of the process feed. When air is mixed with the process feed, a large volume of inert gases, mainly nitrogen, is also introduced to the process. Once the inert gases enter the process stream, they remain in the process stream until after the Fischer-Tropsch process. Thus, most gasification processes require an air separation unit to produce pure oxygen as the oxidant for the gasification process.

The raw synthesis gas produced by the gasifier contains contaminants that are poisons for the Fischer-Tropsch catalyst. Contaminants, such as hydrgoen sulfide and halogenated compounds (that can arise when plastic waste – such as from PVC plastic waste – is part of the feedstock), must be removed from the synthesis gas. It may also be beneficial to remove carbon dioxide and water, depending on the requirements of the design of the Fischer-Tropsch gas loop.

2.3 Feedstocks

Gasification technology offers the important ability to take a wide range of feedstocks and process them into synthesis gas, from which a similarly diverse number of end products are possible. In fact, the technology has advanced to such an extent that gasifiers have been developed to suit all different types of carbonaceous materials such as highly carbonaceous products from crude oil refining – such as crude oil resids, heavy crude oil, tar sand bitumen, and coke from thermal processes – and biomass, as well as various type of waste such as agricultural waste, industrial waste, and municipal waste (Chapter 1). The flexibility stems from the ability of gasification to take any carbon and hydrogen containing feedstock and then thermochemically break down the feedstock to a gas containing simple compounds which are easy to process into several marketable products.

2.3.1 Natural Gas

Synthesis gas can be produced from natural gas, or virtually any hydrocarbon feedstock, by reaction of the feedstock with steam or oxygen. The reaction of the hydrocarbon with oxygen can be represented by the simple equations:

$$CH_4 + H_2O \rightarrow CO + 3H_2$$

$$2CH_4 + O_2 \rightarrow 2CO + 2H_2$$

The formation of synthesis gas is strongly endothermic and requires high temperatures. Steam reforming of natural gas (or shale gas) proceeds in tubular reactors that are heated externally. The process uses nickel catalyst on a special support that is resistant against the harsh process conditions. Waste heat from the oven section is used to preheat gases and to produce steam. This plant generates synthesis gas with hydrgoen-carbon monoxide (H_2/CO) ratios in the range of 3 to 4, and is suitable for hydrogen production.

The partial oxidation of methane (or other hydrocarbon derivatives) is a non-catalytic, large-scale process to make synthesis gas and yields synthesis gas with a hydrogen-carbon monoxide (H_2/CO) ratio on the order of 2. This is an optimal ratio for gas-to-liquids plants. A catalytic version of partial oxidation that is based on short-contact time conversion of methane, hydrocarbon derivatives or biomass on, for example, rhodium catalysts, is suitable for small-scale applications (Zhu *et al.*, 2004; De Campos Roseno *et al.*, 2018).

2.3.2 Crude Oil Resid, Heavy Crude Oil, Extra Heavy Crude Oil, and Tar Sand Bitumen

Gasification is the only technology which makes possible a zero residue target for refineries, contrary to all conversion technologies (including thermal cracking, catalytic cracking, cooking, deasphalting, hydroprocessing, etc.) which can only reduce the bottom volume, with the complication that the residue qualities generally get worse with the degree of conversion (Speight, 2014a, 2017). While it is unlikely that any members of the crude oil family (Chapter 1) with substantial amounts of volatile components will be used directly as a gasification feedstock, the potential is high for the use of the resids from these crude oil as gasification feedstocks.

In fact, the flexibility of gasification permits to handle any type of refinery residue, including crude oil coke, tank bottoms and refinery sludge and make available a range of value added products including electricity, steam, hydrogen and various chemicals based on synthesis gas chemistry: methanol, ammonia, MTBE, TAME, acetic acid, and formaldehyde (Speight, 2008, Chapter 7). The environmental performance of gasification is unmatched. No other technology processing low-value refinery residues can come close to the emission levels achievable with gasification (Speight, 2014a, 2017).

Gasification is also a method for converting crude oil coke and other refinery non-volatile waste streams (often referred to as refinery residuals and including but not limited to atmospheric residuum, vacuum residuum, visbreaker tar, and deasphalter pitch) into power, steam and hydrogen for use in the production of cleaner transportation fuels. The main requirement for a gasification of carbonaceous feedstocks (including coal and biomass) is that it contains both hydrogen and carbon (Table 2.2, Table 2.3). A number of factors have increased the interest in gasification applications in crude oil refinery operations: (i) coking capacity has increased with the shift to heavier, more sour crude oils being supplied to the refiners, (ii) worldwide coking capacity has increased making sale of high sulfur petcoke more difficult, especially for refineries in the United States, (iii) hazardous waste disposal has become a major issue for refiners, especially in the United States, (iv) pressure to reduce emissions of criteria pollutants and greenhouse gases is increasing,

Table 2.2 Examples of the properties of various resids available for on-site gasification.

Crude oil resid	API gravity	Sulfur % w/w	Nitrogen % w/w	Nickel ppm	Vanadium ppm	Carbon residue % w/w*
Arabian Light >650°F	17.7	3.0	0.2	10.0	26.0	7.5
Arabian Light, >1050°F	8.5	4.4	0.5	24.0	66.0	14.2
Arabian Heavy >650°F	11.9	4.4	0.3	27.0	103.0	14.0
Arabian Heavy >1050°F	7.3	5.1	0.3	40.0	174.0	19.0
Alaska, North Slope >650°F	15.2	1.6	0.4	18.0	30.0	8.5
Alaska, North Slope >1050°F	8.2	2.2	0.6	47.0	82.0	18.0
Lloydminster >650°F	10.3	4.1	0.3	65.0	141.0	12.1
Lloydminster >1050F	8.5	4.4	0.6	115.0	252.0	21.4
Kuwait >650°F	13.9	4.4	0.3	14.0	50.0	12.2
Kuwait >1050°F	5.5	5.5	0.4	32.0	102.0	23.1
Tia Juana >650°F	17.3	1.8	0.3	25.0	185.0	9.3
Tia Juana >1050°F	7.1	2.6	0.6	64.0	450.0	21.6

*Conradson

and (v) requirements to produce ultra-low sulfur fuels are increasing the hydrogen needs of the refineries.

The typical gasification system incorporated into the refinery consists of several process plants including the on-site presence of a gasifier, feedstock preparation, an air separation unit, gas cleaning operations, a sulfur recovery unit, and downstream process options depending on the desired products as well as the optional downstream processes for producing power through cogeneration, hydrogen production, and Fischer-Tropsch synthesis (Chapter 10) (Speight, 2013).

Table 2.3 Examples of other refinery feedstocks that are available for on-site gasification.

	Visbreaker bottoms	Deasphalter bottoms	C5 asphaltenes*	Delayed coke
Ultimate Analysis				
Carbon, % w/w	83.1	85.9	80.47	88.6
Hydrogen, % w/w	10.4	9.5	8.45	2.8
Nitrogen, % w/w	0.6	1.4	1.25	1.1
Sulfur, % w/w*	2.4	2.4	1.88	7.3
Oxygen, % w/w	0.5	0.5	7.95	0.0
Ash, % w/w	<0.5	<0.5	<0.5	<1.0
Specific Gravity	1.008	1.036		0.863
API Gravity	8.9	5.1		

*The use of heptane as the precipitant yields an asphaltene fraction that is different from the pentane-insoluble material as exemplified by differences in the elemental analysis. For example, the H/C ratio of the heptane-asphaltene fraction is lower than the H/C ratio of the pentane asphaltene fraction, indicating a higher degree of aromaticity in the heptane asphaltene fraction. The N/C, O/C, and S/C atomic ratios are usually higher in the heptane asphaltene fraction, indicating higher proportions of the heteroelements in this material. Nevertheless, each fraction is suitable for use as a feedstock for a gasification process.

The benefits of the addition of a gasification system in a refinery to process crude oil coke or other residuals include: (i) production of power, steam, oxygen and nitrogen for refinery use or sale, (ii) source of synthesis gas for hydrogen to be used in refinery operations as well as for the production of low-boiling refinery products through Fischer-Tropsch synthesis, (iii) increased efficiency of power generation, improved air emissions, and reduced waste stream versus combustion of crude oil coke or residues or incineration, (iv) no off-site transportation or storage for crude oil coke or residuals, and (v) the potential to dispose of waste streams including hazardous materials.

Gasification can provide high-purity hydrogen for a variety of uses within the refinery (Speight, 2014a, 2017). Hydrogen is used in refineries to remove sulfur, nitrogen, and other impurities from intermediate to finished product streams and in hydrocracking operations for the conversion of high-boiling distillates and oils into low-boiling products, naphtha, kerosene, and diesel fuel. Hydrocracking and severe hydrotreating require hydrogen which is at least 99% v/v, while less severe hydrotreating can work with gas streams containing 90% v/v pure hydrogen.

Electric power and high-pressure steam can be generated via gasification of crude oil coke and residuals to drive mostly small and intermittent loads such as compressors, blowers, and pumps. Steam can also be used for process heating, steam tracing, partial pressure reduction in fractionation systems, and stripping low-boiling components to stabilize process streams.

Carbon soot is produced during gasification – briefly, soot is a mass of impure carbon particles resulting from the incomplete combustion of hydrocarbons and is more typically a product of the gas-phase combustion process. The soot is transferred to the feedstock by contacting, in sequence, the quench water blowdown with naphtha, and then the naphtha-soot slurry with a fraction of the feed. The soot mixed with the feed is finally recycled into the gasifier, thus achieving 100% w/w conversion of carbon to gas.

2.3.3 Refinery Coke

Coke is the residue left by the destructive distillation (thermal cracking) of crude oil residua. That formed in catalytic cracking operations is usually non-recoverable, as it is often irreversibly adsorbed on to the catalyst and is employed as fuel for the process when it is burned from the catalyst. The composition of crude oil coke varies with the source of the crude oil, but in general, large amounts of high-molecular-weight complex hydrocarbon derivatives (rich in carbon but correspondingly poor in hydrogen) make up a high proportion. The solubility of crude oil *coke* in carbon disulfide has been reported to be as high as 50 to 80%, but this is in fact a misnomer, since the coke is the insoluble, honeycomb material that is the end product of thermal processes.

Crude oil coke is employed for a number of purposes, but its chief use is in the manufacture of carbon electrodes for aluminum refining, which requires a high-purity carbon – low in ash and sulfur free; the volatile matter must be removed by calcining. In addition to its use as a metallurgical reducing agent, crude oil coke is employed in the manufacture of carbon brushes, silicon carbide abrasives, and structural carbon (e.g., pipes and Rashig rings), as well as calcium carbide manufacture from which acetylene is produced:

$$\text{Coke} \rightarrow \text{CaC}_2$$

$$\text{CaC}_2 + \text{H}_2\text{O} \rightarrow \text{HC}\equiv\text{CH}$$

However, with the progressive increase in the amount of coke produced from the heavier (more viscous) refinery feedstocks, there is a renewed interest in the utilization of coke as a feedstock for a gasification process. This is one of the attractive options and is gaining increasing attention to convert the petcoke to value-added products. The process offers the refiners a variety of product slates mainly via the production of synthesis gas. The products include steam, hydrogen, electricity, chemicals (such as methanol, CH_3OH, and ammonia, NH_3), and liquid fuels by way of the Fischer-Tropsch synthesis (Murthy *et al.*, 2013).

2.3.4 Coal

The best example of a carbonaceous feedstock for a gasification process is coal, which is often used as a gasification feedstock in a refinery provided suitable coal resources are within economic reach. In fact, as a result of the rapid increase in the use of coal from the 15th century onwards (Nef, 1957; Taylor and Singer, 1957; Speight, 2013) it is not surprising the concept of using coal to produce a flammable gas, especially the use of the water and hot coal (van Heek and Muhlen, 1991), became commonplace (Elton, 1958). However, there are four main types of coal (Table 2.4) and, for maximum process efficiency, the properties of each type of coal must be given due consideration before selection of the requisite gasification process parameters.

Table 2.4 The four main types (ranks) of coal*.

Type/rank	Properties
Lignite	Sometimes called brown coal.
	The youngest of the coal types, and has the lowest energy content. Contains between 25 and 35% w/w carbon.
	High moisture content coal.
	Typically burned in power plants for electrical production.
Sub-bituminous coal	Has a higher energy content than lignite.
	Contains between 35-45% w/w carbon.
Bituminous coal	Contains anywhere from 45 to 86% w/w carbon.
	A higher heating value than sub-bituminous coal.
	Used for electrical production.
	Plays a large role in the steel and iron industries.
Anthracite	Contains 86 to 97% w/w carbon.
	Has a slightly lower heating value than bituminous coal.

*The term low-rank coal refers to the lowest heating value and higher moisture content lignite and sub-bituminous coals.

The production of gas from coal has been a vastly expanding area of coal technology and, as a result, the characteristics of rank, mineral matter, particle size, and reaction conditions are all recognized as having a bearing on the outcome of the process; not only in terms of gas yields but also on gas properties (Massey, 1974; van Heek and Muhlen, 1991). The products from the gasification of coal may be of low, medium, or high heat-content (high-Btu) content as dictated by the process as well as by the ultimate use for the gas (Figure 2.1) (Fryer and Speight, 1976; Mahajan and Walker, 1978; Anderson and Tillman, 1979; Cavagnaro, 1980; Bodle and Huebler; Argonne, 1990; Baker and Rodriguez, 1990; Probstein and Hicks, 1990; Lahaye and Ehrburger, 1991; Matsukata *et al.*, 1992; Speight, 2013).

Coal is a fossil fuel formed in swamp ecosystems where plant remains were saved from oxidation and biodegradation by being covered with water and mud. Coal is a combustible organic sedimentary rock (composed primarily of carbon, hydrogen and oxygen as well as other minor elements including sulfur) formed from ancient vegetation and consolidated between other rock strata to form coal seams. The harder forms can be regarded as organic metamorphic rocks (e.g., anthracite coal) because of a higher degree of maturation.

Coal remains in adequate supply and at current rates of recovery and consumption, the world global coal reserves have been variously estimated to have a reserves/production ratio of at least 155 years. However, as with all estimates of resource longevity, coal longevity is subject to the assumed rate of consumption remaining at the current rate of consumption and, moreover, to technological developments that dictate the rate at which the coal can be mined. But most importantly, coal is a fossil fuel and an *unclean* energy source that will only add to global warming. In fact, the next time electricity is advertised as a clean energy source, just consider the means by which the majority of electricity is produced – almost 50% of the electricity generated in the United States derives from coal (EIA, 2007; Speight, 2013).

Coal occurs in different forms or *types* (Speight, 2013). Variations in the nature of the source material and local or regional variations in the coalification processes cause the vegetal matter to evolve differently. Various classification systems thus exist to define the different types of coal. Chemically, coal is a hydrogen-deficient hydrocarbon with an atomic hydrogen-to-carbon ratio near 0.8, as compared to crude oil hydrocarbon derivatives, which have an atomic hydrogen-to-carbon ratio approximately equal to 2, and methane (CH_4) that has an atomic carbon-to-hydrogen ratio equal to 4. For this reason, any process used to convert coal to alternative fuels must add hydrogen or redistribute the hydrogen in the original coal to generate hydrogen-rich products and coke (Speight, 2013).

The chemical composition of the coal is defined in terms of its proximate and ultimate (elemental) analyses (Speight, 2013). The parameters of proximate analysis are (i) moisture, (ii) volatile matter, (iii) mineral matter, which is determined as combustion ash, and (iv) fixed carbon. Elemental or ultimate analysis encompasses the quantitative determination of carbon, hydrogen, nitrogen, oxygen, and sulfur within the coal. Additionally, specific physical and mechanical properties of coal and particular carbonization properties are also determined.

Carbon monoxide and hydrogen are produced by the gasification of coal in which a mixture of gases is produced. In addition to carbon monoxide and hydrogen, methane and other hydrocarbon derivatives are also produced depending on conditions. Gasification may be accomplished either *in situ* or in processing plants. *In situ* gasification is accomplished by controlled, incomplete burning of a coal bed underground while adding air and steam. The gases are withdrawn and may be burned to produce heat, generate electricity or are utilized as synthesis gas in indirect liquefaction as well as for the production of chemicals.

Producing diesel and other fuels from coal can be performed through the conversion of coal to synthesis gas, a combination of carbon monoxide, hydrogen, carbon dioxide, and methane. Synthesis gas is subsequently reacted through Fischer-Tropsch Synthesis processes to produce hydrocarbon derivatives that can be refined into liquid fuels. By increasing the quantity of high-quality fuels from coal (while reducing costs), research into this process could help in mitigating the dependence on ever-increasingly expensive and depleting stocks of crude oil.

While coal is an abundant natural resource, its combustion or gasification produces both toxic pollutants and greenhouse gases. By developing adsorbents to capture the pollutants (mercury, sulfur, arsenic, and other harmful gases), it is possible not only to reduce the quantity of emitted gases but also to maximize the thermal efficiency of the cleanup. Thus, gasification offers one of the most clean and versatile ways to convert the energy contained in coal into electricity, hydrogen, and other sources of power. Turning coal into synthesis gas isn't a new concept; in fact the basic technology dates back to World War II.

2.3.5 Biomass

As the last two to three decades have evolved, biomass has been considered as any renewable feedstock which is in principle *carbon neutral* (while the plant is growing, it uses the energy of the sun to absorb the same amount of carbon from the atmosphere as it releases into the atmosphere).

Biomass is a broad term used to describe any organic material or resource which is derived from plant or animal matter, and primarily used as fuel. This includes wood, food crops, grass and woody plants, agriculture and forestry residues, and organic components of municipal and industrial wastes. Biomass excludes organic material which has been transformed via geological processes to fossil fuels such as natural gas, crude oil, and coal. The chemical composition of biomass varies substantially, because of the broad range of materials involved, but the main components are moisture, carbohydrates or sugars, lignin, and mineral matter which becomes ash upon combustion or gasification.

There are two methods for converting biomass into high-value products: (i) biochemical conversion and (ii) thermochemical conversion. Biochemical conversion involves the use of biological processes to convert biomass into biofuels, chemicals and electrical power. In the case of ethanol production, enzymes and/or chemical processes are used to extract sugars from the biomass, which can then be converted to ethanol via fermentation. Thermochemical conversion, either gasification using less than stoichiometric oxygen or pyrolysis (the gasification of biomass in the absence of oxygen), uses heat and pressure to convert biomass to liquid fuels, chemicals and electrical power. Combustion is an option for conversion of biomass to electrical power; however, the synthesis gas produced by gasification is much easier and economical to clean than are the exhaust gases produced by combustion. This results in gasification providing better environmental performance, including a cheaper method of capturing carbon dioxide. In addition, synthesis gas produced by gasification can also be processed into a variety of marketable products, where combustion is limited to electrical production via the steam cycle.

The two main advantages that gasification has over biochemical conversion processes are (i) the speed with which the end product is produced (minutes for gasification compared to days for biochemical conversion) and (ii) the ability of the gasification process to extract the energy held in lignin, the harder structural part of the biomass. Fermentation methods currently are unable to extract the energy stored in the lignin; however, this does present the possibility of using gasification as a waste treatment method for materials that cannot be fermented at a biochemical conversion facility.

Through gasification, biomass is converted into a gas consisting of carbon monoxide, carbon dioxide, hydrogen, and other compounds by applying heat and pressure in the presence of steam and a small amount of oxygen, typical of gasification of any organic material. Biomass produces hydrocarbon compounds in the gas, especially in the absence of oxygen, requiring an extra step to remove them with a catalyst downstream of the gasifier. The water gas shift reaction is then used to convert some of the carbon monoxide with water to form more carbon dioxide and hydrogen:

$$CO + H_2O \rightleftharpoons CO_2 + H_2$$

Generally, biomass does not gasify as easily as coal but most of the main type of gasifiers can be designed to handle biomass or a combination of biomass and coal, but each has positives and negatives and the selection depends on the specific feedstock and the desired product or products.

Raw materials that can be used to produce biomass derived fuels are widely available; they come from a large number of different sources and in numerous forms (Rajvanshi, 1986). The main basic sources of biomass include: (i) wood, including bark, logs, sawdust, wood chips, wood pellets and briquettes, (ii) high-yield energy crops, such as wheat, grown specifically for energy applications, (iii) agricultural crops and residues (e.g., straw), and (iv) industrial waste, such as wood pulp or paper pulp. For processing, a simple form of biomass such as untreated and unfinished wood may be converted into a number of physical forms, including pellets and wood chips, for use in biomass boilers and stoves.

Biomass includes a wide range of materials that produce a variety of products which are dependent upon the feedstock (Balat, 2011; Demirbaş, 2011; Ramroop Singh, 2011; Speight, 2011a). In addition, the heat content of the different types of biomass widely varies and have to be taken into consideration when designing any conversion process (Jenkins and Ebeling, 1985). Thermal conversion processes use heat as the dominant mechanism to convert biomass into another chemical form. The basic alternatives of combustion, torrefaction, pyrolysis, and gasification are separated principally by the extent to which the chemical reactions involved are allowed to proceed (mainly controlled by the availability of oxygen and conversion temperature) (Speight, 2011a).

Energy created by burning biomass (fuel wood), also known as dendrothermal energy, is particularly suited for countries where fuel wood grow more rapidly, e.g., tropical countries. There is a number of other less common, more experimental or proprietary thermal processes that may offer benefits including hydrothermal upgrading (HTU) and hydroprocessing. Some have been developed to be compatible with high moisture content biomass (e.g., aqueous slurries) and allow them to be converted into more convenient forms. Some of the applications of thermal conversion are combined heat and power (CHP) and co-firing. In a typical dedicated biomass power plant, efficiencies range from 7 to 27%. In contrast, biomass co-firing with coal, typically occurs at efficiencies close to those of coal combustors (30 to 40%) (Baxter, 2005; Liu *et al.*, 2011).

Many forms of biomass contain a high percentage of moisture (along with carbohydrates and sugars) and mineral constituents – both of which can influence the economics and viability of a gasification process. The presence of high levels of moisture in biomass reduces the temperature inside the gasifier, which then reduces the efficiency of the gasifier. Many biomass gasification technologies therefore require dried biomass to reduce the moisture content prior to feeding into the gasifier. In addition, biomass can come in a range of sizes. In many biomass gasification systems, biomass must be processed to a uniform size or shape to be fed into the gasifier at a consistent rate as well as to maximize gasification efficiency.

Biomass such as wood pellets, yard and crop waste and "energy crops" including switch grass and waste from pulp and paper mills can also be employed to produce bioethanol and synthetic diesel. Biomass is first gasified to produce synthesis gas and then subsequently converted via catalytic processes to the aforementioned downstream products. Biomass can also be used to produce electricity – either blended with traditional feedstocks, such as coal, or as a single feedstock. Also, most biomass gasification systems use air instead of oxygen for gasification reactions (which is typically used in large-scale industrial and power gasification plants). Gasifiers that use oxygen require an air separation unit to provide the gaseous/liquid oxygen; this is usually not cost-effective at the smaller scales used in biomass gasification plants. Air-blown gasifiers utilize oxygen from air for gasification processes.

The use of biomass to offset use of fossil fuels can lower greenhouse gas emissions. Burning biomass for energy does create carbon dioxide; however, the same amount of carbon dioxide is recaptured in future biomass growth. So as long as there is no carbon dioxide released into the atmosphere during production of the biomass (growth, harvest, and processing), the process sums to zero net carbon dioxide to the atmosphere. Also, if biomass is left to decompose naturally, or openly burned, a large portion of the carbon in the biomass is converted to methane (CH_4). The percentage of carbon in the biomass which is converted to methane is 50% in the case of rotting, and 5 to 10% in the case of open burning. These processes are actually worse in regard to greenhouse gas emissions than is biomass gasification, which converts nearly all of the carbon contained in the biomass to carbon dioxide, since methane is a much stronger greenhouse gas than carbon dioxide.

In general, biomass gasification plants are comparatively smaller to those of typical coal or crude oil coke plants used in the power, chemical, fertilizer and refining industries. As such, they are less expensive to build and have a smaller environmental footprint. While a large industrial gasification plant may take up 150 acres of land and process 2,500 to 15,000 tons per day of feedstock (e.g., coal or crude oil coke), smaller biomass plants typically process 25 to 200 tons of feedstock per day and take up less than 10 acres.

Finally, while biomass may seem to some observers to be the answer to the global climate change issue, advantages and disadvantages of biomass as feedstock must be considered carefully (Table 2.5). In addition, while taking the issues of global climate change into account, it must not be ignored that the Earth is in an interglacial period when warming will take place. The extent of this warming is not known – no one was around to measure the temperature change in the last interglacial period – and by the same token the contribution of anthropological sources to global climate change cannot be measured accurately.

The potential variability of biomass feedstocks, longer-term changes in refuse and the size limitation of a power plant using only waste and/or biomass can be overcome

Table 2.5 Advantages and disadvantages of using biomass as a gasification feedstock.

Advantages	Theoretically inexhaustible fuel source.
	Minimal environmental impact when direct combustion of plant mass is not used to generate energy (i.e., fermentation, pyrolysis, etc., are used instead), (iii) alcohol derivatives and other fuels produced by biomass are efficient, viable, and relatively clean-burning.
	Available on a worldwide basis.
Disadvantages	Possible adverse contribution to global climate change and particulate pollution if combusted directly.
	Not always a cheap source of energy, both in terms of producing biomass and the conversion to alcohols or other fuels.
	Life cycle assessments (LCA) should be taken into account to address energy inputs and outputs but there is most likely a net loss of energy when operated on a small scale which requires that energy to grow the plant mass must be taken into account.

combining biomass, refuse, and coal. It also allows benefit from a premium electricity price for electricity from biomass and the gate fee associated with waste. If the power plant is gasification-based, rather than direct combustion, further benefits may be available. These include a premium price for the electricity from waste, the range of technologies available for the gas to electricity part of the process, gas cleaning prior to the main combustion stage instead of after combustion and public image, which is currently generally better for gasification as compared to combustion. These considerations lead to current studies of co-gasification of wastes/biomass with coal (Speight, 2008).

2.3.6 Solid Waste

Waste may be municipal solid waste (MSW) which had minimal presorting, or refuse-derived fuel (RDF) with significant pretreatment, usually mechanical screening and shredding. Other more specific waste sources (excluding hazardous waste) and possibly including crude oil coke, may provide niche opportunities for co-utilization (Bridgwater, 2003; John and Singh, 2011; Arena, 2012; Speight, 2013, 2014b). The traditional waste-to-energy plant, based on mass-burn combustion on an inclined grate, has a low public acceptability despite the very low emissions achieved over the last decade with modern flue gas clean-up equipment. This has led to difficulty in obtaining planning permissions to construct needed new waste-to-energy plants. After much debate, various governments have allowed options for advanced waste conversion technologies (gasification, pyrolysis and anaerobic digestion), but will only give credit to the proportion of electricity generated from non-fossil waste.

Municipal solid waste is a readily available, low-cost fuel, with a high organic content when processed to suit the particular gasification process being used. Several plants in Japan and Europe already employ gasification technology for treatment of municipal solid waste. Metal and glass must be removed from the municipal solid waste as it is preprocessed into refuse-derived fuel in order to increase the heating value of the feedstock and avoid gasifier operational problems. In communities with recycling programs, costs associated with removing these materials will be minimized, giving waste gasification the greatest opportunity for success. The systems used for the production of refuse-derived fuel usually use a combination of size reduction, screening, magnetic separation and density separation to remove the non-combustible materials (such as metal and glass) from the municipal solid waste.

The principle behind waste gasification and the production of gaseous fuels is that waste contains carbon and it is this carbon that is converted to gaseous products via gasification chemistry. Thus when waste is fed to a gasifier, water, and volatile matter are released and a char residue is left to react further. Use of waste materials as co-gasification feedstocks may attract significant disposal credits (Ricketts *et al.*, 2002). Cleaner biomass materials are renewable fuels and may attract premium prices for the electricity generated. Availability of sufficient fuel locally for an economic plant size is often a major issue, as is the reliability of the fuel supply. Use of more-predictably available feedstock alongside these fuels overcomes some of these difficulties and risks. However, the issues associated with gasification of municipal solid waste include, like the gasification of any mixed feedstock, feedstock homogeneity, for many gasifiers, feedstock heterogeneity and process scale up can lead to a number of mechanical problems, shutdowns, sintering and hot spots leading to corrosion and failure of the reactor wall (most of the processes proposed for waste gasification do not include a separation process).

Furthermore, the disposal of municipal and industrial waste has become an important problem because the traditional means of disposal, landfill, are much less environmentally acceptable than previously. Much stricter regulation of these disposal methods will make the economics of waste processing for resource recovery much more favorable. One method of processing waste streams is to convert the energy value of the combustible waste into a fuel. One type of fuel attainable from waste is a low heating value gas, usually 100 to 150 Btu/scf, which can be used to generate process steam or to generate electricity. Co-processing such waste with coal is also an option (Speight, 2008, 2013a, 2014b). However, co-gasification technology varies, being usually site specific and high feedstock dependent (Ricketts et al., 2002).

One of the major challenges to the gasification process of landfill waste is that such waste has high moisture content and is heterogeneous in nature. Particle size and the presence of a number of components in the waste, such as sulphur, chlorides or metal vary considerably. The interconnected properties of heating value and moisture content play an important role. Hence, pre-preparation must be carefully considered in any waste gasification process. There are a number of different approaches to pre-preparation. Most of these involve mechanical shredding and metals removal using magnetic and electric devices.

Analyses of the composition of municipal solid waste indicate that plastics do make up measurable amounts (5 to 10% w/w or more) of solid waste streams. Many of these plastics are worth recovering as energy. In fact, many plastics, particularly the poly-olefin derivatives, have high calorific values and simple chemical constitutions of primarily carbon and hydrogen. As a result, waste plastics are ideal candidates for the gasification process. Because of the myriad of sizes and shapes of plastic products size reduction is necessary to create a feed material of a size less than 2 inches in diameter. Some forms of waste plastics such as thin films may require a simple agglomeration step to produce a particle of higher bulk density to facilitate ease of feeding. A plastic, such as high-density polyethylene, processed through a gasifier is converted to carbon monoxide and hydrogen and these materials in turn may be used to form other chemicals including ethylene from which the polyethylene is produced – *closed the loop recycling*.

Recovering energy from municipal solid waste in waste-to-energy (WTE) plants reduces the space required for land filling and offsets the use of fossil fuels for electrical production. When compared to combustion for processing of municipal solid waste, gasification decreases air/water emissions. Within this context, gasification uses oxygen and water vapor to produce a combustible synthesis gas from organic compounds in the municipal solid waste, which can be used to generate electricity, produce chemicals, liquid fuels, hydrogen (H_2), etc. The synthesis gas produced from municipal solid waste by a gasifier is cleaned up more economically and using simpler systems compared to combustion exhaust gases due to the synthesis gas being more condensed. The conversion of energy in gasification is also much more efficient than the thermal conversion offered by combustion. Challenges to the commercialization of the gasification of municipal solid waste include the processing costs of converting municipal solid waste to refuse-derived fuel (RDF) and the formation of tars in the high temperature and pressure environment of the gasifier. Tars can make downstream processing of the synthesis gas more difficult and may result in excessive process train downtime.

The heat content of the refuse-derived fuel depends on the amount of moisture and combustible organic material. Refuse-derived fuel typically has less variability than municipal solid waste which can vary greatly when looking at a small sample. This is important for gasification due to the need to optimize gasifier conditions for specific fuel compositions. To reduce the residence time in the gasifier, the refuse-derived fuel is shredded to a smaller size – the shredding process also serves the purpose of uniformly distributing the various materials, giving the refuse derived a more stable composition, in addition to decreasing the moisture content of the refuse-derived fuel.

The traditional waste-to-energy plant, based on mass-burn combustion on an inclined grate, has a low public acceptability despite the very low emissions achieved over the last decade with modern flue gas clean-up equipment. This has led to difficulty in obtaining planning permissions to construct needed new waste-to-energy plants. After much debate, various governments have allowed options for advanced waste conversion technologies (gasification, pyrolysis and anaerobic digestion), but will only give credit to the proportion of electricity generated from non-fossil waste.

Co-utilization of waste and biomass with coal may provide economies of scale that help achieve the above identified policy objectives at an affordable cost. In some countries, governments propose co-gasification processes as being *well suited for community-sized developments* suggesting that waste should be dealt with in smaller plants serving towns and cities, rather than moved to large, central plants (satisfying the so-called *proximity principle*).

In fact, neither biomass nor wastes are currently produced, or naturally gathered at sites in sufficient quantities to fuel a modern large and efficient power plant. Disruption, transport issues, fuel use, and public opinion all act against gathering hundreds of megawatts (MWe) at a single location. Biomass or waste-fired power plants are therefore inherently limited in size and hence in efficiency (labor costs per unit electricity produced) and in other economies of scale. The production rates of municipal refuse follow reasonably predictable patterns over time periods of a few years. Recent experience with the very limited current *biomass for energy* harvesting has shown unpredictable variations in harvesting capability with long periods of zero production over large areas during wet weather. The situation is very different for coal, which is generally mined or imported and thus large quantities are available from a single source or a number of closely located sources, and supply has been reliable and predictable. However, the economics of new coal-fired power plants of any technology or size have not encouraged any new coal-fired power plant in the gas generation market.

The potential unreliability of biomass, longer-term changes in refuse and the size limitation of a power plant using only waste and/or biomass can be overcome combining biomass, solid waste (refuse), and coal. The use of combined feedstocks also allows benefit from a premium electricity price for electricity from biomass and the gate fee associated with waste. If the power plant is gasification-based, rather than direct combustion, further benefits may be available. These include a premium price for the electricity from waste, the range of technologies available for the gas to electricity part of the process, gas cleaning prior to the main combustion stage instead of after combustion and public image, which is currently generally better for gasification as compared to combustion. These considerations lead to current studies of co-gasification of biomass and/or solid waste as combined feedstocks with coal (Speight, 2008, 2013; Luque and Speight, 2015).

For large-scale power generation (>50 MWe), the gasification field is dominated by plant based on the pressurized, oxygen-blown, entrained flow or fixed-bed gasification of fossil fuels. Entrained gasifier operational experience to date has largely been with well-controlled fuel feedstocks with short-term trial work at low co-gasification ratios and with easily handled fuels.

Use of waste materials as co-gasification feedstocks may attract significant disposal credits. Cleaner biomass materials are renewable fuels and may attract premium prices for the electricity generated. Availability of sufficient fuel locally for an economic plant size is often a major issue, as is the reliability of the fuel supply. Use of more-predictably available coal alongside these fuels overcomes some of these difficulties and risks. Coal could be regarded as the stand-in which keeps the plant running when the fuels producing the better revenue streams are not available in sufficient quantities.

Coal characteristics are very different to the alternate sources of hydrocarbon fuels such as biomass and waste. Hydrogen-to-carbon ratios are higher for younger fuels, as is the oxygen content. This means that reactivity is very different under gasification conditions. Gas cleaning issues can also be very different, with sulfur always a major concern for coal gasification and chlorine compounds and tars more important for waste and biomass gasification. There are no current proposals for adjacent gasifiers and gas cleaning systems, one handling biomass or waste and one coal, alongside each other and feeding the same power production equipment. However, there are some advantages to such a design as compared with mixing fuels in the same gasifier and for the gas cleaning systems.

Electricity production or combined electricity and heat production remain the most likely area for the application of gasification or co-gasification. The lowest investment cost per unit of electricity generated is the use of the gas in an existing large power station. This has been done in several large utility boilers, often with the gas fired alongside the main fuel. This option allows a comparatively small thermal output of gas to be used with the same efficiency as the main fuel in the boiler as a large, efficient steam turbine can be used. It is anticipated that addition of gas from a biomass or wood gasifier into the natural gas feed to a gas turbine to be technically possible but there will be concerns as to the balance of commercial risks to a large power plant and the benefits of using the gas from the gasifier.

The use of fuel cells with gasifiers is frequently discussed but the current cost of fuel cells is such that their use for mainstream electricity generation is uneconomic. Furthermore, the disposal of municipal and industrial waste has become an important problem because the traditional means of disposal, landfill, are much less environmentally acceptable than previously. Much stricter regulation of these disposal methods will make the economics of waste processing for resource recovery much more favorable. In fact, one method of processing waste streams is to convert the energy value of the combustible waste into a fuel. One type of fuel attainable from waste is a low heating value gas, usually 100-150 Btu/scf, which can be used to generate process steam or to generate electricity (Gay *et al.*, 1980). Co-processing such waste with coal is also an option (Speight, 2008, 2013; Luque and Speight, 2015).

2.3.7 Black Liquor

Another waste that is not often recognized as a source of energy – in the current context a potential source of synthesis gas – is the waste from pulping processes. As an example, black liquor is the spent liquor from the Kraft process in which pulpwood is converted into paper pulp by removing lignin and hemicellulose constituents as well as other extractable materials from wood to free the cellulose fibers. The equivalent spent cooking liquor in the sulfite process is usually called *brown liquor*, but the terms *red liquor*, *thick liquor*, and *sulfite liquor* are also used. Approximately seven units of black liquor are produced in the manufacture of one unit of pulp (Biermann, 1993).

Black liquor is the spent liquor from the Kraft process in which pulpwood is converted into paper pulp by removing lignin and hemicellulose constituents as well as other extractable materials from wood to free the cellulose fibers. The present-day chemical pulping process uses a complex combustion system called a recovery boiler to generate process heat and electricity as well as to recover the processing chemicals in an almost closed cycle. The recovery boiler is a very complex device, which is actually operated as a gasifier-combustor. After evaporation of the majority of the water, the very high solids black liquor is sprayed onto a mass of char in the bottom of the boiler. Black liquor comprises an aqueous solution of lignin residues, hemicellulose, and the inorganic chemical used in the process and 15% w/w solids of which 10% w/w are inorganic and 5% w/w are organic. Typically, the organic constituents in black liquor are 40 to 45% w/w soaps, 35 to 45% w/w lignin, and 10 to 15% w/w other (miscellaneous) organic materials.

The organic constituents in the black liquor are made up of water/alkali soluble degradation components from the wood. Lignin is partially degraded to shorter fragments with sulfur contents in the order of 1 to 2% w/w and sodium content at approximately 6% w/w of the dry solids. Cellulose (and hemicellulose) is degraded to aliphatic carboxylic acid soaps and hemicellulose fragments. The extractable constituents yield *tall oil soap* and crude turpentine. The tall oil soap may contain up to 20% w/w sodium. Residual lignin components currently serve for hydrolytic or pyrolytic conversion or combustion. Alternative, hemicellulose constituents may be used in fermentation processes.

Gasification of black liquor has the potential to achieve higher overall energy efficiency as compared to those of conventional recovery boilers, while generating an energy-rich synthesis gas. The synthesis gas can then be burned in a gas turbine combined cycle system (*BLGCC* – black liquor gasification combined cycle – and similar to *IGCC*, integrated gasification combined cycle) to produce electricity or converted (through catalytic processes) into chemicals or fuels (e.g., methanol, dimethyl ether, Fischer-Tropsch hydrocarbon derivatives and diesel fuel).

The organic constituents in the black liquor are made up of water/alkali soluble degradation components from the wood. Lignin is partially degraded to shorter fragments with sulfur contents in the order of 1 to 2% w/w and sodium content at approximately 6% w/w of the dry solids. Cellulose (and hemicellulose) is degraded to aliphatic carboxylic acid soaps and hemicellulose fragments. The extractable constituents yield *tall oil soap* and crude turpentine. The tall oil soap may contain up to 20% w/w sodium. Lignin components currently serve for hydrolytic or pyrolytic conversion or combustion. Alternative, hemicellulose constituents may be used in fermentation processes.

In another aspect, lignin pyrolysis produces reducing gases and char which react with the spent pulping chemicals to produce sodium carbonate (Na_2CO_3) and sodium sulfide (Na_2S). Other minerals in the feedstock appear as non-usable chemical ash and have to be removed from the cycle. The gas product from the char bed passes to an oxidizing zone in the furnace where the gas is combusted to produce process steam (and electricity) as well as provide radiant heat back to the char bed for the reduction chemistry to take place.

The product chemicals are molten, drained from the char bed to collectors, and then poured into water to produce green liquor.

Thus, the pulp and paper industry offers unique opportunities for the production of synthesis gas insofar as an important part of many pulp and paper plants is the chemicals recovery cycle where black liquor is combusted in boilers. Substituting the boiler by a gasification plant with additional biofuel and electricity production is very attractive, especially when the old boiler has to be replaced. The equivalent spent cooking liquor in the sulfite process is usually called *brown liquor*, but the terms *red liquor*, *thick liquor*, and *sulfite liquor* are also used. Approximately seven units of black liquor are produced in the manufacture of one unit of pulp (Biermann, 1993).

2.3.8 Mixed Feedstocks

Any carbonaceous feedstock may be co-gasified with waste or biomass for environmental, technical or commercial reasons (Chapter 5, Chapter 6, Chapter 7). It allows larger, more efficient plants than those sized for grown biomass or arising waste within a reasonable transport distance; specific operating costs are likely to be lower and fuel supply security is assured.

Although the gasification processes used for individual feedstocks are relatively straightforward, process efficiency depends for the most part on the unique characteristics of each feedstock. In addition, the non-uniformity of the feedstocks and the variability of the specific compositions over time require flexible and robust gasifiers. For example, consideration should be given to the individual constituents of each feedstock to determine if there are any reactions between the constituents that can have an adverse effect on the process. An example is the composition of any mineral matter that may cause the constituents to interact to form a troublesome slag that may cause harm to the gasifier.

However caution is advised when using mixed feedstocks for gasification or, for that matter, for any conversion process. It has been the method in the past (and even continued to be the method in some cases) to assess the properties of the mixed feedstock by calculating an average value for one or more of the properties of the mixed feedstock. This is a dangerous practice because it ignores the potential for interaction of the contents of each feedstock that can cause changes in the chemistry of the process as well as a loss of process efficiency.

One approach to avoid a significant decrease in conversion efficiency is to develop a formulation of mixed feedstocks in order to produce a more consistent material. Formulation combines various preprocessed resources and/or additives to produce a feedstock that provides process consistency.

Feedstock formulation is not a new concept and has been used for decades by the coal industry. For example, different grades of coal are blended for power generation to reduce the sulfur content and the nitrogen content of the feedstock. In the current context, biomass blending feedstocks refers to the combination of multiple sources of the same biomass resource to average out compositional and moisture variations, whereas aggregation refers to the combination of different raw or preprocessed biomass resources to produce a single, consistent feedstock with desirable properties. Examples include mixing blended corn stover with blended switchgrass; mixing blended wheat straw with blended softwood residuals; and mixing blended Miscanthus with blended rice hulls. This strategy allows the desirable characteristics of many types of feedstocks to be combined to achieve a better feedstock than any of the feedstocks alone (Shi *et al.*, 2013; Lu and Berge, 2014).

Examples to emphasize the above considerations are presented below and relate to (i) the gasification of biomass and coal, (ii) the gasification of biomass and municipal solid waste.

2.3.8.1 Biomass and Coal

The gasification of *biomass and coal* (Chapter 4, Chapter 6) blends is of considerable current interest because of the reduction of the usual high yield of tar products that result from biomass gasification. Various operations involved in the biomass-coal-gasification process such as the typically high moisture-content biomass is usually not just dried, but also subject to torrefaction which involves heating to temperatures typically ranging between 200 and 320°C 390 to 610°F) in the absence of oxygen, at which point the biomass undergoes a mild form of pyrolysis) and possibly compacted – this improves the quality as a feedstock for the process. Also, size reduction of both the biomass and the coal to uniformly sized particles is required for optimum gasification.

The product gas compositions are influenced by both the type of biomass co-gasified, as well as its proportion in the feed mixture. Generally, higher H_2 content results from greater biomass inclusion; in particular, lignin in woody biomass seems to boost H_2 yield in syngas. A wide range of proportions of coal and biomass may be possible for given applications, but the optimum is a complex function of the type of coal used, type(s) of biomass, gasifier type and operating conditions, desired syngas composition, etc., not to mention the available quantities of the biomass which may be considerably less than the coal available.

Furthermore, the cleanup of synthesis gas derived from mixed feedstocks may be more complicated than for gasification or of the individual feedstocks because the species present in each raw feedstock as well as those (environmentally unfriendly) species that are present in elevated amounts from, say, biomass gasification (such as tars and alkalis) may need to be addressed.

2.3.8.2 Biomass and Municipal Solid Waste

The gasification of *biomass and municipal solid waste* differ in many ways from the gasification of crude oil coke or conversion of natural gas to synthesis gas. While the gasification technologies used with biomass (Chapter 6) or municipal or municipal solid waste (Chapter 7) are relatively straightforward, performance depends greatly on the unique characteristics of the feedstock. These feedstocks have much higher moisture content and less heating value by volume than coal. In addition, the non-uniformity of the feedstocks and the variability of the specific compositions over time require flexible and robust gasifiers.

Co-gasification technology varies, being usually site specific and feedstock dependent. In fact, biomass and municipal solid waste feedstocks are highly variable feedstocks that present issues for feed systems as these feedstocks are largely heterogeneous in their delivered state. Some biomass, such as sawdust from lumber mills, can be in a condition suitable for many existing feed systems most municipal solid wastes require extensive preparation or feed system customization. Biomass and municipal solid waste may also have characteristics such as higher moisture content which may necessitate pre-gasification drying. The mineral matter content of each feedstock can also vary widely and the gasifier must be able to handle variable (even high) levels of mineral matter and the ensuing ash.

At the largest scale, the plant may include the well-proven fixed-bed and entrained-flow gasification processes. At smaller scales, emphasis is placed on technologies which appear closest to commercial operation. Pyrolysis and other advanced thermal conversion processes are included where power generation is practical using the on-site feedstock produced. However, the needs to be addressed are (i) core fuel handling and gasification/pyrolysis technologies, (ii) fuel gas clean-up, and (iii) conversion of fuel gas to electric power (Ricketts et al., 2002).

2.4 Influence of Feedstock Quality

The influence of physical process parameters and the effect of feedstock type on gasification include, for example, the fact that the reactivity of coal generally decreases with increase in rank (from lignite to subbituminous coal to bituminous coal anthracite). Furthermore, the smaller the particle size, the more contact area between the coal and the reaction gases, leading to a more rapid reaction. For medium-rank coal and low-rank coal, reactivity increases with an increase in pore volume and surface area, but for coal having a carbon content greater than 85% w/w these factors have no effect on reactivity. In fact, in high-rank coal, pore sizes are so small that the reaction is diffusion controlled. Other feedstocks (such as crude oil residua and biomass) are so variable that gasification behavior and products vary over a wide range. The volatile matter produced during the thermal reactions varies widely and the ease with which tar products are formed as part of the gaseous products makes gas cleanup more difficult.

The mineral matter content of the feedstock also has an impact on the composition of the produced synthesis gas. Gasifiers may be designed to remove the produced ash in solid or liquid (slag) form. In fluidized or fixed-bed gasifiers, the ash is typically removed as a solid, which limits operational temperatures in the gasifier to well below the ash melting point. In other designs, particularly slagging gasifiers, the operational temperatures are designed to be above the ash melting temperature. The selection of the most appropriate gasifier is often dependent on the melting temperature and/or the softening temperature of the ash and the feedstock which is to be used at the facility.

Gasification reactors are very susceptible to ash production and properties. Ash can cause a variety of problems particularly in up or downdraught gasifiers. Slagging or clinker formation in the reactor, caused by melting and agglomeration of ashes, at the best will greatly add to the difficulty of gasifier operation. If no special measures are taken, slagging can lead to excessive tar formation and/or complete blocking of the reactor. A worst case is the possibility of air-channeling which can lead to a risk of explosion, especially in updraft gasifiers. Whether or not slagging occurs depends on the ash content of the fuel, the melting characteristics of the ash, and the temperature pattern allowed by gasifier design. Local high temperatures in voids in the fuel bed in the oxidation zone, caused by bridging in the bed, may cause slagging even using fuels with a high ash melting temperature.

Generally, slagging is not observed with fuels having mineral matter ash contents less than below 5 to 6% w/w. Severe slagging can be expected for fuels having mineral matter contents in excess of 12% w/w/. For fuels with mineral matter contents between 6 and 12% w/w, the slagging behavior depends to a large extent on the mineral matter composition – reflected in the ash melting temperature – which is influenced by the presence of trace elements giving rise to the formation of low melting point eutectic mixtures.

Updraft and downdraught gasification reactors are able to operate with slagging fuels if specially modified (continuously moving grates and/or external pyrolysis gas combustion). Cross draught gasification reactors, which work at temperatures on the order of 1500°C (2700°F) and higher, need special safeguards with respect to the mineral matter content of the fuel. Fluidized bed reactors, because of their inherent capacity to control the operating temperature, suffer less from ash melting and fusion problems.

High moisture content of the feedstock lowers internal gasifier temperatures through evaporation and the endothermic reaction of steam and char. Usually, a limit is set on the moisture content of feedstock supplied to the gasifier, which can be met by drying operations if necessary. For a typical fixed bed gasifier and moderate carbon content and mineral matter content of the feedstock, the moisture limit may be on the order of 35% w/w. Fluidized-bed and entrained-bed gasifiers have a lower tolerance for moisture, limiting the moisture content to approximately 5 to 10% w/w of the feedstock. Oxygen supplied to the gasifiers must be increased with an increase in mineral matter content (ash production) or moisture content in the feedstock.

Depending on the type of feedstock being processed and the analysis of the gas product desired, pressure also plays a role in product definition (Speight, 2011a, 2013). In fact, some (or all) of the following processing steps will be required: (i) pretreatment of the feedstock, (ii) primary gasification, (iii) secondary gasification of the carbonaceous residue – char – from the primary gasifier, (iv) removal of carbon dioxide, hydrogen sulfide, and other acid gases; (v) shift conversion for adjustment of the carbon monoxide/hydrogen mole ratio to the desired ratio, and (vi) catalytic methanation of the carbon monoxide/hydrogen mixture to form methane. If high heat-content (high-Btu) gas is desired, all of these processing steps are required since gasifiers do not yield methane in the concentrations required (Speight, 2008, 2011a, 2013).

Thus, the reactivity of the feedstock is an important factor in determining the design of the reactor because feedstock reactivity, which determines the rate of reduction of carbon dioxide to carbon monoxide in the reactor, influences reactor design insofar as it dictates the height needed in the reduction zone. In addition, certain operational design characteristics of the reactor system (load following response, restarting after temporary shutdown) are affected by the reactivity of the char produced in the reactor. There is also a relationship between feedstock reactivity and the number of active places on the char surface, these being influenced by the morphological characteristics as well as the geological age of the fuel. The grain size and the porosity of the char produced in the reduction zone influence the surface available for reduction and, therefore, the rate of the reduction reactions which are facilitated by reactor design.

In terms of feedstock quality, mixed feedstock must be given careful attention by virtue of the composition – related to the amount of each feedstock component in the mixture – as well as the potential interaction of the components with each other during the initial heating stage just prior to the gasification stage in the gasifier.

Both fixed-bed and fluidized-bed gasifiers have been used in co-gasification of coal and biomass – these include a downdraft fixed-bed gasifier (Kumabe *et al.*, 2007; Speight,

2011a). However, operational problems when a fluidized-bed gasifier was employed that included (i) defluidization of the fluidized-bed gasifier caused due to agglomeration of low melting point ash present in the biomass, and (ii) clogging of the downstream pipes due to excessive tar accumulation (Pan *et al.*, 2000; Vélez *et al.*, 2009). In addition, cogasification and co-pyrolysis of birch wood and coal in an updraft fixed-bed gasifier as well as in a fluidized-bed gasifier has yielded overhead products with to 6.0% w/w tar content while the fixed-bed reactor gave tar yields on the order of 25 to 26% w/w for cogasification of coal and silver birch wood mixtures (1:1 w/w ratio) at 1000°C (1830°F) (Collot *et al.*, 1999).

From the perspective of the efficient operation of the reactor, the presence of mineral matter has a deleterious effect on fluidized-bed reactors. The low melting point of ash formed from the mineral matter present in woody biomass can lead to agglomeration which influences the efficiency of the fluidization – the ash can cause sintering, deposition, and corrosion of the gasifier construction metal. In addition, biomass containing alkali oxides and salts can cause clinkering/slagging problems (McKendry, 2002).

2.5 Gasification Processes

All of the main types of gasifiers (Chapter 3) can be adapted to be used with various waste types as feedstock, but plasma gasification is becoming a technology of the near future, especially in regard to the treatment of municipal solid waste. In the plasma gasification process, the plasma (ionized gas at high temperature which conducts a strong electrical current) allows extremely high gasification temperatures of 4000°C (7200°F) to over 7000°C (12600°F). These high temperatures completely break down toxic compounds to their elemental constituents, making them easily neutralized, and the gas is mixed with oxygen and steam inside the gasifier. The organic compounds in the fuel are converted to synthesis gas, similar to the other gasification technologies, and any residual materials are captured in a rock-like mass which is highly resistant to leaching. With this technology, all known contaminants can be easily contained, making it ideal for municipal solid waste applications where feedstock composition is sometimes unclear. Also, after the initial electricity required at startup for the plasma gasifier, these systems are self-sustained by running off the electricity produced by firing the synthesis gas in a gas turbine.

Thus, gasification processes can accept a variety of feedstocks but the reactor must be selected on the basis of feedstock properties and the behavior of the feedstock in the process. Gasification processes are segregated according to bed types, which differ in their ability to accept (and use) caking coals and are generally divided into four categories based on reactor (bed) configuration: (i) fixed bed, (ii) fluidized bed, (iii) entrained bed, and (iv) molten salt.

In a fixed-bed process, the coal is supported by a grate and combustion gases (steam, air, oxygen, etc.) pass through the supported coal whereupon the hot produced gases exit from the top of the reactor. Heat is supplied internally or from an outside source, but caking coals cannot be used in an unmodified fixed-bed reactor.

The fluidized-bed system uses finely sized coal particles and the bed exhibits liquid-like characteristics when a gas flows upward through the bed. Gas flowing through the coal produces turbulent lifting and separation of particles and the result is an expanded bed having greater coal surface area to promote the chemical reaction, but such systems have a limited ability to handle caking coals.

An entrained-bed system uses finely sized coal particles blown into the gas steam prior to entry into the reactor and combustion occurs with the coal particles suspended in the gas phase; the entrained system is suitable for both caking and non-caking coals. The molten salt system employs a bath of molten salt to convert coal (Cover *et al.*, 1973; Howard-Smith and Werner, 1976; Speight, 2013).

The aim of underground (or *in situ*) gasification of coal is the conversion into combustible gases by combustion of a coal seam in the presence of air, oxygen, or oxygen and steam. Thus, seams that were considered to be inaccessible, unworkable, or uneconomical to mine could be put to use. In addition, strip mining and the accompanying environmental impacts, the problems of spoil banks, acid mine drainage, and the problems associated with use of high-ash coal are minimized or even eliminated.

The principles of underground gasification are very similar to those involved in the above-ground gasification of coal. The concept involves the drilling and subsequent linking of two boreholes so that gas will pass between the two (King and Magee, 1979). Combustion is then initiated at the bottom of one bore-hole (injection well) and is maintained by the continuous injection of air. In the initial reaction zone (combustion zone), carbon dioxide is generated by the reaction of oxygen (air) with the coal:

$$[C]_{coal} + O_2 \rightarrow CO_2$$

The carbon dioxide reacts with coal (partially devolatilized) further along the seam (reduction zone) to produce carbon monoxide:

$$[C]_{coal} + CO_2 \rightarrow 2CO$$

In addition, at the high temperatures that can frequently occur, moisture injected with oxygen or even moisture inherent in the seam may also react with the coal to produce carbon monoxide and hydrogen:

$$[C]_{coal} + H_2O \rightarrow CO + H_2$$

The gas product varies in character and composition but usually falls into the low-heat (low Btu) category ranging from 125 to 175 Btu/ft^3 (King and Magee, 1979).

2.5.1 Feedstock Pretreatment

While feedstock pretreatment for introduction into the gasifier is often considered to be a physical process in which the feedstock is prepared for gasifier – typically as pellets or finely ground feedstock – there are chemical aspects that must also be considered. Some feedstocks, especially certain types of coal, display caking, or agglomerating, characteristics when heated (Speight, 2013) and these coal types are usually not amenable to treatment by gasification processes employing fluidized-bed or moving-bed reactors; in fact, caked coal is difficult to handle in fixed-bed reactors. The pretreatment involves a mild oxidation treatment which destroys the caking characteristics of coals and usually consists of low-temperature heating of the coal in the presence of air or oxygen.

While this may seemingly be applicable to coal gasification only, this form of coal pretreatment is particularly important when a non-coal feedstock is co-gasified with coal. Co-gasification of other feedstocks, such as coal and especially biomass, with crude oil coke offers a bridge between the depletion of crude oil stocks when coal is used as well as a supplementary feedstock based on renewable energy sources (biomass). These options can contribute to reduce the crude oil dependency and carbon dioxide emissions since biomass is known to be neutral in terms of carbon dioxide emissions. The high reactivity of biomass and the accompanying high production of volatile products suggest that some synergetic effects might occur in simultaneous thermochemical treatment of petcoke and biomass, depending on the gasification conditions such as: (i) feedstock type and origin, (ii) reactor type, and (iii) process parameters (Penrose *et al.*, 1999; Gray and Tomlinson, 2000; McLendon *et al.*, 2004; Lapuerta *et al.*, 2008; Fermoso *et al.*, 2009; Shen *et al.*, 2012; Khosravi and Khadse, 2013; Speight, 2013, 2014a, 2014b; Luque and Speight, 2015).

The first step in the process is feedstock preparation which may include unit operations such as (i) size reduction, (ii) screening, and (iii) slurrying. The design of the feedstock preparation section depends on the properties of the raw material feed and on the requirements of the gasification technology that is selected. For example, preparing any form of crude oil resid, biomass, or waste for gasification is very different from the preparation of bituminous coal for the gasifier, and the type of gasifier suitable for the gasification may also (or more than likely) be feedstock dependent (Chapter 12).

For example, carbonaceous fuels are gasified in reactors, a variety of gasifiers such as the fixed- or moving-bed, fluidized-bed, entrained-flow, and molten-bath gasifiers have been developed (Shen *et al.*, 2012; Speight, 2014b). If the flow patterns are considered, the fixed-bed and fluidized-bed gasifiers intrinsically pertain to a countercurrent reactor in that fuels are usually sent into the reactor from the top of the gasifier, whereas the oxidant is blown into the reactor from the bottom. With regard to the entrained-flow reactor, it is necessary to pulverize the feedstock (such as coal and petcoke). On the other hand, when the feedstock is sent into an entrained-flow gasifier, the fuels can be in either form of dry feed or slurry feed. In general, dry-feed gasifiers have the advantage over slurry-feed gasifiers in that the former can be operated with lower oxygen consumption. Moreover, dry-feed gasifiers have an additional degree of freedom that makes it possible to optimize synthesis gas production (Shen *et al.*, 2012).

2.5.2 Feedstock Devolatilization

The devolatilization (or pyrolysis) process commences at approximately 200 to 300°C (390 to 570°F), depending upon the nature and properties of the feedstock. Volatiles are released and a carbonaceous residue (char) is produced, resulting in up to 70% weight loss for many feedstocks. The process determines the structure and composition of the char, which will then undergo gasification reactions.

In a gasifier, the feedstock particle is exposed to high temperatures generated from the partial oxidation of the carbon. As the particle is heated, any residual moisture (assuming that the feedstock has been pre-fired) is driven off and further heating of the particle begins to drive off the volatile gases. Discharge of the volatile products will generate a wide spectrum of hydrocarbon derivatives ranging from carbon monoxide and methane to long-chain hydrocarbon derivatives comprising tars, creosote, and high-boiling oil. The complexity of the products will also affect the progress and rate of the reaction when each product is produced by a different chemical process at a different rate. At a temperature above 500°C (930°F) the conversion of the feedstock to char and ash and char is completed. In most of the early gasification processes, this was the desired byproduct but for gas generation the char provides the necessary energy to effect further heating and – typically, the char is contacted with air or oxygen and steam to generate the product gases. Furthermore, with an increase in heating rate, feedstock particles are heated more rapidly and are burned in a higher temperature region, but the increase in heating rate has almost no substantial effect on the mechanism (Irfan, 2009).

2.5.3 Char Gasification

The gasification process occurs as the char reacts with gases such as carbon dioxide and steam to produce carbon monoxide and hydrogen. Also, corrosive ash elements such as chloride and potassium may be refined out by the gasification process, allowing the high temperature combustion of the gas from otherwise problematic feedstocks. Although the initial gasification stage is completed in seconds or even less at elevated temperature, the subsequent gasification of the char produced at the initial gasification stage is much slower, requiring minutes or hours to obtain significant conversion under practical conditions and reactor designs for commercial gasifiers are largely dependent on the reactivity of the char, which in turn depends on the nature of feedstock. The reactivity of char also depends upon parameters of the thermal process required to produce the char from the original feedstock. The rate of gasification of the char decreases as the process temperature increases due to the decrease in active surface area of char. Therefore a change of char preparation temperature may change the chemical nature of char, which in turn may change the gasification. The reactivity of char may be influenced by catalytic effect of mineral matter in the char.

Heat and mass transfer processes in fixed- or moving-bed gasifiers are affected by complex solids flow and chemical reactions. Moving-bed gasifiers are countercurrent flow reactors in which the feedstock enters at the top of the reactor and oxygen (air) enters at the bottom of the reactor. Because of the countercurrent flow arrangement of the reactor, the heat of reaction from the gasification reactions serves to pre-heat the coal before it enters the gasification reaction zone. Consequently, the temperature of the synthesis gas exiting the gasifier is significantly lower than the temperature needed for complete conversion of the feedstock. However, coarsely crushed feedstock may settle while undergoing (i) thermal drying, (ii) pyrolysis-devolatilization, (iii) gasification, and (iv) reduction. In addition, the particles change in diameter, shape, and porosity – non-ideal behavior may result from bridges, gas bubbles, channeling, and a variable void fraction may also change heat and mass transfer characteristics.

2.5.4 General Chemistry

Gasification involves the thermal decomposition of feedstock and the reaction of the feedstock carbon and other pyrolysis products with oxygen, water, and fuel gases such as methane. The presence of oxygen, hydrogen, water vapor, carbon oxides, and other compounds in the reaction atmosphere during pyrolysis may either support or inhibit numerous reactions with carbonaceous feedstocks and with the products evolved. The distribution of weight and chemical composition of the products are also influenced by the prevailing conditions (i.e., temperature, heating rate, pressure, and residence time) and, last but by no means least, the nature of the feedstock (Speight, 2014a, 2014b, 2017).

Generally, gasification involves two distinct stages that are both feedstock and reactor dependent: (i) devolatilization to produce a semi-char which, the rate of devolatilization has passed a maximum the semi-char is converted to char by elimination of hydrogen followed by (ii) gasification of the char, which is specific to the reactor and the conditions of the reaction.

Chemically, gasification involves the thermal decomposition of the feedstock and the reaction of the feedstock carbon and other pyrolysis products with oxygen, water, and fuel gases such as methane and is represented by a sequence of simple chemical reactions (Table 14.2). However, the gasification process is often considered to involve two distinct chemical stages: (i) devolatilization of the feedstock to produce volatile matter and char, (ii) followed by char gasification, which is complex and specific to the conditions of the reaction – both processes contribute to the complex kinetics of the gasification process (Sundaresan and Amundson, 1978).

Gasification of a carbonaceous material in an atmosphere of carbon dioxide can be divided into two stages: (i) pyrolysis and (ii) gasification of the pyrolytic char. In the first stage, pyrolysis (removal of moisture content and devolatilization) occurs at a comparatively lower temperature. In the second stage, gasification of the pyrolytic char is achieved by reaction with oxygen/carbon dioxide mixtures at high temperature. In nitrogen and carbon dioxide environments from room temperature to 1000°C (1830°F), the mass loss rate of pyrolysis in nitrogen may be significant differently (sometime lower, depending on the feedstock) to mass loss rate in carbon dioxide, which may be due (in part) to the difference in properties of the bulk gases.

Using coal as an example (Chapter 4), gasification in an atmosphere of oxygen/carbon dioxide environment is almost the same as gasification in an atmosphere of oxygen/nitrogen at the same oxygen concentration but this effect is a little bit delayed at high temperature. This may be due to the lower rate of diffusion of oxygen through carbon dioxide and the higher specific heat capacity of carbon dioxide. However, with an increase in the concentration of oxygen, the mass loss rate of coal also increases and, hence, shortens the burn out time of coal. The optimum value of oxygen/carbon dioxide for the reaction of oxygen with the functional groups that are present in the coal feedstock is on the order of 8% v/v.

If air is used for combustion, the product gas will have a heat content of ca. 150-300 Btu/ft^3 (depending on process design characteristics) and will contain undesirable constituents such as carbon dioxide, hydrogen sulfide, and nitrogen. The use of pure oxygen, although expensive, results in a product gas having a heat content on the order of 300 to 400 Btu/ft^3 with carbon dioxide and hydrogen sulfide as byproducts (both of which can be removed from low or medium heat-content, low-Btu or medium-Btu) gas by any of several available processes (Speight, 2013, 2014a, 2017).

If a high heat-content (high-Btu) gas (900 to 1000 Btu/ft^3) is required, efforts must be made to increase the methane content of the gas. The reactions which generate methane are all exothermic and have negative values, but the reaction rates are relatively slow and catalysts may therefore be necessary to accelerate the reactions to acceptable commercial rates. Indeed, the overall reactivity of the feedstock and char may be subject to catalytic effects.

It is also possible that the mineral constituents of the feedstock (such as the mineral matter in coal and biomass) may modify the reactivity by a direct catalytic effect (Davidson, 1983; Baker and Rodriguez, 1990; Mims, 1991; Martinez-Alonso and Tascon, 1991).

In the process, the feedstock undergoes three processes in its conversation to synthesis gas – the first two processes, pyrolysis and combustion, occur very rapidly. In pyrolysis, char is produced as the feedstock heats up and volatiles are released. In the combustion process, the volatile products and some of the char reacts with oxygen to produce various products (primarily carbon dioxide and carbon monoxide) and the heat required for subsequent gasification reactions. Finally, in the gasification process, the feedstock char reacts with steam to produce hydrogen (H_2) and carbon monoxide (CO).

Combustion:

$$2C_{feedstock} + O_2 \rightarrow 2CO + H_2O$$

Gasification:

$$C_{feedstock} + H_2O \rightarrow H_2 + CO$$

$$CO + H_2O \rightarrow H_2 + CO_2$$

The resulting synthesis gas is approximately 63% v/v carbon monoxide, 34% v/v hydrogen, and 3% v/v carbon dioxide. At the gasifier temperature, the ash and other feedstock mineral matter liquefies and exits at the bottom of the gasifier as slag, a sand-like inert material that can be sold as a co-product to other industries (e.g., road building). The synthesis gas exits the gasifier at pressure and high temperature and must be cooled prior to the synthesis gas cleaning stage.

Although processes that use the high temperature to raise high-pressure steam are more efficient for electricity production, full-quench cooling, by which the synthesis gas is cooled by the direct injection of water, is more appropriate for hydrogen production. Full-quench cooling provides the necessary steam to facilitate the water gas shift reaction, in which carbon monoxide is converted to hydrogen and carbon dioxide in the presence of a catalyst:

Water Gas Shift Reaction:

$$CO + H_2O \rightarrow CO_2 + H_2$$

This reaction maximizes the hydrogen content of the synthesis gas, which consists primarily of hydrogen and carbon dioxide at this stage. The synthesis gas is then scrubbed of particulate matter and sulfur is removed via physical absorption (Speight, 2013, 2014a, 2017). The carbon dioxide is captured by physical absorption or a membrane and either vented or sequestered.

Thus, in the initial stages of gasification, the rising temperature of the feedstock initiates devolatilization and the breaking of weaker chemical bonds to yield volatile tar, volatile oil, phenol derivatives, and hydrocarbon gases. These products generally react further in the gaseous phase to form hydrogen, carbon monoxide, and carbon dioxide. The char (fixed carbon) that remains after devolatilization reacts with oxygen, steam, carbon dioxide, and hydrogen. Overall, the chemistry of gasification is complex but can be conveniently (and simply) represented by the following reactions:

$$C + O_2 \rightarrow CO_2 \qquad \Delta H_r = -393.4 \text{ MJ/kmol} \qquad (2.1)$$

$$C + \tfrac{1}{2}O_2 \rightarrow CO \qquad \Delta H_r = -111.4 \text{ MJ/kmol} \qquad (2.2)$$

$$C + H_2O \rightarrow H_2 + CO \qquad \Delta H_r = 130.5 \text{ MJ/kmol} \qquad (2.3)$$

$$C + CO_2 \leftrightarrow 2CO \qquad \Delta H_r = 170.7 \text{ MJ/kmol} \qquad (2.4)$$

$$CO + H_2O \leftrightarrow H_2 + CO_2 \qquad \Delta H_r = -40.2 \text{ MJ/kmol} \qquad (2.5)$$

$$C + 2H_2 \rightarrow CH_4 \qquad \Delta H_r = -74.7 \text{ MJ/kmol} \qquad (2.6)$$

The designation C represents carbon in the original feedstock as well as carbon in the char formed by devolatilization of the feedstock. Reactions (2.1) and (2.2) are exothermic oxidation reactions and provide most of the energy required by the endothermic gasification reactions (2.3) and (2.4). The oxidation reactions occur very rapidly, completely consuming all of the oxygen present in the gasifier, so that most of the gasifier operates under reducing conditions. Reaction (2.5) is the water-gas shift reaction, in which water (steam) is converted to hydrogen – this reaction is used to alter the hydrogen/carbon monoxide ration when synthesis gas is the desired product, such as for use in Fischer-Tropsch processes. Reaction (2.6) is favored by high pressure and low temperature and is, thus, mainly important in lower temperature gasification systems. Methane formation is an exothermic reaction that does not consume oxygen and, therefore, increases the efficiency of the gasification process and the final heat content of the product gas. Overall, approximately 70% of the heating value of the product gas is associated with the carbon monoxide and hydrogen but this varies depending on the gasifier type and the process parameters (Speight, 2011a; Chadeesingh, 2011; Speight, 2013).

In essence, the direction of the gasification process is subject to the constraints of thermodynamic equilibrium and variable reaction kinetics. The combustion reactions (reaction of the feedstock or char with oxygen) essentially go to completion. The thermodynamic equilibrium of the rest of the gasification reactions are relatively well defined and collectively have a major influence on thermal efficiency of the process as well as on the gas composition. Thus, thermodynamic data are useful for estimating key design parameters for a gasification process, such as: (i) calculating of the relative amounts of oxygen and/or steam required per unit of feedstock, (ii) estimating the composition of the produced synthesis gas, and (iii) optimizing process efficiency at various operating conditions.

Other deductions concerning gasification process design and operations can also be derived from the thermodynamic understanding of its reactions. Examples include: (i) production of synthesis gas with low methane content at high temperature, which requires an amount of steam in excess of the stoichiometric requirement, (ii) gasification at high temperature, which increases oxygen consumption and decreases the overall process efficiency,

(iii) production of synthesis gas with a high methane content, which requires operation at low temperature (approximately 700°C, 1290°F) but the methanation reaction kinetics will be poor without the presence of a catalyst.

Relative to the thermodynamic understanding of the gasification process, the kinetic behavior is much more complex. In fact, very little reliable global kinetic information on gasification reactions exists, partly because it is highly dependent on (i) the chemical nature of the feed, which varies significantly with respect to composition, mineral impurities, (ii) feedstock reactivity, and (iii) process conditions. In addition, physical characteristics of the feedstock (or char) also play a role in phenomena such boundary layer diffusion, pore diffusion and ash layer diffusion which also influence the kinetic outcome. Furthermore, certain impurities, in fact, are known to have catalytic activity on some of the gasification reactions which can have further influence on the kinetic imprint of the gasification reactions.

With some feedstocks, the higher the amounts of volatile material produced in the early stages of the process the higher the heat content of the product gas. In some cases, the highest gas quality may be produced at the lowest temperatures but when the temperature is too low, char oxidation reaction is suppressed and the overall heat content of the product gas is diminished. All such events serve to complicate the reaction rate and make derivative of a global kinetic relationship applicable to all types of feedstock subject to serious question and doubt.

Depending on the type of feedstock being processed and the analysis of the gas product desired, pressure also plays a role in product definition. In fact, some (or all) of the following processing steps will be required: (i) pretreatment of the feedstock, (ii) primary gasification of the feedstock, (iii) secondary gasification of the carbonaceous residue from the primary gasifier; (iv) removal of carbon dioxide, hydrogen sulfide, and other acid gases; (v) shift conversion for adjustment of the carbon monoxide/hydrogen mole ratio to the desired ratio; and (vi) catalytic methanation of the carbon monoxide/hydrogen mixture to form methane. If high heat-content (high-Btu) gas is desired, all of these processing steps are required since gasifiers do not typically yield methane in the significant concentration.

2.5.5 Stage-by-Stage Chemistry

Though there is a considerable overlap of the processes, each can be assumed to occupy a separate zone where fundamentally different chemical and thermal reactions take place. The gasification technology package consists of a fuel and ash handling system, gasification system – reactor, gas cooling and cleaning system. There are also auxiliary systems, namely the water treatment plant to meet the requirements of industry and pollution control board. The prime mover for power generation consists of either a diesel engine or a spark-ignited engine coupled to an alternator. In the case of a thermal system, the end use device is a standard industrial burner.

2.5.5.1 *Primary Gasification*

Primary gasification involves thermal decomposition of the raw feedstock via various chemical processes (Table 2.6) and many schemes involve pressures ranging from atmospheric

Table 2.6 Types of reactions that occur in a gasifier.

$2C + O_2 \rightarrow 2CO$
$C + O_2 \rightarrow CO_2$
$C + CO_2 \rightarrow 2CO$
$CO + H_2O \rightarrow CO_2 + H_2$ (shift reaction)
$C + H_2O \rightarrow CO + H_2$ (water gas reaction)
$C + 2H_2 \rightarrow CH_4$
$2H_2 + O_2 \rightarrow 2H_2O$
$CO + 2H_2 \rightarrow CH_3OH$
$CO + 3H_2 \rightarrow CH_4 + H_2O$ (methanation reaction)
$CO_2 + 4H_2 \rightarrow CH_4 + 2H_2O$
$C + 2H_2O \rightarrow 2H_2 + CO_2$
$2C + H_2 \rightarrow C_2H_2$
$CH_4 + 2H_2O \rightarrow CO_2 + 4H_2$

to 1000 psi. Air or oxygen may be admitted to support combustion to provide the necessary heat. The product is usually a low heat content (low-Btu) gas ranging from a carbon monoxide/hydrogen mixture to mixtures containing varying amounts of carbon monoxide, carbon dioxide, hydrogen, water, methane, hydrogen sulfide, nitrogen, and typical tar-like products of thermal decomposition of carbonaceous feedstocks are complex mixtures and include hydrocarbon oils and phenolic products (Dutcher *et al.*, 1983; Speight, 2011a, 2013, 2014b).

Devolatilization of the feedstock occurs rapidly as the temperature rises above 300°C (570°F). During this period, the chemical structure is altered, producing solid char, tar products, condensable liquids, and low molecular weight gases. Furthermore, the products of the devolatilization stage in an inert gas atmosphere are very different from those in an atmosphere containing hydrogen at elevated pressure. In an atmosphere of hydrogen at elevated pressure, additional yields of methane or other low molecular weight gaseous hydrocarbon derivatives can result during the initial gasification stage from reactions such as: (i) direct hydrogenation of feedstock or semi-char because of any reactive intermediates formed and (ii) the hydrogenation of other gaseous hydrocarbon derivatives, oils, tars, and carbon oxides. Again, the kinetic picture for such reactions is complex due to the varying composition of the volatile products which, in turn, are related to the chemical character of the feedstock and the process parameters, including the reactor type.

A solid char product may also be produced, and may represent the bulk of the weight of the original feedstock, which determines (to a large extent) the yield of char and the composition of the gaseous product.

2.5.5.2 Secondary Gasification

Secondary gasification usually involves the gasification of char from the primary gasifier, which is typically achieved by reaction of the hot char with water vapor to produce carbon monoxide and hydrogen:

$$C_{char} + H_2O \rightarrow CO + H_2$$

The reaction requires heat input (endothermic) for the reaction to proceed in its forward direction. Usually, an excess amount of steam is also needed to promote the reaction. However, excess steam used in this reaction has an adverse effect on the thermal efficiency of the process. Therefore, this reaction is typically combined with other gasification reactions in practical applications. The hydrogen-carbon monoxide ratio of the product synthesis gas depends on the synthesis chemistry as well as process engineering.

The mechanism of this reaction section is based on the reaction between carbon and gaseous reactants, not for reactions between feedstock and gaseous reactants. Hence the equations may over-simply the actual chemistry of the steam gasification reaction. Even though carbon is the dominant atomic species present in feedstock, feedstock is more reactive than pure carbon. The presence of various reactive organic functional groups and the availability of catalytic activity via naturally occurring mineral ingredients can enhance the relative reactivity of the feedstock – for example anthracite, which has the highest carbon content among all ranks of coal (Speight, 2013), is most difficult to gasify or liquefy.

After the rate of devolatilization has passed a maximum another reaction becomes important – in this reaction the semi-char is converted to char (sometimes erroneously referred to as *stable char*) primarily through the evolution of hydrogen. Thus, the gasification process occurs as the char reacts with gases such as carbon dioxide and steam to produce carbon monoxide and hydrogen. The resulting gas (producer gas or synthesis gas) may be more efficiently converted to electricity than is typically possible by direct combustion. Also, corrosive elements in the ash may be refined out by the gasification process, allowing high temperature combustion of the gas from otherwise problematic feedstocks (Speight, 2011a, 2013, 2014b).

Oxidation and gasification reactions consume the char and the oxidation and the gasification kinetic rates follow Arrhenius type dependence on temperature. On the other hand, the kinetic parameters are feedstock specific and there is no true global relationship to describe the kinetics of char gasification – the characteristics of the char are also feedstock specific. The complexity of the reactions makes the reaction initiation and the subsequent rates subject to many factors, any one of which can influence the kinetic aspects of the reaction.

Although the initial gasification stage (devolatilization) is completed in seconds or even less at elevated temperature, the subsequent gasification of the char produced at the initial gasification stage is much slower, requiring minutes or hours to obtain significant conversion under practical conditions and reactor designs for commercial gasifiers are largely dependent on the reactivity of the char and also on the gasification medium (Johnson, 1979; Sha, 2005). Thus, the distribution and chemical composition of the products are also influenced by the prevailing conditions (i.e., temperature, heating rate, pressure, residence time, etc.) and, last but not least, the nature of the feedstock. Also, the presence of oxygen, hydrogen, water vapor, carbon oxides, and other compounds in the reaction atmosphere during pyrolysis may either support or inhibit numerous reactions with feedstock and with the products evolved.

The reactivity of char produced in the pyrolysis step depends on the nature of the feedstock and increases with oxygen content of the feedstock but decreases with carbon content. In general, char produced from a low-carbon feedstock is more reactive than char produced from a high-carbon feedstock. The reactivity of char from a low-carbon feedstock may be influenced by catalytic effect of mineral matter in char. In addition, as the carbon content of the feedstock increases, the reactive functional groups present in the feedstock decrease and the char becomes more aromatic and cross-linked in nature (Speight, 2013). Therefore, char obtained from high-carbon feedstock contains a lesser number of functional groups and higher proportion of aromatic and cross-linked structures, which reduce reactivity. The reactivity of char also depends upon thermal treatment it receives during formation from the parent feedstock – the gasification rate of char decreases as the char preparation temperature increases due to the decrease in active surface areas of char. Therefore, a change of char preparation temperature may change the chemical nature of char, which in turn may change the gasification rate.

Typically, char has a higher surface area compared to the surface area of the parent feedstock, even when the feedstock has been pelletized, and the surface area changes as the char undergoes gasification – the surface area increases with carbon conversion, reaches maximum and then decreases. These changes in turn affect gasification rates – in general, reactivity increases with the increase in surface area. The initial increase in surface area appears to be caused by cleanup and widening of pores in the char. The decrease in surface area at high carbon conversion may be due to coalescence of pores, which ultimately leads to collapse of the pore structure within the char.

Heat transfer and mass transfer processes in fixed- or moving-bed gasifiers are affected by complex solids flow and chemical reactions. Coarsely crushed feedstock settles while undergoing heating, drying, devolatilization, gasification and combustion. Also, the feedstock particles change in diameter, shape, and porosity – non-ideal behavior may result from certain types of chemical structures in the feedstock, gas bubbles, and channel and a variable void fraction may also change heat and mass transfer characteristics.

An important factor is the importance of the pyrolysis temperature as a major factor in the thermal history, and consequently in the thermodynamics of the feedstock char. However, the thermal history of a char should also depend on the rate of temperature rise to the pyrolysis temperature and on the length of time the char is kept at the pyrolysis temperature (soak time), which might be expected to reduce the residual entropy of the char by employing a longer soak time.

Alkali metal salts are known to catalyze the steam gasification reaction of carbonaceous materials, including coal. The process is based on the concept that alkali metal salts (such as potassium carbonate, sodium carbonate, potassium sulfide, sodium sulfide, and the like) will catalyze the steam gasification of feedstocks. The order of catalytic activity of alkali metals on the gasification reaction is:

Cesium (Cs) > rubidium (Rb) > potassium (K) > sodium (Na) > lithium (Li)

Catalyst amounts on the order of 10 to 20% w/w potassium carbonate will lower bituminous coal gasifier temperatures from 925°C (1695°F) to 700°C (1090°F) and that the catalyst can be introduced to the gasifier impregnated on coal or char. In addition, tests with potassium carbonate showed that this material also acts as a catalyst for the methanation reaction. In addition, the use of catalysts can reduce the amount of tar formed in the process. In the case of catalytic steam gasification of coal, carbon deposition reaction may affect catalyst life by fouling the catalyst active sites. This carbon deposition reaction is more likely to take place whenever the steam concentration is low.

Ruthenium-containing catalysts are used primarily in the production of ammonia. It has been shown that ruthenium catalysts provide five to 10 times higher reactivity rates than other catalysts. However, ruthenium quickly becomes inactive due to its necessary supporting material, such as activated carbon, which is used to achieve effective reactivity. However, during the process, the carbon is consumed, thereby reducing the effect of the ruthenium catalyst.

Catalysts can also be used to favor or suppress the formation of certain components in the gaseous product by changing the chemistry of the reaction, the rate of reaction, and the thermodynamic balance of the reaction. For example, in the production of synthesis gas (mixtures of hydrogen and carbon monoxide), methane is also produced in small amounts. Catalytic gasification can be used to either promote methane formation or suppress it.

2.5.5.3 Water Gas Shift Reaction

The water gas shift reaction (shift conversion) is necessary because the gaseous product from a gasifier generally contains large amounts of carbon monoxide and hydrogen, plus lesser amounts of other gases. Carbon monoxide and hydrogen (if they are present in the mole ratio of 1:3) can be reacted in the presence of a catalyst to produce methane. However, some adjustment to the ideal (1:3) is usually required and, to accomplish this, all or part of the stream is treated according to the waste gas shift (shift conversion) reaction. This involves reacting carbon monoxide with steam to produce a carbon dioxide and hydrogen whereby the desired 1:3 mole ratio of carbon monoxide to hydrogen may be obtained:

$$CO(g) + H_2O(g) \rightarrow CO_2(g) + H_2(g)$$

Even though the water gas shift reaction is not classified as one of the principal gasification reactions, it cannot be omitted in the analysis of chemical reaction systems that involve synthesis gas. Among all reactions involving synthesis gas, this reaction equilibrium is least sensitive to the temperature variation – the equilibrium constant is least strongly dependent on the temperature. Therefore, the reaction equilibrium can be reversed in a variety of practical process conditions over a wide range of temperature.

The water gas shift reaction in its forward direction is mildly exothermic and although all of the participating chemical species are in gaseous form, the reaction is believed to be heterogeneous insofar as the chemistry occurs at the surface of the feedstock and the reaction is actually catalyzed by carbon surfaces. In addition, the reaction can also take place homogeneously as well as heterogeneously and a generalized understanding of the water gas shift reaction is difficult to achieve. Even the published kinetic rate information is not immediately useful or applicable to a practical reactor situation.

Synthesis gas from a gasifier contains a variety of gaseous species other than carbon monoxide and hydrogen. Typically, they include carbon dioxide, methane, and water (steam).

Depending on the objective of the ensuing process, the composition of synthesis gas may need to be preferentially readjusted. If the objective of the gasification process is to obtain a high yield of methane, it would be preferred to have the molar ratio of hydrogen to carbon monoxide at 3:1:

$$CO(g) + 3H_2(g) \rightarrow CH_4(g) + H_2O(g)$$

On the other hand, if the objective of generating synthesis gas is the synthesis of methanol via a vapor-phase low-pressure process, the stoichiometrically consistent ratio between hydrogen and carbon monoxide would be 2:1. In such cases, the stoichiometrically consistent synthesis gas mixture is often referred to as *balanced gas*, whereas a synthesis gas composition that is substantially deviated from the principal stoichiometry of the reaction is called *unbalanced gas*. If the objective of synthesis gas production is to obtain a high yield of hydrogen, it would be advantageous to increase the ratio of hydrogen to carbon monoxide by further converting carbon monoxide (and water) into hydrogen (and carbon dioxide) via the water gas shift reaction.

The water gas shift reaction is one of the major reactions in the steam gasification process, where both water and carbon monoxide are present in ample amounts. Although the four chemical species involved in the water gas shift reaction are gaseous compounds at the reaction stage of most gas processing, the water gas shift reaction, in the case of steam gasification of feedstock, predominantly takes place on the solid surface of feedstock (heterogeneous reaction). If the product synthesis gas from a gasifier needs to be reconditioned by the water gas shift reaction, this reaction can be catalyzed by a variety of metallic catalysts.

Choice of specific kinds of catalysts has always depended on the desired outcome, the prevailing temperature conditions, composition of gas mixture, and process economics. Typical catalysts used for the reaction include catalysts containing iron, copper, zinc, nickel, chromium, and molybdenum.

2.5.5.4 Carbon Dioxide Gasification

The reaction of carbonaceous feedstocks with carbon dioxide produces carbon monoxide (*Boudouard reaction*) and (like the steam gasification reaction) is also an endothermic reaction:

$$C(s) + CO_2(g) \rightarrow 2CO(g)$$

The reverse reaction results in carbon deposition (carbon fouling) on many surfaces including the catalysts and results in catalyst deactivation.

This gasification reaction is thermodynamically favored at high temperatures (>680°C, >1255°F), which is also quite similar to the steam gasification. If carried out alone, the reaction requires high temperature (for fast reaction) and high pressure (for higher reactant concentrations) for significant conversion but as a separate reaction a variety of factors come into play: (i) low conversion, (ii) slow kinetic rate, and (iii) low thermal efficiency.

Also, the rate of the carbon dioxide gasification of a feedstock is different from the rate of the carbon dioxide gasification of carbon. Generally, the carbon-carbon dioxide reaction follows a reaction order based on the partial pressure of the carbon dioxide that is approximately 1.0 (or lower) whereas the feedstock-carbon dioxide reaction follows a reaction order based on the partial pressure of the carbon dioxide that is 1.0 (or higher). The observed higher reaction order for the feedstock reaction is also based on the relative reactivity of the feedstock in the gasification system.

2.5.5.5 Hydrogasification

Not all high heat-content (high-Btu) gasification technologies depend entirely on catalytic methanation and, in fact, a number of gasification processes use hydrogasification, that is, the direct addition of hydrogen to feedstock under pressure to form methane.

$$C_{char} + 2H_2 \rightarrow CH_4$$

The hydrogen-rich gas for hydrogasification can be manufactured from steam and char from the hydrogasifier. Appreciable quantities of methane are formed directly in the primary gasifier and the heat released by methane formation is at a sufficiently high temperature to be used in the steam-carbon reaction to produce hydrogen so that less oxygen is used to produce heat for the steam-carbon reaction. Hence, less heat is lost in the low-temperature methanation step, thereby leading to higher overall process efficiency.

Hydrogasification is the gasification of feedstock in the presence of an atmosphere of hydrogen under pressure. Thus, not all high heat-content (high-Btu) gasification technologies depend entirely on catalytic methanation and, in fact, a number of gasification processes use hydrogasification, that is, the direct addition of hydrogen to feedstock under pressure to form methane:

$$[C]_{feedstock} + H_2 \rightarrow CH_4$$

The hydrogen-rich gas for hydrogasification can be manufactured from steam by using the char that leaves the hydrogasifier. Appreciable quantities of methane are formed directly in the primary gasifier and the heat released by methane formation is at a sufficiently high temperature to be used in the steam-carbon reaction to produce hydrogen so that less oxygen is used to produce heat for the steam-carbon reaction. Hence, less heat is lost in the low-temperature methanation step, thereby leading to higher overall process efficiency.

The hydrogasification reaction is exothermic and is thermodynamically favored at low temperatures (<670°C, <1240°F), unlike the endothermic both steam gasification and carbon dioxide gasification reactions. However, at low temperatures, the reaction rate is inevitably too slow. Therefore, a high temperature is always required for kinetic reasons, which in turn requires high pressure of hydrogen, which is also preferred from equilibrium considerations. This reaction can be catalyzed by salts such as potassium carbonate (K_2CO_3), nickel chloride ($NiCl_2$), iron chloride ($FeCl_2$), and iron sulfate ($FeSO_4$). However, use of a catalyst in feedstock gasification suffers from difficulty in recovering and reusing the catalyst and the potential for the spent catalyst becoming an environmental issue.

In a hydrogen atmosphere at elevated pressure, additional yields of methane or other low molecular weight hydrocarbon derivatives can result during the initial feedstock gasification stage from direct hydrogenation of feedstock or semi-char because of active intermediate formed in the feedstock structure after pyrolysis. The direct hydrogenation can also increase the amount of feedstock carbon that is gasified as well as the hydrogenation of gaseous hydrocarbon derivatives, oil, and tar.

The kinetics of the rapid-rate reaction between gaseous hydrogen and the active intermediate depends on hydrogen partial pressure (P_{H_2}). Greatly increased gaseous hydrocarbon derivatives produced during the initial feedstock gasification stage are extremely important in processes to convert feedstock into methane (SNG, synthetic natural gas, substitute natural gas).

2.5.5.6 Methanation

Several exothermic reactions may occur simultaneously within a methanation unit. A variety of metals have been used as catalysts for the methanation reaction; the most common, and to some extent the most effective methanation catalysts, appear to be nickel and ruthenium, with nickel being the most widely used (Cusumano *et al.*, 1978):

Ruthenium (Ru) > nickel (Ni) > cobalt (Co) > iron (Fe) > molybdenum (Mo).

Nearly all the commercially available catalysts used for this process are, however, very susceptible to sulfur poisoning and efforts must be taken to remove all hydrogen sulfide (H_2S) before the catalytic reaction starts. It is necessary to reduce the sulfur concentration in the feed gas to less than 0.5 ppm v/v in order to maintain adequate catalyst activity for a long period of time.

The synthesis gas must be desulfurized before the methanation step since sulfur compounds will rapidly deactivate (poison) the catalysts. A processing issue may arise when the concentration of carbon monoxide is excessive in the stream to be methanated since large amounts of heat must be removed from the system to prevent high temperatures and deactivation of the catalyst by sintering as well as the deposition of carbon. To eliminate this problem, temperatures should be maintained below 400°C (750°F).

The methanation reaction is used to increase the methane content of the product gas, as needed for the production of high-Btu gas.

$$4H_2 + CO_2 \rightarrow CH_4 + 2H_2O$$

$$2CO \rightarrow C + CO_2$$

$$CO + H_2O \rightarrow CO_2 + H_2$$

Among these, the most dominant chemical reaction leading to methane is the first one. Therefore, if methanation is carried out over a catalyst with a synthesis gas mixture of hydrogen and carbon monoxide, the desired hydrogen-carbon monoxide ratio of the feed synthesis gas is around 3:1. The large amount of water (vapor) produced is removed by condensation and recirculated as process water or steam. During this process, most of the exothermic heat due to the methanation reaction is also recovered through a variety of energy integration processes.

Whereas all the reactions listed above are quite strongly exothermic except the forward water gas shift reaction, which is mildly exothermic, the heat release depends largely on the amount of carbon monoxide present in the feed synthesis gas. For each 1% v/v carbon monoxide in the feed synthesis gas, an adiabatic reaction will experience a 60°C (108°F) temperature rise, which may be termed as *adiabatic temperature rise*.

2.5.5.7 Catalytic Gasification

Catalysts are commonly used in the chemical and crude oil industries to increase reaction rates, sometimes making certain previously unachievable products possible (Speight, 2002; Speight, 2014a; Hsu and Robinson, 2017; Speight, 2017). Acids, through donated protons (H^+), are common reaction catalysts, especially in the organic chemical industries. This it is not surprising that catalysts can be used to enhance the reactions involved in gasification and use of appropriate catalysts not only reduces reaction temperature but also improves the gasification rates.

In addition, thermodynamic constraints of the gasification process limit the thermal efficiency are not inherent but the result of design decisions based on available technology, as well as the kinetic properties of available catalysts. The latter limits the yield of methane to that obtainable at global equilibrium over carbon in the presence of carbon monoxide and hydrogen. The equilibrium composition is shown to be independent of the thermodynamic properties of the char or feedstock. These limitations give non-isothermal two-stage processes significant thermodynamic advantages. The results of the analysis suggest directions for modifying present processes to obtain higher thermal efficiencies, and a two-stage process scheme that would have significant advantages over present technologies and should be applicable to a wide range of catalytic and non-catalytic processes (Shinnar *et al.*, 1982; McKee, 1981).

Alkali metal salts of weak acids, such as potassium carbonate (K_2CO_3), sodium carbonate (Na_2CO_3), potassium sulfide (K_2S), and sodium sulfide (Na_2S) can catalyze the carbon-steam gasification reaction. Catalyst amounts on the order of 10 to 20% w/w K_2CO_3 will lower the temperature required for gasification of bituminous coal from approximately 925°C (1695°F) to 700°C (1090°F) and that the catalyst can be introduced to the gasifier impregnated on coal or char.

Disadvantages of catalytic gasification include increased materials costs for the catalyst itself (often rare metals), as well as diminishing catalyst performance over time. Catalysts can be recycled, but their performance tends to diminish with age or by poisoning. The relative difficulty in reclaiming and recycling the catalyst can also be a disadvantage. For example, the potassium carbonate catalyst can be recovered from spent char with a simple water wash, but some catalysts may not be so accommodating. In addition to age, catalysts can also be diminished by poisoning. On the other hand, many catalysts are sensitive to particular chemical species which bond with the catalyst or alter it in such a way that it no longer functions. Sulfur, for example, can poison several types of catalysts including palladium and platinum.

2.5.6 Physical Effects

Depending on the type of feedstock being processed and the analysis of the gas product desired, pressure also plays a role in product definition. In fact, some (or all) of the following processing steps will be required: (i) pretreatment of the feedstock, (ii) primary gasification of the feedstock, (iii) secondary gasification of the carbonaceous residue from the primary gasifier, (iv) removal of carbon dioxide, hydrogen sulfide, and other acid gases, (v) shift conversion for adjustment of the carbon monoxide/hydrogen mole ratio to the desired ratio, and (vi) catalytic methanation of the carbon monoxide/hydrogen mixture to form methane.

Most notable effects in the physical chemistry of the gasification process are those effects due to the chemical character of the feedstock as well as the physical composition of the feedstock (Speight, 2011a, 2013). In more general terms of the character of the feedstock, gasification technologies generally require some initial processing of the feedstock with the type and degree of pretreatment a function of the process and/or the type of feedstock. For example, the Lurgi process will accept lump feedstock (1 in., 25 mm, to 28 mesh), but it must be non-caking with the fines removed – feedstock that show caking or agglomerating tendencies form a plastic mass in the bottom of a gasifier and subsequently plug up the system thereby markedly reducing process efficiency. Thus, some attempt to reduce caking tendencies is necessary and can involve preliminary partial oxidation of the feedstock thereby destroying the caking properties.

Another factor, often presented as very general *rule of thumb*, is that optimum gas yields and gas quality are obtained at operating temperatures of approximately 595 to 650°C (1100 to 1200°F). A gaseous product with a higher heat content (BTU/ft.^{3}) can be obtained at lower system temperatures but the overall yield of gas (determined as the *fuel-to-gas ratio*) is reduced by the unburned char fraction.

The major difference between combustion and gasification from the point of view of the chemistry involved is that combustion takes place under oxidizing conditions, while gasification occurs under reducing conditions. In the gasification process, the feedstock (in the presence of steam and oxygen at high temperature and moderate pressure) is converted to a mixture of product gases. In the initial stages of gasification, the rising temperature of the feedstock initiates devolatilization of the feedstock and the breaking of weaker chemical bonds to yield tar, oil, volatile species, and hydrocarbon gases. These products generally react further to form hydrogen, carbon monoxide, and carbon dioxide. The fixed carbon that remains after devolatilization reacts with oxygen, steam, carbon dioxide, and hydrogen.

Depending on the gasifier technology employed and the operating conditions, significant quantities of water, carbon dioxide, and methane can be present in the product gas, as well as a number of minor and trace components. Under the reducing conditions in the gasifier, most of the sulfur in the fuel sulfur is converted to hydrogen sulfide (H_2S) as well as to smaller yields of carbonyl sulfide (COS). Organically bound nitrogen in the feedstock is generally (but not always) converted to gaseous nitrogen (N_2) – some ammonia (NH_3) and a small amount of hydrogen cyanide (HCN) are also formed. Any chlorine in the feedstock (such as coal) is converted to hydrogen chloride (HCl) with some chlorine present in the particulate matter (fly ash). Trace elements, such as mercury and arsenic, are released during gasification and partition among the different phases, such as fly ash, bottom ash, slag, and product gas.

Fuels for gasification reactors differ significantly in chemical properties, physical properties, and morphological properties and, hence, require different reactor design and operation. It is for this reason that, during more than a century of gasification experience, a large number of different gasifiers has been developed – each reactor designed to accommodate the specific properties of a typical fuel or range of fuels. In short, the gasification reactor that is designed to accommodate all (or most) types of fuels does not exist.

However, before choosing a gasifier for any individual fuel it is important to ensure that the fuel meets the requirements of the gasifier or that it can be treated to meet these requirements. Practical tests are needed if the fuel has not previously been successfully gasified. In other words the fuel must match the gasifier and the gasifier must match the fuel.

2.6 Products

The original concept of the gasification process was to produce a fuel gas for use in homes (including street lighting) and industrial operations. Thus, the gasification of carbonaceous residues is generally aimed to feedstock conversion to gaseous products. In fact, gasification offers one of the most versatile methods (with a reduced environmental impact with respect to combustion) to convert carbonaceous feedstocks into electricity, hydrogen, and other valuable energy products. Depending on the previously described type of gasifier (e.g. air-blown, enriched oxygen-blown) and the operating conditions, gasification can be used to produce a fuel gas that is suitable for several applications.

Gasification agents are typically air, oxygen-enriched air or oxygen and the products of the combustion or gasification oxidation reaction change significantly as the oxygen-to-fuel ratio changes from combustion to gasification conditions, which are dependent upon gasifier design and operation (Luque and Speight, 2015). The mixture under gasifying conditions is fuel-rich and there is not enough oxygen to effect complete conversion of the feedstock, in terms of gas quality. As a result, the feedstock carbon reacts to produce carbon instead of carbon dioxide and the feedstock hydrogen is converted to hydrogen rather than to water. Thus, the quantity and quality of the gas generated in a gasification reactor is influenced not only by the feedstock characteristics but predominantly by the gasifier type and configuration well as by the amount of air, oxygen or steam introduced into the system, which is also influence by the configuration of the gasifier.

At the same time, the fate of the nitrogen and sulfur in the fuel is also dictated by oxygen availability (i.e., the configuration of the gasification reactor). The nitrogen and sulfur in a gasification process has important and environmental consequences. Instead of being converted to the respective oxides, the fuel-bound nitrogen is predominantly converted to molecular nitrogen (N2) and hydrogen cyanide (HCN) while the sulfur in the fuel produces hydrogen cyanide (HCN) and carbonyl sulfide (COS).

Steam is sometimes added for temperature control, heating value enhancement or to permit the use of external heat (*allothermal gasification*). The major chemical reactions break and oxidize hydrocarbon derivatives to give a product gas of carbon monoxide (CO), carbon dioxide (CO_2), hydrogen (H_2), and water (H_2O). Other important components include hydrogen sulfide (H_2S), various compounds of sulfur and carbon, ammonia, low molecular weight hydrocarbon derivatives, and tar.

As a very general *rule of thumb*, optimum gas yields and gas quality are obtained at operating temperatures of approximately 595 to 650°C (1100 to 1200°F). A gaseous product with a higher heat content (BTU/ft.³) can be obtained at lower system temperatures but the overall yield of gas (determined as the *fuel-to-gas ratio*) is reduced by the unburned portion of the feedstock, which usually appears as char.

Gasification for electric power generation enables the use of a common technology in modern gas-fired power plants (*combined cycle*) to recover more of the energy released by burning the fuel. The use of these two types of turbines in the combined cycle system involves (i) a combustion turbine and (ii) a steam turbine. The increased efficiency of the combined cycle for electrical power generation results in a 50% v/v decrease in carbon dioxide emissions compared to conventional coal plants. Gasification units could be modified to further reduce their climate change impact because a large part of the carbon dioxide generated can be separated from the other product gas *before* combustion (for example carbon dioxide can be separated/sequestered from gaseous byproducts by using adsorbents (e.g., MOFs) to prevent its release to the atmosphere).

Gasification has also been considered for many years as an alternative to combustion of solid or liquid fuels. Gaseous mixtures are simpler to clean as compared to solid or high-viscosity liquid fuels. Cleaned gases can be used in internal combustion-based power plants that would suffer from severe fouling or corrosion if solid or low quality liquid fuels were burned inside them.

In fact, the hot synthesis gas produced by gasification of carbonaceous feedstocks can then be processed to remove sulfur compounds, mercury, and particulate matter prior to its use as fuel in a combustion turbine generator to produce electricity. The heat in the exhaust gases from the combustion turbine is recovered to generate additional steam. This steam, along with the steam produced by the gasification process, drives a steam turbine generator to produce additional electricity. In the past decade, the primary application of gasification to power production has become more common due to the demand for high efficiency and low environmental impact.

2.6.1 Gaseous Products

The products from gasification may be of low, medium, or high heat-content (high-Btu) content as dictated by the process as well as by the ultimate use for the gas (Speight, 2008, 2011a, 2013). However, the products of gasification are varied insofar as the gas composition varies with the system employed (Speight, 2013). It is emphasized that the gas product must be first freed from any pollutants such as particulate matter and sulfur compounds before further use, particularly when the intended use is a water gas shift or methanation (Cusumano *et al.*, 1978; Probstein and Hicks, 1990).

Product gases from fixed-bed versus fluidized-bed gasifier configurations vary significantly. Fixed-bed gasifiers are relatively easy to design and operate and are best suited for small to medium-scale applications with thermal requirements of up to several thermal megawatts (megawatts thermal, MWt). For large-scale applications, fixed-bed gasifiers may encounter problems with bridging of the feedstock (especially in the case of biomass feedstocks) and non-uniform bed temperatures. Bridging leads to uneven gas flow, while non-uniform bed temperature may lead to hot spots, ash formation, and slagging. Large-scale applications are also susceptible to temperature variations throughout the gasifier because of poor mixing in the reaction zone.

Pressurized gasification systems lend themselves to economical synthesis gas production and can also be more flexible in production turndown depending on the reactor design. Typically this is the case for both a *pressurized bubbling reactor* and a *circulating fluidized-bed reactor*, while the flexibility of an atmospheric fluidized-bed reactor is typically limited to narrower pressure and production ranges. Both designs are well suited for pressurized synthesis gas production. Pressurized designs require more costly reactors, but the downstream equipment (such as gas cleanup equipment, heat exchangers, synthesis gas reactors) will consist of fewer and less expensive components (Worley and Yale, 2012).

In the process, the feedstock undergoes three processes during the conversation to synthesis gas – the first two processes, pyrolysis and combustion, occur very rapidly. In pyrolysis, char is produced as the feedstock heats up and volatile products are released. In the combustion process, the volatile products and some of the char reacts with oxygen to produce secondary products (primarily carbon dioxide and carbon monoxide) and the heat required for subsequent gasification reactions. Finally, the char reacts with steam to produce hydrogen (H_2) and carbon monoxide (CO).

Combustion:

$$2C_{feedstock} + O_2 \rightarrow 2CO + H_2O$$

Gasification:

$$C_{feedstock} + H_2O \rightarrow H_2 + CO$$

$$CO + H_2O \rightarrow H_2 + CO_2$$

At the gasifier temperature, the ash and other feedstock mineral matter liquefies and exits at the bottom of the gasifier as slag, a sand-like inert material that can be sold as a co-product to other industries (e.g., road building). The synthesis gas exits the gasifier at pressure and high temperature and must be cooled prior to the cleaning stage. Full-quench cooling, by which the synthesis gas is cooled by the direct injection of water, is more appropriate for hydrogen production. The procedure provides the necessary steam to facilitate the water gas shift reaction, in which carbon monoxide is converted to hydrogen and carbon dioxide in the presence of a catalyst:

Water Gas Shift Reaction:

$$CO + H_2O \rightarrow CO_2 + H_2$$

This reaction maximizes the hydrogen content of the synthesis gas, which consists primarily of hydrogen and carbon dioxide at this stage. The synthesis gas is then scrubbed of particulate matter and sulfur is removed via physical absorption (Speight, 2008; Chadeesingh, 2011; Speight, 2013). The carbon dioxide is captured by physical absorption or a membrane and either vented or sequestered.

2.6.1.1 Low Btu Gas

During the production of gas by oxidation of the feedstock with air, the oxygen is not separated from the air and, as a result, the gas product invariably has a low Btu value (low heat-content) on the order of 150 to 300 Btu/ft^3. Low Btu gas is also the usual product of *in situ* gasification of coal (Speight, 2013) which is essentially used as a method for obtaining energy from coal without the necessity of mining the coal, especially if the coal cannot be mined or of mining is uneconomical.

Several important chemical reactions, and a host of side reactions, are involved in the manufacture of low heat-content gas under the high-temperature conditions employed (2011; Speight, 2013). Low heat-content gas contains several components, four of which are always major components present at levels of at least several percent; a fifth component, methane, is marginally a major component.

The nitrogen content of low heat-content gas ranges from somewhat less than 33% v/v to slightly more than 50% v/v and cannot be removed by any reasonable means; the presence of nitrogen at these levels makes the product gas *low heat-content* by definition. The nitrogen also strongly limits the applicability of the gas to chemical synthesis. Two other noncombustible components (water, H_2O, and carbon dioxide, CO) further lower the heating value of the gas; water can be removed by condensation and carbon dioxide by relatively straightforward chemical means.

The two major combustible components are hydrogen and carbon monoxide; the H_2/CO ratio varies from approximately 2:3 to approximately 3:2. Methane may also make an appreciable contribution to the heat content of the gas. Of the minor components hydrogen sulfide is the most significant and the amount produced is, in fact, proportional to the sulfur content of the feed coal. Any hydrogen sulfide present must be removed by one, or more, of several procedures (Mokhatab *et al.*, 2006; Speight, 2019).

Producer gas is a low Btu gas typically obtained from a coal gasifier (fixed-bed) upon introduction of air instead of oxygen into the fuel bed. The composition of the producer gas is approximately 28% v/v carbon monoxide, 55% v/v nitrogen, 12% v/v hydrogen, and 5% v/v methane with some carbon dioxide. *Water gas* is a medium Btu gas which is produced by the introduction of steam into the hot fuel bed of the gasifier. The composition of the gas is approximately 50% v/v hydrogen and 40% v/v carbon monoxide with small amounts of nitrogen and carbon dioxide. Also, low heat-content gas is of interest to industry as a fuel gas or even, on occasion, as a raw material from which ammonia, methanol, and other compounds may be synthesized.

2.6.1.2 Medium Btu Gas

Medium Btu gas (medium heat-content gas) has a heating value in the range 300 to 550 Btu/ft^3) and the composition is much like that of low heat-content gas, except that there is virtually no nitrogen. The primary combustible gases in medium heat-content gas are hydrogen and carbon monoxide (Kasem, 1979). Medium heat-content gas is considerably more versatile than low heat-content gas; like low heat-content gas, medium heat-content gas may be used directly as a fuel to raise steam, or used through a combined power cycle to drive a gas turbine, with the hot exhaust gases employed to raise steam, but medium heat-content gas is especially amenable to synthesize methane (by methanation), higher hydrocarbon derivatives (by Fischer-Tropsch synthesis), methanol, and a variety of synthetic chemicals.

The reactions used to produce medium heat-content gas are the same as those employed for low heat-content gas synthesis, the major difference being the application of a nitrogen barrier (such as the use of pure oxygen) to keep diluent nitrogen out of the system.

In medium heat-content gas, the H_2/CO ratio varies from 2:3 C to 3:1 and the increased heating value correlates with higher methane and hydrogen contents as well as with lower carbon dioxide contents. Furthermore, the very nature of the gasification process used to produce the medium heat-content gas has a marked effect upon the ease of subsequent processing. For example, the CO_2-acceptor product is quite amenable to use for methane production because it has (i) the desired H_2/CO ratio just exceeding 3:1, (ii) an initially high methane content, and (iii) relatively low water and carbon dioxide contents. Other gases may require appreciable shift reaction and removal of large quantities of water and carbon dioxide prior to methanation.

Town gas is also a medium Btu that is produced in the coke ovens and has the approximate composition: 55% v/v hydrogen, 27% v/v methane, 6% v/v carbon monoxide, 10% v/v nitrogen, and 2% v/v carbon dioxide. Carbon monoxide can be removed from the gas by catalytic treatment with steam to produce carbon dioxide and hydrogen.

2.6.1.3 High Btu Gas

High Btu gas (heat-content gas) is essentially pure methane and often referred to as *synthetic natural gas* or *substitute natural gas* (SNG) (Kasem, 1979; c.f. Speight, 1990, 2013). However, to qualify as substitute natural gas, a product must contain at least 95% methane, giving an energy content (heat content) of synthetic natural gas on the order of 980 to 1080 Btu/ft^3).

The commonly accepted approach to the synthesis of high heat-content gas is the catalytic reaction of hydrogen and carbon monoxide:

$$3H_2 + CO \rightarrow CH_4 + H_2O$$

To avoid catalyst poisoning, the feed gases for this reaction must be quite pure and, therefore, impurities in the product are rare. The large quantities of water produced are removed by condensation and recirculated as very pure water through the gasification system. The hydrogen is usually present in slight excess to ensure that the toxic carbon monoxide is reacted; this small quantity of hydrogen will lower the heat content to a small degree.

The carbon monoxide/hydrogen reaction is somewhat inefficient as a means of producing methane because the reaction liberates large quantities of heat. In addition, the methanation catalyst is troublesome and prone to poisoning by sulfur compounds and the decomposition of metals can destroy the catalyst. Hydrogasification may be thus employed to minimize the need for methanation:

$$[C]_{coal} + 2H2 \rightarrow CH_4$$

The product of hydrogasification is far from pure methane and additional methanation is required after hydrogen sulfide and other impurities are removed.

Synthetic natural gas (SNG) is methane obtained from the reaction of carbon monoxide or carbon with hydrogen. Depending on the methane concentration, the heating value can be in the range of high-Btu gases.

2.6.1.4 Synthesis Gas

Synthesis gas is a mixture mainly of hydrogen and carbon monoxide which is comparable in its combustion efficiency to natural gas (Speight, 2008 Chapter 7). This reduces the emissions of sulfur, nitrogen oxides, and mercury, resulting in a much cleaner fuel (Nordstrand *et al.*, 2008; Lee *et al.*, 2006; Sondreal *et al.*, 2004, 2006; Yang *et al.*, 2007; Wang *et al.*, 2008). The resulting hydrogen gas can be used for electricity generation or as a transport fuel. The gasification process also facilitates capture of carbon dioxide emissions from the combustion effluent (see discussion of carbon capture and storage below).

Although synthesis gas can be used as a stand-alone fuel, the energy density of synthesis gas is approximately half that of natural gas and is therefore mostly suited for the production of transportation fuels and other chemical products. Synthesis gas is mainly used as an intermediary building block for the final production of various fuels such as synthetic natural gas, methanol, and synthetic fuel (dimethyl ether – synthesized gasoline and diesel fuel) (Chadeesingh, 2011; Speight, 2013).

The use of synthesis gas offers the opportunity to furnish a broad range of environmentally clean fuels and chemicals and there has been steady growth in the traditional uses of synthesis gas. Almost all hydrogen gas is manufactured from synthesis gas and there has been an increase in the demand for this basic chemical. In fact, the major use of synthesis gas is in the manufacture of hydrogen for a growing number of purposes, especially in crude oil refineries (Speight, 2014a, 2017). Methanol not only remains the second-largest consumer of synthesis gas but has shown remarkable growth as part of the methyl ethers used as octane enhancers in automotive fuels.

The Fischer-Tropsch synthesis remains the third-largest consumer of synthesis gas, mostly for transportation fuels but also as a growing feedstock source for the manufacture of chemicals, including polymers. The hydroformylation of olefins (the Oxo reaction), a completely chemical use of synthesis gas, is the fourth-largest use of carbon monoxide and hydrogen mixtures. A direct application of synthesis gas as fuel (and eventually also for chemicals) that promises to increase is its use for *integrated gasification combined cycle* (IGCC) units for the generation of electricity (and also chemicals) from coal, crude oil coke or high-boiling (high-density) resids. Finally, synthesis gas is the principal source of carbon monoxide, which is used in an expanding list of carbonylation reactions, which are of major industrial interest.

Since the synthesis gas is at high pressure and has a high concentration of carbon dioxide, a physical solvent, can be used to capture carbon dioxide (Speight, 2008, 2013), which is desorbed from the solvent by pressure reduction and the solvent is recycled into the system.

2.6.2 Liquid Products

The production of liquid fuels from coal via gasification is often referred to as the *indirect liquefaction* of coal (Speight, 2013). In these processes, coal is not converted directly into liquid products but involves a two-stage conversion operation in which coal is first converted (by reaction with steam and oxygen) to produce a gaseous mixture that is composed primarily of carbon monoxide and hydrogen (synthesis gas). The gas stream is subsequently purified (to remove sulfur, nitrogen, and any particulate matter) after which it is catalytically converted to a mixture of liquid hydrocarbon products.

The synthesis of hydrocarbon derivatives from carbon monoxide and hydrogen (synthesis gas) (the Fischer-Tropsch synthesis) is a procedure for the indirect liquefaction of coal and other carbonaceous feedstocks (Starch *et al.*, 1951; Batchelder, 1962; Dry, 1976;

88 Synthesis Gas

Anderson, 1984; Speight, 2011a, 2011b). This process is the only coal liquefaction scheme currently in use on a relatively large commercial scale; South Africa is currently using the Fischer-Tropsch process on a commercial scale in its SASOL complex (Singh, 1981).

Thus, coal is converted to gaseous products at temperatures in excess of 800°C (1470°F), and at moderate pressures, to produce synthesis gas:

$$[C]_{coal} + H_2O \rightarrow CO + H_2$$

The gasification may be attained by means of any one of several processes or even by gasification of coal in place (underground, or in situ, gasification of coal).

In practice, the Fischer-Tropsch reaction is carried out at temperatures of 200 to 350°C (390 to 660°F) and at pressures of 75 to 4000 psi. The hydrogen/carbon monoxide ratio is typically on the order of 2/2:1 or 2/5:1. Since up to three volumes of hydrogen may be required to achieve the next stage of the liquids production, the synthesis gas must then be converted by means of the water-gas shift reaction) to the desired level of hydrogen:

$$CO + H_2O \rightarrow CO_2 + H_2$$

After this, the gaseous mix is purified and converted to a wide variety of hydrocarbon derivatives:

$$nCO + (2n + 1)H_2 \rightarrow C_nH_{2n+2} + nH_2O$$

These reactions result primarily in low- and medium-boiling aliphatic compounds suitable for gasoline and diesel fuel.

2.6.3 Tar

Another key contribution to efficient gasifier operation is the need for a tar reformer. Tar reforming occurs when water vapor in the incoming synthesis gas is heated to a sufficient temperature to cause steam reforming in the gas conditioning reactor, converting condensable hydrocarbon derivatives (tars) to non-condensable lower molecular weight molecules. The residence time in the conditioning reactor is sufficient to also allow a water gas shift reaction to occur and generate increased amounts of hydrogen in the synthesis gas.

Thus, tar reforming technologies – which can be thermally driven and/or catalytically driven – are utilized to break down or decompose tars and high-boiling hydrocarbon products into hydrogen and carbon monoxide. This reaction increases the hydrogen-to-carbon monoxide (H_2/CO) ratio of the synthesis gas and reduces or eliminates tar condensation in downstream process equipment. Thermal tar reformer designs are typically fluid-bed or fixed-bed type. Catalytic tar reformers are filled with heated loose catalyst material or catalyst block material and can be fixed or fluid bed designs.

Typically, the tar reformer is a refractory lined steel vessel equipped with catalyst blocks, which may contain a noble metal or a nickel-enhanced material. Synthesis gas is routed to the top of the vessel and flows down through the catalyst blocks. Oxygen and steam are added to the tar reformer at several locations along the flow path to enhance the synthesis gas composition and achieve optimum performance in the reformer. The tar reformer utilizes a catalyst to decompose tars and high boiling hydrocarbon derivatives into hydrogen and carbon monoxide. Without this decomposition the tars and high boiling hydrocarbon derivatives in the synthesis gas will condense as the synthesis gas is cooled in the downstream process equipment. In addition, the tar reformer increases the hydrogen/carbon monoxide ratio for optimal conversion. The synthesis gas is routed from the tar reformer to downstream heat recovery and gas cleanup unit operations.

References

Anderson, L.L., and Tillman, D.A. 1979. *Synthetic Fuels from Coal: Overview and Assessment*. John Wiley and Sons Inc., New York. 33.

Anderson, R.B. 1984. In: *Catalysis on the Energy Scene*. S. Kaliaguine and A. Mahay (editors). Elsevier, Amsterdam, Netherlands. Page 457.

Anthony, D.B., and Howard, J.B. 1976. Coal Devolatilization and Hydrogasification. *AIChE Journal*. 22: 625.

Arena, U. 2012. Process and Technological Aspects of Municipal Solid Waste Gasification. A Review. *Waste Management*, 32: 625-639.

Argonne. 1990. Environmental Consequences of, and Control Processes for, Energy Technologies. Argonne National Laboratory. *Pollution Technology Review No. 181*. Noyes Data Corp., Park Ridge, New Jersey. Chapter 6.

Baker, R.T.K., and Rodriguez, N.M. 1990. In: *Fuel Science and Technology Handbook*. Marcel Dekker Inc, New York. Chapter 22.

Balat, M. 2011. Fuels from Biomass – An Overview. In: *The Biofuels Handbook*. J.G. Speight (Editor). Royal Society of Chemistry, London, United Kingdom. Part 1, Chapter 3.

Batchelder, H.R. 1962. In: *Advances in Petroleum Chemistry and Refining*. J.J. McKetta Jr. (Editor). Interscience Publishers Inc., New York. Volume V. Chapter 1.

Baxter, L. 2005. Biomass-Coal Co-Combustion: Opportunity for Affordable Renewable Energy. *Fuel* 84(10): 1295-1302.

Bhattacharya, S., Md. Mizanur Rahman Siddique, A.H., and Pham, H-L. 1999. A Study in Wood Gasification on Low Tar Production. *Energy*, 24: 285-296.

Biermann, C.J. 1993. *Essentials of Pulping and Papermaking*. Academic Press Inc., New York.

Boateng, A.A., Walawender, W.P., Fan, L.T., and Chee, C.S. 1992. Fluidized-Bed Steam Gasification of Rice Hull. *Bioresource Technology*, 40(3): 235-239.

Bodle, W.W., and Huebler, J. 1981. In: *Coal Handbook*. R.A. Meyers (Editor). Marcel Dekker Inc., New York. Chapter 10.

Brage, C., Yu, Q., Chen, G., and Sjöström, K. 2000. Tar Evolution Profiles Obtained from Gasification of Biomass and Coal. *Biomass and Bioenergy*, 18(1): 87-91.

Brar, J.S., Singh, K., Wang, J., and Kumar, S. 2012. Cogasification of Coal and Biomass: A Review. *International Journal of Forestry Research*, (2012): 1-10.

Bridgwater, A.V. 2003. Renewable Fuels and Chemicals by Thermal Processing of Biomass. *Chem. Eng. Journal*, 91: 87-102.

Cavagnaro, D.M. 1980. *Coal Gasification Technology*. National Technical Information Service, Springfield, Virginia.

Chadeesingh, R. 2011. The Fischer-Tropsch Process. In *The Biofuels Handbook*. J.G. Speight (Editor). The Royal Society of Chemistry, London, United Kingdom. Part 3, Chapter 5, Page 476-517.

Chen, G., Sjöström,, K. and Bjornbom, E. 1992. Pyrolysis/Gasification of Wood in a Pressurized Fluidized Bed Reactor. *Ind. Eng. Chem. Research*, 31(12): 2764-2768.

Chen, C., Horio, M., and Kojima, T. 2000. Numerical Simulation of Entrained Flow Coal Gasifiers. Part II: Effects of Operating Conditions on Gasifier Performance. *Chemical Engineering Science*, 55(18): 3875-3883.

Collot, A.G., Zhuo, Y., Dugwell, D.R., and Kandiyoti, R. 1999. Co-Pyrolysis and Cogasification of Coal and Biomass in Bench-Scale Fixed-Bed and Fluidized Bed Reactors. *Fuel*, 78: 667-679.

Couvaras, G. 1997. Sasol's Slurry Phase Distillate Process and Future Applications. *Proceedings. Monetizing Stranded Gas Reserves Conference, Houston, December*.

Cover, A.E., Schreiner, W.C., and Skaperdas, G.T. 1973. Kellogg's Coal Gasification Process. *Chem. Eng. Progr.* 69(3): 31.

Cusumano, J.A., Dalla Betta, R.A., and Levy, R.B. 1978. Catalysis in Coal Conversion. Academic Press Inc., New York.

Davidson, R.M. 1983. *Mineral Effects in Coal Conversion*. Report No. ICTIS/TR22, International Energy Agency, London, United Kingdom.

De Campos Roseno, K.T., de B. Alves, R.M., Giudici, R., and Schmal, M. 2018. Syngas Production Using Natural Gas from the Environmental Point of View. In: *Biofuels - State of Development*. B. Krzysztof (Editor). IntechOpen, DOI: 10.5772/intechopen.74605. Chapter 13. Page 273-290. https://www.intechopen.com/books/biofuels-state-of-development/syngas-production-using-natural-gas-from-the-environmental-point-of-view

Demirbaş, A. 2011. Production of Fuels from Crops. In: *The Biofuels Handbook*. J.G. Speight (Editor). Royal Society of Chemistry, London, United Kingdom. Part 2, Chapter 1.

Dry, M.E. 1976. Advances in Fischer-Tropsch Chemistry. *Ind. Eng. Chem. Prod. Res. Dev.* 15(4): 282-286.

Dutcher, J.S., Royer, R.E., Mitchell, C.E., and Dahl, A.R. 1983. In: *Advanced Techniques in Synthetic Fuels Analysis*. C.W. Wright, W.C. Weimer, and W.D. Felic (Editors). Technical Information Center, United States Department of Energy, Washington, DC. Page 12.

EIA. 2007. Net Generation of Heat by Energy Sources by Type of Producer. Energy Information Administration, United States Department of Energy, Washington, DC. http://www.eia.doe.gov/cneaf/electricity/epm/table1_1.html

Elton, A. 1958. In: *A History of Technology*. C. Singer, E.J. Holmyard, A.R. Hall, and T.I. Williams (Editors). Clarendon Press, Oxford, United Kingdom. Volume IV. Chapter 9.

Ergudenler, A., and Ghaly, A.E. 1993. Agglomeration of Alumina Sand in a Fluidized Bed Straw Gasifier at Elevated Temperatures. *Bioresource Technology*, 43(3): 259-268.

Fermoso, J., Plaza, M.G., Arias, B., Pevida, C., Rubiera, F., and Pis, J.J. 2009. Co-Gasification of Coal with Biomass and Petcoke in a High-Pressure Gasifier for Syngas Production. *Proceedings. 1st Spanish National Conference on Advances in Materials Recycling and Eco-Energy. Madrid, Spain. November 12-13*.

Fryer, J.F., and Speight, J.G. 1976. *Coal Gasification: Selected Abstract and Titles*. Information Series No. 74. Alberta Research Council, Edmonton, Canada.

Fujita, S., Terunuma, H., Nakamura, M., and Takezawa, N. 1991. Mechanisms of Methanation of CO and CO2 over Ni. *Industrial & Engineering Chemistry Research*, 30: 1146-1151.

Fujita, S., Nakamura, M., Doi, T., and Takezawa, N. 1993. Mechanisms of Methanation of Carbon Dioxide and Carbon Monoxide Over Nickel/Alumina Catalysts. *Applied Catalysis, A: General*, 104: 87-100.

Gabra, M., Pettersson, E., Backman, R., and Kjellström, B. 2001. Evaluation of Cyclone Gasifier Performance for Gasification of Sugar Cane Residue – Part 1: Gasification of Bagasse. *Biomass and Bioenergy*, 21(5): 351-369.

Gay, R.L., Barclay, K.M., Grantham, L.F., and Yosim, S.J. 1980. *Fuel Production from Solid Waste. Symposium on Thermal Conversion of Solid Waste and Biomass.* Symposium Series No. 130, American Chemical Society, Washington, DC. Chapter 17. Page 227-236.

Gray, D., and Tomlinson, G. 2000. Opportunities For Petroleum Coke Gasification Under Tighter Sulfur Limits For Transportation Fuels. *Proceedings. 2000 Gasification Technologies Conference, San Francisco, California. October 8-11.*

Hanaoka, T., Inoue, S., Uno, S., Ogi, T., and Minowa, T. 2005. Effect of Woody Biomass Components on Air-Steam Gasification. *Biomass and Bioenergy*, 28(1): 69-76.

Hickman, D.A., and Schmidt, L.D. 1993. Syngas Formation by Direct Catalytic Oxidation of Methane. *Science.* 259: 343-346.

Hotchkiss, R. 2003. Coal Gasification Technologies. *Proceedings. Institute of Mechanical Engineers Part A*, 217(1): 27-33.

Howard-Smith, I., and Werner, G.J. 1976. *Coal Conversion Technology.* Noyes Data Corp., Park Ridge, New Jersey. Page 71.

Hsu, C.S., and Robinson, P.R. (Editors). 2017. *Handbook of Petroleum Technology.* Springer International Publishing AG, Cham, Switzerland.

Irfan, M.F. 2009. Pulverized Coal Pyrolysis and Gasification in $N_2/O_2/CO_2$ Mixtures by Thermogravimetric Analysis. Research Report. *Novel Carbon Resource Sciences Newsletter*, Kyushu University, Fukuoka, Japan. Volume 2: 27-33.

Ishi, S. 1982. Coal Gasification Technology. *Energy*, 15(7): 40-48, 1982.

Jenkins, B.M., and Ebeling, J.M. 1985. Thermochemical Properties of Biomass Fuels. *California Agriculture* (May-June), Page 14-18.

John, E., and Singh, K. 2011. Properties of Fuels from Domestic and Industrial Waste. In: *The Biofuels Handbook.* J.G. Speight (Editor). The Royal Society of Chemistry, London, United Kingdom. Part 3, Chapter 2, Page 377-407.

Johnson J.L. 1979. *Kinetics of Coal Gasification.* John Wiley and Sons Inc., Hoboken, New Jersey.

Kasem, A. 1979. *Three Clean Fuels from Coal: Technology and Economics.* Marcel Dekker Inc., New ork.

Khosravi1, M., and Khadse, A., 2013. Gasification of Petcoke and Coal/Biomass Blend: A Review. *International Journal of Emerging Technology and Advanced Engineering*, 3(12): 167-173.

King, R.B., and Magee, R.A. 1979. In: *Analytical Methods for Coal and Coal Products.* C. Karr Jr. (Editor). Academic Press Inc., New York. Volume III. Chapter 41.

Ko, M.K., Lee, W.Y., Kim, S.B., Lee, K.W., and Chun, H.S. 2001. Gasification of Food Waste with Steam in Fluidized Bed. *Korean Journal of Chemical Engineering*, 18(6): 961-964.

Kumabe, K., Hanaoka, T., Fujimoto, S., Minowa, T., and Sakanishi, K. 2007. Cogasification of Woody Biomass and Coal with Air and Steam. *Fuel*, 86: 684-689.

Lahaye, J., and Ehrburger, P. (Editors). 1991. *Fundamental Issues in Control of Carbon Gasification Reactivity.* Kluwer Academic Publishers, Dordrecht, Netherlands.

Lapuerta M., Hernández J.J., Pazo A., and López, J. 2008. Gasification and co-gasification of biomass wastes: Effect of the biomass origin and the gasifier operating conditions. *Fuel Processing Technology*, 89(9): 828-837.

Lee, S., Speight, J.G., and Loyalka, S. 2007. *Handbook of Alternative Fuel Technologies.* CRC-Taylor & Francis Group, Boca Raton, Florida.

Lee, S., and Shah, Y.T. 2013. *Biofuels and Bioenergy.* CRC Press, Taylor & Francis Group, Boca Raton, Florida.

Liu, G., Larson, E.D., Williams, R.H., Kreutz, T.G., Guo, X. 2011. Making Fischer-Tropsch Fuels and Electricity from Coal and Biomass: Performance and Cost Analysis. *Energy & Fuels*, 25: 415-437.

Lu, X., and Berge, N.D. 2014. Influence of Feedstock Chemical Composition on Product Formation and Characteristics Derived from the Hydrothermal Carbonization of Mixed Feedstocks. *Bioresource Technology*, 166: 120-131.

Luque, R., and Speight, J.G. (Editors). 2015. *Gasification for Synthetic Fuel Production: Fundamentals, Processes, and Applications*. Woodhead Publishing, Elsevier, Cambridge, United Kingdom.

Lv, P.M., Xiong, Z.H., Chang, J., Wu, C.Z., Chen, Y., and Zhu, J.X. 2004. An Experimental Study on Biomass Air-Steam Gasification in a Fluidized Bed. *Bioresource Technology*, 95(1): 95-101.

Mahajan, O.P., and Walker, P.L. Jr. 1978. In: *Analytical Methods for Coal and Coal Products*. C. Karr Jr. (Editor). Academic Press Inc., New York. Volume II. Chapter 32.

Martinez-Alonso, A., and Tascon, J.M.D. 1991. In: *Fundamental Issues in Control of Carbon Gasification Reactivity*. Lahaye, J., and Ehrburger, P. (Editors). Kluwer Academic Publishers, Dordrecht, Netherlands.

Massey, L.G. (Editor). 1974. *Coal Gasification*. Advances in Chemistry Series No. 131. American Chemical Society, Washington, D.C.

Matsukata, M., Kikuchi, E., and Morita, Y. 1992. A New Classification of Alkali and Alkaline Earth Catalysts for Gasification of Carbon. *Fuel*. 71: 819-823.

McKee, D.W. 1981. *The Catalyzed Gasification Reactions of Carbon. The Chemistry and Physics of Carbon*. P.L. Walker, Jr. and P.A. Thrower (Editors). Marcel Dekker Inc., New York). Volume 16. Page 1.

McKendry, P. 2002. Energy Production from Biomass Part 3: Gasification Technologies. *Bioresource Technology*, 83(1): 55-63.

McLendon T.R., Lui A.P., Pineault R.L., Beer S.K., and Richardson S.W. 2004. High-Pressure Co-Gasification of Coal and Biomass in a Fluidized Bed. *Biomass and Bioenergy*, 26(4): 377-388.

Mims, C.A. 1991. In: *Fundamental Issues in Control of Carbon Gasification Reactivity*. Lahaye, J., and Ehrburger, P. (Editors). Kluwer Academic Publishers, Dordrecht, Netherlands. Page 383.

Mokhatab, S., Poe, W.A., and Speight, J.G. 2006. *Handbook of Natural Gas Transmission and Processing*. Elsevier, Amsterdam, Netherlands.

Murthy, B.N., Sawarkar, A.N., Deshmukh, N.A., Mathew, T., and Joshi, J.B. 2013. Petroleum Coke Gasification: A Review. *Canadian Journal of Chemical Engineering*, (923): 441-468.

Nef, J.U. 1957. In: *A History of Technology*. C. Singer, E.J. Holmyard, A.R. Hall, and T.I. Williams (Editors). Clarendon Press, Oxford, United Kingdom. Volume III. Chapter 3.

Nordstrand D., Duong D.N.B., Miller, B.G. 2008. *Combustion Engineering Issues for Solid Fuel Systems*. Chapter 9 Post-combustion Emissions Control. B.G. Miller and D. Tillman (Editors). Elsevier, London, United Kingdom.

Pakdel, H., and Roy, C. 1991. Hydrocarbon Content of Liquid Products and Tar from Pyrolysis and Gasification of Wood. *Energy & Fuels*, 5: 427-436.

Pan, Y.G., Velo, E., Roca, X., Manyà, J.J., and Puigjaner, L. 2000. Fluidized-Bed Cogasification of Residual Biomass/Poor Coal Blends for Fuel Gas Production. *Fuel*, 79: 1317-1326.

Penrose, C.F., Wallace, P.S., Kasbaum, J.L., Anderson, M.K., and Preston, W.E. 1999. Enhancing Refinery Profitability by Gasification, Hydroprocessing and Power Generation. *Proceedings. Gasification Technologies Conference San Francisco, California. October.*

Probstein, R.F., and Hicks, R.E. 1990. *Synthetic Fuels*. pH Press, Cambridge, Massachusetts. Chapter 4.

Rajvanshi, A.K. 1986. Biomass Gasification. In: *Alternative Energy in Agriculture*, Vol. II. D.Y. Goswami (Editor). CRC Press, Boca Raton, Florida. Page 83-102.

Ramroop Singh, N. 2011. Biofuel. In: *The Biofuels Handbook*. J.G. Speight (Editor). Royal Society of Chemistry, London, United Kingdom. Part 1, Chapter 5.

Rapagnà, N.J., and Latif, A. 1997. Steam Gasification of Almond Shells in a Fluidized Bed Reactor: The Influence of Temperature and Particle Size on Product Yield and Distribution. *Biomass and Bioenergy*, 12(4): 281-288.

Rapagnà, N.J., and, A. Kiennemann, A., and Foscolo, P.U. 2000. Steam-Gasification of Biomass in a Fluidized-Bed of Olivine Particles. *Biomass and Bioenergy*, 19(3): 187-197.

Ricketts, B., Hotchkiss, R., Livingston, W., and Hall, M. 2002. Technology Status Review of Waste/Biomass Co-Gasification with Coal. *Proceedings. Inst. Chem. Eng. Fifth European Gasification Conference. Noordwijk, Netherlands. April 8-10.*

Sha, X. 2005. Coal Gasification. In: *Coal, Oil Shale, Natural Bitumen, Heavy Oil and Peat. Encyclopedia of Life Support Systems (EOLSS)*, Developed under the Auspices of the UNESCO, EOLSS Publishers, Oxford, UK, [http://www.eolss.net].

Shen, C-H., Chen, W-H., Hsu, H-W., Sheu, J-Y., and Hsieh, T-H. 2012. Co-Gasification Performance of Coal and Petroleum Coke Blends in A Pilot-Scale Pressurized Entrained-Flow Gasifier. *Int. J. Energy Res.*, 36: 499-508.

Shi, J., Thompson, V.S., Yancey, N.A. Stavila, V., Simmons, B.A., and Singh, S. 2013. Impact of Mixed Feedstocks and Feedstock Densification on Ionic Liquid Pretreatment Efficiency. *Biofuels*, 4(1): 63-72.

Shinnar, R., Fortuna, G. and Shapira, D. 1982. Thermodynamic and Kinetic Constraints of Catalytic Natural Gas Processes. *Ind. Eng. Chem. Process Design and Development* 21: 728-750.

Silaen, A., and Wang, T. 2008. Effects of Turbulence and Devolatilization Models on Gasification Simulation. *Proceedings. 25th International Pittsburgh Coal Conference, Pittsburgh, Pennsylvania. September 29-October 2.*

Sjöström, K., Chen, G., Yu, Q., Brage, C., and Rosén, C. 1999. Promoted Reactivity of Char in Cogasification of Biomass and Coal: Synergies in the Thermochemical Process. *Fuel*, 78: 1189-1194.

Sondreal, E.A., Benson, S. A., Pavlish, J. H., and Ralston, N.V.C. 2004. An Overview of Air Quality III: Mercury, Trace Elements, and Particulate Matter. *Fuel Processing Technology*. 85: 425-440.

Sondreal, E.A., Benson, S. A., and Pavlish, J.H. 2006. Status of Research on Air Quality: Mercury, Trace Elements, and Particulate Matter. *Fuel Processing Technology*. 65/66: 5-22.

Speight, J.G. 1990. In: *Fuel Science and Technology Handbook*. J.G. Speight (Editor). Marcel Dekker Inc., New York. Chapter 33.

Speight, J.G. 2002. *Chemical Process and Design Handbook*. McGraw-Hill Inc., New York.

Speight, J.G. 2008. *Synthetic Fuels Handbook: Properties, Processes, and Performance*. McGraw-Hill, New York.

Speight, J.G. 2009. *Enhanced Recovery Methods for Heavy Oil and Tar Sands*. Gulf Publishing Company, Houston, Texas.

Speight, J.G. (Editor). 2011a. *The Biofuels Handbook*. Royal Society of Chemistry, London, United Kingdom.

Speight, J.G. 2011b. *The Refinery of the Future*. Gulf Professional Publishing, Elsevier, Oxford, United Kingdom.

Speight, J.G. 2013. *The Chemistry and Technology of Coal* 3rd Edition. CRC Press, Taylor & Francis Group, Boca Raton, Florida.

Speight, J.G. 2014a. *The Chemistry and Technology of Petroleum* 5th Edition. CRC Press, Taylor & Francis Group, Boca Raton, Florida.

Speight, J.G. 2014b. *Gasification of Unconventional Feedstocks*. Gulf Professional Publishing, Elsevier, Oxford, United Kingdom.

Speight, J.G. 2017. *Handbook of Petroleum Refining*. CRC Press, Taylor & Francis Group, Boca Raton, Florida.

Speight, J.G. 2019. *Natural Gas: A Basic Handbook* 2nd Edition. Gulf Publishing Company, Elsevier, Cambridge, Massachusetts.

Storch, H.H., Golumbic, N., and Anderson, R.B. 1951. *The Fischer Tropsch and Related Syntheses*. John Wiley & Sons Inc., New York.

Sundaresan, S., and Amundson, N.R. 1978. Studies in Char Gasification – I: A lumped Model. *Chemical Engineering Science*, 34: 345-354.

Taylor, F.S., and Singer, C. 1957. In: *A History of Technology*. C. Singer, E.J. Holmyard, A.R. Hall, and T.I. Williams (Editors). Clarendon Press, Oxford, United Kingdom. Volume II. Chapter 10.

Van Heek, K.H., Muhlen, H-J. 1991. In: *Fundamental Issues in Control of Carbon Gasification Reactivity*. J. Lahaye and P. Ehrburger (editors). Kluwer Academic Publishers Inc., Netherlands. Page 1.

Vélez, J.F., Chejne, F., Valdés, C.F., Emery, E.J., and Londoño, C.A. 2009. Cogasification of Colombian Coal and Biomass in a Fluidized Bed: An Experimental Study. *Fuel*, 88: 424-430.

Wang, Y., Duan Y., Yang L., Jiang Y., Wu C., Wang Q., and Yang X. 2008. Comparison of Mercury Removal Characteristic between Fabric Filter and Electrostatic Precipitators of Coal-Fired Power Plants. *Journal of Fuel Chemistry and Technology*, 36(1): 23-29.

Yang, H., Xua, Z., Fan, M., Bland, A.E., and Judkins, R.R. 2007. Adsorbents for Capturing Mercury in Coal-Fired Boiler Flue Gas. *Journal of Hazardous Materials*. 146:1-11.

Worley, M., and Yale, J. 2012. *Biomass Gasification Technology Assessment*. Subcontract Report No. NREL/SR-5100-57085. National Renewable Energy Laboratory, Golden, Colorado. November.

Zhu, Q., Zhao, X., and Deng, Y. 2004. Advances in the Partial Oxidation of Methane to Synthesis Gas. *Journal of Natural Gas Chemistry*, 13: 191-203.

3
Gasifier Types and Gasification Chemistry

3.1 Introduction

The gasification of a carbonaceous feedstock is the conversion of the feedstock by any one of a variety of processes to combustible gases (Fryer and Speight, 1976; Radović *et al.*, 1983; Radović and Walker, 1984; Garcia and Radović, 1986; Calemma and Radović, 1991; Kristiansen, 1996; Speight, 2008, 2013a, 2013b). In fact, gasification offers one of the most versatile methods (with a lesser environmental impact than combustion) to convert carbonaceous feedstocks (such as coal, petroleum residua, biomass, and industrial waste) (McKendry, 2002; Jangsawang *et al.*, 2006; Senneca, 2007; Butterman and Castaldi, 2008; Speight, 2008, 2013a, 2013b; 2014; Luque and Speight, 2015) into electricity, hydrogen, and other valuable energy products. Gasification may be one of the most flexible technologies to produce clean-burning as the chemical building block for a wide range of products.

Moreover, gasification is one of the critical technologies that enable hydrogen production from carbonaceous feedstocks (Lee *et al.*, 2007; Speight, 2008, 2011, 2013a, 2013b; 2014). Gasifiers produce synthesis gas that has multiple applications and can be used for hydrogen production, electricity generation and chemical plants. Integrated gasification combined cycle (IGCC) plants utilize the synthesis gas in a combined cycle power plant (gas turbine and steam turbine) to produce electricity (Speight, 2013a, 2013b).

There has been a general tendency to classify gasification processes by virtue of the heat content of the gas which is produced; it is also possible to classify gasification processes according to the type of reactor vessel and whether or not the system reacts under pressure. However, for the purposes of the present text gasification processes are segregated according to the bed types, which differ in their ability to accept (and convert) various types of feedstock (Collot, 2002, 2006).

Although there are many successful commercial coal gasifiers (Speight, 2013a, 2013b; Luque and Speight, 2015), the basic form and concept as well as the details on the design and operation for the commercial coal gasifiers are closely guarded as proprietary information. In fact, the production of gas from carbonaceous feedstocks has been an expanding area of technology. As a result, several types of gasification reactors have arisen and there has been a general tendency to classify gasification processes by virtue of the heat content of the gas which is produced. Furthermore, it is also possible to classify gasification processes according to the type of reactor vessel and whether or not the system reacts under pressure (Collot, 2002, 2006).

Thus, the gasification of a carbonaceous feedstock (or a derivative thereof, such as char) is, essentially, the conversion of carbonaceous material by any one of a variety of chemical processes to produce combustible gases (Higman, and van der Burgt, 2008; Speight, 2008, 2013a). With the rapid increase in the use of coal from the 15th century onwards it is not surprising the concept of using coal to produce a flammable gas that was used for domestic heating, industrial heating, and power generation, especially the use of the water and hot coal, became commonplace in the 19th and 20th centuries (Speight, 2013a, 2013b).

The gasification process includes a series of reaction steps that convert the carbonaceous feedstock (composed of carbon, hydrogen, and oxygen as well as impurities such as sulfur-containing and nitrogen-containing moieties and metallic constituents) into *synthesis gas* (*syngas*, a mixture of carbon monoxide, CO, and hydrogen H_2) and an array of hydrocarbon derivatives. This conversion is generally accomplished by introducing a gasifying agent (air, oxygen, and/or steam) into a reactor vessel containing the carbonaceous feedstock where the temperature, pressure, and flow pattern (moving bed, fluidized, or entrained bed) are controlled. However there are gases other than carbon monoxide and hydrogen and the proportions of the resultant product gases (such as carbon dioxide, CO_2, methane, CH_4, water vapor, H_2O, hydrogen sulfide, H_2S, and sulfur dioxide, SO_2, but including carbon monoxide, CO, and hydrogen, H_2) depends on (i) the type of feedstock, (ii) the composition of the feedstock, (iii) the gasifying agent or the gasifying medium, and (iv) the thermodynamics and chemistry of the gasification reactions as controlled by the process operating parameters (Singh *et al.*, 1980; Shabbar and Janajreh, 2013; Speight, 2013a, 2013b; Luque and Speight, 2015).

The kinetic rates and extents of conversion for the several chemical reactions that are a part of the gasification process are variable and are typically functions of (i) temperature, (ii) pressure, (iii) reactor and configuration, (iv) gas composition, and (v) the nature – chemical composition and properties – of the feedstock being gasified (Johnson, 1979; Penner, 1987; Müller *et al.*, 2003; Slavinskaya *et al.*, 2009; Speight, 2013a, 2013b).

Generally, the reaction rate (i.e., the rate of feedstock conversion) is higher at higher temperatures, whereas reaction equilibrium may be favored at either higher or lower temperatures depending on the specific type of gasification reaction. The effect of pressure on the rate also depends on the specific reaction. Thermodynamically, some gasification reactions such as carbon-hydrogen reaction producing methane are favored at high pressures (>1030 psi) and relatively lower temperatures (760 to 930°C; 1400 to 1705°F), whereas low pressures and high temperatures favor the production of synthesis gas (i.e., carbon monoxide and hydrogen) via steam or carbon dioxide gasification reaction.

Because of the overall complexity of the gasification process, it is necessary to present a description of the chemistry of the gasification reactions, and it is the purpose of this chapter to present descriptions of the various reactions involved in the gasification of carbonaceous materials as well as the various thermodynamic aspects of these reactions which dictate the process parameters used to produce the various gases.

It is the purpose of this chapter to present the different categories of gasification reactors as they apply to various types of feedstocks. Within each category there are several commonly known gasifier types, each of which are designed for a different process (Table 3.1, Table 3.2), some of which are in current use and some of which are in lesser use (Speight, 2013a, 2013b).

3.2 Gasifier Types

Many gasifiers employ a single stage which is simple in design and less expensive with respect to manufacturing pressure vessel. When the feedstock is relatively uniform in

Table 3.1 Categories of gasification processes (Braunstein *et al.*, 1977).

Process types	Unit
Fixed-Bed Processes	Foster Wheeler Stoic Process
	Lurgi Process
	Wellman-Galusha Process
	Woodall-Duckham Process
Fluidized-Bed Processes	Agglomerating Burner Process
	Carbon Dioxide Acceptor Process
	Coalcon Process
	COED/COGAS Process
	Exxon Catalytic Gasification Process
	Hydrane Process
	Hygas Process
	Pressurized Fluid-Bed Process
	Synthane Process
	U-Gas Process
	Winkler Process
Entrained-Bed Processes	Bi-Gas Process
	Combustion Engineering Process
	Koppers-Totzek Process
	Texaco Process
Molten Salt Processes	Atgas Process
	Pullman-Kellogg Process
	Rockgas Process
	Rummel Single-Shaft Process

quality and in other properties, the residence time inside the gasifier will be constant in theory if the constant feeding is guaranteed. When the feedstock and oxygen feeding is uniform, all the times, the performance of the gasifier will be satisfactory, although there would be some mechanical- or components-related problems. The most important factor in operating a gasifier should be the constant rate of feedstock introduction into the unit. Unfortunately, feedstocks such as biomass and waste are becoming more and more heterogeneous. In many feedstocks from widely different origins may be mixed with coal

Table 3.2 Summary of the characteristics of the various types of gasifier.

Updraft fixed-bed gasifier   • The feedstock is fed in at the top of the gasifier, and the air, oxygen or steam intake is at the bottom, hence the feedstock and gases move in opposite directions.   • Some of the resulting char falls and burns to provide heat.   • The methane and tar-rich gas leaves at the top of the gasifier, and the ash falls from the grate for collection at the bottom of the gasifier.
Downdraft fixed-bed gasifier   • The feedstock is fed in at the top of the gasifier and the air, and oxygen or steam intake is also at the top or from the sides, hence the feedstock and gases move in the same direction.   • Some of the feedstock is burnt, falling through the gasifier throat to form a bed of hot charcoal which the gases have to pass through (a reaction zone).   • This ensures a fairly high quality synthesis gas, which leaves at the base of the gasifier, with ash collected under the grate.
Entrained-flow gasifier   • Powdered feedstock is fed into a gasifier with pressurized oxygen and/or steam.   • A turbulent flame at the top of the gasifier burns some of the feedstock, providing large amounts of heat, at high temperature (1200 to 1500°C, 2190 to 2730°F).   • Fast conversion of feedstock into very high quality synthesis gas.   • The ash melts onto the gasifier walls, and is discharged as molten slag.
Bubbling fluidized-bed gasifier   • A bed of fine inert material sits at the gasifier bottom, with air, oxygen or steam being blown upwards through the bed just fast enough to agitate the material.   • Feedstock is fed in from the side, mixes, and combusts or forms synthesis gas which leaves upwards.   • Operates at temperatures below 900°C (1,650°F) to avoid ash melting and sticking. Can be pressurized.
Circulating fluidized-bed gasifier   • A bed of fine inert material has air, oxygen or steam blown upwards through it fast enough to suspend material throughout the gasifier.   • Feedstock is fed in from the side, is suspended, and combusts providing heat, or reacts to form synthesis gas.   • The mixture of synthesis gas and particles are separated using a cyclone, with material returned into the base of the gasifier.   • Operates at temperatures below 900°C (1650°F) to avoid ash melting and sticking. Can be pressurized.
Dual fluidized-bed gasifier   • This system has two chambers – a gasifier and a combustor.   • Feedstock is fed into the circulating fluidized-bed/bubbling fluidized-bed gasification chamber, and converted to nitrogen-free synthesis gas and char using steam.   • The char is burnt in air in the circulating fluidized-bed/bubbling fluidized-bed combustion chamber, heating the accompanying bed particles.   • This hot bed material is then fed back into the gasification chamber, providing the indirect reaction heat.   • Cyclones remove any circulating fluidized bed chamber synthesis gas or flue gas.   • Operates at temperatures below 900°C to avoid ash melting and sticking. Could be pressurized.

(Continued)

Table 3.2 Summary of the characteristics of the various types of gasifier. (*Continued*)

Plasma gasifier
• Untreated feedstock is dropped into the gasifier, coming into contact with an electrically generated plasma, usually at atmospheric pressure and temperatures of 1500 to 5000°C (2730 to 9030°F).
• Organic matter is converted into very high quality synthesis gas, and inorganic matter is vitrified into inert slag.
• Note that plasma gasification uses plasma torches. It is also possible to use plasma arcs in a subsequent process step for synthesis gas clean-up.

(Chapter 5, Chapter 6, Chapter 7) and particle residence time inside the gasifier might not be sufficient to guarantee the full conversion of all of the feedstock. Low reactivity or larger size feedstock particles that are contained could pass through the gasifier without fully reacting. The two-stage gasifier is better able to accommodate the heterogeneous feedstock particles in a single reactor. Introducing the amount of feedstock and oxygen can be manipulated in two separate positions at the gasifier. By adjusting the amounts, hot local temperature is possible in the gasifier that will gasify even the least reactive particles within the feedstock.

Thus, if one single pass through the gasifier is not sufficient to convert all organic components to syngas, unreacted char can be collected and recycled to enhance carbon conversion in excess of 99%. But recycling usually incorporates expensive additional feeding systems and, if possible, it is preferable if the gasifier can achieve complete (100%) carbon conversion in a single pass through the gasifier.

In terms of feedstock variability, several different types of feedstocks that vary widely in composition (Chapter 1) are available for gasification and include crude oil resids, coal (including peat), wood, wood waste (branches, twigs, roots, bark, wood shavings and sawdust) as well as a multitude of agricultural residues (such as maize cobs, coconut shells, coconut husks, cereal straws, rice husks, etc.), domestic waste, and industrial waste. Because the feedstocks differ substantially in the respective chemical, physical and morphological properties, they make different demands on the method of gasification and consequently require different reactor design and/or on the gasification technology. It is for this reason that, during more than a century of gasification experience, a large number of different gasifiers has been developed and marketed of which the principal types are (i) the moving/fixed-bed gasifier, (ii) the entrained-flow gasifier, and (iii) the fluidized-bed gasifier (Figure 3.1) (Ricketts *et al.*, 2002).

Moving-bed (fixed-bed) gasifiers (Figure 3.1a) commonly operate at moderate pressures (350 to 450 psi). Feedstocks in the form of large particles are loaded into the top of the refractory-lined gasifier vessel and move slowly downward through the bed, while reacting with high oxygen content gas introduced at the bottom of the gasifier that is flowing countercurrently upward in the gasifier. The basic configuration is the same as seen in the common blast furnace. Reactions within the gasifier occur in different zones. In the drying zone at the top of the gasifier, the entering coal is heated and dried, while cooling the product gas before it leaves the reactor. The feedstock is further heated and devolatilized by the higher temperature gas as it descends through the carbonization zone. In the next

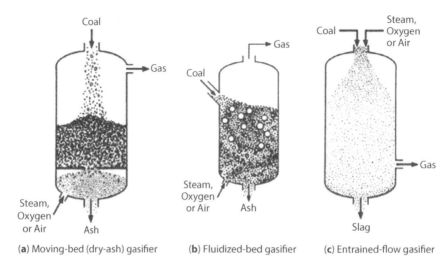

Figure 3.1 The principal types of gasifiers.

zone – the gasification zone – the devolatilized feedstock is gasified by reaction with steam and carbon dioxide. Near the bottom of the gasifier – the combustion zone, which operates at the highest temperature – oxygen reacts with the remaining char.

Moving-bed gasifiers operate in two different modes. In the dry-ash mode of operation (such as in the Lurgi dry-ash gasifier), the temperature is moderated to below the ash-slagging temperature by reaction of the char with excess steam. The ash below the combustion zone is cooled by the entering steam and oxidant (oxygen or air) and produced as a solid ash. In the slagging mode of operation (such as in the British Gas/Lurgi gasifier) much less steam is used, and as the result, a much higher temperature is achieved in the combustion zone, melting the ash and producing slag. The moisture content of the feedstock is the main factor which determines the discharge gas temperature. Typically, the product gas leaving the gasifier is quenched by direct contact with recycle water to condense and remove tars and oils. After quench, heat can be recovered from the gas by generation of low pressure steam.

Fluidized-bed gasifiers (Figure 3.1b) suspend feedstock particles in an oxygen-rich gas so the resulting bed within the gasifier acts as a fluid. These gasifiers employ back-mixing, and efficiently mix fresh feedstock particles with feedstock particles already undergoing gasification. To sustain fluidization, or suspension of feedstock particles within the gasifier, feedstock of small particles sizes (<6 mm) is typically used.

The feedstock enters at the side of the reactor, while steam and oxidant enter near the bottom with enough velocity to fully suspend or fluidize the reactor bed. Due to the thorough mixing within the gasifier, a constant temperature is sustained in the reactor bed. The gasifiers normally operate at moderately high temperature to achieve an acceptable carbon conversion rate (on the order of 90 to 95%) and to decompose most of the tar, oils, and other liquid byproducts. However, the operating temperatures are usually less than the ash fusion temperature so as to avoid clinker formation and the possibility of de-fluidization of the bed. This, in turn means that fluidized-bed gasifiers are best suited to relatively reactive feedstocks, such as biomass.

Some char particles are entrained in the raw syngas as it leaves the top of the gasifier, but are recovered and recycled back to the reactor via a cyclone. Ash particles, removed below the bed, give up heat to the incoming steam and recycle gas. At startup, the bed is heated externally before the feedstock is introduced. Fluidized-bed gasifiers may differ in ash conditions (dry or agglomerated/slagging) and in design configurations for improving char use. Also, depending on the degree of fluidization and bed height, these types of reactors sometimes are also named as circulating fluidized-bed reactors, and/or transport reactors.

In the *entrained-flow gasifier*, the finely ground feedstock and the oxidant (air or oxygen) and/or steam are fed co-currently to the gasifier. This results in the oxidant and steam surrounding or entraining the feedstock particles as they flow through the gasifier in a dense cloud. Entrained-flow gasifiers operate at high temperature and pressure – with extremely turbulent flow – which causes rapid feedstock conversion and allows high throughput. The gasification reactions occur at a very high rate (typical residence time is on the order of few seconds), with high carbon conversion efficiencies (98 to 99.5%). The tar, oil, and other liquids produced from devolatilization of the feedstock inside the gasifier are decomposed into hydrogen (H_2), carbon monoxide (CO) and small amounts of hydrocarbon gases. Entrained-flow gasifiers have the ability to handle practically any coal feedstock and produce a clean, tar-free synthesis gas. Given the high operating temperatures, gasifiers of this type melt the mineral matter ash into vitreous inert slag.

The finely ground feedstock can be fed to the gasifier in either the dry form or in the slurry form. The former (the dry form) uses a lock hopper system, while the latter (the slurry form) relies on the use of high-pressure slurry pumps. The slurry feed is a simpler operation, but it introduces water into the reactor which needs to be evaporated. The result of this additional water is a product synthesis gas with higher hydrgoen-to-carbon monoxide ratio, but with a lower gasifier thermal efficiency.

Although all types of gasifier are designed for accommodating the specific properties of the feedstock, the gasifier that is able to handle all or most feedstocks or feedstock types has not yet been developed.

In fact, compared to a typical fossil fuel, the complex ligno-cellulosic structure of biomass is more difficult to gasify. The nature of the mineral impurities in conjunction with the presence of various inorganic species, as well as sulfur and nitrogen containing compounds, adversely impacts the benign thermal processing of the oxygenated hydrocarbon structure of the biomass. In contract to combustion of biomass feedstocks in which fuel-bound nitrogen and sulfur are converted to NOx and SOx, steam gasification involves thermal treatment under a reducing atmosphere resulting in fuel-bound nitrogen release as molecular nitrogen and fuel-bound sulfur conversion to hydrogen sulfide that is more easily removed by means of adsorption beds (Mokhatab *et al.*, 2006; Speight, 2009, 2013a, 2013b). Unlike combustion, the gasification process is more energy intensive. Careful engineering of the process reactor is necessary if the result is to produce rather than consume a significant amount of energy or power as a result of the thermal treatment.

Thus, a careful section must be made of the type of gasifier, which is feedstock dependent. Four types of gasifier configuration are currently available for commercial use: (i) fixed-bed gasifier, which is subdivided into the countercurrent fixed-bed gasifier and the co-current fixed-bed gasifier, (ii) the fluid-bed gasifier, (iii) the entrained-flow gasifier, and (iv) processes involving the use of molten salt(s) or molten metal(s) (Kohl *et al.*, 1978; Speight, 2011; 2013a, 2013b). All systems show relative advantages and disadvantages with respect to fuel type, application and simplicity of operation, and for this reason each will have its own technical and/or economic advantages in a particular set of circumstances.

However, each type of gasifier may be designed to operate either at atmospheric pressure or at high pressure. In the latter type of operation, the hydrogasification process is optimized and the quality of the product gas (in terms of heat, or Btu, content) is improved (Anthony and Howard, 1976). In addition, the reactor size may be reduced and the need to pressurize the gas before it is introduced into a pipeline is eliminated (if a high heat-content gas is to be the ultimate product). High-pressure systems may have problems associated with the introduction of the feedstock into the reactor. Furthermore, low pressure or atmospheric pressure gasification reactors are frequently designed with an accompanying fuel gas compressor after the synthesis gas clean-up processes.

Each type of gasifier will operate satisfactorily with respect to stability, gas quality, efficiency and pressure losses only within certain ranges of the fuel properties of which the most important are: (i) energy content, (ii) moisture content, (iii) volatile matter production, (iv) mineral matter content – ash forming propensity, (v) ash chemical composition and reactivity, (vi) feedstock reactivity, (vii) feedstock size and size distribution, (viii) bulk density of the feedstock, and (ix) feedstock propensity for char formation. Before choosing a gasifier for any individual fuel it is important to ensure that the fuel meets the requirements of the gasifier or that it can be treated to meet these requirements.

Thus, the gasifier is the core of the gasification system and is a *vessel* where the feedstock reacts with oxygen (or air) at high temperature (Luque and Speight, 2015). In order to accommodate the different feedstocks and different process requirements, there are several gasifier designs (Table 3.2), which are distinguished by (i) the use of wet or dry feedstock, (ii) the use of air or oxygen, (iii) the flow direction within the gasifier – up-flow, down-flow, or circulating flow, and (iv) the cooling process for the synthesis gas and other gaseous products.

3.2.1 Fixed-Bed Gasifier

In a fixed-bed process the feedstock is supported by a grate and combustion gases (such as steam, air, oxygen) pass through the supported feedstock whereupon the hot produced gases exit from the top of the reactor. Heat is supplied internally or from an outside source, but some carbonaceous feedstocks (such as caking coal) cannot be used in an unmodified fixed-bed reactor.

The descending-bed-of-solids system is often referred to as a moving or fixed bed or, on occasion, a countercurrent descending-bed reactor. In the gasifier, the feedstock (approximately 1/8 to 1 in., 3 to 25 mm, diameter) is laid down at the top of a vessel while reactant gases are introduced at the bottom of the vessel and flow at relatively low velocity upward through the interstices between the feedstock lumps. As the feedstock descends it is reacted first by devolatilization using the sensible heat from the rising gas, then hydrogenated by the hydrogen in the reactant gas, and finally burned to an ash. The reactions are, therefore, carried out in a countercurrent fashion (Silaen and Wang, 2008).

Thus, the *countercurrent fixed-bed gasifier* (*updraft gasifier, counterflow gasifier*) consists of a fixed bed of carbonaceous fuel through which the gasification agent (steam, oxygen and/or air) flows in countercurrent configuration. The ash is either removed dry or as a slag. The slagging gasifiers require a higher ratio of steam and oxygen to carbon in order to reach temperatures higher than the ash fusion temperature. The nature of the gasifier means that the fuel must have high mechanical strength and must be non-caking so that it will form a permeable bed, although recent developments have reduced these restrictions to some extent. The throughput for this type of gasifier is relatively low but thermal efficiency is high as the gas exit temperatures are relatively low but, as a result, production of methane and tar is significant at typical operation temperatures.

The main advantage of this gasifier is the effective heat exchange in the reactor. Conventional non-slagging gasifiers adopt fluidized-bed type of reactor. There are indications that entrained-bed type of non-slagging gasifier might provide the advantages of fast reaction and the utilization of inorganics as a fly-ash form, or use the collected fly-ash as a low-grade fuel.

High-temperature synthesis gas, before led out of the gasifier, dries the biomass material as it moves downwards the reactor. By that heat exchange taken place, the raw synthesis gas is cooled significantly on its way through the bulk filling. Synthesis gas temperature at its exit from the reactor is approximately 250°C (480°F – in downdraft gasifiers it is approximately 800°C, 1470°F). Since synthesis gas is exploited in order to dry the incoming feedstock, the system sensitivity to feedstock moisture content is less than other in other gasification reactors. On the other hand, the countercurrent flow of feedstock and synthesis gas results in higher tar content (10 to 20% w/w) in the raw synthesis gas. Other advantages of updraft gasification include: (i) simple, low-cost process, (ii) able to handle feedstocks (such as biomass) with a high moisture and high inorganic content (such as municipal solid waste), (iii) proven technology.

The *co-current fixed-bed* (*downdraft*) gasifier is similar to the countercurrent gasifier, but the gasification agent gas flows in co-current configuration with the fuel (downwards, hence the name downdraft gasifier). Heat needs to be added to the upper part of the bed, either by combusting small amounts of the fuel or from external heat sources. The produced gas leaves the gasifier at a high temperature, and most of this heat is often transferred to the gasification agent added in the top of the bed, resulting in energy efficiency almost equivalent to the countercurrent gasifier. In this configuration, any produced tar must pass through a hot bed of char thereby removing much of the tar from the product slate.

Due to the fact that the gaseous products from the pyrolysis step pass through the oxidation zone, the tar compounds concentration in the raw synthesis gas is less than in the case of updraft gasifiers. These gasifiers are easier to control but are more sensitive to the properties quality of the feedstock. For example, in the case of biomass feedstocks, while updraft gasifiers can process biomass with moisture content up to 50% w/w, in downdraft gasification a moisture content range between 10 and 25% is required.

The advantages of downdraft gasification are: (i) up to 99.9% of the tar formed is consumed, requiring minimal or no tar cleanup, (ii) minerals remain with the char/ash, reducing the need for a cyclone, (iii) proven, simple and low-cost process. However, the disadvantages of downdraft gasification are: (i) the feed should be dried to a low moisture content (<20% w/w moisture), (ii) the synthesis gas exiting the reactor is at high temperature, requiring a secondary heat recovery system, and (iii) 4 to 7% of the carbon remains unconverted.

Crossdraft gasification reactors – which operate well on dry air blast and dry fuel – do have advantages over updraft gasification reactors and downdraft gasifiers v the disadvantages – such as high exit gas temperature, poor carbon dioxide reduction and high gas velocity – but the disadvantages, which are the consequences of the design, outweigh the advantages.

Unlike downdraft and updraft gasifiers, the ash bin, fire and reduction zone in crossdraft gasifiers are separated. This design characteristic limits the type of fuel for operation to low mineral matter fuels such as wood, charcoal and coke. The load following ability of the crossdraft gasifier is quite good due to concentrated partial zones which operates at temperatures up to 2000°C (3600°F). The relatively higher temperature in crossdraft gasification reactor has an effect on gas composition – resulting in high carbon monoxide content and low hydrogen and methane content when dry fuel such as charcoal is used.

In the countercurrent fixed-bed gasifier (*updraft gasifier*) the gasification agent (steam, oxygen and/or air) flows through a fixed bed of waste in countercurrent configuration. The ash is either removed in the dry condition or as a slag. The slagging gasifiers have a lower ratio of steam to carbon, achieving temperatures higher than the ash fusion temperature. The nature of the gasifier means that the fuel must have high mechanical strength and must ideally be non-caking so that it will form a permeable bed, although recent developments have reduced these restrictions to some extent. The throughput for this type of gasifier is relatively low but the thermal efficiency is high as the temperatures in the gas exit are relatively low. However, tar (can be recycled to the gasifier) and methane production can be significant at typical operation temperatures and the product gas must be extensively cleaned before use.

In the fixed-bed (or moving-bed) gasifier a deep bed of waste is present in almost all the volume of the reactor and different zones can be distinguished, with a sequence that depends on the flow direction of the waste and gasification medium. These zones are not physically fixed and move upwards and downwards depending on operating conditions, so that they can be to some extent overlapping. In the updraft reactors the waste is fed in at the top of the gasifier, and the oxidant intake is at the bottom, so that the waste moves countercurrently to the gases, and passes through different zones (drying, pyrolysis, reduction and oxidation) successively. The fuel is dried in the top of the gasifier, so that waste with high moisture content can be used. Some of the resulting char falls and burns to provide heat. The methane and tar-rich gas leave at the top of the gasifier, and the ash falls from the grate for collection at the bottom.

The co-current fixed-bed gasifier (*downdraft gasifier*) is similar to the countercurrent fixed-bed gasifier type except that the gasification agent gas flows in co-current configuration with the descending waste. Heat needs to be added to the upper part of the bed, either by combusting small amounts of the fuel or from external heat sources. The produced gas leaves the gasifier at a high temperature, and most of this heat is often transferred to the gasification agent added in the top of the bed, resulting in energy efficiency on level with the countercurrent type. The gasifier is configured so that the tar product must pass through a hot bed of char and, as a result, the yield of tar is much lower than the tar yield in the countercurrent fixed-bed gasifier.

In the downdraft gasifier, the waste is fed in at the top of the gasifier while the oxidant is introduced from the top or the sides so that waste and gases move in the same direction. It is possible to distinguish the same zones of updraft gasifiers but in a different order. Some of the waste is burned, falling through the gasifier throat to form a bed of hot char which the gases have to pass through. This ensures a high-quality synthesis gas (with relatively low tar content), which leaves at the base of the gasifier, with ash collected under the grate.

One particular updraft gasifier that has commercial potential for gasifying MSW is the high-pressure, oxygen-injected slagging fixed bed. Originally developed for the gasification of coal briquettes, these units operate at a maximum temperature of approximately 1650°C (3000°F), above the grate and at pressures of approximately 450 psi. In theory, the high temperatures crack all tars and other volatiles into non-condensable, light gases. Also under these conditions, the ash becomes molten and is removed, as is occurs in iron blast furnaces.

3.2.2 Fluid-Bed Gasifier

In the *fluidized-bed gasifier* (*fluid-bed gasifier*), the fuel is fluidized in oxygen (or air) and steam and the ash is removed dry or as heavy agglomerates. The temperatures are relatively low in dry ash gasifiers, so the fuel must be highly reactive. Feedstock throughput is higher than for the fixed bed, but not as high as for the entrained flow gasifier. The conversion efficiency is low and a recycle operation or subsequent combustion of solids is necessary to increase conversion. Fluidized-bed gasifiers are most useful for fuels that form highly corrosive ash (such as biomass) that would damage the walls of slagging gasifiers.

The fluidized-bed system uses finely sized feedstock particles and the bed exhibits liquid-like characteristics (in the form of fluid flow) when a gas flows upward through the bed. Gas flowing through the feedstock produces turbulent lifting and separation of particles and the result is an expanded bed having greater feedstock surface area to promote the chemical reaction.

The fluidized-bed system requires the feedstock to be finely ground and the reactant gases are introduced through a perforated deck near the bottom of the vessel. The volume rate of gas flow is such that its velocity is sufficient to suspend the solids but not high enough to blow them out of the top of the vessel. The result is an active boiling bed of solids having very intimate contact with the upward-flowing gas, which gives a very uniform temperature distribution. The solid flows rapidly and repeatedly from bottom to top and back again, while the gas flows rather uniformly upward. The reactor is said to be completely back-mixed and no countercurrent flow is possible. If a degree of countercurrent flow is desired, two or more fluid-bed stages are placed one above the other. Reaction rates are faster than in the moving bed because of the intimate contact between gas and solids and the increased solids surface area due to the smaller particle size.

Compared with the fixed-bed gasifiers, the sequence of reactor processes (drying, pyrolysis, oxidation and reduction) is not obvious at a certain point of the gasifier since they take place in the entire reactor thus resulting to a more homogeneous type of reaction (Collot *et al.*, 1999). This means the existence of more constant and lower temperatures inside the reactor, where no *hot spots* are observed. Due to the lower operating temperatures, ash does not melt and it is more easily removed from the reactor. In addition, sulfur-contain constituents of the feedstock and chlorine-containing constituents of the feedstock can be absorbed in the inert bed material thus eliminating the fouling hazard and reducing the maintenance costs. Another significant difference is that fluidized-bed gasifiers are much less to biomass quality than fixed-bed systems, and they can even operate with mixed biomass feedstock.

One critical advantage of a fluidized-bed gasification system (as opposed to downdraft or fixed-bed system) is the use of multiple feed stocks without experiencing downtime (Capareda, 2011). Another important characteristic of the fluidized-bed system is the ability to operate at various throughputs without having to use a larger diameter unit. This is accomplished by changing the appropriate bed material. By using a larger bed material, more air flow rate is required for fluidization and thus more biomass may need to be fed at higher rates to maintain the same fuel to air ratio as before. The reactor free-board must then be high enough so that bed materials are not blown out of the system. Also, a fluidized-bed gasification reactor is also designed to be accompanied by a cyclone downstream of the gasifier to capture the larger particles that are entrained out of the reactor as a result of the fluidity of the bed and the velocity of the gas rising though the bed. These particles are recycled back into the reactor but overall, the residence time of the feedstock particles in a fluidized-bed gasifier is shorter than that of a moving-bed gasifier.

Fluidized-bed gasifiers come in three basic types: (i) the bubbling fluidized-bed gasifier, (ii) the circulating fluidized-bed gasifier, and (iii) the dual fluidized-bed gasifier. Depending on the manner in which the feedstock is treated and the inflow speed, the fluidized-bed gasifier can be characterized either as a *bubbling fluidized-bed* system or as a *circulating fluidized-bed* system – the circulating fluidized-bed system corresponds to higher velocity of the gasification medium. Uniform bed formation in a fluid-bed reactor is very important for efficient bed utilization and consistent operation during gasification of the feedstock. In order to enhance the mixing and uniformity of a bubbling fluid bed, the feedstock is fed to the bed at multiple feed points around the circumference of the reactor vessel. In addition, the fluidization medium, whether air, oxygen, steam, or some combination of these substances, should be uniform in composition and should be introduced in multiple locations.

In the *bubbling fluidized-bed gasifier*, the feedstock biomass is fed from the side, and/or below the bottom of the bed, and the velocity of the gasifying agent is controlled so that it is just greater than the minimum fluidization velocity of the bed material. The product gas exits from the top of the gasifier and ash is either removed from the bottom or from the product gas using a cyclone.

The *circulating fluidized-bed gasifier* systems use two integrated units. In the first unit (the riser) the bed material is kept fluidized by the gasifying agent, with a higher velocity than that found in a bubbling fluidized bed unit. This allows the bed material to be fluidized to a greater extent than in the bubbling fluidized bed unit and the overall residence time is higher, due to the circulation, which is effected by passing the product gas and entrained bed material through a cyclone which separates the product gas from the bed material which is recirculated back to the riser.

A *bubbling fluid-bed* design is generally more sensitive to bed utilization. The size of the feedstock particles greatly affects the rate of gasification and the ability of the biomass to migrate to the center of the bed in a bubbling fluid-bed design. With small particles, the gasification is very quick, and unburned material might not make it to the center of the bed, resulting in oxygen slip and a void center in the bubbling fluid-bed reactor. If all or a majority of the feedstock quickly gasifies, there will be insufficient char to maintain a uniform bed. For this reason, more detail is required in designing the in-feed system with the proper number of in-feed points and controlling and/or monitoring the size particle distribution of the feedstock material. A bubbling fluid bed will generally require additional feed points that must be balanced for larger particle sizes.

The advantages of the bubbling fluidized-bed gasifier are: (i) yields a uniform product gas, (ii) exhibits a nearly uniform temperature distribution throughout the reactor, (iii) able to accept a wide range of fuel particle sizes, including fines, (iv) provides high rates of heat transfer between inert material, fuel and gas, and (v) high conversion possible with low tar and unconverted carbon. The disadvantages of bubbling fluidized-bed gasification are that a large bubble size may result in gas bypass through the bed.

On the other hand, a *circulating fluid-bed* design operates at a higher velocity and incorporates recycling of the char and bed material, resulting in complete mixing regardless of feedstock size. Generally, the circulating fluid-bed designs are more flexible but are still limited by the amount of very fine material that they can process.

The advantages of the circulating fluidized-bed gasifier are: (i) suitable for rapid reactions, (ii) high heat transport rates possible due to high heat capacity of bed material, and (iii) high conversion rates possible with low tar and unconverted carbon. The disadvantages of the circulating fluidized-bed gasifier are: (i) temperature gradients occur in direction of solid flow, (ii) the size of fuel particles determine minimum transport velocity – high velocities may result in equipment erosion, and (iii) heat exchange is less efficient than bubbling fluidized-bed.

In the *dual fluidized-bed gasifier*, the gasification and the combustion parts of the process are separated using two separate fluidized beds. The feedstock is fed into the base of the gasifier bed, usually fluidized by steam. The second bed acts as a char combustor using air in a fast fluidized bed which heats the bed material. The bed material acts as the heat transfer medium between beds and this avoids gas transfer, allowing a nitrogen-free syngas to be produced; the bed material is separated from the combustion flue gases in a cyclone and recirculated to the gasifier.

A novel reactor design that is particularly appropriate for biomass is the *indirectly heated gasification* technology which utilizes a bed of hot particles (sand), which is fluidized using steam. Solids (sand and char) are separated from the synthesis gas via a cyclone and then transported to a second fluidized-bed reactor. The second bed is air blown and acts as a char combustor, generating a flue gas exhaust stream and a stream of hot particles. The hot (sand) particles are separated from the flue gas and recirculated to the gasifier to provide the heat required for pyrolysis. This approach results in a product gas that is practically nitrogen free and has a heating value of approximately 400 Btu/ft^3) (Turn, 1999).

Another novel design is the new fluidized-bed gasifier with increased gas-solid interaction combining two circulating fluidized-bed reactors (Schmid et al., 2011). The aim of the design is to generate a nitrogen-free product gas with low tar content and low fines (particulate matter) content. The system accomplishes this by division into an air/combustion and a fuel/gasification reactor – the two reactors are interconnected via loop seals to assure the global circulation of bed material.

The fuel/gasification reactor is a circulating fluidized bed but with the special characteristic of almost countercurrent flow conditions for gas phase and solids. The gas velocity and the geometrical properties in the fuel/gasification reactor are chosen in such a way that entrainment of coarse particles is low at the top. Due to the dispersed downward movement of the solids, volatile products are not produced in the upper part of the fuel reactor and the issues related to insufficient gas phase conversion and high tar content are avoided.

Finally, the design of fluidized-bed gasification reactor is extremely important (for all of the reasons given above) because both the axial and radial transport of solids within the bed influence gas-solid contact, the thermal gradient, and the heat transfer coefficient. Segregation in a fluidized bed is affected by the particle density, shape, size, superficial gas velocity, mixture composition, bed aspect ratio (the ratio of the static bed height divided by the dynamic or expanded bed height). Variations in the size, shape and density of the fuel particles can cause severe mixing problems which result in changes in temperature gradients within the reactor, increase tar formation and agglomeration, and decrease the conversion efficiency (Cranfield, 1978; Bilbao, 1988). Effective mixing of fuels of various sizes is needed to maintain uniform temperature and a good mix depends on the relative concentrations of the solids in the bed and the velocity of the gas (Bilbao *et al.*, 1988; Ghaly *et al.*, 1989).

The conversion efficiency in a fixed-bed unit may be low due to elutriation (separation of lighter particles from heavier particles) of carbonaceous material. However, recycle or subsequent combustion of solids can be used to increase conversion. Fluidized-bed gasifiers are most useful for fuels that form highly corrosive ash that would damage the walls of slagging gasifiers. Certain types of waste and biomass fuels generally contain high levels of corrosive ash and the fluid-bed gasifier also is appropriate for go-gasification of these feedstocks.

3.2.3 Entrained-Bed Gasifier

An entrained-bed system (entrained flow system) uses finely sized feedstock particles blown into the gas steam prior to entry into the reactor and combustion occurs with the feedstock particles suspended in the gas phase (Chen *et al.*, 2000). In the *entrained-flow gasifier* (*entrained-bed gasifier*) (Figure 3.1c) a dry pulverized solid, an atomized liquid fuel or a fuel slurry is gasified with oxygen (much less frequent: air) in co-current flow and the gasification reactions take place in a dense cloud of very fine particles. The high temperatures and pressures also mean that a higher throughput can be achieved; however, thermal efficiency is somewhat lower as the gas must be cooled before it can be cleaned with existing technology. The high temperatures also mean that tar and methane are not present in the product gas; however, the oxygen requirement is higher than for the other types of gasifiers.

The entrained-flow reactor requires a smaller particle size of the feedstock than the fluid-bed gasifier so that the feedstock can be conveyed pneumatically by the reactant gases. Velocity of the mixture must be approximately 20 ft/sec (6.1 m/sec) or higher depending upon the fineness of the feedstock. In this case, there is little or no mixing of the solids and gases, except when the gas initially meets the solids. Furthermore, apart from higher temperature, entrained-flow gasification usually takes place at elevated pressure (pressurized entrained-flow gasifiers) reaching operating pressures even up to 40 and 50 bars. The existence of such high temperatures and pressures requires more sophisticated reactor design and construction materials used.

The design of an entrained-flow reactor gives a residence time of the feedstock in the reaction zone to be on the order of seconds, or tens of seconds. This short residence time requires that entrained-flow gasifiers operate at high temperatures to achieve high carbon conversion. Consequently, most entrained-flow gasifiers are designed to use oxygen rather than air and also to operate above the slagging temperature of the feedstock mineral matter.

All entrained-flow gasifiers are designed to remove the major part of the ash as a slag as the operating temperature is well above the ash fusion temperature. A smaller fraction of the ash is produced either as a very fine dry fly ash or as black colored fly ash slurry. Some fuels, in particular certain types of biomass, can form slag that is corrosive for ceramic inner walls that serve to protect the gasifier outer wall. However, some entrained-bed type of gasifiers do not possess a ceramic inner wall but have an inner water or steam cooled wall covered with partially solidified slag. For fuel that produces ash with a high ash fusion temperature, limestone can be mixed with the fuel prior to gasification in order to lower the ash fusion temperature. Typically, the fuel particles must be smaller than for other types of gasifier – the fuel must be pulverized.

Entrained-flow gasifiers are highly efficient and useful for large-scale gasification, and are commonly employed for coal, biomass and refinery residues. Their requirement for highly pulverized fuel particles presents problems when gasifying biomass. On the other hand, gasification in these gasifiers is above 1000°C (1830°F) which aids in cracking tar; they are therefore advantageous for biomass gasification where tar is a serious issue.

The gasification reactions take place in a dense cloud of very fine particles. Most coals are suitable for this type of gasifier because of the high operating temperatures and, this co-gasification of coal with pelletized solid waste is an option. However, the waste feedstock particles must be much smaller than for other types of gasifiers, i.e., the waste must be pulverized, which requires somewhat more energy than for the other types of gasifiers. By far the most energy consumption related to entrained-flow gasification is not the milling of the fuel but the production of oxygen used for the gasification.

The high temperatures and pressures lend themselves to a higher throughput than can be achieved with other gasifiers but thermal efficiency is somewhat lower as the gas must be cooled before it can be cleaned with existing technology. Because of the high temperatures, tar and methane are not present to any great extent (if at all) in the product gas but the oxygen requirement is higher than for the other types of gasifier units. All entrained-flow gasifiers remove the major part of the ash as a slag – the operating temperature is well above the ash fusion temperature.

A smaller fraction of the ash is produced either as a very fine dry fly ash or as black colored fly ash slurry. Some fuels, in particular certain types of waste and biomass, can form slag that is corrosive for ceramic inner walls that serve to protect the gasifier outer wall. However, some entrained-flow types of gasifiers do not possess a ceramic inner wall but have an inner water or steam cooled wall covered and (to some extent) protected by partially solidified slag – these types of gasifiers do not suffer severe adverse effects from corrosive slag. If the waste is likely to produce ash with a very high ash fusion temperature, limestone or dolomite can be mixed with the waste prior to gasification (He *et al.*, 2009) – this usually is sufficient to lower the fusion temperature of the ash.

Most modern large-scale gasification systems utilize the entrained-flow design. However, for municipal solid waste, fixed-bed and fluid-bed designs predominate due to the low reactivity and high moisture content and high mineral matter content (high propensity for ash formation) of the municipal solid waste (CH2MHill, 2009).

3.2.4 Molten Salt Gasifier

Another type of gasifier is the molten media reactor in which the feedstock is dispersed in a molten carrier. The molten salt gasifier (molten metal gasifier) use, as the name implies, a molten medium of an inorganic salt (or molten metal) to generate the heat to decompose the feedstock into products and there are a number of applications of the molten bath gasification (Rummel, 1959).

The carrier generally provides heat for the gasification and can participate in the gasification if it contains catalytically active species such as nickel or iron. The major advantages of molten media reactors are the high heat transfer rates, as the molten media is the heat source, and the ability of the molten media to absorb contaminants in the feedstock, mainly sulphur and alkalis. Molten media reactors can have a high temperature depending on the molten media and can accept most feedstocks. The disadvantage is that the processes must handle the molten media. For molten salts this can become very troublesome as the media are highly corrosive.

A number of different designs have evolved through various stages of development but the basic concept is that instead of using a formed gasifying chamber where the reactions occur in suspension, the feedstocks are gasified in a molten bath of salt or metal. This type of design allows for more complete processing of the feedstock and also allows for a greater variety of feedstocks to be efficiently processed in the same gasifier.

In molten bath gasifiers, crushed feedstock, steam air and/or oxygen are injected into a bath of molten salt, iron, or feedstock ash. The feedstock appears to *dissolve* in the melt where the volatiles crack and are converted into carbon monoxide and hydrogen. The feedstocks carbon reacts with oxygen and steam to produce carbon monoxide and hydrogen. Unreacted carbon and mineral ash float on the surface from which they are discharged.

High temperatures (approximately 900°C, 1650°F and above, depending on the nature of the melt) are required to maintain the bath molten. Such temperature levels favor high reaction rates and throughputs and low residence times. Consequently, tar and volatile oil products are not produced in any great quantity, if at all. Gasification may be enhanced by the catalytic properties of the melt used. Molten salts, which are generally less corrosive and have lower melting points than molten metals, can strongly catalyze the steam-feedstock reaction and lead to very high conversion efficiencies.

In the process, the carbonaceous feedstock devolatilizes with some thermal cracking of the volatile constituents leaving the fixed carbon and sulfur to dissolve in the molten salt (such as an iron salt) whereupon carbon is oxidized to carbon monoxide by oxygen introduced through lances placed at a shallow depth in the bath). The sulfur migrates from the molten salt to the slag layer where it reacts with lime to produce calcium sulfide.

The product gas, which leaves the gasifier at ca. 1425°C (2600°F), is cooled, compressed, and fed to a shift converter where a portion of the carbon monoxide is reacted with steam to attain a carbon monoxide to hydrogen ratio of 1:3. The carbon dioxide so produced is removed and the gas is again cooled and enters a methanator where carbon monoxide and hydrogen react to form methane. Excess water is removed from the methane-rich product and, depending on the type of feedstock used and the extent of purification required – the final gas product may have a heat content of 920 Btu/ft^3.

As another example, the Pullman-Kellogg process involves contacting feedstock with a melt of an inorganic salt such as sodium carbonate to convert the feedstock. In the process, air is bubbled into the bottom of the gasifier through multiple inlet nozzles and the feedstock (typically sized to 1/4 in.; 6 mm) is fed beneath the surface of the molten salt bath using a central feed tube whereupon natural circulation and agitation of the melt disperses the material. The main gasification reaction is a partial oxidation reaction and any volatile matter from the feedstock reacts to produce a fuel gas free of oils, tars, as well as ammonia. A water-gas shift equilibrium exists above the melt and, accordingly, in the reducing environment, carbon dioxide and water concentrations are minimal.

In practice, the molten salt design allows for some of the catalysis process to take place within the gasifier instead of downstream. For example, if the reactor or process design allows the hydrogen and carbon monoxide to be produced in separate distinct streams, the need for post-process separation prior to catalyzing into synthetic fuels will be eliminated.

The molten salt/metal design also allows for a greater variety of co-products to be produced on-site. All gasification methods allow for co-production of various chemicals and gases but the molten metal process adds various metals, such as vanadium and nickel as well as a variety of trace elements, to the mix. Most gasifier feedstocks contain trace metals which can then be extracted in the molten metal process, instead of being disposed of as slag. Also, the design and operation of molten metal reactors is such that the use of a fluxing material, such as lime or limestone, is required. When combined with the silica ash that is generated through normal gasification, the slag produced and removed from the molten metal reactor can be used directly as cement or formed into bricks for construction materials.

3.2.5 Plasma Gasifier

Plasma is a high-temperature, highly ionized (electrically charged) gas capable of conducting electrical current. Plasma technology has a long history of development and has evolved into a valuable tool for engineers and scientists who need to use very high temperatures for new process applications (Messerle and Ustimenko, 2007; Heberlein and Murphy, 2008). Man-made plasma is formed by passing an electrical discharge through a gas such as air or oxygen (O_2). The interaction of the gas with the electric arc dissociates the gas into electrons and ions, and causes its temperature to increase significantly, often (in theory) exceeding 6000°C (10830°F).

There are two basic types of plasma torches, the *transferred torch* and the *non-transferred torch*. The transferred torch creates an electric arc between the tip of the torch and a metal bath or the conductive lining of the reactor wall. In the non-transferred torch, the arc is produced within the torch itself. The plasma gas is fed into the torch and heated, and it then exits through the tip of the torch.

In the plasma-based process, the gasifier is heated by a plasma torch system located near the bottom of the reactor vessel. In the gasifier, the feedstock is charged into a vertical reactor vessel (refractory lined or water-cooled) at atmospheric pressure. A superheated blast of air, which may be enriched with oxygen, is provided to the bottom of the gasifier, at the stoichiometric amount required for gasification. The amount of air fed is controlled so that a low velocity of the upward flowing gas is maintained and the pulverized (small-particle) feedstock can be fed directly into the reactor. Additional air and/or steam can be provided at different levels of the gasifier to assist with the pyrolysis and gasification components of the process. The temperature of the synthesis gas leaving the top of the gasifier is maintained above 1000°C (1830°F) – at this temperature, tar formation is eliminated.

The high operating temperatures decompose the feedstock (and/or all hazardous and toxic components) and dramatically increase the kinetics of the various reactions occurring in the gasification zone, converting all organic materials into hydrogen (H_2) and carbon monoxide (CO). Any residual materials from inorganic constituents and heavy metals will be melted and produced as a vitrified slag which is highly resistant to leaching. *Magmavication* or *vitrification* is the result of the interaction between plasma and inorganic materials, in presence of a coke bed or coke-like products in the cupola or reactor, a vitrified material is produced that can be used in the manufacture of architectural tiles and construction materials (Leal-Quirós, 2004).

Plasma is used in two different ways in the gasification process: (i) as a heat source during gasification and (ii) for tar cracking after standard gasification. Primarily, plasma gasification is employed for the decomposition of toxic organic wastes, along with rubber and plastics, although the first reason and currently the main application for plasma gasification is the treatment of hazardous biomass waste. However, the technology has also gained interest for the production of synthesis gas and electricity generation in recent years as the costs have entered into a commercially competitive range. Also, due to the very high temperatures produced, the plasma process can be employed for toxic wastes, rubber, and treatment of plastic waste.

Serious efforts have been made, with some success, to apply plasma gasification technology to gasification technology and to treat industrial and municipal solid wastes (MSW) over the last two decades. It is believed that the technology can be used as a gasification reactor thereby allowing: (i) greater feedstock flexibility enabling a variety of fuels such as coal, biomass, and municipal solid waste to be used as fuel without the need for pulverizing, (ii) air blowing and thus an oxygen plant is not required, (iii) high conversion (>99%) of carbonaceous matter to synthesis gas, (iv) the absence of tar in the synthesis, (v) the production of high heating value synthesis gas suitable for use in a combustion turbine operation, (vi) production of little or no char, ash or residual carbon, (vii) production of a glassy slag with beneficial value, (viii) high thermal efficiency, and (ix) low carbon dioxide emissions.

In the process, the gasifier is heated by a plasma torch system located near the bottom of the reactor vessel. In the gasifier, the feedstock is charged into a vertical reactor vessel (refractory lined or water-cooled) at atmospheric pressure. A superheated blast of air, which may be enriched with oxygen, is provided to the bottom of the gasifier, at the stoichiometric amount required for gasification. The amount of air fed is such that the superficial velocity of the upward flowing gas is low, and that the pulverized feedstock can be fed directly into the reactor. Additional air and/or steam can be provided at different levels of the gasifier to assist with pyrolysis and gasification. The temperature of the synthesis gas leaving the top of the gasifier is maintained above 1000°C (1830°F). At this temperature, tar formation is eliminated.

Gasification takes place at very high temperatures, driven by the plasma torch system, which is located at the bottom of the gasifier vessel. The high operating temperatures break down the feedstock and/or all hazardous and toxic components into their respective elemental constituents, and dramatically increases the kinetics of the various reactions occurring in the gasification zone, converting all organic materials into hydrogen (H_2) and carbon monoxide (CO). Any residual materials from inorganic constituents of the feedstock (including heavy metals) will be melted and produced as a vitrified slag which is highly resistant to leaching.

The main purported benefits of this process are (i) the yield of gas with a high content of hydrogen and a high content of carbon monoxide, (ii) improved heat content, (iii) low yield of carbon dioxide, and (iv) a low yield of tar. The process can be employed for wet biomass such as sewage sludge which are otherwise difficult to gasify (Sikarwar et al., 2016).

3.2.6 Other Types

The *rotary kiln gasifier* is used in several applications, varying from the industrial waste to cement production and reactor accomplishes two objectives simultaneously: (i) moving solids into and out of a high-temperature reaction zone and (ii) assuring thorough mixing of the solids during reaction. The kiln is typically comprised of a steel cylindrical shell lined with abrasion-resistant refractory – to prevent overheating of the metal – and is usually inclined slightly toward the discharge port. The movement of the solids being processed is controlled by the speed of rotation (approximately 1.5 rpm). In this gasifier, the gasifying agents, air and/or oxygen and steam are introduced along a rotating horizontal cylindrical reactor vessel. Gasification takes place along the length of the vessel in stages until synthesis gas is released from the end while ash drops out. Rotary reactors enable complete mixing of the gasifying agents with air while the process is closely controlled by the rotational speed and air flow. The lower gas temperatures (800 to 900°C, 1470 to 1650°F) – while a high enough temperature to volatilize tar constituents and – allows easier handling of the ash.

The *moving grate gasifier* is based on the system used for waste combustion in a waste-to-energy process. The constant-flow grate feeds the waste feedstock continuously to the incinerator furnace and provides movement of the waste bed and ash residue toward the discharge end of the grate. During the operation there is stoking and mixing of the burning material and this allows some flexibility in the composition of the fuel for the gasifier. The thermal conversion takes place in two stages: (i) the primary chamber for gasification of the waste (typically at an equivalence ratio of 0.5) and (ii) the secondary chamber for high-temperature oxidation of the synthesis gas produced in the primary chamber.

The unit is equipped with a horizontal oil-cooled grate that is divided into several separate sections, each with a separate primary air supply, and a water-cooled guillotine-type controller that is installed at the inlet of the gasification unit to control the thickness of the fuel bed. The oxidation in the secondary chamber is facilitated by multiple injections of air and recycled flue-gas (Grimshaw and Lago, 2010). A distinct benefit of the moving grate gasifiers (like the fluid-bed gasifiers) is that the process can accommodate wet feedstocks (Hankalin *et al.*, 2011).

Large-scale *oxygen gasifiers* may play a prominent role in the conversion of municipal waste. If small oxygen gasifiers and plants are developed (50 tons/day), they could play a crucial role in energy self-sufficient farms, manufacturing ammonia and methanol or gasoline from residues at the farm cooperative level to eliminate the heavy dependence on fossil fuels that makes arms vulnerable to variable fuel costs and uncertain supply.

3.2.7 Gasifier Selection

The most commonly used gasifiers are the fixed-bed gasifier, the fluidized-bed gasifier, and the entrained-flow gasifier. One critical advantage of a fluidized-bed gasification system (as opposed to downdraft or fixed-bed system) is the use of multiple feedstocks without experiencing downtime. Another important characteristic of the fluidized-bed system is the ability to operate at various throughputs without having to use a larger diameter unit. This is accomplished by changing the appropriate bed material – by using a larger bed material, more air flow rate is required for fluidization and thus more biomass may need to be fed at higher rates to maintain the same fuel to air ratio as before. The reactor free-board must then be high enough so that bed materials are not blown out of the system (Capareda, 2011).

A fixed-bed gasifier can be either (i) an updraft gasifier in which the feedstock enters from the top of the unit while the gasifying agent enters from the bottom of the unit or (ii) a downdraft gasifier in which both the feedstock and the gasification agent enter from the top of the unit, with the fuel coming in from a lock-hopper. In the updraft gasifier, the char at the bottom of the bed meets the gasifying agent first, and complete combustion occurs, producing water and carbon dioxide and raising the temperature to approximately 1000°C (1830°F). The hot gases percolate upwards through the bed, driving endothermic reactions with unreacted char to form hydrgoen and carbon monoxide with consequent cooling to approximately 750°C (1380°F). The gases pyrolyze the dry biomass, which is descending, and also (near the top of the reactor) dry the incoming biomass. Updraft gasifiers typically produce between 10 and 20% w/w tar in the produced gas. The allowable level of tar in the product gas depends on the downstream application.

In contrast to an updraft gasifier, in a downdraft gasifier (closed top) the gas flows co-currently with the fuel. A *throated gasifier* has a restriction part-way down the gasifier where air or oxygen is added, and where the temperature rises to a range on the order of 1200 to 1400°C (2190 to 2550°F), and the fuel feedstock is either burned or pyrolyzed. The combustion gases then pass down over the hot char at the bottom of the bed, where they are reduced to hydrogen and carbon monoxide. The high temperature within the throat ensures that the tars formed during pyrolysis are significantly cracked (homogeneous cracking), with further cracking occurring as the gas meets the hot char on the way out of the bed (heterogeneous cracking), leading to a less tar-containing off-gas.

Another interesting and efficient design for fixed bed is the open top fixed-bed reactor, which has been found to be more efficient and reliable especially with high-moisture content feedstock and produces a high-quality gas with low tar content (Sikarwar *et al.*, 2016). The gasifier consists of a vertical tube with an open top and water seal at the bottom. The top third of the reactor is made of stainless steel, with an annular jacket around it. The remaining lower part is made of ceramic material to avoid high-temperature corrosion (>600°C, >1110°F) caused by the different gases prevailing at that point in the gasifier. The hot combustible gases produced are taken to the upper annulus of the gasifier by way of a grate and an insulated pipe. These gases transfer the heat to the feedstock, aid in drying and enhance the thermal efficiency of the process. A recirculating duct connects the upper annular part of the gasifier to the lower part and is insulated with alumino-silicate blankets. Constant homogeneous air flow through the bed resulting in a final fuel-rich state enhances the gasifier performance. Furthermore, a superior quality synthesis gas with lower tar content is obtained on account of gas movement through a deep hot bed of charcoal.

To use the product gas from a biomass gasifier as synthesis gas, there are several properties which have to be taken into account: (i) the hydrogen-to-carbon monoxide ration, (ii) the amount of inert constituents, such as nitrogen, (iii) the amount of methane and higher molecular weight hydrocarbon derivatives, and (iv) the presence of catalyst poisons, such as sulfur-, nitrogen-, and chlorine-containing constituents.

For most synthesis a hydrogen-to-carbon monoxide ration on the order to 2-to-3 is required and this ratio is normally adjusted in a separate catalytic reactor before the synthesis reactor, where some carbon monoxide is converted to hydrogen by the water-gas shift reaction. If the gasifier produces the correct hydrogen-to-carbon monoxide ratio, the exothermal water-gas shift reaction can be avoided, which reduces operation costs and also increases the efficiency.

Impurities such as nitrogen are inert during the synthesis and the concentration has to be as low as possible. The inert constituents do reduce the partial pressure and, hence, reduce the conversion. Also, for synthesis reactions where the product is separated as a liquid and where the unconverted gas is recycled, the inert constituents have to be bled off as they would otherwise be accumulated. Also for the production of synthetic natural gas (SNG, i.e., methane) the inert constituents have to be less than 1% v/v or the heating value of the synthetic natural gas will not meet the specifications for the gas.

Catalyst poisons deactivate the synthesis catalyst and have to be removed to very low levels. The most frequent poisons are sulfur-containing compounds such as hydrogen sulfide (H_2S) carbonyl sulfide (COS), mercaptan derivatives (RSH), or thiophene derivatives. The organic sulfur components are mainly present in fluidized-bed gasifiers, and not in high-temperature gasification. The removal technology has to be adapted to the type of the sulfur components – for example, thiophene deroivatives cannot be removed by the use of zinc oxide (ZnO) adsorbers.

Overall, the selection of the gasification reactor requires careful consideration insofar as the reactor influences the whole conversion chain – the gas treatment options (Chapter 9) have to be adjusted to the type of gasifier and the synthesis should also fit to the properties of the synthesis gas such as, for example, the hydrogen-to-carbon monoxide ration and the amount of inert constituents.

3.3 General Chemistry

In the current context, process chemistry is the chemistry that occurs in a gasification reactor. Also, because some impurities cannot be detected at small scales, a gasification process that works sufficiently on the bench or pilot scale may prove to be inefficient, or even impossible at the demonstration and or the commercial scale. To remove any such doubts about the efficiency of the chemistry at scale-up, the unit operators must be knowledgeable not only about reactions but also about impurities that may develop from side reactions. Thus, process chemistry requires a blend of theoretical and practical knowledge and, in addition to creating the desired product, the process chemist must always keep cost and safety in mind.

Within the gasification process, the chemical reactions of gasification can progress to different extents depending on the gasification conditions (like temperature and pressure) and the feedstock used. Combustion reactions (Table 3.3) can also take place in a gasification process, but, in comparison with conventional combustion which uses a stoichiometric excess of oxidant, gasification typically uses one-fifth to one-third of the theoretical oxidant. This only partially oxidizes the carbon feedstock and as a *partial oxidation* process, the major combustible products of gasification are carbon monoxide (CO) and hydrogen, with only a minor portion of the carbon completely oxidized to carbon dioxide (CO_2). The heat produced by the partial oxidation provides most of the energy required to drive the endothermic gasification reactions.

Table 3.3 Comparison of products from combustion and gasification processes.

	Combustion	Gasification
Carbon	CO_2	CO
Hydrogen	H_2O	H_2
Nitrogen	NO, NO_2	HCN, NH_3 or N_2
Sulfur	SO_2 or SO_3	H_2S or COS
Water	H_2O	H_2

Also, in the low-oxygen, reducing environment of the gasifier, most of the feedstock sulfur is converted to hydrogen sulfide (H_2S), with a small amount forming carbonyl sulfide (COS). Nitrogen chemically bound in the feed generally converts to gaseous nitrogen (N_2), with some ammonia (NH_3), and a small amount forming hydrogen cyanide (HCN). Chlorine is primary converted to hydrogen chloride (HCl). In general, the quantities of sulfur, nitrogen, and chloride in the fuel are sufficiently small that they have a negligible effect on the main synthesis gas components of carbon monoxide (CO) and hydrogen (H_2). Trace elements associated with both organic and inorganic components in the feed, such as mercury, arsenic and other heavy metals, appear in the various ash and slag fractions, as well as in gaseous emissions, and need to be removed from the synthesis gas to further use (Chapter ??).

Chemically, gasification involves the thermal decomposition of feedstock and the reaction of the feedstock carbon and other pyrolysis products with oxygen, water, and fuel gases such as methane (Table 3.4). In fact, the gasification process is often considered to involve two distinct chemical stages: (i) devolatilization of the feedstock to produce volatile matter and char followed by (ii) char gasification, which is complex and specific to the conditions of the reaction – both processes contribute to the complex kinetics of the gasification process (Sundaresan and Amundson, 1978).

Thus, in the initial stages of the gasification process, the rising temperature of the feedstock initiates devolatilization and the breaking of weaker chemical bonds to yield volatile tar, volatile oil, phenol derivatives, and hydrocarbon gases. These products generally react further in the gaseous phase to form hydrogen, carbon monoxide, and carbon dioxide. The char (fixed carbon) that remains after devolatilization reacts with oxygen, steam, carbon dioxide, and hydrogen. Overall, the chemistry of the gasification of carbonaceous feedstocks is complex but can be conveniently (and simply) represented by the following reaction:

$$C + O_2 \rightarrow CO_2 \qquad \Delta H_r = -393.4 \text{ MJ/kmol} \qquad (3.1)$$

$$C + \tfrac{1}{2}O_2 \rightarrow CO \qquad \Delta H_r = -111.4 \text{ MJ/kmol} \qquad (3.2)$$

$$C + H_2O \rightarrow H_2 + CO \qquad \Delta H_r = 130.5 \text{ MJ/kmol} \qquad (3.3)$$

$$C + CO_2 \leftrightarrow 2CO \qquad \Delta H_r = 170.7 \text{ MJ/kmol} \qquad (3.4)$$

$$CO + H_2O \leftrightarrow H_2 + CO_2 \qquad \Delta H_r = -40.2 \text{ MJ/kmol} \qquad (3.5)$$

$$C + 2H_2 \rightarrow CH_4 \qquad \Delta H_r = -74.7 \text{ MJ/kmol} \qquad (3.6)$$

The designation C represents carbon in the original feedstock as well as carbon in the char formed by devolatilization of the feedstock. Reactions (3.1) and (3.2) are exothermic oxidation reactions and provide most of the energy required by the endothermic gasification reactions (3.3) and (3.4). The oxidation reactions occur very rapidly, completely consuming all of the oxygen present in the gasifier, so that most of the gasifier operates under reducing conditions. Reaction (3.5) is the water-gas shift reaction, in which water (steam) is converted to hydrogen – this reaction is used to alter the hydrogen/carbon monoxide ratio when synthesis gas is the desired product, such as for use in Fischer-Tropsch processes. Reaction (3.6), is favored by high pressure and low temperature and is, thus, mainly important in lower-temperature gasification systems. Methane formation is an exothermic reaction that does not consume oxygen and, therefore, increases the efficiency of the gasification process and the final heat content of the product gas. Overall, approximately 70% of the heating value of the product gas is associated with the carbon monoxide and hydrogen but this varies depending on the gasifier type and the process parameters (Chadeesingh, 2011).

In essence, the direction of the gasification process is subject to the constraints of thermodynamic equilibrium and variable reaction kinetics. The combustion reactions (reaction of feedstock or char with oxygen) essentially go to completion. The thermodynamic equilibrium of the rest of the gasification reactions are relatively well defined and collectively

Table 3.4 Coal gasification reactions.

$2C + O_2 \rightarrow 2CO$
$C + O_2 \rightarrow CO_2$
$C + CO_2 \rightarrow 2CO$
$CO + H_2O \rightarrow CO_2 + H_2$ (shift reaction)
$C + H_2O \rightarrow CO + H_2$ (water gas reaction)
$C + 2H_2 \rightarrow CH_4$
$2H_2 + O_2 \rightarrow 2H_2O$
$CO + 2H_2 \rightarrow CH_3OH$
$CO + 3H_2 \rightarrow CH_4 + H_2O$ (methanation reaction)
$CO_2 + 4H_2 \rightarrow CH_4 + 2H_2O$
$C + 2H_2O \rightarrow 2H_2 + CO_2$
$2C + H_2 \rightarrow C_2H_2$
$CH_4 + 2H_2O \rightarrow CO_2 + 4H_2$

have a major influence on thermal efficiency of the process as well as on the gas composition. Thus, thermodynamic data are useful for estimating key design parameters for a gasification process, such as: (i) calculating of the relative amounts of oxygen and/or steam required per unit of feedstock, (ii) estimating the composition of the produced synthesis gas, and (iii) optimizing process efficiency at various operating conditions.

Other deductions concerning gasification process design and operations can also be derived from the thermodynamic understanding of its reactions. Examples include: (i) production of synthesis gas with low methane content at high temperature, which requires an amount of steam in excess of the stoichiometric requirement, (ii) gasification at high temperature, which increases oxygen consumption and decreases the overall process efficiency, (iii) production of synthesis gas with a high methane content, which requires operation at low temperature (approximately 700°C, 1290°F) but the methanation reaction kinetics will be poor without the presence of a catalyst.

Relative to the thermodynamic understanding of the gasification process, the kinetic behavior is much more complex. In fact, very little reliable global kinetic information on gasification reactions exists, partly because it is highly dependent on (i) the process conditions and (ii) the chemical nature of the coal feed, which varies significantly with respect to composition, mineral impurities, and reactivity. In addition, physical characteristics of the feedstock (or the char produced therefrom) also play a role in phenomena such boundary layer diffusion, pore diffusion and ash layer diffusion which also influence the kinetic picture. Furthermore, certain impurities, in fact, are known to have catalytic activity on some of the gasification reactions which can have further influence on the kinetic imprint of the gasification reactions.

3.3.1 Devolatilization

Devolatilization (Chapter 2) occurs rapidly as the feedstock is heated above 400°C (750°F) (Anthony and Howard, 1976; Silaen and Wang, 2008). During this period, the feedstock structure is altered, producing solid char, tars, condensable liquids, and low molecular weight gases. Furthermore, the products of the devolatilization stage in an inert gas atmosphere are very different from those in an atmosphere containing hydrogen at elevated pressure. In a hydrogen atmosphere at elevated pressure, additional yields of methane or other low molecular weight gaseous hydrocarbon can result during the initial feedstock gasification stage from reactions such as: (i) direct hydrogenation of feedstock or the produced char because of active intermediate formed within feedstock structure after feedstock pyrolysis, and (ii) the hydrogenation of other gaseous hydrocarbon derivatives, oils, tars, and carbon oxides. Again, the kinetic picture for such reactions is complex due to the varying composition of the volatile products which, in turn, are related to the character of the feedstock and the process parameters, including the reactor type.

3.3.2 Products

If air is used for combustion, the product gas will have a heat content on the order of 150 to 300 Btu/ft^3 depending on process design characteristics and will contain undesirable constituents such as carbon dioxide, hydrogen sulfide, and nitrogen. The use of pure oxygen results in a product gas having a heat content of 300 to 400 Btu/ft^3 with carbon dioxide and hydrogen sulfide as byproducts, both of which can be removed from low-heat content or medium heat-content – low-Btu gas, medium-Btu gas, or high-Btu gas – by any of several available chemical processes (Table 3.4, Table 3.5) (Mokhatab *et al.*, 2006; Speight, 2013a, 2014).

If high heat-content (high-Btu) gas (900 to 1000 Btu/ft^3) is required, efforts must be made to increase the methane content of the gas. The reactions which generate methane are all exothermic and have negative values, but the reaction rates are relatively slow and catalysts may therefore, be necessary to accelerate the reactions to acceptable commercial rates. Indeed, it is also possible that the mineral constituents of the feedstock and char may modify the reactivity by a direct catalytic mechanism. The presence of oxygen, hydrogen, water vapor, carbon oxides, and other compounds in the reaction atmosphere during pyrolysis may either support or inhibit numerous reactions with the feedstock and with the products evolved.

If high-Btu gas (high heat-content gas; 900 to 1000 Btu/ft^3) is the desired product, efforts must be made to increase the methane content of the gas. The reactions which generate methane are all exothermic and have negative values (Lee, 2007), but the reaction rates are relatively slow and catalysts may, therefore, be necessary to accelerate the reactions to acceptable commercial rates.

3.4 Process Options

Several types of gasifiers are currently available for commercial use and the selection of a suitable gasifier is dependent upon the character (properties) and behavior of the feedstock during the process, which must also include an assessment of the necessary process parameter set in place to achieve the desired goal (Chapter 1, Chapter 2). Furthermore, the output and quality of the gas produced is determined by the equilibrium established during the process when the heat of oxidation (combustion) balances the heat of vaporization and volatilization plus the sensible heat (temperature rise) of the exhaust gases. The quality of the outlet gas (measured in BTU/ft^3) is determined by the amount of volatile gases (such as hydrogen, carbon monoxide, water, carbon dioxide, and methane) in the gas stream.

Table 3.5 Gasification products.

Product	Characteristics
Low-Btu gas (150–300 Btu/scf)	Around 50% nitrogen, with smaller quantities of combustible H_2 and CO, CO_2 and trace gases, such as methane
Medium-Btu gas (300–550 Btu/scf)	Predominantly CO and H_2, with some incombustible gases and sometimes methane
High-Btu gas (980–1080 Btu/scf)	Almost pure methane

3.4.1 Effects of Process Parameters

As anticipated, the quality of the gas generated in a system is influenced by feedstock characteristics, gasifier configuration as well as the amount of air, oxygen or steam introduced into the system. The output and quality of the gas produced is determined by the equilibrium established when the heat of oxidation (combustion) balances the heat of vaporization and volatilization plus the sensible heat (temperature rise) of the exhaust gases. The quality of the outlet gas (BTU/ft.^{3}) is determined by the amount of volatile gases (such as hydrogen, carbon monoxide, water, carbon dioxide, and methane) in the gas stream. With some feedstocks, the higher the amounts of volatile produced in the early stages of the process, the higher the heat content of the product gas. In some cases, the highest gas quality may be produced at lower temperatures. However, char oxidation reaction is suppressed when the temperature is too low, and the overall heat content of the product gas is diminished.

Gasification agents are normally air, oxygen-enriched air or oxygen. Steam is sometimes added for temperature control, heating value enhancement or to allow the use of external heat (*allothermal gasification*). The major chemical reactions break and oxidize hydrocarbon derivatives to give a product gas containing carbon monoxide, carbon dioxide, hydrogen and water. Other important components include hydrogen sulfide, various compounds of sulfur and carbon, ammonia, low-boiling hydrocarbon derivatives and high-boiling hydrocarbon derivatives (tars).

Depending on the employed gasifier technology and operating conditions, significant quantities of water, carbon dioxide and methane can be present in the product gas, as well as a number of minor and trace components. Under reducing conditions in the gasifier, most of the feedstock sulfur converts to hydrogen sulfide (H_2S), but 3-10% converts to carbonyl sulfide (COS). Organically bound nitrogen in the coal feedstock is generally converted to gaseous nitrogen (N_2), but some ammonia (NH_3) and a small amount of hydrogen cyanide (HCN) are also formed. Any chlorine in the coal is converted to hydrogen chloride (HCl), with some chlorine present in the particulate matter (fly ash). Trace elements, such as mercury and arsenic, are released during gasification and partition among the different phases (e.g., fly ash, bottom ash, slag, and product gas). Thus, relative to the chemical and thermodynamic understanding of the gasification process and data derived from thermodynamic studies (van der Burgt, 2008; Shabbar and Janajreh, 2013), the kinetic behavior of coal feedstocks is more complex.

The chemistry of coal gasification is quite complex and, only for discussion purposes can the chemistry be viewed as consisting of a few major reactions which can progress to different extents depending on the gasification conditions (such as temperature and pressure) and the feedstock used. Combustion reactions take place in a gasification process, but, in comparison with conventional combustion which uses a stoichiometric excess of oxidant, gasification typically uses one-fifth to one-third of the theoretical oxidant. This only partially oxidizes the carbon feedstock. As a *partial oxidation* process, the major combustible products of gasification are carbon monoxide (CO) and hydrogen, with only a minor portion of the carbon completely oxidized to carbon dioxide (CO_2). The heat produced by the partial oxidation provides most of the energy required to drive the endothermic gasification reactions. Furthermore, while the basic thermodynamic cycles pertinent to coal gasification have long been established, novel combination and the use of alternative fluids to water/steam offer the prospect of higher process efficiency through use of thermodynamic studies.

Finally, very little reliable kinetic information on coal gasification reactions exists, partly because it is highly depended on the process conditions *and* the nature of the coal feedstock, which can vary significantly with respect to composition, mineral impurities, and reactivity as well as the potential for certain impurities to exhibit catalytic activity on some of the gasification reactions. Indeed, in spite of the efforts of many researchers, kinetic data are far from able to be applied to gasification of coal or char in various processes. All such parameters serve to complicate the reaction rate and make derivative of a global kinetic relationship applicable to all types of coal subject to serious question and doubt.

3.4.2 Effect of Heat Release

The gasification reactor must be configured to accommodate the energy balance of the chemical reactions. During the gasification process, most of the energy bound up in the fuel is not released as heat. In fact, the fraction of the chemical energy of the feedstock, or the heating content (typically shown as Btu/lb or calories per kilogram), which remains in the product gases (especially the sun thesis gas) is an important measure of the efficiency of a gasification process (which is dependent upon the reactor configuration) and is known as the *cold gas efficiency*. Most commercial-scale gasification reactors have a cold gas efficiency on the order of 65% to 80%, or even higher.

Thus, it is important for the reactor to limit the amount of heat that is transferred out of the zone where the gasification reactions are occurring. If not, the temperature within the gasification zone could be too low to allow the reactions to proceed – as an example, a minimum temperature on the order of 1000°C (1830°F) is typically needed to gasify coal. As a result, a gasification reactor is typically refractory-lined with no water cooling to ensure as little heat loss as possible. Gasification reactors also typically operate at elevated pressure (often as high as 900 psia), which allows them to have very compact construction with minimum surface area and minimal heat loss.

3.4.3 Other Effects

In addition to being designed and selected for feedstock type, another design option for the gasification reactor involves the method for cooling the synthesis gas produced by the gasifier. Regardless of the type of gasifier, the exiting synthesis gas must be cooled down to approximately 100°C (212°F) in order to utilize conventional acid gas removal technology. This can be accomplished either by passing the synthesis gas through a series of heat exchangers which recover the sensible heat for use (for example, in the stem cycle an *integrated combined cycle unit*, IGCC unit) or by directly contacting the synthesis gas with relatively cool water (a *quench* operation). The quench operation results in some of the quench water being vaporized and mixed with the synthesis gas. The quenched synthesis gas is saturated with water and must pass through a series of condensing heat exchanges which remove the moisture from the synthesis gas (so it can be recycled to the quench zone).

Quench designs have a negative impact on the heating rate of related equipment (such as the IGCC unit) because the sensible heat of the high temperature synthesis gas is converted to low-level process heat rather than high-pressure steam. However, quench designs have much lower capital costs and can be justified when low-cost feedstock (such as biomass or waste) is available. Quench designs also have an advantage if carbon dioxide capture is desired. The saturated synthesis gas exiting a quench section has near the optimum water/carbon monoxide ratio as the feedstock to a water-gas shift reactor which will convert the carbon monoxide to carbon dioxide. Non-quench designs that require carbon dioxide capture need to add steam to the synthesis gas before it is sent to a water-gas shift reactor.

Finally, a variety of organic feedstocks have been gasified using supercritical water. Under supercritical conditions (T>375°C, >675°F, P>3250 psi) water reacts with biomass to produce a gas rich in hydrogen and methane. The water self-dissociates to form both hydroxonium ions (H_3O^+) and hydroxyl ions (OH^-) that act as catalysts for converting the biomass into gas. The gas composition from supercritical water gasification of biomass below 400°C (<750°F) typically has high content of methane and carbon dioxide but a low content of hydrogen. Above 400°C (<750°F), the hydrgoen content increases and the methane content decreases. The carbon monoxide content during supercritical gasification at a temperature below 600°C (1110°F) is several orders of magnitude lower than the content of hydrogen, carbon dioxide and methane.

As an example of the process, the complete conversion of glucose (22% by weight in water) to a hydrogen-rich synthesis gas can be achieved at a weight hourly space velocity (WHSV) of 22.2 h-1 in supercritical water at 600°C 1110°F), 34.5 MPa. Complete conversions of whole biomass feeds were also achieved at the same temperature and pressure (Xu et al., 1996).

References

Anthony, D.B., and Howard, J.B. 1976. Coal Devolatilization and Hydrogasification. *AIChE Journal*. 22: 625-656.

Butterman, H.C., and Castaldi, M.J. 2008. CO_2 Enhanced Steam Gasification of Biomass Fuels. Paper No. NAWTEC16-1949. *Proceedings. NAWTEC16 – 16th Annual North American Waste-to-Energy Conference, Philadelphia, Pennsylvania. May 19-21.*

Calemma, V. and Radović, L.R. 1991. On the Gasification Reactivity of Italian Sulcis Coal. *Fuel*, 70: 1027.

Capareda, S. 2011. Advances in Gasification and Pyrolysis Research Using Various Biomass Feedstocks. *Proceedings. 2011 Beltwide Cotton Conferences, Atlanta, Georgia, January 4-7.* Page 467-472.

Chadeesingh, R. 2011. The Fischer-Tropsch Process. In *The Biofuels Handbook*. J.G. Speight (Editor). The Royal Society of Chemistry, London, United Kingdom. Part 3, Chapter 5, Page 476-517.

Chen, C., Horio, M., and Kojima, T. 2000. Numerical Simulation of Entrained Flow Coal Gasifiers. Part II: Effects of Operating Conditions on Gasifier Performance. *Chemical Engineering Science*, 55(18): 3875-3883.

Collot, A.G., Zhuo, Y., Dugwell, D.R., and Kandiyoti, R. 1999. Co-pyrolysis and cogasification of coal and biomass in bench-scale fixed-bed and fluidized bed reactors. *Fuel*, 78: 667-679.

Collot, A.G. 2002. *Matching Gasifiers to Coals*. Report No. CCC/65. Clean Coal Centre, International Energy Agency, London, United Kingdom.

Collot, A.G. 2006. Matching Gasification Technologies to Coal Properties. *International Journal of Coal Geology*, 65: 191-212.

Cranfield, R. 1978. Solids Mixing in Fluidized Beds of Large Particles. *AIChE Journal*, 74(176): 54-59.

Fryer, J.F., and Speight, J.G. 1976. *Coal Gasification: Selected Abstract and Titles*. Information Series No. 74. Alberta Research Council, Edmonton, Alberta, Canada.

Garcia, X. and Radović, L.R. 1986. Gasification Reactivity of Chilean Coals. *Fuel* 65: 292.

Ghaly, A.E., Al-Taweel, A.M., Hamdullahpur, F., Ugwu, I. 1989. Physical and Chemical Properties of Cereal Straw as Related to Thermochemical Conversion. *Proceedings. 7th. Bioenergy R&D Seminar.* E.N. Hogan (Editor). Ministry of Energy, Mines, and Resources Ministry, Ottawa, Ontario, Canada. Page 655-661.

Heberlein, J., and Murphy, A.B. 2008. Thermal Plasma Waste Treatment. *J. Phys. D: Appl. Phys.* 41: 1-20.

Higman, C., and Van der Burgt, M. 2003. *Gasification.* 2nd Edition. Gulf Professional Publishers. Gasification. Elsevier, Amsterdam, Netherlands.

Kohl, A.L., Harty, R.B., Johanson, J.G., and Naphthali, L.M. 1978. Molten Salt Coal Gasification Process. *Chem Eng Prog.*, 74: 73.

Kristiansen, A. 1996. *Understanding Coal Gasification.* Report No. IEACR/86. IEA Coal Research, International Energy Agency, London, United Kingdom.

Jangsawang W., Klimanek, A., and Gupta, A.K. 2006. Enhanced Yield of Hydrogen From Wastes Using High Temperature Steam Gasification. *J. of Energy Resources Technology*, 128(3): 79-185.

Johnson J.L. 1979. *Kinetics of Coal Gasification.* John Wiley and Sons Inc., Hoboken, New Jersey.

Lee, S. 2007. Gasification of Coal. In: *Handbook of Alternative Fuel Technologies.* S. Lee, J.G. Speight, and S. Loyalka (Editors). CRC Press, Taylor & Francis Group, Boca Raton, Florida. 2007.

Lee, S., Speight, J.G., and Loyalka, S. 2007. *Handbook of Alternative Fuel Technologies.* CRC Press, Taylor & Francis Group, Boca Raton, Florida.

Luque, R., and Speight, J.G. (Editors). 2015. *Gasification for Synthetic Fuel Production: Fundamentals, Processes, and Applications.* Woodhead Publishing, Elsevier, Cambridge, United Kingdom.

McKendry, P. 2002. Energy Production from Biomass Part 3: Gasification Technologies. *Bioresource Technology*, 83(1): 55-63.

Messerle, V.E., and Ustimenko, A.B. 2007. Solid Fuel Plasma Gasification. Advanced Combustion and Aerothermal Technologies. *NATO Science for Peace and Security Series C. Environmental Security.* Page 141-1256.

Mokhatab, S., Poe, W.A., and Speight, J.G. 2006. *Handbook of Natural Gas Transmission and Processing.* Elsevier, Amsterdam, Netherlands.

Müller, R., von Zedtwitz, P., Wokaun, A., and Steinfeld, A. 2003. Kinetic Investigation on Steam Gasification of Charcoal under High-Flux Radiation. *Chemical Engineering Science*, 58: 5111-5119.

Penner S.S. 1987. *Coal Gasification.* Pergamon Press Inc., New York.

Radović, L.R., Walker, P.L., Jr., and Jenkins, R.G. 1983. Importance of Carbon Active Sites in the Gasification of Coal Chars, *Fuel* 62: 849.

Radović, L.R., and Walker, P.L., Jr. 1984. Reactivities of Chars Obtained as Residues in Selected Coal Conversion Processes. *Fuel Processing Technology*, 8: 149-154.

Ricketts, B., Hotchkiss, R., Livingston, W., and Hall, M. 2002. Technology Status Review of Waste/Biomass Co-Gasification with Coal. *Proceedings. IChemE 5th European Gasification Conference, Noordwijk, The Netherlands. 8-10 April 8-10.*

Rummel, R. 1959. Gasification in a Slag Bath. *Coke Gas*, 21(247) 493-501.

Schmid, J.C., Pfeifer, C., Kitzler, H., Pröll, T., and Hofbauer, H. 2011. A New Dual Fluidized Bed Gasifier Design for Improved In Situ Conversion of Hydrocarbons. *Proceedings. International Conference on Polygeneration Strategies (ICPS 2011), Vienna, Austria. August 30-September 1.*

Senneca, O. 2007. Kinetics of Pyrolysis, Combustion and Gasification of Three Biomass Fuels. *Fuel Processing Technology*, 88: 87-97.

Sha, X. 2005. Coal Gasification. In: *Coal, Oil Shale, Natural Bitumen, Heavy Oil and Peat. Encyclopedia of Life Support Systems (EOLSS),* Developed under the Auspices of the UNESCO, EOLSS Publishers, Oxford, UK; http://www.eolss.net

Shabbar, S., and Janajreh, I. 2013. Thermodynamic Equilibrium Analysis of Coal Gasification Using Gibbs Energy Minimization Method. *Energy Conversion and Management*, 65: 755-763.

Silaen, A., and Wang, T. 2008. Effects of Turbulence and Devolatilization Models on Gasification Simulation. *Proceedings. 25th International Pittsburgh Coal Conference, Pittsburgh, Pennsylvania. September 29-October 2.*

Singh, S.P., Weil, S.A., and Babu, S.P. 1980. Thermodynamic Analysis of Coal Gasification Processes. *Energy*, 5(8-9): 905–914.

Slavinskaya, N.A., Petrea, D.M., and Riedel. U. 2009. Chemical Kinetic Modeling in Coal Gasification Overview. *Proceedings. 5th International Workshop on Plasma Assisted Combustion (IWEPAC). Alexandria, Virginia.*

Speight, J.G. 2008. *Synthetic Fuels Handbook: Properties, Processes, and Performance*. McGraw-Hill, New York.

Speight, J.G. (Editor). 2011. *The Biofuels Handbook*. Royal Society of Chemistry, London, United Kingdom.

Speight, J.G. 2013a. *The Chemistry and Technology of Coal* 3rd Edition. CRC Press, Taylor & Francis Group, Boca Raton, Florida.

Speight, J.G. 2013b. *Coal-Fired Power Generation Handbook*. Scrivener Publishing, Salem, Massachusetts.

Speight, J.G. 2014. *The Chemistry and Technology of Petroleum* 5th Edition. CRC Press, Taylor and Francis Group, Boca Raton, Florida.

Sundaresan, S., and Amundson, N.R. 1978. Studies in Char Gasification – I: A lumped Model. *Chemical Engineering Science*, 34: 345-354.

Turn, S.Q. 1999. Biomass Integrated Gasifier Combined Cycle Technology: Application in the Cane Sugar Industry. *International Sugar Journal*, 101: 1205.

Van der Burgt, M. 2008. In: *Gasification* 2nd Edition. C. Higman and M. van der Burgt (Editors). Gulf Professional Publishing, Elsevier, Amsterdam, Netherlands. Chapter 2.

Xu, X., Matsumura, Y., Stenberg, J., Antal, M.J. Jr. 1996. Carbon-Catalyzed Gasification of Organic Feedstocks in Supercritical Water. *Ind. Eng. Chem. Res.*, 2522-2530.

4
Gasification of Coal

4.1 Introduction

As already defined (Chapter 2), the gasification process is a technology that converts carbon-containing materials into synthesis gas (also often referred to as *syngas*) which, in turn, can be used to produce electricity and other valuable products, such as chemicals, fuels, and fertilizers (Table 4.1). The gasification does not involve combustion, but instead uses little or no oxygen (or air) in a closed reactor to convert carbon-based materials directly into gaseous products. Thus, the process can be employed to recover the energy locked into a variety of carbonaceous materials (such as, in addition to coal, crude oil residua, biomass, and solid waste) thereby converting such materials into valuable products and eliminating the need for incineration or landfilling.

Gasification processes have been in commercial use for more than 200 years when the feedstock was typically coal. During those years, the process evolved using coal as the major feedstock leading to the production of a variety of gaseous products. On the other hand, the focus has shifted somewhat and the current two most significant reasons for the gasification process are (i) the variable but continuing upward price of natural gas and (ii) the variable but continuing upward price of highway transportation fuels. The most significant reason is the need for energy independence. In other words, the use of domestic energy sources not only for electricity production but also for synthetic natural gas (SNG) and liquids for transportation is a must. Gasification is a key fundamental baseline technology for converting carbonaceous feedstocks to valuable products such as hydrogen, synthetic natural gas, transportation fuels, and a variety of chemicals (Breault, 2010).

The gasification process breaks down the carbon-containing materials to the molecular level, so impurities such as nitrogen-containing constituents, sulfur-containing constituents, and (in some cases) mercury can be easily removed and sold as valuable industrial commodities. Any carbon dioxide is emitted as a concentrated gas stream in synthesis gas at high pressure. In this form, it can be captured and sequestered more easily and at lower costs (Mokhatab *et al.*, 2006; Speight, 2019).

Furthermore, since coal has been, and still is, recognized as the most abundant fossil fuel available on Earth, it is not surprising that the gasification technology was developed using coal as the prime feedstock in the 19th century – even low-grade coal can be used for gasification – and, as a result, gasification the technology became of interest in many regions of the world. In fact, processes for producing energy via coal gasification have been in commercial use for more than 180 years. Earlier gasification technologies included the production of town gas for lighting using peat and coal as feedstocks and the production of fuels from wood to power motor vehicles during World War II. Throughout, this period, the gasification process has been most commonly associated with the centuries-old blast furnace,

Table 4.1 Products from synthesis gas.

Synthesis gas				
	Fuel gas			
	Town gas			
	Hydrogen	Ammonia		
		Urea		
	Fischer-Tropsch liquids	Synthetic natural gas		
		Naphtha		
		Kerosene		
		Waxes		
	Methanol	Dimethyl ether	Ethylene	Polyolefins
			Propylene	Polyolefins
		Acetic acid	Methyl acetate	
			Acetate esters	
			Polyvinyl acetate	
			Acetic anhydride	

which is a type of metallurgical furnace traditionally used to extract metals, generally iron, from ore. Blast furnaces have been around for centuries, with the earliest models developed in the 1st century BC. The gasification of coal has been reliably used on a commercial scale for at least 150 years in the fertilizer, and chemical industries, and for more than 100 years in the electric power industry.

Briefly, a blast furnace is a refractory-lined vessel that is used to convert iron ore (typically, magnetite, Fe_3O_4, combined with hematite, Fe_2O_3) into molten iron. Driven by gravity, they work in a countercurrent manner: iron ore, metallurgical coke, and limestone are fed into the top of the furnace and hot air (the *blast*) is blown into the bottom of the furnace. The materials fed into the top of the furnace gradually descend as the hot gases rise through them. In this process, oxidized iron ore is stripped of the oxygen molecules and converted into elemental iron with the use of carbon as a reducing agent. This reduction happens slowly as the iron ore descends through the furnace, emitting gases until it reaches the zone near the bottom of the furnace where the iron ore melts completely. At this point, the liquid iron is removed (*tapped*) from the bottom of the furnace and poured into casts and sent to a steel mill. The low-value gases produced in the blast furnace were typically extracted at the top of the furnace and burned to heat the air blasted into the furnace. Carbon is used as a reducing agent – it combines with the oxygen in the ore to form carbon monoxide (CO) and eventually carbon dioxide (CO_2). The main carbon-bearing material used in traditional blast furnaces is coal, sometimes transformed into coke, which also serves to generate the high temperatures needed to smelt the iron ore.

The town gas, a gaseous product manufactured from coal, containing approximately 50% hydrogen, with the rest comprised of mostly methane and carbon dioxide, with 3% to 6% carbon monoxide, that was produced in the blast furnace supplied lighting and heating for industrializing America and Europe beginning in the early 1800s. The first public street lighting with gas took place in Pall Mall, London, on January 28, 1807. Not long after that, in 1816, Baltimore, Maryland, began the first commercial gas lighting of residences, streets, and businesses. Since that time, gasification has had its ups and downs with more and longer periods of down as communities began to electrify. The few highpoints of gasification during the past 100 years are worthy of identification. Gasification was used extensively during World War II to convert coal into transportation fuels via the Fischer-Tropsch process. It has been used extensively in the last 50 to 60 years to convert coal and heavy oil into hydrogen for the production of ammonia/urea fertilizer. The chemical industry and the refinery industry applied gasification in the 1960s and 1980s, respectively, for feedstock preparation. In the past several decades, years, it has been used by the power industry in integrated gasification combined cycle (IGCC) plants.

The early gasworks used iron retorts to heat the fuel, pyrolyzing it to gas, oils, and coke or charcoal. Later improvements were the use of fireclay and then silica retorts to achieve higher pyrolysis temperatures. The plants operated with a thermal efficiency which converted 70% to 80% of the energy in the fuel to salable products, producing a gas containing 500 Btu/SCF. Another widely used process was the blue water-gas process in which the solid fuel was heated to very high temperatures with a blast of air which formed a low energy gas (100 Btu/ft^3, called "producer gas") for use as fuel for manufacturing processes. When sufficiently hot, the air was cut off and steam was blown in from the opposite end of the vessel which produced a higher energy gas (300 Btu/ft^3). This blue water gas (blue because it burned with a blue flame) could be converted to carbureted water gas by using the high off-gas temperature to crack oils, yielding a higher energy gas (500 Btu/ft^3). Using these processes, the gas industry grew rapidly and by the time of World War Il there were 1,200 plants in the United States producing and selling gas.

Gas has many advantages over solid fuels since it can be distributed relatively easily and the combustion of the gas can be controlled to give high efficiency; it can be burned automatically; and it burns with low emissions, making smokeless cities a distinct possibility. It burns with a higher temperature needed in many industrial processes and no local storage is necessary. It is ideal for cooking and heating in homes and is a necessity for many modern manufacturing processes. A given amount of energy is worth two to four times as much energy in the form of gas as it would be in the form of a solid fuel. In addition., gas can be used to operate spark and diesel engines or turbines to generate power. The use of producer gas to run an engine was first tried in 1881 and by the 1920s portable gas producers were being used to run trucks and tractors in Europe. These gas generators operated on either wood or charcoal and produced a gas with a rather high heat content.

While it was possible to run engines on this gas, it was not convenient, and solid fuels for automotive use did not achieve wide acceptability. There was continued activity aimed at improving gas generators by individual inventors and a few companies until World War II. Commercial installations to run both stationary and mobile engines continued at a low level. The beginning of World War II and the scarcity of liquid fuels in Europe intensified the search for domestically available fuels and resulted in a great surge of activity in designing and installing gas generators. Gas generators were also used on tractors, boats, motorcycles, and even on railway shunting engines.

Techniques were developed for converting both diesel and spark ignition engines to generator gas operation. These engines operated reliably, although there was a derating of power output to approximately 75% of the gasoline rating, and considerable additional maintenance of filters, coolers, and the generator itself was required of the operator. It required 20 lb of wood to replace 1 gallon of gasoline. The end of the war brought renewed supplies of liquid fossil fuels and a rapid reconversion of vehicles to diesel and gasoline. Since the war a few generators have been in operation, primarily in underdeveloped countries.

Through the history of gas generation and use, coal has been the lead feedstock for the gasification of carbonaceous materials. It is for this reason that it is necessary to include a description of the processes used for coal gasification that have evolved into process for the gasification of other carbonaceous feedstocks. This chapter presents an assessment of the technologies that have evolved for the gasification of coal with a description of the current state of the art of the various technologies. From this, it will be possible to assess the technologies for application to other carbonaceous feedstocks, such as non-coal feedstocks as well as mixed feedstocks.

4.2 Coal Types and Properties

The chemical conversion of coal to gaseous products was first used to produce gas for lighting and heat in the United Kingdom more than 200 years ago. The gasification of coal or a derivative (i.e., char produced from coal) is, essentially, the conversion of coal (by any one of a variety of processes) to produce combustible gases (Fryer and Speight, 1976; Radović et al., 1983; Radović and Walker, 1984; Garcia and Radović, 1986; Calemma and Radović, 1991; Kristiansen, 1996; Speight, 2008). With the rapid increase in the use of coal from the 15th century onwards (Nef, 1957; Taylor and Singer, 1957) it is not surprising the concept of using coal to produce a flammable gas became commonplace (Elton, 1958).

In the historical context, the early gasworks used iron retorts to heat the fuel, pyrolyzing it to gas, oils, and coke or charcoal. Later improvements were the use of fireclay and then silica retorts to achieve higher pyrolysis temperatures. The plants operated with a thermal efficiency which converted 70% to 80% of the energy in the fuel to salable products, producing a gas containing 500 Btu/ft^3.

Another widely used process was the *blue water-gas process* in which the feedstock (typically coal) was heated to very high temperatures with a blast of air (referred to as the *blow*), which formed a low-energy gas (100 Btu/ft^3; called producer gas) for use as fuel for manufacturing processes. When sufficiently hot, the air was cut off and steam was blown in from the opposite end of the vessel (the *run*) which produced a higher energy gas (300 Btu/ft^3). This blue water gas (which burned with a blue flame, hence the name) could be converted to *carbureted water gas* by using the high off-gas temperature to crack oils, yielding a gas with 500 Btu/ft^3. Using these processes, the gas industry grew rapidly and by the time of World War II there were 1,200 plants in the United States producing and selling gas. With the coming of pipeline networks in the 1930s natural gas gradually replaced manufactured gas, and these plants have almost all closed down.

Depending on the type of gasifier (e.g., air-blown, enriched oxygen-blown) and the operating conditions (Chapter 2), gasification can be used to produce a fuel gas that is suitable for several applications. Coal gasification for electric power generation enables the use of a technology common in modern gas-fired power plants, the use of *combined cycle* technology to recover more of the energy released by burning the fuel.

As a very general *rule of thumb*, optimum gas yields and gas quality are obtained at operating temperatures of approximately 595 to 650°C (1100 to 1200°F). A gaseous product with a higher heat content (BTU/ft^3) can be obtained at lower system temperatures but the overall yield of gas (determined as the *fuel-to-gas ratio*) is reduced by the unburned char fraction.

The influence of physical process parameters and the effect of coal type on coal conversion is an important part of any process where coal is used as a feedstock, especially with respect to coal combustion and coal gasification (Speight, 2013a, 2013b). The reactivity of coal generally decreases with increase in rank (from lignite to subbituminous coal to bituminous coal anthracite). Furthermore, the smaller the particle size, the more contact area between the coal and the reaction gases causing faster reaction. For medium-rank coal and low-rank coal, reactivity increases with an increase in pore volume and surface area, but these factors have no effect on reactivity for coals having carbon content greater than 85% w/w. In fact, in high-rank coals, pore sizes are so small that the reaction is diffusion controlled.

The volatile matter produced by the coal during thermal reactions varies widely for the four main coal ranks and is low for high-rank coals (such as anthracite) and higher for increasingly low-rank coals (such as lignite) (Speight, 2013a, 2013b). The more reactive coals produce higher yields of gas and volatile products as well as lower yields of char. Thus, for high-rank coals, the utilization of char within the gasifier is much more of an issue than for lower-rank coal. However, the ease with which they are gasified leads to high levels of tar in the gaseous products which makes gas cleanup more difficult.

The mineral matter content of the coal does not have much impact on the composition of the gas product. Gasifiers may be designed to remove the produced ash in solid or liquid (slag) form (Chapter 2). In fluidized- or fixed-bed gasifiers, the ash is typically removed as a solid, which limits operational temperatures in the gasifier to well below the ash melting point. In other designs, particularly slagging gasifiers, the operational temperatures are designed to be above the ash melting temperature. The selection of the most appropriate gasifier is often dependent on the melting temperature and/or the softening temperature of the ash and the type of coal that is to be used at the facility.

In fact, coal which displays caking, or agglomerating, characteristics when heated (Speight, 2013a) is not usually amenable to use as feedstock for gasification processes that employ a fluidized-bed gasifier or a moving-bed gasifier; in fact, caking coal is difficult to handle in fixed-bed reactors. Pretreatment of the caking coal by a mild oxidation process (typically consisting of low-temperature heating of the coal in the presence of air or oxygen) destroys the caking characteristics of the coal.

High moisture content of the feedstock lowers internal gasifier temperatures through evaporation and the endothermic reaction of steam and char. Usually, a limit is set on the moisture content of coal supplied to the gasifier, which can be met by coal drying operations if necessary. For a typical fixed-bed gasifier and moderate rank and ash content of the coal, this moisture limit in the coal limit is on the order of 35% w/w. Fluidized-bed gasifiers and entrained-bed gasifiers have a lower tolerance for moisture, limiting the moisture content to approximately 5 to 10% w/w a similar coal feedstock. Oxygen supplied to the gasifiers must be increased with an increase in mineral matter content (ash production) or moisture content in the coal.

In regard to the maceral content, differences have been noted between the different maceral groups with inertinite being the most reactive. In more general terms of the character of the coal, gasification technologies generally require some initial processing of the coal feedstock with the type and degree of pretreatment a function of the process and/or the type of coal. For example, the Lurgi process will accept *lump* coal [1 inch (25 mm) to 28 mesh], but it must be noncaking coal with the fines removed. The caking, agglomerating coals tend to form a plastic mass in the bottom of a gasifier and subsequently plug up the system thereby markedly reducing process efficiency.

With some coal feedstocks, the higher the amounts of volatile matter produced in the early stages of the gasification process the higher the heat content of the product gas. In some cases, the highest gas quality may be produced at the lowest temperatures but when the temperature is too low, char oxidation reaction is suppressed and the overall heat content of the product gas is diminished.

Coals of the western United States tend to have lower heating values, lower sulfur contents, and higher moisture contents relative to bituminous coals from the eastern United States. The efficiency loss associated with high moisture and ash content coals is more significant for slurry-feed gasifiers. Consequently, dry-feed gasifiers, such as the Shell gasifier, may be more appropriate for low-quality coals. There is also the possibility that western coals can be combined with petroleum coke in order to increase the heating value and decrease the moisture content of the gasification feedstock.

4.3 Gas Products

The products from the gasification of coal may be of low, medium, or high heat-content (high-Btu) content as dictated by the process as well as by the ultimate use for the gas (Chapter 1) (Fryer and Speight, 1976; Mahajan and Walker, 1978; Anderson and Tillman, 1979; Cavagnaro, 1980; Bodle and Huebler; Argonne, 1990; 1981; Baker and Rodriguez, 1990; Probstein and Hicks, 1990; Lahaye and Ehrburger, 1991; Matsukata *et al.*, 1992). Furthermore, variation in coal quality has an impact on the heating value of the product gas as well as the conditions in the gasifier (i.e., temperature, heating rate, pressure, and residence time) (Speight, 2013a, 2013b).

The gasification process involves two distinct stages: (1) *coal devolatilization* followed by (2) *char gasification*, which is specific to the conditions of the reaction. Both stages have an effect on the yield and quality of the product gas.

Depending on the type of coal being processed and the analysis of the gas product desired, pressure also plays a role in product definition (Speight, 2013a). In fact, some (or all) of the following processing steps will be required: (1) pretreatment of the coal (if caking is a problem); (2) primary gasification of the coal; (3) secondary gasification of the carbonaceous residue from the primary gasifier; (4) removal of carbon dioxide, hydrogen sulfide, and other acid gases; (5) shift conversion for adjustment of the carbon monoxide/hydrogen mole ratio to the desired ratio; and (6) catalytic methanation of the carbon monoxide/hydrogen mixture to form methane. If high heat-content (high-Btu) gas is desired, all of these processing steps are required since coal gasifiers do not yield methane in the concentrations required (Mills, 1969; Cusumano *et al.*, 1978).

4.3.1 Coal Devolatilization

Devolatilization occurs as the coal is heated above 400°C (750°F) and during this period, the coal structure is altered, producing solid char, tars, condensable liquids, and light gases. The devolatilization products formed in an inert gas atmosphere are very different from those in an atmosphere containing hydrogen at elevated pressure. After devolatilization char then gasifies at a lower rate, the specific reactions that take place during this second stage depend on the gasification medium.

After the rate of devolatilization has passed a maximum, another reaction occurs in which the semi-char is converted to char primarily through the evolution of hydrogen. In a hydrogen atmosphere at elevated pressure, additional yields of methane or other low molecular weight gaseous hydrocarbon can result during the initial coal gasification stage from reactions such as: (1) direct hydrogenation of coal or semi-char because of active intermediate formed in coal structure after coal pyrolysis, and (2) the hydrogenation of other gaseous hydrocarbons, oils, tars, and carbon oxides.

4.3.2 Char Gasification

Char gasification occurs as the char reacts with gases such as carbon dioxide (CO_2) and steam (H_2O) to produce carbon monoxide (CO) and hydrogen (H_2):

$$2C + CO_2 + H_2O \rightarrow 3CO + H2$$

The resulting gas (producer gas or synthesis gas) may be more efficiently converted to electricity than is typically possible by direct combustion of coal. Also, corrosive ash elements such as chloride and potassium may be refined out by the gasification process, allowing high temperature combustion of the gas from otherwise problematic coal feedstocks.

Although the devolatilization reaction is completed in short order (typically in seconds) at elevated temperature, the subsequent gasification of the char is much slower, requiring minutes or hours to obtain significant conversion under practical conditions – in fact reactor design for commercial gasification processes is largely dependent on the reactivity of the char.

The reactivity of char produced in the pyrolysis step depends on the nature of parent coal and increases with oxygen content of parent coal but decreases with carbon content. In general, char produced from low-rank coal is more reactive than char produced from high-rank coal. The reactivity of char from low-rank coal may be influenced by catalytic effect of mineral matter in char. In addition, as the carbon content of coal increases, the reactive functional groups present in coal decrease and the coal substance becomes more aromatic and cross-linked in nature (Speight, 2013a). Therefore char obtained from high-rank coal contains a lesser number of functional groups and higher proportion of aromatic

132 SYNTHESIS GAS

and cross-linked structures, which reduce reactivity. The rate of gasification of the char decreases as the process temperature increases due to the decrease in active surface area of char. Therefore a change of char preparation temperature may change the chemical nature of char, which in turn may change the gasification rate (Johnson, 1979; Penner, 1987; Speight, 2013a).

4.3.3 Gasification Chemistry

Coal gasification occurs under reducing conditions – coal (in the presence of steam and oxygen at high temperature and moderate pressure) is converted to a mixture of product gases. The chemistry of coal gasification can be conveniently (and simply) represented by the following reactions:

$$C + O_2 \rightarrow CO_2 \qquad \Delta Hr = -393.4 \text{ MJ/kmol} \qquad (1)$$

$$C + \tfrac{1}{2} O_2 \rightarrow CO \qquad \Delta Hr = -111.4 \text{ MJ/kmo} \qquad (2)$$

$$C + H_2O \rightarrow H_2 + CO \qquad \Delta Hr = 130.5 \text{ MJ/kmol} \qquad (3)$$

$$C + CO_2 \leftrightarrow 2CO \qquad \Delta Hr = 170.7 \text{ MJ/kmol} \qquad (4)$$

$$CO + H_2O \leftrightarrow H_2 + CO_2 \qquad \Delta Hr = -40.2 \text{ MJ/kmol} \qquad (5)$$

$$C + 2H_2 \rightarrow CH_4 \qquad \Delta Hr = -74.7 \text{ MJ/kmol} \qquad (6)$$

Reactions (1) and (2) are exothermic oxidation reactions and provide most of the energy required by the endothermic gasification reactions (3) and (4). The oxidation reactions occur very rapidly, completely consuming all of the oxygen present in the gasifier, so that most of the gasifier operates under reducing conditions. Reaction (5) is the water-gas shift reaction, in which water (steam) is converted to hydrogen – this reaction is used to alter the hydrogen/carbon monoxide ration when synthesis gas is the desired product, such as for use in Fischer-Tropsch processes. Reaction (6) is favored by high pressure and low temperature and is, thus, mainly important in lower temperature gasification systems. Methane formation is an exothermic reaction that does not consume oxygen and, therefore, increases the efficiency of the gasification process and the final heat content of the product gas. Overall, approximately 70% of the heating value of the product gas is associated with the carbon monoxide and hydrogen but this can be higher depending upon the gasifier type (Chapter 2) (Chadeesingh, 2011).

Depending on the gasifier technology employed and the operating conditions (Chapter 2), significant quantities of water, carbon dioxide, and methane can be present in the product gas, as well as a number of minor and trace components. Under the reducing conditions in the gasifier, most of the organically bound sulfur in the coal feedstock is converted to hydrogen sulfide (H_2S), but a small amount (3 to 10% w/w) is converted to carbonyl sulfide (COS). Organically bound nitrogen in the coal feedstock is generally converted to gaseous nitrogen (N_2), but small amounts of ammonia (NH_3) and hydrogen cyanide (HCN) are also formed. Any chlorine in the coal (which typically originates from time in the coal seam) is converted to hydrogen chloride (HCl) with some chlorine present in the particulate matter (fly ash). Trace elements, such as mercury and arsenic, are released during gasification and partition among the different phases, such as fly ash, bottom ash, slag, and product gas.

4.3.4 Other Process Options

In addition to the generic reactor designs of a gasification process (Chapter 2), there are several other design options that the gasification process, each of which can have noticeable impacts on the downstream processes. For example, atmospheric and/or pressurized gasifiers can operate at either atmospheric pressure or at pressures as high as (900 psi). Low-pressure or atmospheric-pressure gasifiers will require a fuel gas compressor after the clean-up of the synthesis gas (Chapter 9). In addition, high-pressure gasifiers can also have a positive impact on the performance of the synthesis clean-up section. If capture of carbon dioxide is required, the high-pressure gasifier operation will improve the performance of physical absorption processes that can remove carbon dioxide from the synthesis gas.

Dry feed or slurried feed feedstock is typically introduced into a pressurized gasifier either pneumatically as a dry solid or pumped as feedstock-water slurry. Slurry-feed systems result in less efficient conversion of the feedstock to synthesis gas because some of the synthesis gas must be combusted in order to generate the heat needed to vaporize the water in the slurry. Consequently, the synthesis gas produced by a slurry-fed gasifier typically contains more carbon dioxide than the synthesis gas produced by a dry-fed gasifier.

The oxidant for the gasification process can be either air-blown or oxygen-blown. Air-blown gasifiers produce a much lower calorific value synthesis gas than oxygen-blown gasifiers. The nitrogen in the air typically dilutes the synthesis gas by a factor of 3 compared to oxygen-blown gasification and therefore, while a synthesis gas calorific value on the order of 300 Btu/ft^3 might be typical from an oxygen-blown gasifier, an air-blown gasifier will typically produce synthesis gas with a calorific value on the order of 100 Btu/ft^3. Because the nitrogen in air must be heated to the gasifier exit temperature by burning some of the synthesis gas, air-blown gasification is more favorable for gasifiers which operate at lower temperatures (i.e., non-slagging).

4.3.4.1 Hydrogasification

Not all high heat-content (high-Btu) gasification technologies depend entirely on catalytic methanation and, in fact, a number of gasification processes use hydrogasification, that is, the direct addition of hydrogen to coal under pressure to form methane (Anthony and Howard, 1976).

$$C_{char} + 2H_2 \rightarrow CH_4$$

The hydrogen-rich gas for hydrogasification can be manufactured from steam by using the char that leaves the hydrogasifier. Appreciable quantities of methane are formed directly in the primary gasifier and the heat released by methane formation is at a sufficiently high temperature to be used in the steam-carbon reaction to produce hydrogen, which then requires less oxygen to produce heat for the steam-carbon reaction.

4.3.4.2 Catalytic Gasification

Catalysts are commonly used in the chemical and petroleum industries to increase reaction rates, sometimes making certain previously unachievable products possible (Speight, 2002; Speight, 2014, 2017, 2019). Use of appropriate catalysts not only reduces reaction temperature but also improves the gasification rates. In addition, catalysts also reduce tar formation (Shinnar *et al.*, 1982; McKee, 1981). Catalysts can also be used to favor or suppress the formation of certain components in the gaseous product. For example, in the production of synthesis gas (mixtures of hydrogen and carbon monoxide), methane is also produced in small amounts. Catalytic gasification can be used to either promote or suppress methane formation.

Alkali metal salts of weak acids, such as potassium carbonate (K_2CO_3), sodium carbonate (Na_2CO_3), potassium sulfide (K_2S), and sodium sulfide (Na_2S) can catalyze steam gasification of coal. Catalyst amounts on the order of 10 to 20% w/w potassium carbonate (K_2CO_3) will lower bituminous coal gasifier temperatures from 925°C (1695°F) to 700°C (1090°F) and that the catalyst can be introduced to the gasifier impregnated on coal or char.

Ruthenium-containing catalysts are used primarily in the production of ammonia. It has been shown that ruthenium catalysts provide five to 10 times higher reactivity rates than other catalysts. However, ruthenium quickly becomes inactive due to its necessary supporting material, such as activated carbon, which is used to achieve effective reactivity. However, during the process, the carbon is consumed, thereby reducing the effect of the ruthenium catalyst.

Disadvantages of catalytic gasification include increased materials costs for the catalyst itself (often rare metals), as well as diminishing catalyst performance over time. Catalysts can be recycled, but their performance tends to diminish with age or by poisoning. The relative difficulty in reclaiming and recycling the catalyst can also be a disadvantage. For example, the potassium carbonate catalyst can be recovered from spent char with a simple water wash, but some catalysts may not be so accommodating. In addition to age, catalysts can also be diminished by poisoning. On the other hand, many catalysts are sensitive to particular chemical species which bond with the catalyst or alter it in such a way that it no longer functions. Sulfur, for example, can poison several types of catalysts including palladium and platinum.

4.3.4.3 Plasma Gasification

Plasma is a high-temperature, highly ionized (electrically charged) gas capable of conducting electrical current and this type of gasification technology is sufficiently different from the type mentioned elsewhere (Chapter 3). Plasma technology has a long history of development and has evolved into a valuable tool for engineers and scientists who need to use very high temperatures for new process applications (Kalinenko *et al.*, 1993; Messerle and Ustimenko, 2007). Man-made plasma is formed by passing an electrical discharge through a gas such as air or oxygen (O_2). The interaction of the gas with the electric arc dissociates the gas into electrons and ions, and causes its temperature to increase significantly, often (in theory) exceeding 6000°C (10830°F).

Plasma technology has the following potential benefits over a typical coal gasification plant: (1) greater feedstock flexibility enabling coal, coal fines, mining waste, lignite, and other opportunity fuels (such as biomass and municipal solid waste) to be used as fuel without the need for pulverizing, (2) air blown and thus an oxygen plant is not required, (3) high conversion (>99%) of carbonaceous matter to synthesis gas, (4) absence of tar in the synthesis, (5) capable of producing high heating value synthesis gas suitable for use in a combustion turbine operation, (6) no char, ash or residual carbon, (7) only producing a glassy slag with beneficial value, (8) high thermal efficiency, and (9) low carbon dioxide emissions.

In the process, the gasifier is heated by a *plasma torch* system located near the bottom of the gasifier. The coal feedstock is charged into the vertical gasifier (refractory lined or water-cooled) at atmospheric pressure. A superheated blast of air, which may be enriched with oxygen, is provided to the bottom of the gasifier, at the stoichiometric amount required for gasification. The amount of air fed is such that the superficial velocity of the upward flowing gas is low, and that the pulverized feedstock can be fed directly into the reactor. Additional air and/or steam can be provided at different levels of the gasifier to assist with pyrolysis and gasification. The temperature of the synthesis gas leaving the top of the gasifier is maintained above 1000°C (1830°F). At this temperature, tar formation is eliminated.

4.3.5 Process Optimization

The output and quality of the gas produced is determined by the equilibrium established when the heat of oxidation (combustion) balances the heat of vaporization and volatilization plus the sensible heat (temperature rise) of the exhaust gases. The quality of the outlet gas (BTU/ft^3) is determined by the amount of volatile gases (such as hydrogen, carbon monoxide, water, carbon dioxide, and methane) in the gas stream.

In a gasifier, the coal particle is exposed to high temperatures generated from the partial oxidation of the carbon. As the particle is heated, any residual moisture (assuming that the coal has been pre-fired) is driven off and further heating of the particle begins to drive off the volatile gases. Discharge of these volatiles will generate a wide spectrum of hydrocarbons ranging from carbon monoxide and methane to long-chain hydrocarbons comprising tars, creosote, and heavy oil. At temperatures above 500°C (930°F) the conversion of the coal to char and ash and char is completed. In most of the early gasification processes, this was the desired byproduct but for gas generation the char provides the necessary energy to effect further heating and – typically, the char is contacted with air or oxygen and steam to generate the product gases.

Gasification of coal/char in a carbon dioxide atmosphere can be divided into two stages, the first stage due to pyrolysis (removal of moisture content and devolatilization) which is comparatively at lower temperature and char gasification by different oxygen/carbon dioxide mixtures at high temperature. In nitrogen and carbon dioxide environments from room temperature to 1000°C (1830°F), the mass loss rate of coal pyrolysis in nitrogen is lower than that of carbon dioxide and may be due to the difference in properties of the bulk gases. The gasification process of pulverized coal in oxygen/carbon dioxide environment is almost the same as compared with that in oxygen/nitrogen at the same oxygen concentration but this effect is a little bit delayed at high temperature. This may be due to the lower rate of diffusion of oxygen through carbon dioxide and the higher specific heat capacity of carbon dioxide. However, with the increase of oxygen concentration the mass loss rate of coal also increases and hence it shortens the burnout time of coal. The optimum value oxygen/

136 Synthesis Gas

carbon dioxide ratio for the reaction of oxygen with the functional group present in the coal sample was found to be about 8%.

The combination of pyrolysis and gasification process can be a unique and fruitful technique as it can save the prior use of gasifying medium and the production of fresh char simultaneously in one process. With the increase of heating rate, coal particles are faster heated in a short period of time and burnt in a higher temperature region, but the increase in heating rate has almost no substantial effect on the combustion mechanism of coal. The increase of heating rate causes a decrease in activation energy value (Irfan, 2009).

A final design option in terms of process efficiency involves the method for cooling the synthesis gas produced by the gasifier. Regardless of the type of gasifier, the exiting synthesis gas must be cooled down to approximately 100°C (212°F) in order to utilize conventional acid gas removal technology (Chapter 9). This can be accomplished either by passing the synthesis gas through a series of heat exchangers which recover the sensible heat for use in the steam cycle of the integrated gasification combined cycle (IGCC) system, or by directly contacting the synthesis gas with relatively cool water. This latter process (the quench process) results in some of the quench water being vaporized and mixed with the synthesis gas. The quenched synthesis gas is saturated with water and must pass through a series of condensing heat exchanges which remove the moisture from the synthesis gas (so it can be recycled to the quench zone). The quench system offers an advantage if carbon dioxide capture is desired. The saturated synthesis gas exiting a quench section has near the optimum gas ratio for feedstock into a water-gas shift catalyst which will convert the carbon monoxide to carbon dioxide. Non-quench designs that require carbon dioxide capture will have to add steam to the synthesis gas before it is sent to a water-gas shift reactor.

4.4 Product Quality

The original concept of the process was the production of a fuel gas which is still the operative principle in many gas works no matter what the feedstock. Briefly, a gasworks is an industrial plant for the production of flammable gas, including (but not used as such at the time) synthesis gas. Many of these plants have been made redundant in the developed world by the use of natural gas, though the gasometers may still be used for storage space. Early gasworks were built for factories in the Industrial Revolution from about 1805 as a light source and for industrial processes requiring gas, and for lighting in country houses from about 1845.

The products of the gasworks (which used coal as the feedstock to the gasifier) were varied insofar as the gas composition varied with the type of coal and the gasification system employed. Furthermore, the quality of gaseous product(s) had to be improved by removal of any pollutants such as particulate matter and sulfur compounds before further use, particularly when the intended use is a water gas shift or methanation (Cusumano *et al.*, 1978; Probstein and Hicks, 1990; Speight, 2013a, 2013b).

4.4.1 Low Btu Gas

Low Btu gas (low heat-content gas) is the product when the oxygen is not separated from the air and, as a result, the gas product invariably has a low heat-content (150 to 300 Btu/ft^3).

Several important chemical reactions (Table 4.2), and a host of side reactions, are involved in the manufacture of low heat-content gas under the high temperature conditions employed. Low Btu gas (low heat-content gas) contains several components (Table 4.3). In medium heat-content gas, the H_2/CO ratio varies from 2:3 to approximately 3:1 and the increased heating value correlates with higher methane and hydrogen contents as well as with lower carbon dioxide content.

The nitrogen content of low heat-content gas ranges from somewhat less than 33% v/v to slightly more than 50% v/v and cannot be removed by any reasonable means, which limits the applicability of the gas to chemical synthesis. Two other noncombustible components (water, H_2O, and carbon dioxide, CO_2) further lower the heating value of the gas; water can be removed by condensation and carbon dioxide by relatively straightforward chemical means.

Table 4.2 Coal gasification reactions.

$2 C + O_2 \rightarrow 2 CO$
$C + O_2 \rightarrow CO_2$
$C + CO_2 \rightarrow 2 CO$
$CO + H_2O \rightarrow CO_2 + H_2$ (shift reaction)
$C + H_2O \rightarrow CO + H_2$ (water gas reaction)
$C + 2 H_2 \rightarrow CH_4$
$2 H_2 + O_2 \rightarrow 2 H_2O$
$CO + 2 H_2 \rightarrow CH_3OH$
$CO + 3 H_2 \rightarrow CH_4 + H_2O$ (methanation reaction)
$CO_2 + 4 H_2 \rightarrow CH_4 + 2 H_2O$
$C + 2 H_2O \rightarrow 2 H_2 + CO_2$
$2 C + H_2 \rightarrow C_2H_2$
$CH_4 + 2 H_2O \rightarrow CO_2 + 4 H_2$

Table 4.3 Coal gasification products.

Product	Characteristics
Low-Btu gas (150–300 Btu/scf)	Around 50% nitrogen, with smaller quantities of combustible H_2 and CO, CO_2 and trace gases, such as methane
Medium-Btu gas (300–550 Btu/scf)	Predominantly CO and H_2, with some incombustible gases and sometimes methane
High-Btu gas (980–1080 Btu/scf)	Almost pure methane

The two major combustible components are hydrogen and carbon monoxide; the hydrogen/carbon monoxide ratio varies from approximately 2:3 to about 3:2. Methane may also make an appreciable contribution to the heat content of the gas. Of the minor components hydrogen sulfide is the most significant and the amount produced is, in fact, proportional to the sulfur content of the feed coal. Any hydrogen sulfide present must be removed by one, or more, of several available on-stream commercial processes (Speight, 2013a, 2014, 2019).

4.4.2 Medium Btu Gas

Medium Btu gas (medium heat-content gas) has a heating value in the range 300 to 550 Btu/ft^3 and the composition is much like that of low heat-content gas, except that there is virtually no nitrogen. The primary combustible gases in medium heat-content gas are hydrogen and carbon monoxide (Kasem, 1979).

Medium heat-content gas is considerably more versatile than low heat-content gas; like low heat-content gas, medium heat content gas may be used directly as a fuel to raise steam, or used through a combined power cycle to drive a gas turbine, with the hot exhaust gases employed to raise steam, but medium heat content gas is especially amenable to synthesize methane (by methanation), higher hydrocarbons (by Fischer-Tropsch synthesis), methanol, and a variety of synthetic chemicals (David and Occelli, 2010; Chadeesingh, 2011).

The reactions used to produce medium heat-content gas are the same as those employed for low heat-content gas synthesis, the major difference being the application of a nitrogen barrier (such as the use of pure oxygen) to keep diluent nitrogen out of the system.

4.4.3 High Btu Gas

High Btu Gas (High heat-content gas) is essentially pure methane and often referred to as synthetic natural gas or substitute natural gas (SNG) (Kasem, 1979; Speight, 1990, 2013a). However, to qualify as substitute natural gas, a product must contain at least 95% methane; the energy content of synthetic natural gas is 980 to 1080 Btu/ft^3. The commonly accepted approach to the synthesis of high heat-content gas is the catalytic reaction of hydrogen and carbon monoxide.

$$3H_2 + CO \rightarrow CH_4 + H_2O$$

The hydrogen is usually present in slight excess to ensure that the toxic carbon monoxide is reacted; this small quantity of hydrogen will lower the heat content to a small degree.

The carbon monoxide/hydrogen reaction is somewhat inefficient as a means of producing methane because the reaction liberates large quantities of heat. In addition, the methanation catalyst is troublesome and prone to poisoning by sulfur compounds and the decomposition of metals can destroy the catalyst. Thus, hydrogasification may be employed to minimize the need for methanation.

$$C_{coal} + 2H_2 \rightarrow CH_4$$

The product of hydrogasification is *not* pure methane and additional methanation is required after hydrogen sulfide and other impurities are removed.

4.4.4 Methane

Several exothermic reactions may occur simultaneously within a methanation unit (Seglin, 1975). A variety of metals have been used as catalysts for the methanation reaction; the most common, and to some extent the most effective methanation catalysts, appear to be nickel and ruthenium, with nickel being the most widely used (Seglin, 1975; Cusumano et al., 1978; Tucci and Thompson, 1979; Watson, G.H., 1980). The synthesis gas must be desulfurized before the methanation step since sulfur compounds will rapidly deactivate (poison) the catalysts (Cusumano et al., 1978). A problem may arise when the concentration of carbon monoxide is excessive in the stream to be methanated since large amounts of heat must be removed from the system to prevent high temperatures and deactivation of the catalyst by sintering as well as the deposition of carbon (Cusumano et al., 1978) – to eliminate carbon deposition, process temperatures should be maintained below 400°C (750°F).

4.4.5 Hydrogen

Hydrogen is also by coal gasification (Johnson et al., 2007). Although several gasifier types exist (Chapter 2), entrained-flow gasifiers are considered most appropriate for producing both hydrogen and electricity from coal since they operate at temperatures high enough (approximately 1500°C, 2730°F) to enable high carbon conversion and prevent downstream fouling from tars and other residuals.

In the process, the coal undergoes three processes in its conversation to synthesis gas – the first two processes, pyrolysis and combustion, occur very rapidly. In pyrolysis, char is produced as the coal heats up and volatiles are released. In the combustion process, the volatile products and some of the char reacts with oxygen to produce various products (primarily carbon dioxide and carbon monoxide) and the heat required for subsequent gasification reactions. Finally, in the gasification process, the coal char reacts with steam to produce hydrogen (H_2) and carbon monoxide (CO).

$$2C_{coal} + O_2 \rightarrow 2CO + H_2O$$

$$C_{coal} + H_2O \rightarrow H_2 + CO$$

$$CO + H_2O \rightarrow H_2 + CO_2$$

The resulting synthesis gas is approximately 63% v/v carbon monoxide, 34% v/v hydrogen, and 3% v/v carbon dioxide. At the gasifier temperature, the ash and other coal mineral matter liquefies and exits at the bottom of the gasifier as slag, a sand-like inert material that can be sold as a co-product to other industries (such as the road building industry). The synthesis gas exits the gasifier at pressure and high temperature and must be cooled prior to the synthesis gas cleaning stage.

Although processes that use the high temperature to raise high-pressure steam are more efficient for electricity production (Speight, 2013b), full-quench cooling, by which the synthesis gas is cooled by the direct injection of water, is more appropriate for hydrogen production and provides the necessary steam to facilitate the catalytic water gas shift reaction:

$$CO + H_2O \rightarrow CO_2 + H_2$$

Unlike pulverized coal combustion plants in which expensive emissions control technologies are required to scrub contaminants from large volumes of flue gas, smaller and less expensive emissions control technologies are appropriate for coal gasification plants since the clean-up occurs in the synthesis gas. The synthesis gas is at high pressure and contains contaminants at high partial pressures, which facilitates gas cleaning.

As with other processes, the characteristics of the coal feedstock (e.g., heating value and ash, moisture, and sulfur content) have a substantial impact on plant efficiency and emissions. As a result, the cost of producing hydrogen from coal gasification can vary substantially depending on the proximity to appropriate coal types.

4.4.6 Other Gases

There is a series of products that are called by older (even archaic) names that should also be mentioned here as clarification: (1) producer gas, (2) water gas, (3) town gas, and (4) synthetic natural gas.

Producer gas is a low Btu gas obtained from a coal gasifier (fixed-bed) upon introduction of air instead of oxygen into the fuel bed. The composition of the producer gas is approximately 28% v/v carbon monoxide, 55% v/v nitrogen, 12% v/v hydrogen, and 5% v/v methane with some carbon dioxide.

Water gas is a medium Btu has which is produced by the introduction of steam into the hot fuel bed of the gasifier. The composition of the gas is approximately 50% v/v hydrogen and 40% v/v carbon monoxide with small amounts of nitrogen and carbon dioxide.

Town gas is a medium Btu has that is produced in the coke ovens and has the approximate composition: 55% v/v hydrogen, 27% v/v methane, 6% v/v carbon monoxide, 10% v/v nitrogen, and 2% v/v carbon dioxide. Carbon monoxide can be removed from the gas by catalytic treatment with steam to produce carbon dioxide and hydrogen.

Synthetic natural gas (SNG) is methane obtained from the reaction of carbon monoxide or carbon with hydrogen. Depending on the methane concentration, the heating value can be in the range of high-Btu gases.

4.5 Chemicals Production

The coal carbonization industry was established initially as a means of producing coke and but a secondary industry emerged (in fact, became necessary) to deal with the secondary or byproducts (namely, gas, ammonia liquor, crude benzole, and tar) produced during carbonization (Table 4.4) (Speight, 2013a).

4.5.1 Coal Tar Chemicals

Coal tar is black or dark brown colored liquid or a high-viscosity semi-solid which is one of the byproducts formed when coal is carbonized. Coal tars are complex and variable mixtures

Table 4.4 Products (% w/w) from coal carbonization.

Product	Low temperature	High temperature
Gas	5.0	20.0
Liquor	15.0	2.0
Light oils	2.0	0.5
Tar	10.0	4.0
Coke	70.0	75.0

of polycyclic aromatic hydrocarbons, phenols, and heterocyclic compounds. Because of its flammable composition coal tar is often used for fire boilers, in order to create heat. Before any heavy oil flows easily they must be heated.

By comparison, coal tar creosote is a distillation product of coal tar and consists of aromatic hydrocarbons, anthracene, naphthalene, and phenanthrene derivatives. At least 75% of the coal tar creosote mixture is polycyclic aromatic hydrocarbons (PAHs). Unlike the coal tars and coal tar creosotes, coal tar pitch is a residue produced during the distillation of coal tar. The pitch is a shiny, dark brown to black residue which contains polycyclic aromatic hydrocarbons and their methyl and poly-methyl derivatives, as well as heteronuclear aromatic compounds.

Primary distillation of crude tar produces pitch (residue) and several distillate fractions, the amounts and boiling ranges of which are influenced by the nature of the crude tar (which depends upon the coal feedstock) and the processing conditions. For example, in the case of the tar from continuous vertical retorts, the objective is to concentrate the tar acids (phenol derivatives, cresol derivatives, and xylenol derivatives) into carbolic oil fractions. On the other hand, the objective with coke oven tar is to concentrate the naphthalene and anthracene components into naphthalene oil and anthracene oil, respectively.

The first step in refining benzole is steam distillation which is employed to remove compounds boiling below benzene. To obtain pure products, the benzole can be distilled to yield a fraction containing benzene, toluene, and xylene(s) (BTX). Benzene is used in the manufacture of numerous products including nylon, gammexane, polystyrene, phenol, nitrobenzene, and aniline. On the other hand, toluene is a starting material in the preparation of saccharin, trinitrotoluene (TNT), and polyurethane foams. The xylenes present in the light oil are not always separated into the individual pure isomers since xylene mixtures can be marketed as specialty solvents. Higher-boiling fractions of the distillate from the tar contain pyridine bases, naphtha, and coumarone resins. Other tar bases occur in the higher-boiling range and these are mainly quinoline, iso-quinoline, and quinaldine.

Pyridine has long been used as a solvent, in the production of rubber chemicals, textile water-repellant agents and in the synthesis of drugs. The derivatives 2-benzylpyridine and 2- aminopyridine, are used in the preparation of antihistamines. Another market for pyridine is in the manufacture of the non-persistent herbicides *diquat* and *paraquat*. Alpha-picoline (2-picoline; 2-methylpryridine) is used for the production of 2-vinylpyridine which, when copolymerized with butadiene and styrene, produces a used as a latex adhesive which is used in the manufacture of car tires. Other uses are in the preparation of 2-beta-methoxyethylpyridine (known as Promintic, an anthelmintic for cattle) and in the synthesis of a 2-picoline quaternary compound (Amprolium) which is used against coccidiosis in young poultry. Beta-picoline (3-picoline; 3-methylpryridine) can be oxidized to nicotinic acid, which with the amide form (nicotinamide) belongs to the vitamin B complex; both products are widely used to fortify human and animal diets. Gama-picoline (4-picoline; 4-methylpyridine) is an intermediate in the manufacture of isonicotinic acid hydrazide (Isoniazide) which is a tuberculostatic drug. 2,6-Lutidine (2,6-dimethylpyridine) can be converted to dipicolinic acid which is used as a stabilizer for hydrogen peroxide and peracetic acid. Solvent naphtha and high-boiling naphtha are the mixtures obtained when the 150 to 200°C (300 to 390°F) fraction, after removal of tar acids and tar bases, is fractionated. These naphtha fractions are used as solvents.

The tar-acid free and tar-base free coke oven naphtha can be fractionated to give a narrow-boiling fraction (170 to 185°C; 340 to 365°F) containing coumarone and indene. This is treated with strong sulfuric acid to remove unsaturated components and is then washed and redistilled. The concentrate is heated with a catalyst (such as a boron fluoride/phenol complex) to polymerize the indene and part of the coumarone. Unreacted oil is distilled off and the resins obtained vary from pale amber to dark brown in color. They are used in the production of flooring tiles and in paints and polishes.

Naphthalene and several tar acids are the important products extracted from volatile oils from coal tar. It is necessary to first extract the phenolic compounds from the oils and then to process the phenol-depleted oils for naphthalene recovery.

Tar acids are produced by extraction of the oils with aqueous caustic soda at a temperature sufficient to prevent naphthalene from crystallizing. The phenols react with the sodium hydroxide to give the corresponding sodium salts as an aqueous extract known variously as crude sodium phenate, sodium phenolate, sodium carbolate, or sodium cresylate. The extract is separated from the phenol-free oils which are then taken for naphthalene recovery.

Naphthalene is probably the most abundant component in high-temperature coal tars. The primary fractionation of the crude tar concentrates the naphthalene into oils which, in the case of coke-oven tar, contain the majority (75 to 90% w/w) of the total naphthalene. After separation, naphthalene can be oxidized to produce phthalic anhydride which is used in the manufacture of alkyd and glyptal resins and plasticizers for polyvinyl chloride and other plastics. The main chemical extracted on the commercial scale from the higher-boiling oils (b.p. 250°C, 480°F) is crude anthracene. The majority of the crude anthracene is used in the manufacture of dyes after purification and oxidation to anthraquinone.

Creosote is the residual distillate oils obtained when the valuable components, such as naphthalene, anthracene, tar acids, and tar bases have been removed from the corresponding fractions. Creosote is a brownish-black/yellowish-dark green oily liquid with a characteristic sharp odor, obtained by the fractional distillation of crude coal tars. The approximate distillation range is 200 to 400°C (390 to 750°F) (ITC, 1990). The chemical composition of creosotes is influenced by the origin of the coal and also by the nature of the distilling process; as a result, the creosote components are rarely consistent in their type and concentration.

As a corollary to this section where the emphasis has been on the production of bulk chemicals from coal, a tendency-to-be-forgotten item must also be included. That is the mineral ash from coal processes. Coal minerals are a very important part of the coal matrix and offer the potential for the recovery of valuable inorganic materials (Speight, 2013a).

However, there is another aspect of the mineral content of coal that must be addressed, and that relates to the use of the ash as materials for roadbed stabilization, landfill cover, cementing (due to the content of pozzolanic materials), and wall construction.

4.5.2 Fischer-Tropsch Chemicals

Fischer-Tropsch chemicals are those chemicals produced by conversion of the synthesis gas mixture (carbon monoxide, CO, and hydrogen, H2) to higher molecular weight liquid fuels and other chemicals (Chapter 10) (Penner, 1987; Chadeesingh, 2011; Speight, 2013a). In principle, synthesis gas can be produced from any hydrocarbon feedstock. These include (i) natural gas, (ii) naphtha, (iii) crude oil resid, (iv) crude oil coke, (v) coal, and (vi) biomass.

The synthesis of hydrocarbons from carbon monoxide and hydrogen (synthesis gas) (the Fischer-Tropsch synthesis) is a procedure for the indirect liquefaction of coal (Anderson, 1984; Dry and Erasmus, 1987). This process is the only coal liquefaction scheme currently in use on a relatively large commercial scale; South Africa is currently using the Fischer-Tropsch process on a commercial scale in its SASOL complex (Singh, 1981), although Germany produced roughly 156 million barrels of synthetic petroleum annually using the Fischer-Tropsch process during World War II.

In the Fischer-Tropsch process, coal is converted to gaseous products at temperatures in excess of 800°C (1470°F), and at moderate pressures, to produce synthesis gas.

$$C + H_2O \rightarrow CO + H_2$$

In practice, the Fischer-Tropsch reaction is generally carried out at temperatures in the range 200 to 350°C (390 to 660°F) and at pressures of 75 to 4000 psi; the hydrogen/carbon monoxide ratio is usually at ca. 2.2:1 or 2.5:1. Since up to three volumes of hydrogen may be required to achieve the next stage of the liquids production, the synthesis gas must then be converted by means of the water-gas shift reaction to the desired level of hydrogen after which the gaseous mix is purified (acid gas removal, etc.) and converted to a wide variety of hydrocarbons.

$$CO + H_2O \rightarrow CO_2 + H_2$$

$$CO + (2n+1)H_2 \rightarrow C_nH_{2n+2} + H_2O$$

These reactions result primarily in low- and medium-boiling aliphatic compounds; present commercial objectives are focused on the conditions that result in the production of n-hydrocarbons as well as olefins and oxygenated materials (Speight, 2013a).

4.5.2.1 Fischer-Tropsch Catalysts

Catalysts play a major role in synthesis gas conversion reactions. For hydrocarbon and synthesis of higher molecular weight alcohols, dissociation of carbon monoxide is a necessary reaction condition. For methanol synthesis, the carbon monoxide molecule remains intact. Hydrogen has two roles in catalytic synthesis gas synthesis reactions. In addition to being a reactant needed for hydrogenation of carbon monoxide, it is commonly used to reduce the metalized synthesis catalysts and activate the metal surface. A variety of catalysts can be used for the Fischer–Tropsch process, but the most common are the transition metals cobalt, iron, and ruthenium. Nickel can also be used, but tends to favor methane formation (*methanation*).

Cobalt-based catalysts are highly active, although iron may be more suitable for low-hydrogen-content synthesis gases such as those derived from coal due to its promotion of the water-gas-shift reaction. In addition to the active metal the catalysts typically contain a number of promoters, such as potassium and copper.

Group 1 alkali metals (including potassium) are poisons for cobalt catalysts but are promoters for iron catalysts. Catalysts are supported on high-surface-area binders/supports such as silica, alumina, and zeolites (Spath and Dayton, 2003). Cobalt catalysts are more active for Fischer-Tropsch synthesis when the feedstock is natural gas. Natural gas has a high hydrogen to carbon ratio, so the water-gas-shift is not needed for cobalt catalysts. Iron catalysts are preferred for lower-quality feedstocks such as coal or biomass. Unlike the other metals used for this process (Co, Ni, Ru), which remain in the metallic state during synthesis, iron catalysts tend to form a number of phases, including various oxides and carbides during the reaction. Control of these phase transformations can be important in maintaining catalytic activity and preventing breakdown of the catalyst particles.

Fischer-Tropsch catalysts are sensitive to poisoning by sulfur-containing compounds. The sensitivity of the catalyst to sulfur is greater for cobalt-based catalysts than for their iron counterparts. Promoters also have an important influence on activity. Alkali metal oxides and copper are common promoters, but the formulation depends on the primary metal, iron vs. cobalt (Spath and Dayton, 2003). Alkali oxides on cobalt catalysts generally cause activity to drop severely even with very low alkali loadings. C_{5+} and carbon dioxide selectivity increase while methane and C_2 to C_4 selectivity decrease. In addition, the olefin to paraffin ratio increases.

4.5.2.2 Product Distribution

The product distribution of hydrocarbons formed during the Fischer–Tropsch process follows an Anderson-Schulz-Flory distribution (Spath and Dayton, 2003):

$$W_n/n = (1-\alpha)^2 \alpha^{n-1}$$

W_n is the weight fraction of hydrocarbon molecules containing n carbon atoms, α is the chain growth probability or the probability that a molecule will continue reacting to form a longer chain. In general, α is largely determined by the catalyst and the specific process conditions. According to the above equation, methane will always be the largest single product; however by increasing α close to one, the total amount of methane formed can be minimized compared to the sum of all of the various long-chained products. Therefore, for production of liquid transportation fuels it may be necessary to crack the Fischer-Tropsch longer chain products.

It has been proposed that zeolites or other catalyst substrates with fixed sized pores that can restrict the formation of hydrocarbons longer than some characteristic size (usually n<10). This would tend to drive the reaction to minimum methane formation without producing the waxy products.

4.6 Advantages and Limitations

The production of gas from coal has been a vastly expanding area of coal technology for power generation and is, in reality, another form of coal-fired power generation in which coal is used as the feedstock to produce the hot gases to drive the turbines. As with combustion processes, coal characteristics such as rank, mineral matter, particle size, and reaction conditions are all recognized as having a bearing on the outcome of the gasification process – not only in terms of gas yields but also on gas properties (Massey, 1974; Hanson *et al.*, 2002).

Coal gasification technology offers the poly-generation: co-production of electric power, liquid fuels, chemicals, hydrogen and from the synthesis gas generated from gasification. Chemical gasification plants based on entrained-flow and more especially on moving-bed technologies are at present operating all over the world with the biggest plants located in South Africa (Sasol) (Speight, 2008, 2013a). In addition, gasification is an important step of the indirect liquefaction of coal for production of liquid fuels (Speight, 2008, 2013a). Another advantage is the ability of the gasifier technology to accommodate feedstock other than coal either separately or as a blend with coal (Speight, 2013a, 2013b).

One of the major environmental advantages of coal gasification is the opportunity to remove impurities such as sulfur and mercury and soot before burning the fuel, using readily available chemical engineering processes. In addition the ash produced is in a vitreous or glasslike state which can be recycled as concrete aggregate, unlike pulverized coal combustion plants which generate ash that must be landfilled, potentially contaminating groundwater.

The increased efficiency of the "combined cycle" for electrical power generation results in a 50% decrease in emissions of carbon dioxide compared to conventional coal plants. As the technology required to develop economical methods of carbon sequestration, the removal of carbon dioxide from combustion byproducts to prevent its release to the atmosphere, coal gasification units could be modified to further reduce their climate change impact because a large part of the carbon dioxide generated can be separated from the synthesis gas before combustion.

However, coal gasification, while providing a route to deriving energy from coal which facilitates the removal of ash and sulfur, has two major disadvantages: (1) the process consumes large quantities of water, especially significant in arid western states where some of the largest coal reserves are located, and (2) the process is less efficient than direct combustion. Some reactors provide limited optimization of either process efficiency or water consumption. Performance optimization is both application- and site-specific, and the choice of a coal gasification system depends to a large extent on the requirements and locations of the end-use markets.

References

Anderson, L.L., and Tillman, D.A. 1979. *Synthetic Fuels from Coal: Overview and Assessment*. John Wiley & Sons Inc., New York.

Anderson, R.B. 1984. In *Catalysis on the Energy Scene*. S. Kaliaguine and A. Mahay (Editors). Elsevier Science Publishers, Amsterdam, Netherlands. Page 457.

Anthony, D.B., and Howard, J.B. 1976. Coal Devolatilization and Hydrogasification. *AIChE Journal*. 22: 625.

Argonne. 1990. Environmental Consequences of, and Control Processes for, Energy Technologies. Argonne National Laboratory. *Pollution Technology Review No. 181*. Noyes Data Corp., Park Ridge, New Jersey. Chapter 5.

Baker, R.T.K., and Rodriguez, N.M. 1990. Coal. In: *Fuel Science and Technology Handbook*. Marcel Dekker Inc, New York. Chapter 22.

Bodle, W.W., and Huebler, J. 1981. In: *Coal Handbook*. R.A. Meyers (Editor). Marcel Dekker Inc., New York. Chapter 10.

Breault, R.W. 2010. Gasification Processes Old and New: A Basic Review of the Major Technologies. *Energies*, 3: 216-240.

Calemma, V. and Radović, L.R. 1991. On the Gasification Reactivity of Italian Sulcis Coal. *Fuel*, 70: 1027.

Cavagnaro, D.M. 1980. *Coal Gasification Technology*. National Technical Information Service, Springfield, Virginia.

Chadeesingh, R. 2011. The Fischer-Tropsch Process. In *The Biofuels Handbook*. J.G. Speight (Editor). The Royal Society of Chemistry, London, United Kingdom. Part 3, Chapter 5, Page 476-517.

Cusumano, J.A., Dalla Betta, R.A., and Levy, R.B. 1978. *Catalysis in Coal Conversion*. Academic Press Inc., New York.

Dry, M.E., and Erasmus, H.B. De W. 1987. Update of the Sasol Synfuels Process. *Ann. Rev. Energy*. 12: 21.

Elton, A. 1958. In *A History of Technology*. C. Singer, E.J. Holmyard, A.R. Hall, and T.I. Williams (Editors). Clarendon Press, Oxford, England. Volume IV. Chapter 9.

Fryer, J.F., and Speight, J.G. 1976. *Coal Gasification: Selected Abstract and Titles*. Information Series No. 74. Alberta Research Council, Edmonton, Canada.

Garcia, X., and Radović, L.R. 1986. Gasification Reactivity of Chilean Coals. *Fuel* 65: 292.

Hanson, S., Patrick, J.W., and Walker, A. 2002. The Effect of Coal Particle Size on Pyrolysis and Steam Gasification. *Fuel*, 81: 531-537.

Irfan, M.F. 2009. Research Report: Pulverized Coal Pyrolysis & Gasification in $N_2/O_2/CO_2$ Mixtures by Thermo-gravimetric Analysis. *Novel Carbon Resource Sciences Newsletter*, Kyushu University, Fukuoka, Japan. Volume 2: 27-33.

Johnson, J.L., 1979. *Kinetics of Coal Gasification*. John Wiley & and Sons Inc., New York.

Johnson, N., Yang, C., and Ogden, J. 2007. Hydrogen Production via Coal Gasification. Advanced Energy Pathways Project, Task 4.1 Technology Assessments of Vehicle Fuels and Technologies, Public Interest Energy Research (PIER) Program, California Energy Commission. May.

Kalinenko, R.A., Kuznetsov, A.P., Levitsky, A.A., Messerle, V.E., Mirokhin, Yu.A., Polak, L.S., Sakipov, Z.B., Ustimenko, A.B. 1993. Pulverized Coal Plasma Gasification. *Plasma Chemistry and Plasma Processing*, 3(1): 141-167.

Kasem, A. 1979. *Three Clean Fuels from Coal: Technology and Economics*. Marcel Dekker Inc., New York.

Kristiansen, A. 1996. *Understanding Coal Gasification*. IEA Coal Research report IEACR/86. International Energy Agency, London, United Kingdom.

Lahaye, J., and Ehrburger, P. (Editors). 1991. *Fundamental Issues in Control of Carbon Gasification Reactivity*. Kluwer Academic Publishers, Dordrecht, Netherlands.

Mahajan, O.P., and Walker, P.L. Jr. 1978. In *Analytical Methods for Coal and Coal Products*. C. Karr Jr. (Editor). Academic 6 Press Inc., New York. Volume II. Chapter 32.

Massey, L.G. (Editor). 1974. Coal Gasification. Advances in Chemistry Series No. 131. American Chemical Society, Washington, D.C. Massey, L.G. 1979. In *Coal Conversion Technology*. C.Y.

Wen and E.S. Lee (Editors). Addison-Wesley Publishers Inc., Reading, Massachusetts. Page 313.

Matsukata, M., Kikuchi, E., and Morita, Y. 1992. A New Classification of Alkali and Alkaline Earth Catalysts for Gasification of Carbon. *Fuel.* 71: 819-823

McKee, D.W. 1981. *The Catalyzed Gasification Reactions of Carbon. The Chemistry and Physics of Carbon.* P.L. Walker, Jr. and P.A. Thrower (Editors). Marcel Dekker Inc., New York). Volume 16. Page 1.

Messerle, V.E., and Ustimenko, A.B. 2007. Solid Fuel Plasma Gasification. Advanced Combustion and Aerothermal Technologies. *NATO Science for Peace and Security Series C. Environmental Security.* Page 141-1256.

Mills, G.A. 1969. Conversion of Coal to Gasoline. *Ind. Eng. Chem.,* 61(7): 6-17.

Mokhatab, S., Poe, W.A., and Speight, J.G. 2006. *Handbook of Natural Gas Transmission and Processing.* Elsevier, Amsterdam, Netherlands.

Nef, J.U. 1957. In: *A History of Technology.* C. Singer, E.J. Holmyard, A.R. Hall, and T.I. Williams (Editors). Clarendon Press, Oxford, England. Volume III. Chapter 3.

Penner, S.S. 1987. *Coal Gasification.* Pergamon Press Limited, New York.

Probstein, R.F., and Hicks, R.E. 1990. *Synthetic Fuels.* pH Press, Cambridge, Massachusetts. Chapter 4.

Radović, L.R., Walker, P.L., Jr., and Jenkins, R.G. 1983. Importance of Carbon Active Sites in the Gasification of Coal Chars, *Fuel* 62: 849.

Radović, L.R., and Walker, P.L., Jr. 1984. Reactivities of Chars Obtained as Residues in Selected Coal Conversion Processes, *Fuel Processing Technology*, 8: 149.

Seglin, L. (Editor). 1975. *Methanation of Synthesis Gas.* Advances in Chemistry Series No. 146. American Chemical Society, Washington, DC.

Shinnar, R., Fortuna, G. and Shapira, D. 1982. Thermodynamic and Kinetic Constraints of Catalytic Synthetic Natural Gas Processes. *Ind. Eng. Chem. Process Design and Development*, 21: 728-750.

Spath, P.L., and Dayton, D.C. 2003. *Preliminary Screening – Technical and Economic Assessment of Synthesis Gas to Fuels and Chemicals with Emphasis on the Potential for Biomass-Derived Syngas.* Report No. NREL/TP-510-3492. Contract No. DE-AC36-99-GO10337. National Renewable Energy Laboratory, Golden, Colorado.

Speight, J.G. 1990. In *Fuel Science and Technology Handbook.* J.G. Speight (Editor). Marcel Dekker Inc., New York. Chapter 33.

Speight, J.G. 2002. *Chemical Process and Design Handbook.* McGraw-Hill Inc., New York.

Speight, J.G. 2008. *Synthetic Fuels Handbook: Properties, Processes, and Performance.* McGraw-Hill, New York.

Speight, J.G. 2013a. *The Chemistry and Technology of Coal.* 3rd Edition. CRC Press, Taylor and Francis Group, Boca Raton, Florida.

Speight. J.G. 2013b. *Coal-Fired Power Generation Handbook.* Scrivener Publishing, Salem, Massachusetts, 2013.

Speight, J.G. 2014. *The Chemistry and Technology of Petroleum* 5th Edition. CRC Press, Taylor & Francis Group, Boca Raton, Florida.

Speight, J.G. 2019. *Natural Gas: A Basic Handbook* 2nd Edition. Gulf Publishing Company, Elsevier, Cambridge, Massachusetts.

Taylor, F.S., and Singer, C. 1957. In: *A History of Technology.* C. Singer, E.J. Holmyard, A.R. Hall, and T.I. Williams (Editors). Clarendon Press, Oxford, England. Volume II. Chapter 10.

Tucci, E.R., and Thompson, W.J. 1979. Monolith Catalyst Favored for Methylation. *Hydrocarbon Processing.* 58(2): 123-126.

Watson, G.H. 1980. *Methanation Catalysts.* Report ICTIS/TR09 International Energy Agency, London, United Kingdom.

5
Gasification of Heavy Feedstocks

5.1 Introduction

Gasification of heavy (viscous) feedstocks (*hydrocarbonaceous materials, residua, process residues, process bottoms, high-viscosity waste streams*), which are non-volatile materials that are not truly hydrocarbon derivatives insofar as they contain elements other than carbon and hydrogen) involves, like all gasification processes, complete thermal decomposition of the feedstock into gaseous products (Wolff and Vliegenthart, 2011; Speight, 2014). The term *heavy hydrocarbon derivatives* (or high molecular weight hydrocarbon derivatives) is often applied to residua but is, in fact, an incorrect term insofar as the residua are not true hydrocarbon derivatives – the so-called *hydrocarbon derivatives* in residua contain elements other than carbon and hydrogen (Table 5.1) and are produced in the distillation section of the refinery (Figure 5.1). In addition, other refinery products, such as visbreaker bottoms (the non-volatile products from a visbreaking unit) and coke (Table 5.2), may also be used for the production of synthesis gas.

Typically, like all gasification processes, the process is carried out at high temperature (>1000°C, >1830°F) producing synthesis gas (*syngas*), some carbon black, and ash as major products; the amount of ash depends upon the amount of mineral matter in the feedstock. Integrated gasification combined cycle (IGCC) is an alternative process for residua conversion and is a known and used technology within the refining industry for (i) hydrogen production, (ii) fuel gas production, and (iii) power generation which, when coupled with efficient gas cleaning methods has minimum effect on the environment (low SOx and NOx) (Wolff, 2007; Goldhammer *et al.*, 2008; Speight, 2013c, 2013d).

The ability of the gasification process to handle heavy crude oil, extra heavy oil, and tar sand bitumen or any refinery bottom streams enhances the economic potential of most refineries and oil fields. Upgrading heavy crude oil – either in the oil field at the source or residua in the refinery – is (and will continue to be) an increasingly prevalent means of extracting maximum value from each barrel of oil produced (Speight, 2011a, 2014). Upgrading can convert marginal heavy crude oil into low-boiling, higher value crude, and can convert high-boiling, sour refinery bottoms into valuable transportation fuels. On the other hand, most upgrading techniques leave behind an even heavier residue and the costs deposition of such a byproduct may approach the value of the production of liquid fuels and other saleable products. In short, the gasification of residua, crude oil coke, or other high-boiling feedstocks to generate synthesis gas produces a clean fuel for firing in a gas turbine. Gasification is (i) a well-established technology, (ii) has broad flexibility of feedstocks and operation, and (iii) is the most environmentally friendly route for handling these feedstocks for power production.

Within the refinery, residuum coking and solvent deasphalting have been used for several decades to upgrade bottoms streams to intermediate products that may be processed to

Table 5.1 Examples of the properties of various resids.

Crude oil resid	API gravity	Sulfur % w/w	Nitrogen % w/w	Nickel ppm	Vanadium ppm	Carbon residue % w/w*
Arabian Light >650°F	17.7	3.0	0.2	10.0	26.0	7.5
Arabian Light, >1050°F	8.5	4.4	0.5	24.0	66.0	14.2
Arabian Heavy >650°F	11.9	4.4	0.3	27.0	103.0	14.0
Arabian Heavy >1050°F	7.3	5.1	0.3	40.0	174.0	19.0
Alaska, North Slope >650°F	15.2	1.6	0.4	18.0	30.0	8.5
Alaska, North Slope >1050°F	8.2	2.2	0.6	47.0	82.0	18.0
Lloydminster >650°F	10.3	4.1	0.3	65.0	141.0	12.1
Lloydminster >1050F	8.5	4.4	0.6	115.0	252.0	21.4
Kuwait >650°F	13.9	4.4	0.3	14.0	50.0	12.2
Kuwait >1050°F	5.5	5.5	0.4	32.0	102.0	23.1
Tia Juana >650°F	17.3	1.8	0.3	25.0	185.0	9.3
Tia Juana >1050°F	7.1	2.6	0.6	64.0	450.0	21.6

*Conradson

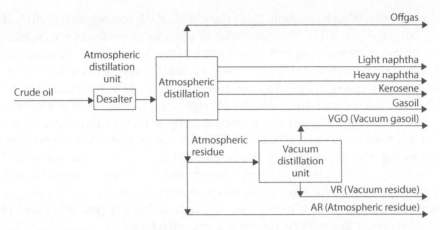

Figure 5.1 The distillation section of a refinery.

Table 5.2 Examples of other refinery feedstocks that are available for on-site gasification.

Ultimate Analysis	Visbreaker bottoms	Deasphalter bottoms	C5 Asphaltenes*	Delayed coke
Carbon, % w/w	83.1	85.9	80.47	88.6
Hydrogen, % w/w	10.4	9.5	8.45	2.8
Nitrogen, % w/w	0.6	1.4	1.25	1.1
Sulfur, % w/w*	2.4	2.4	1.88	7.3
Oxygen, % w/w	0.5	0.5	7.95	0.0
Ash, % w/w	<0.5	<0.5	<0.5	<1.0
Specific Gravity	1.008	1.036		0.863
API Gravity	8.9	5.1		

*The use of heptane as the precipitant yields an asphaltene fraction that is different from the pentane-insoluble material as exemplified by differences in the elemental analysis. For example, the H/C ratio of the heptane-asphaltene fraction is lower than the H/C ratio of the pentane asphaltene fraction, indicating a higher degree of aromaticity in the heptane asphaltene fraction. The N/C, O/C, and S/C atomic ratios are usually higher in the heptane asphaltene fraction, indicating higher proportions of the heteroelements in this material. Nevertheless, either fraction is suitable for use as a feedstock for a gasification process.

produce transportation fuels (Parkash, 2003; Gary et al., 2007; Speight, 2011a, 2014; Hsu and Robinson, 2017; Speight, 2017). The installation of a gasifier in a refinery is a realistic option for the conversion of high-boiling feedstocks leading to the production of added-value. In fact, the flexicoking process used a gasifier as an integral part of the system to convert excess coke to fuel gas (Gray and Tomlinson, 2000; Parkash, 2003; Gary et al., 2007; Sutikno and Turini, 2012; Speight, 2011a, 2014; Hsu and Robinson, 2017; Speight, 2017). Thus, by integrating the gasifier as a fully functional process option with gasification, important synergies may be realized and include: (i) increased crude and fuel flexibility, (ii) enhanced profitability through reduced capital and operating cost, (iii) lower environmental emissions, and (iv) increased reliability and efficiency of utilities. Indeed, the integration between bottoms processing units and gasification can serve as a springboard for other economically enhancing integration. The integration of gasification with existing or new hydroprocessing unit, and power generation unit, presents some unique synergies that will enhance the efficiency of refinery.

The production of high-quality fuels will result in a higher demand for related hydrogen and conversion technologies. Furthermore, the trend towards low-sulfur fuels and changes in the product mix of refineries will affect technology choice and needs. For example, the current desulfurization and conversion technologies use relatively large amounts of hydrogen (which is an energy-intensive product) and increased hydrogen consumption will lead to increased energy use and operation expenses, unless more efficient technologies for hydrogen production are developed.

The demand for high-value crude oil products will maximize production of transportation fuels at the expense of both residua and low-boiling gases. Hydroprocessing of residua will be widespread rather than appearing on selected refineries. At the same time, hydrotreated residua will be the common feedstocks for fluid catalytic cracking units. Additional conversion capacity will be necessary to process increasingly heavier crudes and meet a reduced demand for residua.

Thus, the gasification of such feedstocks to produce hydrogen and/or power will be an attractive option for refiners (Campbell, 1997; Dickenson et al., 1997; Gross and Wolff, 2000; Speight, 2011a, 2014, 2017). The premise that the gasification section of a refinery will be the *garbage can* for deasphalter residues, high-sulfur coke, as well as other refinery wastes is worthy of consideration (Yuksel Orhan et al., 2014).

The gasification process typically involve a partial oxidation (POX) reaction which is a chemical reaction that occurs when a less-than-stoichiometric amount of air is reacted with the feedstock. The product is typically a hydrogen-rich synthesis gas. The thermal partial oxidation process (TPOX process) involves reaction temperatures that are dependent on the oxidant-fuel ration (air-fuel ratio or oxygen-fuel ratio). However, typical reaction temperatures are on the order of 1200°C (2190°F) and higher. In the catalytic partial oxidation process (COPX process), the use of a catalyst reduced the required reaction temperature to approximately 800 to 900°C (1470 to 1650°F). The catalytic partial oxidation process can be employed if the sulfur content is less than 50 ppm. A higher sulfur content can result in catalyst poisoning and, thus, the thermal partial oxidation process is used for the high-sulfur feedstocks.

5.2 Heavy Feedstocks

Heavy feedstocks are materials such as crude oil residua, heavy crude oils, extra heavy oil, tar sand bitumen, and crude oil coke that have low volatility – in fact many such materials have no volatility. However, for the purposes of this chapter non-volatile products from other sources such as the products from coal and oil shale are not included. Furthermore it is preferable to use the term *heavy feedstocks* (or *hydrocarbonaceous materials*) because of the presence of elements other than carbon and hydrogen and they are not true hydrocarbon derivatives. Thus, the term hydrocarbonaceous materials includes (for the purposes of this chapter): (i) crude oil residua, (ii) heavy crude oil, (iii) extra heavy oil, (iv) tar sand bitumen, and (v) other feedstocks such as crude oil coke – all of which can be used as feedstocks for the gasification process (Table 5.3) (Parkas, 2003; Gary *et al.*, 2007; Speight, 2011a, 2014; Hsu and Robinson, 2017; Speight, 2017).

5.2.1 Crude Oil Residua

A crude oil *resid* (*residuum*, pl. *resids*, *residua*) is the non-volatile residue obtained from crude oil after non-destructive distillation has removed all the volatile constituents of the feedstock. The temperature of the distillation unit is usually (but not always because residence time in the hot zone is also a factor) maintained below 350°C (660°F) since the rate of thermal decomposition of crude oil constituents is minimal below this temperature but the rate of thermal decomposition of crude oil constituents is substantial above 350°C (660°F).

Resids are black, viscous materials and are obtained by distillation of a crude oil under atmospheric pressure (atmospheric residuum) or under reduced pressure (vacuum residuum). They may be liquid at room temperature (generally atmospheric residua) or almost solid (generally vacuum residua) depending upon the nature of the crude oil. When a residuum is obtained from crude oil and thermal decomposition has commenced, it is more usual to refer (incorrectly) to this product as *pitch* (Speight, 2014). The differences between parent crude oil and the residua are due to the relative amounts of various constituents present, which are removed or remain by virtue of their relative volatility.

The chemical composition of a residuum form is complex. Physical methods of fractionation usually indicate high proportions of asphaltene constituents and resins, even in

Table 5.3 Options for resid processing in a refinery.

Resid	Gasification		
	Solvent deasphalting	Deasphalted oil	Visbreaking
			Delayed coking
			Fluid catalytic cracking
		Deasphalter bottoms	Visbreaking
			Delayed coking
			Fuel oil blend stock
			Road asphalt
			Gasification

amounts up to 50% (or higher) of the residuum (Gary et al., 2007; Speight, 2011a, 2014; Hsu and Robinson, 2017; Speight, 2017). In addition, the presence of ash-forming metallic constituents, including such organometallic compounds as those of vanadium and nickel as well as other metal constituents is also a distinguishing feature of residua and the non-volatile feedstocks. Furthermore, the deeper the *cut* into the crude oil, the greater is the concentration of sulfur and metals in the residuum and the greater the deterioration in physical properties of the residuum (Parkash, 2003; Gary et al., 2007; Speight, 2011a, 2014; Hsu and Robinson, 2017; Speight, 2017).

Resids are typically gasified to produce fuel gas. As an example, the resid from a solvent deasphalting unit is gasified by partial oxidation method under pressure of approximately 570 psi and at a temperature between 1300 and 1500°C (2370°F and 2730°F) (Bernetti et al., 2000). The high temperature generated gas flows into the specially designed waste heat boiler, in which the hot gas is cooled and high-pressure saturated steam is generated. The gas from the waste heat boiler is then heat exchanged with the fuel gas and flows to the carbon scrubber, where unreacted carbon particles are removed from the generated gas by water scrubbing.

The gas from the carbon scrubber is further cooled by the fuel gas and boiler feed water and led into the sulfur compound removal section, where hydrogen sulfide (H_2S) and carbonyl sulfide (COS) are removed from the gas to obtain clean fuel gas. This clean fuel gas is heated with the hot gas generated in the gasifier and finally supplied to the gas turbine at a temperature of 250 to 300°C (480 to 570°F).

The exhaust gas from the gas turbine having a temperature of approximately 550 to 600°C (1020 to 1110°F) flows into the heat recovery steam generator consisting of five heat exchange elements. The first element is a superheater in which the combined stream of the high-pressure saturated steam generated in the waste heat boiler and in the second element (high pressure steam evaporator) is super-heated. The third element is an economizer, the fourth element is a low pressure steam evaporator and the final or the fifth element is a de-aerator heater. The off-gas from heat recovery steam generator having a temperature of approximately 130°C (265°F) is emitted into the air via stack.

In order to decrease the nitrogen oxide (NO_x) content in the flue gas, two methods can be applied. The first method is the injection of water into the gas turbine combustor. The second method is to selectively reduce the nitrogen oxide content by injecting ammonia gas in the presence of de-NO_x catalyst that is packed in a proper position of the heat recovery steam generator. The latter is more effective than the former to lower the nitrogen oxide emissions to the air.

In the Hybrid gasification process, a slurry of resid and coal oil is injected into the gasifier where it is pyrolyzed in the upper part of the reactor to produce gas and char. The chars produced are then partially oxidized to ash. The ash is removed continuously from the bottom of the reactor. In this process, vacuum residue and coal are mixed together into slurry to produce clean fuel gas. The slurry fed into the pressurized gasifier is thermally cracked at a temperature of 850 to 950°C (1560 to 1740°F) and is converted into gas, tar and char. The mixture oxygen and steam in the lower zone of the gasifier gasify the char. The gas leaving the gasifier is quenched to a temperature of 450°C (840°F) in the fluidized-bed heat exchanger, and is then scrubbed to remove tar, dust and steam at around 200°C (390°F).

The resid-coal slurry is gasified in the fluidized-bed gasifier. The charged slurry is converted to gas and char by thermal cracking reactions in the upper zone of the fluidized bed. The produced char is further gasified with steam and oxygen that enter the gasifier just below the fluidizing gas distributor. Ash is discharged from the gasifier and indirectly cooled with steam and then discharged into the ash hopper. It is burned with an incinerator to produce process steam. Coke deposited on the silica sand is removed in the incinerator.

5.2.2 Heavy Crude Oil

When crude oil occurs in a reservoir that allows the crude material to be recovered by pumping operations as a free-flowing dark to light colored liquid, it is often referred to as *conventional crude oil* (or *conventional petroleum*). Heavy crude oil a *type* of crude oil that is different from the conventional crude oil insofar as it is more difficult to recover from the reservoir. Heavy crude oil has a much higher viscosity (and lower API gravity) than conventional crude oil, and primary recovery of heavy crude oil requires thermal stimulation of the reservoir (Speight, 2008, 2009, 2013a, 2013b, 2014).

The definition of heavy crude oil is commonly (but incorrectly) based on the API gravity or viscosity. In fact, for many years, crude oil and heavy crude oil were very generally (but arbitrarily without much scientific foundation) defined in terms of physical properties. For example, heavy crude oils were considered to be those crude oils that had gravity somewhat less than 20° API with the heavy crude oils falling into the API gravity range 10 to 15°. For example, Cold Lake heavy crude oil has an API gravity equal to 12° and extra heavy crude oils and as tar sand bitumen, usually have an API gravity in the range 5 to 10° (Athabasca bitumen = 8° API) under ambient conditions. Residua would vary depending upon the temperature at which distillation was terminated but usually vacuum residua are in the range 2 to 8° API (Speight, 2000; Parkash, 2003; Ancheyta and Speight, 2007; Speight, 2014).

5.2.3 Extra Heavy Crude Oil

Extra heavy crude oil is also a material that suffers from the use of arbitrary nomenclature. Extra heavy oil occurs in the solid or near-solid state and is generally has mobility under reservoir conditions – this may be due to the temperature of the reservoir/deposit rather than the ambient properties of the material. In fact, he term *extra heavy oil* is a recently evolved term (related to viscosity) of little scientific meaning and often initiates confusion as it is incorrectly used to refer to tar sand bitumen. While this type of oil may resemble tar sand bitumen and does not flow easily, extra heavy oil is generally recognized as having mobility in the reservoir compared to tar sand bitumen, which is typically incapable of mobility (free flow) under reservoir conditions. For example, the tar sand bitumen located in Alberta, Canada, is not mobile in the deposit and requires extreme methods of recovery to recover the bitumen while much of the extra heavy oil located in the Orinoco belt of Venezuela requires recovery methods that are less extreme because of the mobility of the material in the reservoir (Speight, 2009, 2013a, 2013b, 2014).

5.2.4 Tar Sand Bitumen

Bitumen (also, on occasion, referred to as *native asphalt*, and *extra heavy oil*) is a naturally occurring material that is found in tar sand deposits (*oil sand deposits* in Canada) where the permeability is low and passage of fluids through the deposit can only be achieved by prior application of fracturing techniques. Tar sand bitumen is a high-boiling material with little, if any, material boiling below 350°C (660°F) and the properties may resemble those of an atmospheric residuum. However, to get beyond the use of one or two properties to define tar sand bitumen, *tar sands* have been defined in the United States (FE-76-4) more correctly and from a functional aspect as:

> ...*the several rock types that contain an extremely viscous hydrocarbon which is not recoverable in its natural state by conventional oil well production methods including currently used enhanced recovery techniques. The hydrocarbon-bearing rocks are variously known as bitumen-rocks oil, impregnated rocks, oil sands, and rock asphalt.*

Furthermore, the recovery of the bitumen depends to a large degree on the collective composition and. generally, the bitumen found in tar sand deposits is an extremely viscous material that is *immobile under reservoir conditions* and cannot be recovered through a well by the application of secondary or enhanced recovery techniques (Speight 2013a, 2013b, 2014). However, the term *tar sand* is actually a misnomer; more correctly, the name *tar* is usually applied to the high-boiling product remaining after the destructive distillation of coal or other organic matter (Speight, 2013c). The bitumen in tar sand formations requires a high degree of thermal stimulation for recovery to the extent that some thermal decomposition may have to be induced. Current recovery operations of bitumen in tar sand formations involve use of a mining technique and non-mining techniques are continually being developed (Speight, 2009, 2013a, 2013b, 2014).

It is incorrect to refer to native bituminous materials as *tar* or *pitch*. Although the word tar is descriptive of the black, high-boiling bituminous material, it is best to avoid its use with respect to natural materials and to restrict the meaning to the volatile or near-volatile products produced in the destructive distillation of such organic substances as coal (Speight, 2013c). Thus, alternative names, such as *bituminous sand* or *oil sand*, are gradually finding usage, with the former name (bituminous sands) more technically correct. The term *oil sand* is also used in the same way as the term *tar sand*, and these terms are used interchangeably throughout this text.

5.2.5 Other Feedstocks

Other gasification feedstocks are variable and will depend upon the location of the refinery into which the gasifier has been integrated. Such feedstock may arise from fossil fuel and from non-fossil fuel sources (Speight, 2008, 2011a, 2011b). Thus, a wide variety of feedstocks can be considered for gasification, ranging from solids to liquids to gaseous streams. Although when the feed is a gas or liquid, the operation is frequently referred to as partial oxidation (POX) and the partial oxidation of gases and liquids is very similar to the gasification of solids.

The major requirement for a suitable feedstock is that it contains a significant content of carbon and hydrogen. Solid feedstocks include coal, petcoke, biomass, and other solid waste streams. There are many hydrocarbon-containing gas and liquid streams that may be used as a feedstock for gasification. However, the streams most commonly employed are generally low-value byproducts or waste streams generated by the various processes. These feedstocks are often mixed feedstocks where two or more streams have been blended prior to introduction into the gasifier. In such cases, as has already been explained (Chapter 2), caution is advised when using mixed feedstocks for gasification or, for that matter, for any conversion process. It has been the method in the past (and even continued to be the method in some cases) to assess the properties of the mixed feedstock by calculating an average value for one or more of the properties of the mixed feedstock. This is a dangerous practice because it ignores the potential for interaction of the contents of each feedstock that can cause changes in the chemistry of the process as well as a loss of process efficiency.

With resids as gasification feedstocks, it is necessary to determine whether or not the resid constituents are unlikely to form a separate phase (of asphaltene-type material or the products of the reacted asphaltene constituents) as the feedstock passes through hot pipes and into the gasifier. Such a phenomenon may result in blocked pipes and (at best) a decrease in process efficiency or (at worst) a shutdown of the process.

In order to avoid a significant decrease in conversion efficiency when mixed feedstocks are employed, a formulation of mixed feedstocks is developed in order to produce a more consistent material (Chapter 2). Formulation combines various preprocessed resources and/or additives to produce a feedstock that provides process consistency.

5.2.5.1 Crude Oil Coke

The other feedstock worthy of note that is used as a feedstock for the gasification processes is crude oil coke (often referred to as petcoke), which is the residue left by the destructive distillation of crude oil residua (Gray and Tomlinson, 2000; Speight, 2008, 2014). That formed in catalytic cracking operations is usually non-recoverable, as it is often employed as fuel for the process. The composition of crude oil coke varies with the source of the crude oil, but in general, large amounts of high-molecular-weight complex hydrocarbon derivatives (rich in carbon but correspondingly poor in hydrogen) make up a high proportion. The solubility of crude oil *coke* in carbon disulfide has been reported to be as high as 50 to 80% w/w, but this is in fact a misnomer, since the coke is the insoluble, honeycomb material that is the end product of thermal processes.

Three physical structures of coke can be produced by delayed coking: (i) shot coke, (ii) sponge coke, or (iii) needle coke.

Shot coke is an abnormal type of coke resembling small balls. Due to mechanisms not well understood the coke from some coker feedstocks forms into small, tight, non-attached clusters that look like pellets, marbles or ball bearings. It usually is a very hard coke, i.e., low Hardgrove grindability index (Speight, 2013, 2014). Such coke is less desirable to the end users because of difficulties in handling and grinding. It is believed that feedstocks high in asphaltene constituents and low API favor shot coke formation. Blending aromatic materials with the feedstock and/or increasing the recycle ratio reduces the yield of shot coke. Fluidization in the coke drums may cause formation of shot coke. Occasionally, the smaller *shot coke* may agglomerate into ostrich egg-sized pieces. While shot coke may look like it is entirely made up of shot, most shot coke is not 100% shot.

Sponge coke is the common type of coke produced by delayed coking units (Parkash, 2003; Gary *et al.*, 2007; Speight, 2011a, 2014; Hsu and Robinson, 2017; Speight, 2017). It is in a form that resembles a sponge and has been called honeycombed. Sponge coke, mostly used for anode-grade is dull and black, having porous, amorphous structure.

Needle coke (*acicular coke*) is a special quality coke produced from aromatic feedstocks; it is silver-gray, having crystalline broken needle structure, and is believed to be chemically produced through cross linking of condensed aromatic hydrocarbon derivatives during coking reactions. It has a crystalline structure with more unidirectional pores and is used in the production of electrodes for the steel and aluminum industries and is particularly valuable because the electrodes must be replaced regularly.

Crude oil coke is employed for a number of purposes, but its chief use (subject to composition and properties) is in the manufacture of carbon electrodes for aluminum refining, which requires a high-purity carbon – low in ash and sulfur free; the volatile matter must be removed by calcining. In addition to its use as a metallurgical reducing agent, crude oil coke is employed in the manufacture of carbon brushes, silicon carbide abrasives, and structural carbon (e.g., pipes and Rashig rings), as well as calcium carbide manufacture from which acetylene is produced:

$$\text{Coke} \rightarrow \text{CaC}_2$$

$$\text{CaC}_2 + \text{H}_2\text{O} \rightarrow \text{HC} \equiv \text{CH}$$

Coke that is unsuitable for any of the above uses is used either as a fuel for the refinery or as a source of synthesis gas and hydrogen. In either case the presence of nitrogen, oxygen, sulfur, and metals in the coke feedstock requires that the gaseous products be subject to thorough gas-cleaning methods (Speight, 2014).

In the gasification process, several processes are combined to ensure the efficiency of the gasification and the production of the product gases (Table 5.4).

5.2.5.2 Solvent Deasphalter Bottoms

The deasphalting unit (deasphalter) is a unit in a crude oil refinery for bitumen upgrader that separates an asphalt-like product from crude oil, heavy crude oil, extra heavy oil, or tar sand bitumen. The deasphalter unit is usually placed after the vacuum distillation tower where, by the use of a low-boiling liquid hydrocarbon solvent (such as propane or butane under pressure), the insoluble asphalt-like product (*deasphalter bottoms*) is separated from the feedstock – the other output from the deasphalter is deasphalted oil (DAO).

The solvent deasphalting process has been employed for more than six decades to separate high molecular weight fractions of crude oil boiling beyond the range of economical commercial distillation. The earliest commercial applications of solvent deasphalting used liquid propane as the solvent to extract high-quality lubricating oil bright stock from vacuum residue. The process has been extended to the preparation of catalytic cracking feeds, hydrocracking feeds, hydrodesulfurization feedstocks, and asphalts. The latter product (asphalt, also called *deasphalter bottoms*) is used for (i) road asphalt manufacture, (ii) refinery fuel, or (iii) gasification feedstock for hydrogen production.

In fact, the combination of ROSE solvent deasphalting and gasification has been commercially proven at the ERG Petroli refinery (Bernetti *et al.*, 2000). The combination is very synergistic and offers a number of advantages including a low-cost feedstock to the gasifier, thus enhancing the refinery economics, and converts low-value feedstock to high-value products such as power, steam, hydrogen, and chemical feedstock. The process also improves the economics of the refinery by eliminating/reducing the production of low-value fuel oil and maximizing the production of transportation fuel.

Table 5.4 The gasification process flow for crude oil coke.

Stage	Description
1.	Petroleum coke is crushed and mixed with a flux and water to produce a slurry.
2.	Slurry is pumped at high pressure into the gasifiers.
3.	Slurry and high purity oxygen react at elevated temperature and pressure to form synthesis gas and slag (melted ash and flux).
4.	Slag is quenched with water, crushed, washed and exported for sale or disposal.
5.	Remaining synthesis gas is mainly comprised of CO, H2, water and CO2, with small levels of methane, argon, nitrogen and sulfur species.
6.	Synthesis gas passes through water quench and water scrubber to remove entrained particulates before being split into two streams, one for methanol and the other for hydrogen production.
7.	Both streams pass-through an acid gas removal to remove sulfur compounds and carbon dioxide.
8.	Sulfur removed in the process is converted to sulfuric acid in the wet sulfuric acid.
9.	Carbon dioxide is compressed and dehydrated for sale.
10.	The resulting clean synthesis gas is ready to be used to produce a variety of products, notably hydrogen and methanol.

5.3 Synthesis Gas Production

High-boiling feedstocks are gasified and the produced gas is purified to clean fuel gas (Gross and Wolff, 2000), for example:

$$\text{Residuum} \rightarrow CO + CO_2 + H_2 + SO_x + NO_x + \text{particulate matter} \rightarrow CO + H_2$$

Synthesis gas

As an example, solvent deasphalter residuum is gasified by partial oxidation method under pressure of approximately 570 psi and at a temperature between 1300 and 1500°C (2370°F and 2730°F) (Bernetti *et al.*, 2000). The high temperature generated gas flows into a waste heat boiler, in which the hot gas is cooled and high-pressure saturated steam is generated. The gas from the waste heat boiler is then heat exchanged with the fuel gas and flows to the carbon scrubber, where particulate matter is removed from the generated gas by water scrubbing. The gas from the carbon scrubber is further cooled by the fuel gas and boiler feed water and led into the sulfur compound removal section, where hydrogen sulfide (H_2S) and carbonyl sulfide (COS) are removed from the gas to obtain clean fuel gas – the presence of these two gases is not always observed and is dependent upon the operational parameters of the gasification process. If the gas is designated as fuel gas, the clean gas is heated with the hot gas generated in the gasifier and finally supplied to the gas turbine at a temperature of 250 to 300°C (480 to 570°F).

The exhaust gas from the gas turbine having a temperature of approximately 550 to 600°C (1020 to 1110°F) flows into the heat recovery steam generator consisting of five heat exchange elements. The first element is a superheater in which the combined stream of the high-pressure saturated steam generated in the waste heat boiler and in the second element (high pressure steam evaporator) is super-heated. The third element is an economizer, the fourth element is a low pressure steam evaporator and the final or the fifth element is a de-aerator heater. The off-gas from heat recovery steam generator having a temperature of approximately 130°C (265°F) is emitted into the air via stack.

In order to decrease the nitrogen oxide (NO_x) content in the flue gas, two methods can be applied. The first method is the injection of water into the gas turbine combustor. The second method is to selectively reduce the nitrogen oxide content by injecting ammonia gas in the presence of de-NO_x catalyst that is packed in a proper position of the heat recovery steam generator. The latter is more effective that the former to lower the nitrogen oxide emissions to the air.

The process for producing synthesis gas typically has three components: (i) synthesis gas generation, (ii) waste heat recovery, and (iii) gas processing. Within each of these components, there are several options. For example, synthesis gas can be generated to yield a range of compositions ranging from high-purity hydrogen to high-purity carbon monoxide. Two major routes can be utilized for high-purity gas production: (i) pressure-swing adsorption, (ii) utilization of a cryogenic procedure, where separation is achieved using low temperatures, and (iii) the use of permeable membrane technology is also seeing more common use (Speight, 2007, 2014).

5.3.1 Partial Oxidation Technology

Partial oxidation is the most commonly used process for the gasification of heavy crude oils and other refinery residues, although virtually all mixtures – even volatile products – are suitable feedstocks (Liebner, 2000). However, aside from special applications, gasification is a *bottom-of-the-barrel process* converting feedstocks continuing sulfur, nitrogen, and to a clean synthesis gas consisting mainly of hydrogen and carbon monoxide. In fact, gasification is replacing direct combustion due to environmental regulations, since ash removal and flue gas clean-up are more difficult and expensive than synthesis gas cleaning at elevated pressures.

The main advantages derived from application of gasification in a refinery are: (i) the capability of processing low-quality, highly viscous and heavy feedstocks, and in addition – mostly in quench gasifiers – emulsions (tank sludge), slurries (coke) and other liquid wastes, (ii) the capability of processing high-sulfur feedstocks because of the almost complete removal of sulfur compounds in the gas treating unit downstream of the gasifier, (iii) the possibility of producing hydrogen for the various conversion and upgrading processes of the refinery, with increased production of gas oil, which is a desirable product/feedstock for other refinery units, and (iv) the many outlets for synthesis gas, such as hydrogen for the refinery or for export, electricity via the integrated gasification combined cycle (IGCC) process and the production of chemicals such as ammonia, methanol, acetic acid, and oxo-alcohols.

Partial oxidation, POX (POx), reactions occur when a sub-stoichiometric fuel-air mixture is partially combusted in a reformer. The general reaction equation without catalyst (*thermal partial oxidation*, TPOX, TPOx) can be represented as:

$$C_nH_m + (2n + m)/2 O_2 \rightarrow nCO + (m/2) H_2O$$

There are no exact stoichiometric reactions because of the variable composition of the feedstocks to the gasifier. To produce such an equation would only serve to mislead any potential kinetic studies.

A TPOX reactor is similar to the autothermal reactor (ATR) with the main difference being no catalyst is used. The feedstock, which may include steam, is mixed directly with oxygen by an injector which is located near the top of the reaction vessel. Both partial oxidation as well as reforming reactions occurs in the combustion zone below the burner.

On the other hand, very high temperatures, approximately 1300°C (2370°F), are required to achieve near complete reaction. This necessitates the consumption of some of the hydrogen and a greater than stoichiometric consumption of oxygen, i.e., oxygen-rich conditions. Capital costs are high on account of the need to remove soot and acid gases from the synthesis gas. Operating expenses are also high due to the need for oxygen at high pressure.

A possible means of improving the efficiency of synthesis gas production is via catalytic partial oxidation process (CPOX process). The catalytic partial oxidation process has several advantages over steam reforming, especially the higher energy efficiency. The reaction is in fact, not endothermic as is the case with steam reforming, but rather slightly exothermic. Further, an hydrogen/carbon monoxide ratio of close to 2.0, i.e., the ideal ratio for the Fischer-Tropsch and methanol synthesis, is produced by this technology. The process can occur by either of two routes: (i) direct or (ii) indirect.

The *direct catalytic partial oxidation process* occurs through a mechanism involving only surface reaction on the catalyst, the *direct* route produces synthesis gas according to the following reaction:

$$2CH_4 + O_2 \rightarrow 2CO + 4H_2$$

On the other hand, the *indirect catalytic partial oxidation process* comprises total combustion of methane to carbon dioxide and water, followed by steam reforming and the water-gas shift reaction. Here, equilibrium conversions can be greater than 90% at ambient pressure. However, in order for an industrial process for this technology to be economically viable, an operating pressure of more than 300 psi would be required. Unfortunately, under such pressures, equilibrium conversions are lower. Further, an operational problem arises on account of the highly exothermic combustion step, which makes for problematic temperature control of the process and the possibility of temperature runaways.

It must be noted that in most studies of catalytic partial oxidation in microreactors, in most to nearly all cases the conversion occurred via the indirect route. It is apparent that only the direct mechanism is likely to occur at short contact times. However, product yields higher than equilibrium values are obtained when high flow rates through fixed-bed reactors are used.

The principal advantage of the partial oxidation process is the ability of the process to accommodate a variety of feedstocks – these can comprise high molecular-weight organic feedstocks, such as crude oil coke (Gunardson and Abrardo, 1999). Additionally, since emissions of nitrogen oxides (NOx) and sulfur oxides (SOx) are minimal, the technology can be considered to be environmentally acceptable.

5.3.1.1 Shell Gasification Process

The *Shell Gasification process* (a *partial oxidation process*) is a flexible process for generating synthesis gas, principally hydrogen and carbon monoxide, for the ultimate production of high-purity high-pressure hydrogen, ammonia, methanol, fuel gas, town gas or reducing gas by reaction of gaseous or liquid hydrocarbon derivatives with oxygen, air or oxygen-enriched air. The most important step in converting high-boiling feedstocks to industrial gas is the partial oxidation of the oil using oxygen with the addition of steam. The most important step is the partial oxidation of the oil using oxygen with the addition of steam.

The gasification process takes place in a refractory-lined reactor at temperatures on the order of 1400°C (2550°F) and pressures between 30 to 1140 psi. The chemical reactions in the gasification reactor proceed without catalyst to produce gas containing carbon amounting to some 0.5 to 2% w/w by weight, based on the feedstock. The carbon is removed from the gas with water, extracted in most cases with feed oil from the water and returned to the feed oil. The high reformed gas temperature is utilized in a waste heat boiler for generating steam. The steam is generated at 850 to 1565 psi, some of which is used as process steam and for oxygen and oil preheating.

5.3.1.2 Texaco Process

The *Texaco gasification process* (a *partial oxidation gasification process*) for generating synthetic gas, principally hydrogen and carbon monoxide. The characteristic of the process is to inject feedstock together with carbon dioxide, steam or water into the gasifier. Therefore, feedstocks such as (i) residua, (ii) solvent deasphalted residua, or (iii) crude oil coke produced by any coking process can be used as feedstock for this gasification process. The product gas from this gasification process can be used for the production of high-purity high-pressurized hydrogen, ammonia, and methanol. The heat recovered from the high temperature gas is used for the generation of steam in the waste heat boiler. Alternatively the less expensive quench type configuration is preferred when high-pressure steam is not needed or when a high degree of shift is needed in the downstream carbon monoxide converter.

In the partial oxidation process, the gasification reaction is a partial oxidation of hydrocarbon derivatives to carbon monoxide and hydrogen and can be represented in simple chemical terms:

$$C_xH_{2y} + x/2O_2 \rightarrow xCO + yH_2$$

$$C_xH_{2y} + xH_2O \rightarrow xCO + (x+y)H_2$$

The gasification reaction is instantly completed, thus producing gas mainly consisting of carbon monoxide and hydrogen – the high-temperature gas leaving the reaction chamber of the gas generator enters a quenching chamber usually linked to the bottom of the gas generator and is quenched to 200 to 260°C (390 to 500°F) with water.

5.3.1.3 Phillips Process

In the process, crude oil coke is mixed with water to make a pumpable slurry which is then fed to a two-stage gasifier. The slurry reacts readily with the oxygen in the first stage of the gasifier to form hydrogen, carbon monoxide, carbon dioxide and methane. The high temperature in the first-stage ensures the conversion of all feedstock materials and traps inorganic materials like ash and metals in a glassy matrix resembling coarse sand. This sand-like material (slag) is inert and has an array of uses in the construction industry.

The hot synthesis gas from the horizontal first stage enters the vertical, second stage of the gasifier, where additional slurry is added to increase the energy content of the gas. This two-stage design increases efficiencies, particularly for low reactivity fuels such as crude oil coke. Hot synthesis gas is then cooled in a heat recovery system, producing high-pressure steam in a fire tube boiler.

The dry system improves efficiency over wet systems, removing more particulates and avoiding black-water problems which lead to equipment wear, and minimizes water consumption and wastewater generation. Sulfur in the synthesis gas is recovered and converted to elemental sulfur, which can be sold in agricultural and other markets. Maximizing sulfur recovery at over 99% of that found in the feedstock, the process recycles all unconverted gases from the tail gas of the sulfur recovery unit to the second stage of the gasifier.

The clean synthesis gas can be further processed, shifting the synthesis gas equilibrium for additional hydrogen production. Required hydrogen purity standards are achieved through standard pressure-swing adsorption design. The downstream hydrogen production process units also will facilitate the capture of carbon dioxide, which can then be compressed and used for enhanced oil recovery or other beneficial uses, or placed in geologic storage. Steam production is achieved through heat recovery steam generators as needed for power or steam export to the host facility.

5.3.2 Catalytic Partial Oxidation

A possible means of improving the efficiency of synthesis gas production from high-boiling feedstocks is by means of *catalytic partial oxidation* (CPOX, CPOx) technology. This technology has several advantages over steam reforming, especially the higher energy efficiency. The reaction is in fact, not endothermic as is the case with steam reforming, but is more on the exothermic side. Typical operating conditions provided are the temperature range of 700 to 1000°C (1290 to 1830°F), atmospheric pressure, oxygen-to-carbon (O/C) ratio = 1.2. The carbon monoxide-hydrogen ratio (CO/H_2) approximately equal to 2.0 is the ideal ratio for the Fischer-Tropsch process and methanol synthesis, and is produced by this technology.

The catalytic partial oxidation process is an attractive option to produce hydrogen and carbon monoxide from hydrocarbon feedstocks. The process can also be followed by the water gas shift reaction to produce a pure hydrogen stream for use in a variety of refining operations (Parkash, 2003; Gary *et al.*, 2007; Speight, 2011a, 2014; Hsu and Robinson, 2017; Speight, 2017).

Since there is no need to feed water, as in steam reforming and oxidative-steam reforming, the reactors for the process reactors are smaller and more easily integrated into reforming systems, as well as remote and distributed power applications. In addition, the exothermic nature of this process reduces the heat needed to be supplied to the system, making it ideal for integration with high-temperature fuel cells. Challenges for the process include catalyst deactivation from high-temperature sintering, metal vaporization, support degradation, coke formation, and poisoning from sulfur and other contaminants.

The reactions occurring in the catalytic partial oxidation process are complex but the overall reaction can be described by the following equation:

$$CH_4 + 1/2 O_2 \leftrightarrows CO + 2H_2$$

$$2CH_4 + O_2 \leftrightarrows 2CO + 4H_2$$

However, several side reactions (again using methane as the example feedstock) occur. The most significant side reactions are the water gas shift reaction, the steam reforming reaction, and the carbon dioxide reforming reaction, which help determine the final synthesis gas composition which will vary based on the process feedstock and the process (reaction) parameters.

$$CH_4 + 2O_2 \rightarrow CO_2 + 2H_2O \quad \text{Combustion} \tag{5.1}$$

$$CH_4 + O_2 \rightarrow CO_2 + 2H_2 \tag{5.2}$$

$$CO + H_2O \leftrightarrow CO_2 + H_2 \quad \text{Water gas shift} \tag{5.3}$$

$$CH_4 + H_2O \leftrightarrow CO + 3H_2 \quad \text{Steam reforming} \tag{5.4}$$

$$CH_4 + CO_2 \leftrightarrow 2CO + 2H_2 \quad \text{Carbon dioxide reforming} \tag{5.5}$$

$$CO + H_2 \leftrightarrow C + H_2O \tag{5.6}$$

$$CH_4 \leftrightarrow C + 2H_2 \quad \text{Methane decomposition} \tag{5.7}$$

$$2CO \leftrightarrow CO_2 + C \quad \text{Boudouard reaction} \tag{5.8}$$

$$2CO + 5O_2 \rightarrow 2CO_2 \tag{5.9}$$

$$2H_2 + O_2 \rightarrow 2H_2O$$

5.4 Products

Products from synthesis gas processes can range from (i) synthesis gas, in a range of carbon monoxide and hydrogen mixtures, (ii) high-purity hydrogen (iii) high-purity carbon monoxide, and (iv) high-purity carbon dioxide.

If hydrogen is the desired product for refinery operations (Wolff, 2007; Sutikno and Turini, 2012; Speight, 2014) the carbon monoxide-hydrogen ratio can approach infinity by conversion of all of the carbon monoxide to carbon dioxide. By contrast, on the other end, the ratio cannot be adjusted to zero (i.e., 100% v/v hydrogen) because water is always produced with the hydrogen. In fact, a general rule of thumb exists in terms of the hydrogen and carbon monoxide produced by the different gasification processes:

Gasification Process	H_2/CO ratio
Steam reforming	3.0 – 5.0
Steam reforming plus oxygen secondary reforming	2.5 – 4.0
Autothermal reforming (ATR)	1.6 – 2.65
Partial oxidation (POx)	1.6 – 1.9

However, in practice the options are not limited to the ranges shown but rather even greater hydrogen-carbon monoxide ratios can be observed if adjustments are made to the process, such as the inclusion of a shift converter to effect near-equilibrium water-gas-shift conversion or by adjusting the amount of steam.

5.4.1 Gas Purification and Quality

Purities in excess of 99.5% v/v of either hydrogen or carbon monoxide produced from synthesis gas can be achieved if required by the refinery (Chapter 9). Four of the major process technologies available are:

1. *Cryogenics plus methanation*: which utilizes a cryogenic process whereby carbon monoxide is liquefied (if necessary, in a number of steps) to produce hydrogen with a purity of on the order of 98% v/v. The condensed carbon monoxide, which often contains methane, is distilled to produce to streams (i) pure carbon monoxide and (ii) a mixture of carbon monoxide and methane. The carbon monoxide and methane stream can be used as fuel. The hydrogen stream is routed to a shift converter where any remaining carbon monoxide is converted to carbon dioxide and hydrogen. The carbon dioxide is removed and any further carbon monoxide or carbon dioxide can be removed by methanation and the resulting hydrogen stream can have purity as high as 99.7% v/v.
2. *Cryogenics plus pressure-swing adsorption*, which utilizes a similar sequential liquefaction of carbon monoxide until hydrogen having a purity of approximately 98% v/v is achieved. Again, the carbon monoxide stream can be distilled to remove methane until it is essentially pure. The hydrogen stream is then processed through multiple swings (depending on the hydrogen purity required) of pressure-swing adsorption cycles until the hydrogen purity as high as 99.999% v/v is produced.
3. *Methane-wash cryogenic process*, in which liquid carbon monoxide is absorbed into a liquid methane stream so that the hydrogen stream contains only ppm levels of carbon monoxide but contains approximately 5 to 8% v/v methane –

hence the hydrogen stream purity is limited to approximately 95% v/v. The liquid carbon monoxide-/methane stream can be distilled to produce pure carbon monoxide as well as a carbon monoxide-methane stream which can be used as fuel gas.

4. *COsorb Process*, which utilizes copper ions (cuprous aluminum chloride, $CuAlCl_4$) in toluene to form a chemical complex with the carbon monoxide thereby separating it from the product gas steam. This process can capture approximately 96% v/v of the carbon monoxide to produce a stream having purity greater than 99% v/v. However, water, hydrogen sulfide and other trace constituents, which can poison the copper catalyst, must be removed prior to introduction of the product gas into the reactor. Furthermore, a hydrogen stream of only up to 97% v/v purity is obtained. However, while the efficiency of cryogenic separation decreases with a decrease in the carbon monoxide content of the feedstock gas, the COsorb process is a more efficient process for treating feedstock gas with low carbon monoxide content.

5.4.2 Process Optimization

Process optimization includes the development of technologies to facilitate cost-effective gasification of all hydrocarbonaceous feedstocks produced by a refinery as well as biomass and coal. Thus, use of high-pressure feed systems and use of development of technologies for co-feeding mixtures to high-pressure gasifiers are necessary options.

In addition, the use of the refinery gasifier to process biomass and waste and reduce the refinery footprint as well as product marketable products must not be ignored. Using biomass as the example, biomass gasification has emerged as a viable option for decentralized power generation, especially in developing countries. Another potential use of producer gas from biomass gasification is in terms of feedstock for Fischer-Tropsch synthesis – a process for manufacture of synthetic gasoline and diesel (Buragohain *et al.*, 2010).

5.5 Advantages and Limitations

One of the most compelling challenges of the 21st century is finding a way to meet national and global energy needs. Crude oil refineries can help meet this challenge while generating more economic value by adopting a gasification process, although there may always be competition for gasification in the processes that convert the high-boiling feedstocks into added-value products, such as liquid fuels.

Accordingly, there are clear benefits of adding a gasification system to a refinery (Speight, 2011a), such as (i) production of power, steam, oxygen and nitrogen for refinery use or sale – refineries typically convert resids and waste or residue into asphalt or bitumen, products from which they may derive very little economic value – gasification technology converts this waste into valuable commodities, such as power, steam, oxygen, hydrogen and nitrogen, which are used in everyday refinery operations; (ii) increased efficiency of power generation, improved air emissions, and reduced waste stream versus combustion of crude oil coke or residua or incineration; and (iii) there is the potential to provide the high-purity hydrogen that is used in a variety of refinery operations such as removal of impurities through hydrotreating and hydrocracking processes.

5.5.1 Other Uses of Residua

The residua (heavy feedstocks, bottoms, hydrocarbonaceous feedstocks) can be routed to other conversion units or blended to high-boiling industrial fuel and/or asphalt. The high-boiling feedstocks typically have a relatively low economic value – often they are of lower value than the original crude oil. Most refineries convert, or upgrade, the low-value high-boiling feedstocks into more valuable low-boiling products (such as gasoline, jet fuel, and diesel fuel).

Thus high-boiling feedstock upgrading creates a need for additional bottom of the barrel processing, both for expansion and for yield improvement. Traditionally, this would automatically call for the addition of atmospheric distillation and/or vacuum distillation units as a starting point. However, there are alternative processing schemes for processing the vacuum or atmospheric residues to maximizing the value of the heavier crude oils.

Thus, hydrogen management has become a priority for current and future refinery operations as consumption continues to rise for greater hydrotreating processes, and processing of heavier and higher sulfur crude oils. In many cases the hydrogen network is limiting refinery throughput and operating margins. The current main source for hydrogen is the steam methane reforming (SMR) of refinery off-gases and natural gas, an inefficient and cost-incurring process.

5.5.2 Gasification in the Future Refinery

The worldwide trends in the crude oil supply indicate a continuous increase of the heavy crudes and, as a result, the increase in the yield of distillation residues is complemented by an increase in their sulfur content. Additional distillates are produced by upgrading the residues. The upgrading step generates final residues, such as visbreaking tar, coke and asphalt which are produced by visbreaking, coking and deasphalting, respectively. The final residues can be converted to usable products such as hydrogen, steam, electricity, ammonia and chemicals. For this purpose, gasification has emerged as the technology of choice because of its superior environmental performance when compared with the competing means for residue utilization. Also, refinery sludges can be co-gasified with the final residues and as such, be converted to usable products. If integrated with the petroleum refinery, gasification can diminish any environmental problems associated with residue and sludge disposal.

There are several options available to refiners for the residue disposal. This includes either using the residues without further processing or converting them into higher demand products using carbon rejection and/or hydrogen addition technologies. In some regions of the world, residues can be blended with a gas oil and sold as a heavy fuel oil. However, this outlet is drying out because of a continuous increase in the sulphur and metal contents of the residues. Gasification was identified to be an answer to problems with the disposal of crude oil resids. This technology can convert any solid or semi-solid carbonaceous material to synthesis gas. The raw gas produced in the first step is cleaned to remove sulfur compounds and particulates. The former are converted into pure sulphur in the Claus plant, while particulates are recycled to the gasifier. Metals (vanadium and nickel) are concentrated in the slag/ash, saleable for metal recovery. Thus, no solid byproduct is left for disposal. The clean synthesis gas can be converted to valuable products such as electricity, steam, hydrogen and chemicals. Gasification technology is commercially proven and used in many parts of the world.

As refineries continue to evolve (Furimsky, 1999; Speight, 2011a), the *panacea* (rather than a *Pandora's box*) for a variety of feedstocks could well be the *gasification refinery* (Table 5.3) which is capable of supplying the traditional refined products, but also meeting much more severe specifications, and petrochemical intermediates such as olefins, aromatics, hydrogen and methanol (Table 5.5) (Penrose *et al.*, 1999; Phillips and Liu, 2002; Breault, 2010; Speight, 2011). Furthermore, the integrated gasification combined cycle (IGCC) option can

Table 5.5 Illustration of the production of petrochemical starting materials from natural gas and crude oil.

Feedstock	Process	Product
Petroleum	Distillation	Light ends
		methane
		ethane
		propane
		butane
	Catalytic cracking	ethylene
		propylene
		butylenes
		higher olefins
	Catalytic reforming	benzene
		toluene
		xylenes
	coking	ethylene
		propylene
		butylenes
		higher olefins
Natural gas	refining	methane
		ethane
		propane
		butane

be used to raise power from feedstocks such as vacuum residua and cracked residua and (in addition to the production of synthesis gas), a major benefit of the integrated gasification combined cycle concept is that power can be produced with the lowest sulfur oxide (SOx) and nitrogen oxide (NOx) emissions of any liquid/solid feed power generation technology.

In fact, the future of the crude oil refining industry will be primarily on processes for the production of improved quality products. Thus, the refinery of the future will have a gasification section devoted to the conversion of coal and biomass to hydrocarbon derivatives by the Fischer-Tropsch process – perhaps even with rich oil shale added to the gasifier feedstock. Many refineries already have gasification capabilities but the trend will increase to the point (over the next two to three decades) where nearly all refineries recognize the need to construct a gasification section to handle residua and a variety of other feedstocks. Biomass, liquids from coal, and liquids from oil shale will increase in importance and such feedstocks (i) will be sent to refineries or (ii) processed at a remote location and then blended with refinery stocks are options for future development and the nature of the feedstocks. Above all, such feedstock must be compatible with refinery feedstocks and not cause fouling (which can lead to process or even refinery shut-down) in any form (Speight, 2011a, 2011b).

Process unit and refinery economics/operations computer models will be optimized, with integration into plant operations via process computer controls. Alternate fuels for power generation will continue to push crude processing toward higher-value products, such as transportation fuels and chemicals. Otherwise extra heavy crude oil and tar sand bitumen that are considered uneconomical to transport to a refinery will be partially refined at their source to facilitate transport; and there will be a new emphasis on partial or full upgrading *in situ* during recovery operations (Speight, 2009, 2014). In addition, involvement of alternative energy sources with crude oil, which leads to the concept of *alternative energy systems* (Szklo and Schaeffer, 2005) in which crude oil refining is integrated with the production of energy from other energy sources.

Thus, *refinery flexibility* will be a key target, especially when related to the increased use of renewable energy sources. And a promising start can be made to such flexibility by incorporating gasification technology into the refinery system as an equal partner in energy production.

In summary, gasification is the only technology which makes possible for the refineries the zero residue target, contrary to all conversion technologies such as thermal cracking, coking, catalytic cracking, deasphalting, hydroprocessing which can only reduce the volume of bottoms, with the complication that the residue qualities generally get worse with the degree of conversion. The flexibility of gasification permits refineries to handle any kind of refinery residue, including crude oil coke and tank bottoms as well as refinery sludge and make available a range of value-added products, electricity, steam, hydrogen and various chemicals based on synthesis gas chemistry. The environmental performance of gasification is unmatched since no other technology processing low-value refinery residues can come close to the emission levels achievable with gasification.

References

Ancheyta, J., and G. Speight, J.G. 2007. *Hydroprocessing of Heavy Oils and Residua.* CRC Press, Taylor & Francis Group, Boca Raton, Florida.

Bernetti, A., De Franchis, M., Moretta, J.C., and Shah, P.M. 2000. Solvent Deasphalting and Gasification: A Synergy. *Petroleum Technology Quarterly*, Q4: 1-7. www.digitalrefining.com/article/1000690

Breault, R.W. 2010. Gasification Processes Old and New: A Basic Review of the Major Technologies. *Energies,* 3: 216-240.

Buragohain, B., Mahanta, P. and Moholkar, V.S. 2010. Thermodynamic Optimization of Biomass Gasification for Decentralized Power Generation and Fischer-Tropsch Synthesis. *Energy*, 35(6): 2557-2579.

Campbell, W.M. 1997. In: *Handbook of Petroleum Refining Processes* 2nd Edition. R.A. Meyers (Editor). McGraw-Hill, New York. Chapter 6.1.

Dickenson, R.L., Biasca, F.E., Schulman, B.L., and Johnson, H.E. 1997. Refiner Options for Converting and Utilizing Heavy Fuel Oil. *Hydrocarbon Processing.* 76(2): 57.

Furimsky, E. 1999. Gasification in Petroleum Refinery of 21st Century. Revue Institut Français de Petrole, *Oil & Gas Science and Technology*, 54(5): 597-618.

Gary, J.G., Handwerk, G.E., and Kaiser, M.J. 2007. *Petroleum Refining: Technology and Economics*, 5th Edition. CRC Press, Taylor & Francis Group, Boca Raton, Florida.

Goldhammer, B.P., Blume, A.M., and Yeung, T.W. 2008. Gasification/IGCC to Improve Refinery Operations. Gas, 1-8. Also: www.digitalrefining.com/article/1000647

Gray, D., and Tomlinson, G. 2000. Opportunities for Petroleum Coke Gasification under Tighter Sulfur Limits for Transportation Fuels. *Proceedings. 2000 Gasification Technologies Confernece, San Francisco, California. October 8-11.*

Gross, M., and Wolff, J. 2000. Gasification of Residue as a Source of Hydrogen for the Refining Industry in India. *Proceedings. Gasification Technologies Conference. San Francisco, California. October 8-11.*

Gunardson, H.H., and Abrardo, J.M. 1999. Produce CO-Rich Synthesis Gas. *Hydrocarbon Processing*, 78(4): 87-93.

Hsu, C.S., and Robinson, P.R. (Editors). 2017. *Handbook of Petroleum Technology*. Springer International Publishing AG, Cham, Switzerland.

Liebner, W. 2000. Gasification by Non-Catalytic Partial Oxidation of Refinery Residues. In: *Modern Petroleum Technology*. A.G. Lucas (Editor). John Wiley & Sons Inc., Hoboken, New Jersey.

Parkash, S. 2003. *Refining Processes Handbook*. Gulf Professional Publishing, Elsevier, Amsterdam, Netherlands.

Penrose, C.F., Wallace, P.S., Kasbaum, J.L., Anderson. M.K., and Preston, W.E. 1999. Enhancing Refinery Profitability by Gasification. *Proceedings. Gasification Technologies Conference, San Francisco, California. October 2-5.*

Phillips, G., and Liu. F. 2002. Advances in Residue Upgrading Technologies Offer Refiners Cost-Effective Options for Zero Fuel Oil Production. *Proceedings. 2002 European Refining Technology Conference, Paris, France. November.*

Rostrup-Neilsen, J.R. 1984. Sulfur-Passivated Nickel Catalysts for Carbon-Free Steam Reforming of Methane. *J. Catal.*, 85: 31-43.

Rostrup-Neilsen, J.R. 1993. Production of Synthesis Gas. *Catalysis Today*, 19: 305-324.

Speight, J.G. 2000. *The Desulfurization of Heavy Oils and Residua*. 2nd Edition. Marcel Dekker Inc., New York.

Speight. J.G. 2007. *Natural Gas: A Basic Handbook*. GPC Books, Gulf Publishing Company, Houston, Texas.

Speight, J.G. 2008. *Synthetic Fuels Handbook: Properties, Processes, and Performance*. McGraw-Hill, New York.

Speight, J.G. 2009. *Enhanced Recovery Methods for Heavy Oil and Tar Sands*. Gulf Publishing Company, Houston, Texas.

Speight, J.G. 2011a. *The Refinery of the Future*. Gulf Professional Publishing, Elsevier, Oxford, United Kingdom.
Speight, J.G. (Editor). 2011b. *The Biofuels Handbook*. The Royal Society of Chemistry, London, United Kingdom.
Speight. J.G. 2013a. *Oil Sand Production Processes*. Gulf Professional Publishing, Elsevier, Oxford, United Kingdom.
Speight, J.G. 2013b. *Heavy Oil Production Processes*. Gulf Professional Publishing, Elsevier, Oxford, United Kingdom.
Speight, J.G. 2013c. *The Chemistry and Technology of Coal* 3rd Edition. CRC Press, Taylor and Francis Group, Boca Raton, Florida, 2007.
Speight, J.G. 2013d. *Coal-Fired Power Generation Handbook*. Scrivener Publishing, Salem Massachusetts.
Speight, J.G. 2014. *The Chemistry and Technology of Petroleum* 5th Edition. CRC Press, Taylor and Francis Group, Boca Raton, Florida.
Speight, J.G. 2017. *Handbook of Petroleum Refining*. CRC Press, Taylor and Francis Group, Boca Raton, Florida.
Sutikno, T., and Turini, K. 2012. Gasifying Coke to Produce Hydrogen in Refineries. *Petroleum Technology Quarterly*, Q3: 1-4. Also: www.digitalrefining.com/article/1000550
Szklo, A., and Schaeffer, R. 2005. Alternative energy sources or integrated alternative energy systems? Oil as a Modern Lance of Peleus for the Energy Transition. *Energy*, 31: 2513-2522.
Udengaard, N.R., Bak-Hansen, J. H., Hanson, D. C., and Stal, J.A. 1992. Sulfur Passivated Reforming Process Lowers Syngas H2/CO Ratio. *Oil & Gas Journal*, 90(10): 62.
Wolff, J. 2007. Gasification Technologies for Hydrogen Manufacturing. *Petroleum Technology Quarterly*, Q2: 1-8. www.digitalrefining.com/article/1000670
Wolff, J., and Vliegenthart, E. 2011. Gasification of Heavy Ends. *Petroleum Technology Quarterly*, Q2: 1-5. www.digitalrefining.com/article/10003988
Yuksel Orhan, O., İs, G., Alper, E. McApline, K., Daly, S., Sycz, M., and Elkamel, A. 2014. Gasification of Oil Refinery Waste for Power and Hydrogen Production. *Proceedings. 2014 International Conference on Industrial Engineering and Operations Management, Bali, Indonesia. January 7-9.* http://iieom.org/ieom2014/pdfs/314.pdf

6

Gasification of Biomass

6.1 Introduction

The two most important thermal conversion systems applicable to biomass resources and biomass wastes are (i) pyrolysis and (ii) gasification. Pyrolysis is a thermal conversion process in a complete absence of oxidant. Products include synthesis gas (carbon monoxide and hydrogen) and methane as well as other low molecular weight hydrocarbons derivative), liquid bio-oil, and solid bio-char. On the other hand, biomass gasification is the conversion of biomass to gaseous products in the presence of a limited (controlled) amount of oxygen (Pandey, 2009; Speight, 2011a).

Biomass gasification has been the focus of research in recent years to estimate efficiency and performance of the gasification process using various types of biomass such as sugarcane residue (Gabra *et al.*, 2001), rice hulls (Boateng *et al.*, 1992), pine sawdust (Lv *et al.*, 2004), almond shells (Rapagnà *et al.*, 2000), almond (S. Rapagnà and Latif, 1997), and food waste (Ko *et al.*, 2001) as well as woody biomass (Hanaoka *et al.*, 2005). Thus, in the current context, the gasification of biomass is a widely used thermochemical process for obtaining products with more value and potential applications than the raw material itself (Hu *et al.*, 2012).

The gasification process is an extension of the pyrolysis process insofar as the thermal process is optimized to give the highest yield gas rather than to produce bio-oil and/or biochar. In any case, bio-oil and biochar can be used for the production of synthesis gas through the agency of the gasification process. The technology used for the gasification process is a tried-and-true technology for coal (Speight, 2013a) and other carbonaceous feedstocks such as crude oil residua (Speight, 2014b). The majority of gasifiers are partial oxidation reactors, in which just sufficient air or oxygen is introduced to burn part of the input biomass to provide the heat for pyrolysis and gasification (Adhikari *et al.*, 2015). In cases where the oxidant is air, the product gas is diluted by the nitrogen present, and although air is 79% nitrogen, the stoichiometry of partial oxidation is such that the final product gas has approximately 50% v/v nitrogen as a diluent.

As a result the gas heating value of the fuel gas derived from air driven partial oxidation gasifiers is relatively low. The value can be higher if the feedstock is very dry, thus minimizing the heat demand for the process and the amount of oxidant required. The product gas has a wide range of applications, from power generation to chemicals production. The power derived from the gasification of carbonaceous feedstocks followed by the combustion of the product gas(es) is considered to be a source of renewable energy of derived gaseous products (Table 6.1) that are generated from a carbonaceous source (e.g., biomass) (Pakdel and Roy, 1991; Speight, 2008, 2014b; Adhikari *et al.*, 2015).

Table 6.1 Gasification products.

Product	Properties
Low-Btu gas	150 to 300 Btu/ft^3
	Approximately 50% v/v nitrogen
	Smaller amounts of carbon monoxide and hydrogen
	Some carbon dioxide
	Trace amounts of methane
Medium-Btu gas	300 to 550 Btu/ft^3
	Predominantly carbon monoxide and hydrogen
	Small amounts of methane
	Some carbon dioxide
High-Btu gas	980 to 1080 Btu/ft^3
	Predominantly methane - typically >85% v/v

If the requirement of the process is a high heat-content gas, the production of products such as methane, ethane, and propane further increases the calorific value. The use of pure oxygen as the gasification agent eliminates the nitrogen diluent and can produce medium calorific value gaseous products. An alternative strategy is to carry out the gasification process by means of indirect heat input. In this case the product stream is even higher in calorific value, as neither nitrogen nor the carbon dioxide produced from the combustion *in-situ* of the partial oxidation processes is present in the product gas stream. The challenges to achieve a clean and useable fuel gas have been addressed through gasifier design and post-gasification processing to remove tar and particulate contaminants in the gas stream.

Tar is a significant potential problem in the gasification of biomass as it can lead to equipment blockages, increased maintenance and makes operation difficult. It is usually a thick, dark-colored liquid with a low condensation temperature and hence can lead to blockages in downstream equipment. In particular, low-boiling hydrocarbon derivatives (C_{2-6}) can actually avoid condensation and instead form tarry aerosols, which in turn degrade the quality of outlet gas and potentially make it unsuitable for use in high-purity applications, e.g., applications other than boilers.

In terms of tar production, during the heating of biomass, the order of mass loss is as follows: first, moisture evaporation takes place from 30 to 120°C (86 to 250°F), followed by hemicellulose breakdown between 150 to 220°C (300 to 430°F), lignin breakdown at temperatures on the order of 220 to 400°C (430 to 750°F) and cellulose being the most stable decomposes at temperatures on the order of around 315 to 450°C (600 to 840°F). Lignin degrades over the widest range due to the diverse range of chemical structures and bonds present. Devolatilization and biomass conversion can occur across a range of temperatures for different kinds of biomass; this is commonly investigated using thermogravimetric analysis (TGA). The typical components of tar are (Sikarwar *et al.*, 2016):

Benzene (38% w/w)
Toluene (14.5% w/w)
Single-ring aromatic hydrocarbon derivatives (14% w/w)
Naphthalene (9.5% w/w)
Two-ring aromatic hydrocarbon derivatives (8% w/w)
Heterocyclic compounds (6.5% w/w)
Phenolic compounds (4.5% w/w)
Three-ring aromatic hydrocarbon derivatives (3.5% w/w)
Four-ring aromatic hydrocarbon derivatives (1% w/w)

The composition and amount of tar depends on the gasifier type and gasification temperature. Tar from the fluidized-bed gasifier can be characterized as secondary and tertiary tar. For example, increasing the temperature of the gasification process typically reduces the amount of tar (the primary tar) but increases the share of higher molecular weight (and higher-boiling) constituents (the secondary tar). The effect of pressure on tar composition is variable and process dependent. However, it has been observed that an increase in pressure may drive the composition of the tar to the more higher-boiling constituents.

The minimization of tar formation can be achieved in two ways: (i) by degradation of the tar inside the gasifier, which is referred to as primary tar reduction and requires an appropriate selection of functional parameters discussed earlier such as the design of the gasifier and the use of catalysts during gasification, (ii) by thermal, mechanical or catalytic cracking of tar in a separate step after the biomass gasification, which is referred to as secondary tar reduction where the tar is deposited further downstream.

In terms of the use of biomass feedstocks and, hence, the production of synthesis gas, the process has grown from a predominately coal conversion process used for making *town gas* for industrial lighting to an advanced process for the production of multi-product, carbon-based fuels from a variety of feedstocks such as the biomass alone, biomass mixed with coal, and biomass mixed with the viscous products from crude oil refining and any other carbonaceous feedstocks (Figure 6.1) (Kumar *et al.*, 2009; Speight, 2013a, 2014a, 2014b; Luque and Speight, 2015; Sikarwar *et al.*, 2016, 2017).

Gasification is a key process for the thermo-chemical conversion of biomass. In the presence of a gasifying agent (such as oxygen or air), biomass is converted to a multifunctional gaseous mixture, usually called bio-synthesis gas or bio-synthesis gas, which can be used for the production of energy (heat and/or electricity generation), chemicals (ammonia), and biofuels. Furthermore, a solid residue after biomass conversion (char) is generally found (Bridgwater, 2003; Rauch *et al.*, 2014; Molino *et al.*, 2016; Sikarwar *et al.*, 2017). Synthesis gas consists of a mixture of carbon monoxide, hydrogen, methane, carbon dioxide (the primary components) and water, hydrogen sulfide, ammonia (secondary components), as well as tarry constituents and other trace species with a composition dependent on feedstock type and characteristics, operating conditions (i.e., the gasifying agents, gasifier temperature and pressure, type of bed materials), and gasification technology (Molino *et al.*, 2013; Asadullah, 2014; Rodríguez-Olalde *et al.*, 2015; Ahmad *et al.*, 2016).

Gasification is an appealing process for the utilization of relatively inexpensive feedstocks, such as biomass and waste, that might in the case of waste be sent to a landfill (where the production of methane – a so-called greenhouse gas – will be produced and often escapes into the atmosphere) or combusted which may not (depending upon the

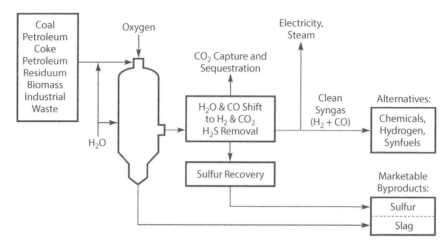

Figure 6.1 The gasification process can accommodate a variety of carbonaceous feedstocks.

feedstock) be energy efficient. Because of the development of the process using coal as the feedstock (Speight, 2013), biomass is often considered to be an unconventional carbonaceous feedstock which also includes heavy oil, extra heavy oil, tar sand bitumen, and crude oil residua (Speight, 2014b).

A biomass gasifier needs uniform-sized and dry fuel for smooth and trouble-free operation. Most gasifier systems are designed either for woody biomass (or dense briquettes made from loose biomass) or for loose, pulverized biomass (Chen et al., 1992; Bhattacharya et al., 1999; Hanaoka et al., 2005). The feedstock requirements vary depending upon the biomass fed to the gasifier, for example: (i) for woody biomass: pieces smaller than 5 to 10 cm (2 to 4 inches) in any dimension, depending on design and bulk density of wood or briquettes, and (ii) for loose biomass: pulverized biomass, depending on design, moisture content up to 15 to 25%, depending on gasifier design, and a mineral matter content (manifested as product ash) below 5% w/w is preferred; with a maximum limit of 20% w/w.

Overall, use of a gasification technology (Speight, 2013a, 2014b) with the necessary gas cleanup options can have a smaller environmental footprint and lesser effect on the environment than landfill operations or combustion of the biomass or organic waste. In fact, there are strong indications that gasification is a technically viable option for the waste conversion, including residual waste from separate collection of municipal solid waste. The process can meet existing emission limits and can have a significant effect on the reduction of landfill disposal using known gasification technologies to produce a synthesis gas product than can be employed as a fuel, a petrochemical feedstock as well as for use as a feedstock to the Fischer-Tropsch process (Arena, 2012; Fabry et al., 2013; Speight, 2014b; Luque and Speight, 2015).

However, it is compared to a typical fossil fuel (if there is such a fuel), the complex ligno-cellulosic structure of biomass is more difficult to gasify or combust. The nature of the mineral impurities in conjunction with the presence of various inorganic species, as well as sulfur and nitrogen containing compounds, adversely impacts the benign thermal processing of the oxygenated hydrocarbon structure of the biomass. While combustion of biomass feedstocks results in fuel-bound nitrogen and sulfur (being converted to NOx and SOx, respectively) steam gasification involves thermal treatment under a reducing atmosphere resulting in fuel-bound nitrogen release as molecular nitrogen (N_2) and fuel-bound sulfur conversion to hydrgoen sulfide (H_2S) that is more easily removed by means of adsorption beds. Unlike combustion, the gasification process is more energy intensive. Careful engineering of the process is necessary if the result is to produce rather than consume a significant amount of energy or power as a result of the thermal treatment (Butterman and Castaldi, 2008).

6.2 Gasification Chemistry

The gasification process consists of the conversion of the organic carbon in the biomass into various constituents that make up the synthesis gas constituents by reacting the biomass at high temperatures (>700°C, 1290°F), without combustion using a controlled amount of oxygen and/or steam (Marano, 2003; Lee, 2007; Lee *et al.*, 2007; Higman and Van der Burgt, 2008; Speight, 2008; Sutikno and Turini, 2012; Speight, 2013a, 2014b; Luque and Speight, 2015). The products include gaseous hydrocarbon derivatives (methane, ethane, and propane), carbon monoxide, carbon dioxide, and hydrogen. Other products may also be produced depending upon the composition of the feedstock.

Biomass gasification consists of many overlapping processes: drying, pyrolysis and partial oxidation. The feasible gasification routes involve the process of producing solid, liquid or gaseous fuels or valuable chemicals by transforming biomass in an O_2-deficient environment. The process can be categorized as mild pyrolysis, slow pyrolysis or as fast pyrolysis. A very simple way of representing the gasification reaction is:

$$\text{Biomass} \rightarrow H_2 + CO + CO_{2(g)} + \text{hydrocarbons}_{(g)} + \text{tar}_{(l)} + \text{char}_{(s)}$$

Fast pyrolysis is considered to be an efficient thermochemical route to transform biomass into liquid fuel, followed by partial oxidation and subsequent gasification to yield the desired products. Elevated temperatures, high heating rates and long volatile stage residence times are vital for synthesis gas production or hydrogen production (Chapter 11). Since the liquid yield through fast pyrolysis is high, the gaseous yield is reduced. This approach does not allow the production directly of usable fuels (or feedstock for blending), but produces pyrolytic oils requiring further upgrading owing to their high acidity. Gasification is an alternative thermo-chemical route to treat biomass, which reduces these limitations.

It is important to distinguish the gasification process from a pyrolysis process. The main difference between pyrolysis and gasification is the absence of a gasifying agent in a pyrolysis process (Chapter 5). Pyrolysis is a thermal degradation of organic compounds, at a range of temperatures on the order of 300 to 900°C (570 to 1650°F), under oxygen-deficient circumstances to produce various types of products such as gases (often referred to as synthesis gas), a liquid product (referred to as bio-oil), and a solid product (referred to as biochar).

On the other hand, the gasification process involves thermal cracking of solid or high-boiling liquid carbonaceous material (such as bio-oil biomass, biomass char) into a combustible gas mixture, mainly composed of hydrogen, carbon monoxide (CO), carbon dioxide (CO_2), and methane (CH_4) and other gases with some byproducts (solid char or slag, oils, and water). The synthesis gas product has a chemical composition and properties that are largely affected by (i) the type and composition of the biomass feedstock, (ii) the operational conditions throughout pyrolysis and gasification such as reactor temperature, residence time, pressure, and (iii) the reactor geometry.

In the current context, and in addtion to the potential reactions that can occur in the gasifier (Table 6.2), the gasification process is often considered to involve two distinct physico-chemical stages: (i) devolatilization of the feedstock to produce volatile matter and char, (ii) followed by char gasification, which is complex and specific to the conditions of the reaction – both processes contribute to the complex kinetics of the gasification process (Sundaresan and Amundson, 1978; Luque and Speight, 2015).

6.2.1 General Aspects

Generally, the gasification of carbonaceous feedstocks (such as biomass, waste, heavy oil, extra heavy oil, tar sand bitumen, crude oil residua) includes a series of reaction steps that convert the feedstock into *synthesis gas* (carbon monoxide, CO, plus hydrogen, H_2) or, if biomass is the feedstock, into bio-synthesis gas and other gaseous products (Table 6.2). The gaseous products and the proportions of these product gases (such as methane, CH_4, and other hydrocarbon derivatives, water vapor (H_2O), carbon dioxide, (CO_2), carbon monoxide (CO), hydrogen (H_2), hydrogen sulfide (H_2S), and sulfur dioxide, (SO_2) depends on: (i) the type of feedstock, (ii) the chemical composition of the feedstock, (iii) the gasifying agent or gasifying medium, as well as (iv) the thermodynamics and chemistry of the gasification reactions as controlled by the process operating parameters (Singh *et al.*, 1980;

Table 6.2 Reactions that occur during gasification of a carbonaceous feedstock.

$2C + O_2 \rightarrow 2CO$
$C + O2 \rightarrow CO_2$
$C + CO_2 \rightarrow 2CO$
$CO + H_2O \rightarrow CO_2 + H_2$ (shift reaction)
$C + H_2O \rightarrow CO + H_2$ (water gas reaction)
$C + 2H_2 \rightarrow CH_4$
$2H_2 + O_2 \rightarrow 2H_2O$
$CO + 2H2 \rightarrow CH_3OH$
$CO + 3H2 \rightarrow CH_4 + H_2O$ (methanation reaction)
$CO_2 + 4H_2 \rightarrow CH_4 + 2H_2O$
$C + 2H_2O \rightarrow 2H_2 + CO_2$
$2C + H_2 \rightarrow C_2H_2$
$CH_4 + 2H_2O \rightarrow CO_2 + 4H_2$

Pepiot *et al.*, 2010; Shabbar and Janajreh, 2013; Speight, 2013a, 2013b, 2014b; Luque and Speight, 2015). In addition, the kinetic rates, the extent of the conversion for any feedstock (irrespective of the composition), as well as the several chemical reactions that are a part of the gasification process are variable and are typically functions of (i) temperature, (ii) pressure, and (iii) reactor configuration, the (iv) gas composition of the product gases, and (v) whether or not the product gases influence the outcome of the reaction (Johnson, 1979; Penner, 1987; Müller *et al.*, 2003; Slavinskaya *et al.*, 2009; Speight, 2013a, 2013b, 2014b; Luque and Speight, 2015).

At a temperature in excess of 500°C (930°F) the conversion of the feedstock to char and ash and char is completed. In most of the early gasification processes, this was the desired byproduct but for gas generation the char provides the necessary energy to effect further heating and – typically, the char is contacted with air or oxygen and steam to generate the product gases. Furthermore, with an increase in heating rate, feedstock particles are heated more rapidly and are burned in a higher temperature region, but the increase in heating rate may have very little effect on the mechanism (Demirbaş, 2004; Tinaut *et al.*, 2008; Irfan, 2009).

Most notable effects in the physical chemistry of the gasification process are those effects due to the chemical character of the feedstock as well as the physical composition of the feedstock (Speight, 2011a, 2013a, 2014a, 2014b). In more general terms of the character of the feedstock, gasification technologies generally require some initial processing of the feedstock with the type and degree of pretreatment a function of the process and/or the type of feedstock (Table 6.3, Table 6.4). This is especially true for biomass where the feedstock can have several constituents that strongly influence the course and outcome of the conversion process (Chapter 2).

Table 6.3 Acidic and alkaline methods for biomass treatment.

Method	Conditions	Outcome
Acid based methods	Low pH using an acid (H_2SO_4, H_3PO_4)	Hydrolysis of the hemicellulose to monomer sugars
		Minimizes the need for hemicellulases
Neutral conditions		Steam pretreatment and hydrothermolysis
		Solubilizes most of the hydrocarbons by conversion to acetic acid
		Does not usually result in total conversion to monomer sugars
		Requires hemicellulases acting on soluble oligomers
Alkaline methods		Leaves a part of the hydrocarbon in the solid fraction
		Requires hemicellulases acting hydrocarbons

Table 6.4 Summation of the methods for the pretreatment of biomass feedstocks.

Physical methods	Miscellaneous methods
Milling:	Explosion:*
-Ball milling	- Steam, NH_3, CO_2, SO_2, Acids, Alkali
-Two-roll milling	- NaOH, NH_3, $(NH_4)_2SO_3$
-Hammer milling	Acid:
Irradiation:	- Sulfuric, hydrochloric, and phosphoric acids
-Gamma-ray irradiation	Gas:
-Electron-beam irradiation	- Chlorine dioxide, nitrous oxide, sulfur dioxide
-Microwave irradiation	Oxidation:
Other methods:	- Hydrogen peroxide
- Hydrothermal	- Wet oxidation
- High pressure steaming	- Ozone
- Extrusion	Solvent extraction of lignin:
- Pyrolysis	- Ethanol-water extraction
	- Benzene-water extraction
	- Butanol-water extraction
	Organic solvents Ionic liquids

* The feedstock material is subjected to the action of steam and high-pressure carbon dioxide before being discharged through a nozzle.

Another factor, often presented as very general *rule of thumb*, is that optimum gas yields and gas quality are obtained at operating temperatures of approximately 595 to 650°C (1100 to 1200°F). A gaseous product with a higher heat content (BTU/ft³) can be obtained at lower system temperatures but the overall yield of gas (determined as the *fuel-to-gas ratio*) is reduced by the unburned char fraction.

With some feedstocks, the higher the amounts of volatile material produced in the early stages of the process the higher the heat content of the product gas. However, in the case of biomass, there is the potential for carbon dioxide to be an initial product of the process which will serve to lower the heat content of the product gas. Also, when the process temperature is not sufficiently high, the char oxidation reaction is suppressed and the overall heat content of the product gas is diminished. All such events serve to complicate the reaction rate and make the attempted derivation of a kinetic relationship that can be applied to all types of feedstock (and gasifier configurations) subject to serious question and doubt.

Pressure also plays a role in the type of product produced and is, as might be anticipated, dependent on the type of feedstock being processed and the gas product that is desired. In fact, some (or all) of the following processing steps will be required: (i) pretreatment of the feedstock, (ii) primary gasification of the feedstock, (iii) secondary gasification of the carbonaceous residue from the primary gasifier; (iv) removal of carbon dioxide, hydrogen sulfide, and other acid gases, (v) shift conversion for adjustment of the carbon monoxide/hydrogen mole ratio to the desired ratio, and (vi) catalytic methanation of the carbon monoxide/hydrogen mixture to form methane.

While pretreatment of the biomass feedstock for introduction into the gasifier is often considered to be a physical process in which the feedstock is prepared for gasifier – typically as pellets or finely ground feedstock – there are chemical aspects that must also be considered (Chapter 2). Co-gasification of other feedstocks, such as coal and with the residuum or coke from crude oil refining (often referred to as petcoke) offers a bridge between the depletion of crude oil stocks when coal, residua, and/or petcoke are used (Speight, 2014b).

The high reactivity of biomass and the accompanying high production of volatile products suggest that some synergetic effects might occur in simultaneous thermochemical treatment of coal, residua, or petcoke, depending on the gasification conditions such as: (i) feedstock type and origin, (ii) reactor type, and (iii) process parameters (Penrose *et al.*, 1999; Gray and Tomlinson, 2000; McLendon *et al.*, 2004; Lapuerta *et al.*, 2008; Fermoso *et al.*, 2009; Shen *et al.*, 2012; Khosravi and Khadse, 2013; Speight, 2013a, 2014a, 2014b; Luque and Speight, 2015).

For example, carbonaceous fuels are gasified in reactors, a variety of gasifiers such as the fixed- or moving-bed, fluidized-bed, entrained-flow, and molten-bath gasifiers have been developed that have differing feedstock requirements (Table 6.5) (Beenackers, 1999; Shen *et al.*, 2012; Speight, 2014b; Luque and Speight, 2015). If the flow patterns are considered, the fixed-bed and fluidized-bed gasifiers intrinsically pertain to a countercurrent reactor in that fuels are usually sent into the reactor from the top of the gasifier, whereas the oxidant is blown into the reactor from the bottom.

6.2.2 Reactions

The gasification process is not a single-step process, but involves a variety of reactions (Table 6.2), and like any gasification process, the gasification of biomass (including bio-oil and biochar) is a complex thermal process that depends on the pyrolysis mechanism to generate gaseous precursors, which in the presence of reactive gases such as oxygen and steam convert the majority of the biomass into a synthesis gas, which is mainly composed of carbon monoxide and hydrogen, and used to produce chemicals and liquid fuels over catalysts (Speight, 2016, 2017; 2019). The distribution of weight and chemical composition of the products are also influenced by the prevailing conditions (i.e., temperature, heating rate, pressure, and residence time) and, last but by no means least, the nature of the feedstock (Speight, 2014a, 2014b).

If air is used for combustion, the product gas will have a heat content of approximately 150-300 Btu/ft^3 (depending on process design characteristics) and will contain undesirable constituents such as carbon dioxide, hydrogen sulfide, and nitrogen. The use of pure oxygen, although expensive, results in a product gas having a heat content on the order of 300 to 400 Btu/ft^3 with carbon dioxide and hydrogen sulfide as byproducts (both of which

Table 6.5 Characteristics of the different types of gasifiers.

Gasifier type	Fuel properties
Fixed/moving bed	Particle size: 1 to 10 cm
	Mechanically stable fuel particles (unblocked passage of gas through the bed)
	Pellets or briquettes preferred
	Updraft configuration more tolerant to biomass moisture content (up to 40 to 50% w/w)
	Drying occurs as biomass moves down the gasifier
Fluidized bed	Ash melting temperature of fuel: higher limit for operating temperature.
	Fuel particle size relatively small to ensure good contact with bed material; typically <40 mm
	Good fuel flexibility due to high thermal inertia of the bed.
Entrained bed	Fuel particle size: < 50 micrometers.
	Pulverized for high fuel conversion in short residence times
	Low moisture content
	Ash melting behavior can influence for reactor/process design.

can be removed from low or medium heat-content, low-Btu or medium-Btu) gas by any of several available processes (Speight, 2013a, 2014a).

If a high heat-content (high-Btu) gas (900 to 1000 Btu/ft^3) is required, efforts must be made to increase the methane content of the gas. The reactions which generate methane are all exothermic and have negative values, but the reaction rates are relatively slow and catalysts may, therefore, be necessary to accelerate the reactions to acceptable commercial rates. Indeed, the overall reactivity of the feedstock and char may be subject to catalytic effects insofar as the mineral constituents of the feedstock (such as the mineral matter in biomass) may modify the reactivity by a direct catalytic effect (Davidson, 1983; Baker and Rodriguez, 1990; Mims, 1991; Martinez-Alonso and Tascon, 1991).

Biomass begins to rapidly decompose with heat once its temperature rises above 240°C (465°F) and the biomass breaks down into a combination of gases, liquids, and solid (char). In the process, the feedstock undergoes three steps in the conversion to gas – the first two steps, pyrolysis and combustion, occur very rapidly – all of which are highly dependent upon the properties of the biomass (Table 6.6). However, there is also a preliminary step in which any moisture in the biomass is removed before it enters the pyrolysis step. All of the moisture needs to be (or will be) removed from the biomass at some point (typically between 100 and 150°C, 212 and 300°F) in the higher temp processes. Moisture removal (often referred to as drying) is one of the major issues that has to be solved for successful gasification of biomass. The high moisture content of the feedstock and/or poor handling of the moisture internally, is one of the most common reasons for failure to produce clean gas in the process.

Table 6.6 Biomass properties that influence the gasification process.

Chemical properties	Elemental composition	C,H,N,O,S (% w/w)
	Proximate analysis	Moisture
		Fixed carbon
		Volatile matter
		Mineral matter
Physical properties	Density	
	Porosity	
	Particle size	
	Particle shape	
Thermal properties	Heating value	
	Elemental composition	
	Types of minerals	Mineral ash
		Ash fusion temperature

In the pyrolysis step, char is produced as the feedstock heats up and volatiles are released. In the combustion step, the volatile products and some of the char reacts with oxygen to produce various products (primarily carbon dioxide and carbon monoxide) and the heat required for subsequent gasification reactions. Finally, in the gasification step, the feedstock char reacts with steam to produce hydrogen (H_2) and carbon monoxide (CO).

Drying:

$$\text{Biomass (wet)} \rightarrow \text{biomass (dry)} + H_2O$$

Pyrolysis:

$$\text{Biomass} \rightarrow \text{volatiles} + \text{char}$$

Combustion:

$$2C_{\text{volatiles + char}} + O_2 \rightarrow 2CO + H_2O$$

Gasification:

$$C_{\text{char}} + H_2O \rightarrow H_2 + CO$$

$$CO + H_2O \rightarrow H_2 + CO_2$$

Thus, in the initial stages of gasification, the rising temperature of the feedstock initiates devolatilization and the breaking of weaker chemical bonds to yield volatile tar, volatile oil, phenol derivatives, and hydrocarbon gases. These products generally react further in the gaseous phase to form hydrogen, carbon monoxide, and carbon dioxide. The char (fixed carbon) that remains after devolatilization reacts with oxygen, steam, carbon dioxide, and hydrogen.

In essence, the direction of the gasification process is subject to the constraints of thermodynamic equilibrium and variable reaction kinetics. The combustion reactions (reaction of the feedstock or char with oxygen) essentially go to completion. The thermodynamic equilibrium of the rest of the gasification reactions are relatively well defined and collectively have a major influence on thermal efficiency of the process as well as on the gas composition. Thus, thermodynamic data are useful for estimating key design parameters for a gasification process, such as: (i) calculating of the relative amounts of oxygen and/or steam required per unit of feedstock, (ii) estimating the composition of the produced bio-synthesis gas, and (iii) optimizing process efficiency at various operating conditions.

Relative to the thermodynamic understanding of the gasification process, the kinetic behavior is much more complex. In fact, very little reliable kinetic information (for general application) on gasification reactions exists, partly because it is highly dependent on (i) the chemical nature of the feed, which varies significantly with respect to composition, mineral impurities, (ii) feedstock reactivity, and (iii) process parameters, such as temperature, pressure, and residence time. In addition, physical characteristics of the feedstock (or char) also play a role in phenomena such as boundary layer diffusion, pore diffusion and ash layer diffusion which also influence the kinetic outcome. Furthermore, certain impurities, in fact, are known to have catalytic activity on some of the gasification reactions which can have further influence on the kinetic imprint of the gasification reactions.

Alkali metal salts are known to catalyze the steam gasification reaction of carbonaceous materials, including biomass – hence the need for pretreatment of biomass to remove such metals (Chapter 2). The process is based on the concept that alkali metal salts (such as potassium carbonate, sodium carbonate, potassium sulfide, sodium sulfide, and the like) will catalyze the steam gasification of feedstocks. The order of catalytic activity of alkali metals on the gasification reaction is:

Cesium (Cs) > rubidium (Rb) > potassium (K) > sodium (Na) > lithium (Li)

6.2.2.1 Water Gas Shift Reaction

If a high-Btu gas is required, the option is application of the water gas shift reaction, in which carbon monoxide is converted to hydrogen and carbon dioxide in the presence of a catalyst:

Water Gas Shift Reaction:

$$CO + H_2O \rightarrow CO_2 + H_2$$

This reaction maximizes the hydrogen content of the gas and the product gas is then scrubbed of particulate matter and sulfur is removed via physical absorption (Speight, 2013a, 2014a). The carbon dioxide is captured by physical absorption or a membrane and either vented or sequestered.

The water gas shift reaction in its forward direction is mildly exothermic and although all of the participating chemical species are in gaseous form, the reaction is believed to be heterogeneous insofar as the chemistry occurs at the surface of the feedstock and the reaction is actually catalyzed by carbon surfaces. In addition, the reaction can also take place homogeneously as well as heterogeneously and a generalized understanding of the water gas shift reaction is difficult to achieve. Even the published kinetic rate information is not immediately useful or applicable to a practical reactor situation.

The water gas shift reaction is one of the major reactions in the steam gasification process, where both water and carbon monoxide are present in ample amounts. Although the four chemical species involved in the water gas shift reaction are gaseous compounds at the reaction stage of most gas processing, the water gas shift reaction, in the case of steam gasification of feedstock, predominantly takes place on the solid surface of feedstock (heterogeneous reaction). If the bio-synthesis gas from a gasifier needs to be reconditioned by the water gas shift reaction, this reaction can be catalyzed by a variety of metallic catalysts.

Choice of specific kinds of catalysts has always depended on the desired outcome, the prevailing temperature conditions, composition of gas mixture, and process economics. Typical catalysts used for the reaction include catalysts containing iron, copper, zinc, nickel, chromium, and molybdenum.

6.2.2.2 Carbon Dioxide Gasification

The reaction of carbonaceous feedstocks with carbon dioxide produces carbon monoxide (*Boudouard reaction*) and (like the steam gasification reaction) is also an endothermic reaction:

$$C(s) + CO_2(g) \rightarrow 2CO(g)$$

The reverse reaction results in carbon deposition (carbon fouling) on many surfaces including the catalysts and results in catalyst deactivation. This reaction may be valuable when biomass is the feedstock because of the potential for the production of carbon dioxide directly from the biomass by the decompositon of carboxylic acid functions:

$$[\equiv C\text{-}CO_2H] \rightarrow CO_2 + [\equiv CH]$$

This gasification reaction is thermodynamically favored at high temperatures (>680°C, >1255°F), which is also quite similar to the steam gasification. If carried out alone, the reaction requires high temperature (for fast reaction) and high pressure (for higher reactant concentrations) for significant conversion but as a separate reaction a variety of factors come into play: (i) low conversion, (ii) slow kinetic rate, and (iii) low thermal efficiency.

Also, the rate of the carbon dioxide gasification of a feedstock is different to the rate of the carbon dioxide gasification of carbon. Generally, the carbon-carbon dioxide reaction follows a reaction order based on the partial pressure of the carbon dioxide that is approximately 1.0 (or lower) whereas the feedstock-carbon dioxide reaction follows a reaction order based on the partial pressure of the carbon dioxide that is 1.0 (or higher). The observed higher reaction order for the feedstock reaction is also based on the relative reactivity of the feedstock in the gasification system.

6.2.2.3 Hydrogasification

Hydrogasification is the gasification of feedstock, such as biomass, in the presence of an atmosphere of hydrogen under pressure. Not all high heat-content (high-Btu) gasification technologies depend entirely on catalytic methanation and, in fact, a number of gasification processes use hydrogasification, that is, the direct addition of hydrogen to feedstock under pressure to form methane.

$$C_{char} + 2H_2 \rightarrow CH_4$$

The hydrogen-rich gas for hydrogasification can be manufactured from steam and char from the hydrogasifier. Appreciable quantities of methane are formed directly in the primary gasifier and the heat released by methane formation is at a sufficiently high temperature to be used in the steam-carbon reaction to produce hydrogen so that less oxygen is used to produce heat for the steam-carbon reaction. Hence, less heat is lost in the low-temperature methanation step, thereby leading to higher overall process efficiency.

The hydrogasification reaction is exothermic and is thermodynamically favored at relatively low temperature (<670°C, <1240°F), unlike the endothermic both steam gasification and carbon dioxide gasification reactions. However, at low temperatures, the reaction rate is inevitably too slow. Therefore, a high temperature is always required for kinetic reasons, which in turn requires high pressure of hydrogen, which is also preferred from equilibrium considerations. This reaction can be catalyzed by salts such as potassium carbonate (K_2CO_3), nickel chloride ($NiCl_2$), iron chloride ($FeCl_2$), and iron sulfate ($FeSO_4$). However, use of a catalyst in feedstock gasification suffers from difficulty in recovering and reusing the catalyst and the potential for the spent catalyst becoming an environmental issue.

6.2.2.4 Methanation

Several exothermic reactions may occur simultaneously within a methanation unit. A variety of metals have been used as catalysts for the methanation reaction; the most common, and to some extent the most effective methanation catalysts, appear to be nickel and ruthenium, with nickel being the most widely used (Cusumano et al., 1978):

Ruthenium (Ru) > nickel (Ni) > cobalt (Co) > iron (Fe) > molybdenum (Mo).

Nearly all the commercially available catalysts used for this process are, however, very susceptible to sulfur poisoning and efforts must be taken to remove all hydrogen sulfide (H_2S) before the catalytic reaction starts. It is necessary to reduce the sulfur concentration in the feed gas to less than 0.5 ppm v/v in order to maintain adequate catalyst activity for a long period of time.

The gas must be desulfurized before the methanation step since sulfur compounds will rapidly deactivate (poison) the catalysts. A processing issues may arise when the concentration of carbon monoxide is excessive in the stream to be methanated since large amounts of heat must be removed from the system to prevent high temperatures and deactivation of the catalyst by sintering as well as the deposition of carbon. To eliminate this problem, temperatures should be maintained below 400°C (750°F).

The methanation reaction is used to increase the methane content of the product gas, as needed for the production of high-Btu gas.

$$4H_2 + CO_2 \rightarrow CH_4 + 2H_2O$$

$$2CO \rightarrow C + CO_2$$

$$CO + H_2O \rightarrow CO_2 + H_2$$

Among these, the most dominant chemical reaction leading to methane is the first one. Therefore, if methanation is carried out over a catalyst with a bio-synthesis gas mixture of hydrogen and carbon monoxide, the desired hydrogen-carbon monoxide ratio of the feedstock gas is on the order of 3:1. The high proportion of water (vapor) produced is removed by condensation and recirculated as process water or steam. During this process, most of the exothermic heat due to the methanation reaction is also recovered through a variety of energy integration processes.

Whereas all the reactions listed above are quite strongly exothermic except the forward water gas shift reaction, which is mildly exothermic, the heat release depends largely on the amount of carbon monoxide present in the feedstock gas. For each 1% v/v carbon monoxide in the feedstock gas, an adiabatic reaction will experience a 60°C (108°F) temperature rise, which may be termed as *adiabatic temperature rise*.

6.3 Gasification Processes

The typical gasification system consists of several process plants including (i) a feedstock preparation area, (ii) the type of gasifier, (iii) a gas cleaning section, and (iv) a sulfur recovery unit, as well as (v) downstream process options that are dependent on the nature of the products. By way of further explanation, the gas generated from biomass by gasification processes can be differentiated in direct (autothermal) and indirect (or allothermal) processes. For biomass applications the direct processes are typically operated with air as the gasifying agent.

Typically, gasification is divided into four steps: drying (endothermic step), pyrolysis (endothermic step), oxidation (exothermic stage), and reduction (endothermic stage). In the drying step, when the biomass (which typically contains 10 to 355 w/w moisture) is heated to about 100°C (212°F), the moisture is converted into steam. After drying, as heating continues, the biomass undergoes pyrolysis which involves burning biomass completely without supplying any oxygen. As a result, the biomass is decomposed or separated into gases, liquids, and solids. During oxidation, which takes place at about 700 to 1400°C (1290 to 2550°F), the solid (carbonized) biomass, reacts with the oxygen in the air to produce carbon dioxide and heat:

$$C + O_2 \rightarrow CO_2 + \text{heat}$$

At higher temperatures and under reducing conditions, when insufficient oxygen is available, the following reactions take place forming carbon dioxide, hydrogen, and methane:

$$C + CO_2 \rightarrow 2CO$$

$$C + H_2O \rightarrow CO + H_2$$

$$CO + H_2O \rightarrow CO_2 + H_2$$

$$C + 2H_2 \rightarrow CH_4$$

Tar-reforming can also be added as a step to produce low molecular weight hydrocarbon derivatives from higher molecular weight constituents of the tarry products. Although the chemistry of biomass gasification is complex (Table 6.7), the overall process can be represented by a simple equation. Thus:

$$\text{Biomass} \rightarrow CO + H2 + CO2 + CH4 + H2O + H2S + NH3 + CxHy + \text{Tar} + \text{Char}$$

6.3.1 Gasifiers

A gasifier differs from a combustor in that the amount of air or oxygen available inside the gasifier is carefully controlled so that only a relatively small portion of the fuel burns completely. Thus, the gasifier contains separate sections for gasification and combustion. The gasification section consists of three parts: (i) the riser, (ii) the settling chamber, and (iii) the downcomer whereas the combustion section contains only one part, the combustor.

In the process, biomass is fed into the riser along with a small amount of superheated steam. Hot bed material (typically 925°C, 1695°F, sand or olivine of 0.2 to 0.3 mm particle size) enters the riser from the combustor through a hole in the riser (opposite of biomass feeding point). The bed material heats the biomass to 850°C (1550°F) causing the biomass particles to undergo conversion into gas. The volume created by the gas from the biomass results in the creation of a turbulent fluidization regime in the riser causing carry-over of the bed material together with the degasified biomass particles (char). The

Table 6.7 Predominant reactions occurring during biomass gasification.

Sub-process	Reaction
Pyrolysis	Biomass → CO + H_2 + CO + CH_4 + H_2O + Tar + Char
Oxidation	Char + O_2 → CO_2 (Char Oxidation)
	2C + O_2 → 2CO (Partial Oxidation)
	$2H_2 + 2O_2$ → $2H_2O$ (Hydrogen Oxidation)
Reduction	C + CO_2 ↔ 2CO (Boudouard Reaction)
	C + H_2O ↔ CO + H2 (Reforming of Char)
	CO + H_2O ↔ CO_2 + H_2 (Water Gas Shift (WGS) Reaction)
	C + $2H_2$ ↔ CH_4 (Methanation Reaction)
	CH_4 + H_2O ↔ CO + $3H_2$ (Steam Reforming of Methane)
	CH_4 + CO_2 ↔ 2CO + $2H_2$ (Dry Reforming of Methane)
Tar Reforming	Tar + H_2O → H_2 + CO_2 + CO + CxHy (Steam Reforming of Tar)

settling chamber reduces the vertical velocity of the gas causing the larger solids (bed material and char) to separate from the gas and to pass into the downcomer. The gas stream leaves the reactor from the top and is sent to the cooling and gas cleaning section in order to remove contaminants such as dust, tar, chloride and sulfur (Chapter 7) before the catalytic conversion of the gas into biomethane.

Four types of gasifier are currently available for commercial use: (i) the countercurrent fixed bed, (ii) co-current fixed bed, (iii) the fluidized bed, and (iv) the entrained flow (Speight, 2008, 2013a; Luque and Speight, 2015).

In a *fixed-bed process*, the feedstock is supported by a grate and combustion gases (steam, air, oxygen, etc.) pass through the supported feedstock whereupon the hot produced gases exit from the top of the reactor. Heat is supplied internally or from an outside source, but some feedstocks cannot be used in an unmodified fixed-bed reactor. Due to the liquid-like behavior, the fluidized-beds are very well mixed, which effectively eliminates the concentration and temperature gradients inside the reactor. The process is also relatively simple and reliable to operate as the bed acts as a large thermal reservoir that resists rapid changes in temperature and operation conditions. The disadvantages of the process include the need for recirculation of the entrained solids carried out from the reactor with the fluid, and the nonuniform residence time of the solids that can cause poor conversion levels. The abrasion of the particles can also contribute to serious erosion of pipes and vessels inside the reactor (Kunii and Levenspiel 1991) – another reason why feedstock pretreatment (Chapter 2) is essential.

Fixed-bed gasifiers are classified as updraft and downdraft gasifiers (Molino, 2018). In the former, biomass is supplied from the top, while the gasifying agent (air or oxygen) is supplied from the bottom (countercurrent). In the latter type of gasifier, the biomass and the gasifying agent are introduced from the top (co-current). For updraft reactors, the sequence of the biomass is drying, pyrolysis, and reduction, finally arriving at the combustion zone, with synthesis gas drawn out from the top. In the case of downdraft reactors, both biomass and oxygen or air are supplied in the drying zone, going the through pyrolysis, combustion, and reduction, with synthesis gas drawn out from the bottom (Dogru *et al.*, 2002).

The *countercurrent fixed bed* (*updraft*) gasifier consists of a fixed bed of carbonaceous fuel (such as biomass) through which the *gasification agent* (steam, oxygen and/or air) flows in countercurrent configuration. An updraft gasifier has distinctly defined zones for partial combustion, reduction, pyrolysis, and drying. The gas leaves the gasifier reactor together with the products of pyrolysis from the pyrolysis zone and steam from the drying zone. The resulting combustible gas is rich in higher molecular weight hydrocarbon derivatives (tars) and, therefore, has a higher calorific value, which makes updraft gasifiers more suitable where the gas is to be used for heat production. The mineral ash is either removed dry or as a slag. In fixed-bed gasifiers biomass raw material is fed to the reactor from its top, through an opening on the reactor's head and moves downwards by gravity. Depending on the way that the gasification medium (air, oxygen or steam) is introduced in the reactor, fixed-bed systems are divided into updraft gasifiers and downdraft gasifiers.

The *co-current fixed-bed (downdraft) gasifier* is similar to the countercurrent type, but the gasification agent gas flows in co-current configuration with the fuel (downwards, hence the name downdraft gasifier). The term co-current is used because air moves in the same direction as that of fuel, downwards. A downdraft gasifier is so designed that tar product, which are produced in the pyrolysis zone, travel through the combustion zone, where the tar is decomposed into lower molecular weight products or burned. As a result, the mixture of gases in the exit stream is relatively clean. The position of the combustion zone is thus a critical element in the downdraft gasifier – the main advantage being that it produces gas with low tar content. The operating temperature varies from a minimum of 900°C (1650°F) to a maximum of 1000 to 1050°C (1830 to 1920°F). In a *cross-draft gasifier*, air enters from one side of the gasifier reactor and leaves from the other. Also, cross-draft gasifiers do not need a grate; the ash falls to the bottom and does not come in the way of normal operation.

In the *fluidized-bed* gasifier, the biomass is fluidized in oxygen (or air) and steam and the biomass is brought into an inert bed of fluidized material (such as sand or char) – the air is distributed through nozzles located at the bottom of the bed. The fluidized bed reactors are diverse, but characterized by the fluid behavior of the catalyst (Dry, 2001). Typically, the feedstock is introduced into the fluidized system either above-bed or directly into the bed, depending upon the size and density of the feedstock and how it is affected by the bed velocities. During normal operation, the bed media is maintained at a temperature between 550 and 1000°C (1020 and 1830°F). When the biomass is introduced under such temperature conditions, the drying and pyrolysis steps proceed rapidly, driving off all gaseous portions of the fuel at relatively low temperature. The remaining char is oxidized within the bed to provide the heat source for the drying and devolatilizing reactions to continue. Fluidized-bed gasifiers produce a uniformly high (800 to 1000°C, 1470 to 1830°F) bed temperature. A fluidized-bed gasifier works as a hot bed of sand particles agitated constantly by air. Fluidized-bed gasifiers are most useful for fuels that form highly corrosive ash that would damage the walls of slagging gasifiers. The ash is removed dry or as high molecular weight agglomerated materials – a disadvantage of biomass feedstocks is that they generally contain high levels of corrosive ash.

Compared with the fixed-bed gasifiers, in the fluidized-bed gasifier, the sequence of processes (drying, pyrolysis, oxidation and reduction) is not obvious at a certain point of the gasifier since they take place in the entire reactor thus resulting to a more homogeneous type of reaction. This means the existence of more constant and lower temperatures inside the reactor, where no "hot spots" are observed. Due to the lower operating temperatures, ash does not melt, so it is easier removed from the reactor. At the same time, sulfur and chloride contaminants can be absorbed in the inert bed material thus eliminating the fouling hazard and reducing the maintenance costs. Another significant difference is that fluidized-bed gasifiers are much less to biomass quality than fixed-bed systems, and they can even operate with mixed biomass feedstock.

In the *entrained-flow* gasifier, a dry pulverized solid, an atomized liquid fuel or a fuel slurry is gasified with oxygen (much less frequent: air) in co-current flow. The high temperatures and pressures also mean that a higher throughput can be achieved but thermal efficiency is somewhat lower as the gas must be cooled before it can be sent to a gas processing facility. All entrained-flow gasifiers remove the major part of the ash as a slag as the operating temperature is well above the ash fusion temperature. The operating temperatures are on the order of 1200 to 1600°C (2190 to 2910°F) and the pressure is 300 to 1200 psi. Entrained-flow gasifiers are suitable for use with biomass feedstocks as long as the feedstock (low moisture) has a low content of mineral matter (manifested as a low ash in the product). Due to the short residence time (0.5 to 4.0 seconds), high temperatures are required for such gasifiers, and as a result, in entrained-flow gasifiers, the gas contains only small quantities of tar.

The main feature of entrained-flow gasification is the requirement for a very fine biomass grain size, even smaller than 0.1 mm which allows the biomass easier to be carried out by the gasification medium. Biomass can be introduced in the reactor either at a dry form or even as slurry (pulverized biomass mixed with water). The retention time of the feedstock inside the reactor is only a few seconds and thus in order to achieve high conversion rates, higher temperatures are applied (between 1200 and 1500°C, 2190 and 2730°F). At these temperature levels, the resulting ash melts, cools down and eventually accumulates as slag. Apart from higher temperature, entrained-flow gasification usually takes place at elevated pressure (pressurized entrained-flow gasifiers) reaching operating pressures even up to 600 and 750 psi. The existence of such high temperatures and pressures requires more sophisticated reactor design and construction materials used.

Entrained-flow gasifiers are classified into two types: (i) top-fed gasifiers and (ii) side-fed gasifiers. A top-fed gasifier is a vertical cylinder reactor where fine particles (pulverized) and the gasification agent are co-currently fed from the top in the form of a jet and the feedstock conversion is achieved by use of an inverted burner. The product gas is taken from the side of the lower section while slag is extracted from the bottom of the reactor. On the other hand, in a side-fed gasifier, the pulverized fed and the gasification agent are co-currently fed by nozzles installed in the lower reactor, resulting in an appropriate mixing of biomass and gasification agent. The gas is extracted from the top of the gasifier and the slag from the bottom. Both configurations are highly efficient, with a standard operating temperature (1300 to 1500°C, 2370 to 2730°F)] and pressure 300 to 1000 psi.

Also, with regard to the entrained-flow reactor, it may be necessary to pulverize the feedstock (such as biomass char). On the other hand, when the feedstock is sent into an entrained-flow gasifier, the biomass can be in either form of dry feed or slurry feed if, for example, bio-oil and a solid feedstock (including biomass and biomass char) are co-gasified. In general, dry-feed gasifiers have the advantage over slurry-feed gasifiers in that the former can be operated with lower oxygen consumption. Moreover, dry-feed gasifiers have an additional degree of freedom that makes it possible to optimize bio-synthesis gas production (Shen *et al.*, 2012).

While there are many alternate uses for the synthesis gas produced by gasification, and a combination of products/utilities can be produced in addition to power. A major benefit of the integrated gasification combined cycle concept is that power can be produced with the lowest sulfur oxide (Sox) and nitrogen oxide (NOx) emissions of any liquid/solid feed power generation technology.

Finally, and by way of explanation, an *indirect gasifier* requires an external source of heat to be transferred to the biomass to pyrolyze the feedstock under conditions of high severity, i.e., long residence time at high temperature. In the case of a biomass feedstock, the process will provide a char stream using either slow or fast pyrolysis to temperatures of 600 to 750°C (1110 to 1380°F), in amounts representing between 12% and 25% of the biomass feedstock.

6.3.2 Fischer-Tropsch Synthesis

Bio-synthesis gas is used as a raw material in different thermochemical processes for the production of second-generation biofuels, both liquid (such as methanol, ethanol, dimethyl ether (DME), and Fischer-Tropsch diesel) and gaseous (such as hydrogen and synthetic natural gas (SNG). In particular, the type and composition of the biomass and the production process strongly influences the composition and heating value of the gaseous products (Table 6.7) (Sikarwar *et al.*, 2016). The production of liquid biofuel as an energy carrier could be very cost-effective because it would take the same infrastructure, storage system, and transportation used for liquefied petroleum gas.

The synthesis reaction is dependent on a catalyst, mostly an iron or cobalt catalyst where the reaction takes place. There is either a low-temperature Fischer-Tropsch process (often represented as the LTFT process) or high-temperature Fischer-Tropsch process (often represented as the HTFT process), with temperatures ranging between 200 to 240°C (390 to 465°F) for the low-temperature Fischer-Tropsch process and 300 to 350°C (570 to 660°F) for the high-temperature Fischer-Tropsch process. The high-temperature Fischer-Tropsch process uses an iron-based catalyst, and the low-temperature Fischer-Tropsch process either an iron-based catalyst or a cobalt-based catalyst. The different catalysts include also nickel-based catalyst and ruthenium-based catalysts.

Finally, it is worthy of note that clean synthesis gas can also be used (i) as chemical *building blocks* to produce a broad range of chemicals using processes well established in the chemical and petrochemical industry), (ii) as a fuel producer for highly efficient fuel cells (which run off the hydrogen made in a gasifier) or perhaps in the future, hydrogen turbines and fuel cell-turbine hybrid systems, and (iii) as a source of hydrogen that can be separated from the gas stream and used as a fuel or as a feedstock for refineries (which use the hydrogen to upgrade crude oil products) (Speight, 2016, 2017, 2019).

6.3.3 Feedstocks

For many decades, coal has been the primary feedstock for gasification units but because of recent concerns about the use of fossil fuels and the resulting environmental pollutants, irrespective of the various gas cleaning processes and gasification plant environmental cleanup efforts, there is a move to feedstocks other than coal for gasification processes (Speight, 2013a, 2014b). But more pertinent to the present text, the gasification process can also use carbonaceous feedstocks which would otherwise have been discarded and unused, such as waste biomass and other similar biodegradable wastes. In this respect, biomass can be used to the fullest potential for the production of hydrogen since the refining industry has seen fit to use viscous feedstock gasification as a source of hydrogen for the past several decades (Speight, 2016, 2014a).

The advantage of the gasification process when a carbonaceous feedstock (a feedstock containing carbon) or hydrocarbonaceous feedstock (a feedstock containing carbon and hydrogen) is employed is that the product of focus – bio-synthesis gas – is potentially more useful as an energy source and results in an overall cleaner process. However, the reactor must be selected on the basis of feedstock properties and predicted behavior in the process.

6.3.3.1 Biomass

Biomass can be considered as any renewable feedstock which is in principle *carbon neutral* (while the plant is growing, it uses the sun's energy to absorb the same amount of carbon from the atmosphere as it releases into the atmosphere) (Speight, 2008, 2011a).

Raw materials that can be used to produce biomass derived fuels are widely available; they come from a large number of different sources and in numerous forms (Rajvanshi, 1986; Speight, 2008, 2011a). The main basic sources of biomass include: (i) wood, including bark, logs, sawdust, wood chips, wood pellets and briquettes, (ii) high-yield energy crops, such as wheat, grown specifically for energy applications, (iii) agricultural crops and residues (e.g., straw), and (iv) industrial waste, such as wood pulp or paper pulp. These different forms of biomass include a wide range of materials that produce a variety of products which are dependent upon the feedstock (Bhattacharya *et al.*, 1999; Balat, 2011; Demirbaş, 2011; Ramroop Singh, 2011; Speight, 2011a). In addition, the heat content of the different types of biomass widely varies and has to be taken into consideration when designing any conversion process (Jenkins and Ebeling, 1985).

Many forms of biomass contain a high percentage of moisture (along with carbohydrates and sugars) and mineral constituents (Chapter 2) – both of which can influence the economics and viability of a gasification process. The presence of high levels of moisture in biomass reduces the temperature inside the gasifier, which then reduces the efficiency of the process. Many biomass gasification technologies therefore require dried biomass to reduce the moisture content prior to feeding into the gasifier. In addition, biomass can come in a range of sizes. In many biomass gasification systems, biomass must be processed to a uniform size or shape to be fed into the gasifier at a consistent rate as well as to maximize gasification efficiency.

Biomass such as wood pellets, yard and crop waste and energy crops including switch grass and waste from pulp and paper mills can also be employed to produce bioethanol and synthetic diesel. Biomass is first gasified to produce bio-synthesis gas and then subsequently converted via catalytic processes to the aforementioned downstream products. Biomass can also be used to produce electricity – either blended with traditional feedstocks, such as coal or by itself (Shen et al., 2012; Khosravi1 and Khadse, 2013a; Speight, 2014b).

Finally, while biomass may seem to some observers to be the answer to the global climate change issue, the types (Table 6.8) and the composition (Table 6.9) of the biomass as well as the advantages and disadvantages (Table 6.10) of biomass as feedstock must be considered carefully (Molino et al., 2016). Also, while taking the issues of global climate change into account, it must not be ignored that the Earth is in an interglacial period when warming will take place. The true extent of this warming is not known – no one was around to measure the temperature change in the last interglacial period – and by the same token the contribution of anthropological sources to global climate change (through the emissions of carbon dioxide in the past and in the present) cannot be measured accurately because, for example, of the mobility of carbon dioxide in ice (Speight and Islam, 2016; Radovanović and Speight, 2018).

6.3.3.2 *Gasification of Biomass with Coal*

Recently, co-gasification of various biomass and coal mixtures has attracted a great deal of interest from the scientific community (Usón et al., 2004). However, biomass and coal require drying and size reduction before they can be fed into a gasifier. Size reduction is needed to obtain appropriate particle sizes; drying is required to achieve a moisture content suitable for gasification operations. In addition, densification of the biomass may be done to make pellets and improve density and material flow in the feeder areas. Both fixed-bed and fluidized-bed gasifiers have been used in cogasification of biomass with coal (McKendry, 2002).

Feedstock combinations including Japanese cedar wood and coal (Kumabe et al., 2007), coal and sawdust (Vélez et al., 2009), coal and pine chips (Pan et al., 2000), coal and silver birch wood (Collot et al., 1999), and coal and birch wood (Brage et al., 2000) have been reported in gasification practice. Co-gasification of coal and biomass has some synergy – the process not only produces a low carbon footprint on the environment, but also improves the H_2/CO ratio in the produced gas which is required for liquid fuel synthesis (Sjöström et al., 1999; Kumabe et al., 2007). In addition, the inorganic matter present in biomass catalyzes the gasification of coal. However, co-gasification processes require custom fittings and optimized processes for the coal and region-specific wood residues.

While cogasification of coal and biomass is advantageous from a chemical viewpoint, some practical problems are present on upstream, gasification, and downstream processes. On the upstream side, the particle size of the coal and biomass is required to be uniform for optimum gasification. In addition, moisture content and pretreatment (torrefaction) are very important during upstream processing. Also, biomass decomposition occurs at a lower temperature than coal and therefore different reactors compatible to the feedstock mixture are required (Speight, 2011; Brar et al., 2012; Speight, 2013a, 2013b). Furthermore, feedstock and gasifier type along with operating parameters not only decide product gas composition but also dictate the amount of impurities to be handled downstream.

Table 6.8 Biomass classification, sub-classification, and potential use.

Classification	Sub-classification	Examples	Use
Terrestrial	–	Forest biomass	Ideal for gasification due to high cellulose and hemicellulose content
		Grasses	Unsuitable for gasification due to high moisture; suitable for fermentation
		Energy crops	Suitable for power generation through biological treatment
		Cultivated crops	Some crops are ideal for gasification while others are consumed directly by humans and animals
Marine	–	Algae	Suitable for biological treatment because of high moisture content
		Water plant	Ideal for biological treatment
Waste	Municipal waste	MSW, biosolids, sewage, landfill gas	Suitable for plasma gasification
	Agricultural solid waste	Livestock and manures, agriculture crop residue bark, leaves, floor residues	Most suitable for composting and other biological treatments
	Forestry residues		Ideal for gasification; pretreatment may be required
	Industrial waste	Demolition wood, sawdust, waste oil	Commonly employed for gasification; also used in biological treatments

Table 6.9 Composition of various types of biomass.

Type of biomass	Cellulose (% w/w)	Hemicellulose (% w/w)	Lignin (% w/w)	Other (% w/w)
Softwood	41	24	28	7
Hardwood	39	35	20	7
Wheat straw	40	28	17	15
Rice straw	30	25	12	33
Bagasse	38	39	20	3
Oak wood	34.5	18.6	28	–
Pine wood	42.1	17.7	25	–
Birch wood	35.7	25.1	19.3	–
Spruce wood	41.1	20.9	28	–
Sunflower seed hull	26.7	18.4	27	–
Coconut shell	24.2	24.7	34.9	–
Almond shell	24.7	27	27.2	–
Poultry litter	27	17.8	11.3	20
Deciduous plant	42	25	21.5	11.5
Coniferous plant	42	26	30	2
Willow plant	50	19	25	6
Larch plant	26	27	35	12

However, first and foremost, coal and biomass require drying and size reduction before they can be fed into a gasifier. Size reduction is needed to obtain appropriate particle sizes; however, drying is required to achieve moisture content suitable for gasification operations. In addition, biomass densification may be conducted to prepare pellets and improve density and material flow in the feeder areas.

It is recommended that biomass moisture content should be less than 15% w/w prior to gasification. High moisture content reduces the temperature achieved in the gasification zone, thus resulting in incomplete gasification. Forest residues or wood has a fiber saturation point at 30 to 31% moisture content (dry basis) (Brar *et al.*, 2012). Compressive and shear strength of the wood increases with decreased moisture content below the fiber saturation point. In such a situation, water is removed from the cell wall leading to shrinkage. The long-chain molecules constituents of the cell wall move closer to each other and bind more tightly. A high level of moisture, usually injected in form of steam in the gasification zone, favors formation of a water-gas shift reaction that increases hydrogen concentration in the resulting gas.

Table 6.10 The advantages and disadvantages of using biomass as a feedstock for energy production and chemicals production.

Advantages
Theoretically inexhaustible fuel source.
Minimal environmental impact when processes such as fermentation and pyrolysis are used.
Alcohols and other fuels produced by biomass are efficient, viable, and relatively clean-burning.
Biomass is available on a worldwide basis.
Disadvantages
Could contribute to global climate change and particulate pollution when direct combustion is employed.
Production of biomass and the technological conversion to alcohols or other fuels can be expensive.
Life cycle assessments (LCA) should be considered to address energy input and output.
Possibly a net loss of energy when operated on a small scale – energy is required to grow the biomass.

The torrefaction process is a thermal treatment of biomass in the absence of oxygen, usually at 250 to 300°C (480 to 570°F) to drive off moisture, decompose hemicellulose completely, and partially decompose cellulose (Speight, 2011a). Torrefied biomass has reactive and unstable cellulose molecules with broken hydrogen bonds and not only retains 79 to 95% of feedstock energy but also produces a more reactive feedstock with lower atomic hydrogen-carbon and oxygen-carbon ratios to those of the original biomass. Torrefaction results in higher yields of hydrogen and carbon monoxide in the gasification process.

Biomass fuel producers, coal producers and, to a lesser extent, waste companies are enthusiastic about supplying co-gasification power plants and realize the benefits of co-gasification with alternate fuels (Speight, 2008, 2011a; Lee and Shah, 2013; Speight, 2013a, 2013b). The benefits of a co-gasification technology involving coal and biomass include the use of a reliable coal supply with gate-fee waste and biomass which allows the economies of scale from a larger plant to be supplied just with waste and biomass. In addition, the technology offers a future option of hydrogen production and fuel development in refineries. In fact, oil refineries and petrochemical plants are opportunities for gasifiers when the hydrogen is particularly valuable (Speight, 2011b, 2014).

While upstream processing is influential from a material handling point of view, the choice of gasifier operation parameters (temperature, gasifying agent, and catalysts) dictate the product gas composition and quality. Biomass decomposition occurs at a lower temperature than coal and therefore different reactors compatible to the feedstock mixture are required (Brar *et al.*, 2012). Furthermore, feedstock and gasifier type along with operating parameters not only decide product gas composition but also dictate the amount of impurities to be handled downstream. Downstream processes need to be modified if coal is co-gasified with biomass. Heavy metals and other impurities such as sulfur-containing compounds and mercury present in coal can make synthesis gas difficult to use and unhealthy for the environment. Alkali present in biomass can also cause corrosion problems at high temperatures in downstream pipes. An alternative option to downstream gas cleaning would be to process the feedstock to remove mercury and sulfur prior to introduction into the gasifier.

Finally, the presence of mineral matter in the coal-biomass feedstock is not appropriate for fluidized-bed gasification. Low melting point of ash present in woody biomass leads to agglomeration which causes defluidization of the ash and sintering, deposition as well as corrosion of the gasifier construction metal bed (Vélez et al., 2009). Biomass containing alkali oxides and salts are likely to produce clinkering/slagging problems from ash formation (McKendry, 2002). Thus, it is imperative to be aware of the melting of biomass ash, its chemistry within the gasification bed (no bed, silica/sand, or calcium bed), and the fate of alkali metals when using fluidized-bed gasifiers.

6.3.3.3 Gasification of Biomass with Other Feedstocks

Co-utilization of waste and biomass with coal may provide economies of scale that help achieve the above identified policy objectives at an affordable cost. In some countries, governments propose co-gasification processes as being *well suited for community-sized developments*, suggesting that waste should be dealt with in smaller plants serving towns and cities, rather than moved to large, central plants (satisfying the so-called *proximity principle*).

Co-gasification technology varies, being usually site specific and high feedstock dependent. At the largest scale, the plant may include the well-proven fixed-bed and entrained-flow gasification processes. At smaller scales, emphasis is placed on technologies which appear closest to commercial operation. Pyrolysis and other advanced thermal conversion processes are included where power generation is practical using the on-site feedstock produced. However, the needs to be addressed are (i) core fuel handling and gasification/pyrolysis technologies, (ii) fuel gas cleanup, and (iii) conversion of fuel gas to electric power (Ricketts et al., 2002).

The use of waste as gasification feedstock or as a co-gasification feedstock can mitigate an important problem. The disposal of municipal and industrial wastes has become an important problem because the traditional means of disposal, landfill, has become environmentally much less acceptable than previously. New, much stricter regulation of these disposal methods will make the economics of waste processing for resource recovery much more favorable. One method of processing waste streams is to convert the energy value of the combustible waste into a fuel. One type of fuel attainable from wastes is a low heating value gas heat content on the order of 100 to 150 Btu/ft^3), which can be used to generate process steam or to generate electricity.

The ability of a refinery to efficiently accommodate heavy feedstock streams (such as heavy oil, extra heavy oil, residua, deasphalter bottoms, visbreaker bottoms, and tar sand bitumen) enhances the economic potential of the refinery and the development of heavy oil and tar sand resources. A refinery with the flexibility to meet the increasing product specifications for fuels through the ability to upgrade heavy feedstocks is an increasingly attractive means of extracting maximum value from each barrel of oil produced. Upgrading can convert marginal heavy crude oil into light, higher-value crude, and can convert heavy, sour refinery bottoms into valuable transportation fuels. On the downside, most upgrading processes also produce an even heavier residue whose disposition costs may approach the value of the upgrade itself.

6.4 Gas Production and Products

The gasification of a carbonaceous feedstock (i.e., char produced from the feedstock such as char from the pyrolysis of biomass) is the conversion of the feedstock (by any one of a variety of processes) to produce gaseous products that are combustible as well as a wide range of chemical products from synthesis gas (Figure 6.2). The products from the gasification of the process may be of low, medium, or high heat-content (high-Btu) content as dictated by the process as well as by the ultimate use for the gas (Baker and Rodriguez, 1990; Probstein and Hicks, 1990; Lahaye and Ehrburger, 1991; Matsukata *et al.*, 1992; Speight, 2013a).

There are many useful products from the gasification of biomass, which include: (i) synthesis gas, an important aspect in the present context, (ii) heat, (iii) power, (iv) biofuels, and (v) biochar. The synthesis gas can be further processed by means of the Fischer-Tropsch process into methanol, dimethyl ether and other chemical feedstocks (Chapter 12). Generally, biomass feedstocks are classified into four main groups: woody biomass, herbaceous biomass, marine biomass, and manure (Sikarwar *et al.*, 2016). The gasifier is usually designed to generate a given product (Chapter 3) but the feedstock material is an important parameter to specify and optimize where possible (Luque and Speight, 2015; Sikarwar *et al.*, 2016, 2017).

6.4.1 Gas Production

Large-scale biomass gasifiers employ one of two types of fluidized-bed configurations: (i) bubbling fluidized bed and circulating fluidized bed or (ii) a combination of both indirectly fired to maintain the bed temperature below the ash fusion temperature of the

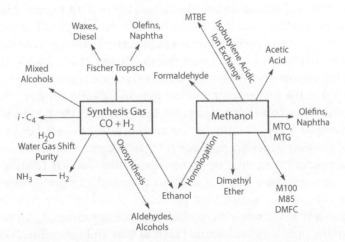

Figure 6.2 Potential products from heavy feedstock gasification.

biomass ash. A bubbling fluidized bed (Chapter 3) consists of fine, inert particles of sand or alumina, which are selected based on their suitability of physical properties such as size, density and thermal characteristics. The gas flow rate is chosen to maintain the bed in a fluidization condition, which enters at the bottom of the vessel. The dimension of the bed at some height above the distributor plate is increased to reduce the superficial gas velocity below the fluidization velocity to maintain inventory of solids and to act as a disengaging zone. A cyclone is used to trap the smaller size particle that exit the fluidized bed to either return fines to the bed or to remove ash-rich fines from the system.

Biomass is introduced either through a feed chute to the top of the bed or deep inside the bed. The deeper introduction of biomass into the bed of inert solids provides sufficient residence time for fines that would otherwise be entrained in the fluidizing gas. The biomass organic constituents pyrolytically vaporize and are partially combusted in the bed. The exothermic combustion provides the heat to maintain the bed at temperature and to volatilize additional biomass. The bed needs to be preheated to the start-up temperature using hydrocarbon resources such as natural gas, fuel oil, either by direct firing or by indirect heating. After the bed reaches the biomass ignition temperature, Biomass is slowly introduced into the bed to raise the bed temperature to the desired operating temperature which is normally in the range of 700 to 900°C (1290 to 1650°F). The bed temperature is governed by the desire to obtain complete devolatilization versus the need to maintain the bed temperature below the biomass ash fusion temperature. The advantages of the fluidized-bed gasifiers are (i) yields a uniform product gas, (ii) the ability to accept a wide range of fuel particles sizes, including fines, (iii) a nearly uniform temperature distribution throughout the reactor, (iv) provides a high rate of heat transfer between inert material and biomass, aiding high conversion, with low tar production. The disadvantage being, formation of large bubbles at higher gas velocities, which bypass the bed reducing the high rate of heat/mass transfer significantly.

Autothermal gasifiers provide the necessary heat of reaction by means of partial oxidation within the gasification reactor. If air is used as oxidizing agent during the process, the product gas contains a high amount of nitrogen. Thus, for synthesis gas production either pure oxygen (in entrained-flow reactors) or mixtures of oxygen and steam (in fluidized-bed reactors) are used as gasification agent. The advantage of autothermal gasification is the direct heating of the reactants and therefore more efficient energy utilization.

Allothermal (or indirect) gasification is characterized by the separation of the processes of heat production and heat consumption (Pfeifer *et al.*, 2009). The allothermal gasification facility almost always consists of two reactors, connected by an energy flow. Biomass is gasified in the first reactor and the remaining solid residue (char) or product gas is combusted in the second reactor to produce the heat for the first process. The transport of the heat can be achieved either by circulating a bed material or by heat exchangers. Allothermal gasifiers generally produce two gas streams: a medium calorific product gas (gasification reactor) with little or no nitrogen and a flue gas (combustion reactor). The production of a nitrogen-free gas without the need of pure oxygen is one of the advantages over autothermal gasification processes. Another advantage is the complete carbon conversion and that there is no problematic waste produced. All carbon-containing streams from the product gas cleaning (such as dust and tar removal) can be recycled to the combustion zone and converted to heat, which is used for the gasification reactions (Tijmensen *et al.*, 2002).

In fixed-bed fluidized-bed, and entrained-flow gasifiers (Chapter 3), the differences in the designs cause differences in composition of the gas which is also influenced by other factors, including feedstock composition, reactor type, and operating parameters (temperature, pressure, oxygen fuel ratio). During oxygen gasification, the combustion products (carbon dioxide and water) are in the product gas and take part in the chemical reactions, mainly in the water-gas shift reaction. On the other hand, higher amount of hydrogen in the product gas can be found during the steam gasification. The hydrogen that occurs in the product gas does not originate only from the fuel as in the case of oxygen gasification but also from the steam. The tar content depends not only on the type of gasifier but also mainly on the operation temperature. In entrained-flow gasification, where the temperature is above 1000°C (1830°F), there is no tar produced, whereas in fluidized-bed gasification, where the temperature is below 1000°C (1830°F), tar is produced and has to be removed from the product gas.

6.4.2 Gaseous Products

The products of gasification are varied insofar as the gas composition varies with the system employed (Speight, 2013a). It cannot be overemphasized that the gas product must be first freed from any pollutants such as particulate matter and sulfur compounds before further use, particularly when the intended use is a water gas shift or methanation (Cusumano *et al.*, 1978; Probstein and Hicks, 1990).

In all cases, the gas exiting the gasifier contains impurities that need to be removed for the advanced use of biomass gasification gas. For example, if the gas is to be used in a Fischer-Tropsch process, the gas must be ultra-clean. Any particulate matter in the gas exiting the gasifier can be removed by a hot gas filter, whereas the tar constituents and low-boiling hydrocarbon derivatives can be converted by reforming to synthesis gas. For tar removal, another possibility is scrubbing by an organic solvent.

However, the gas stream will also, more than likely, contain other impurities, such as hydrogen sulfide, carbonyl sulfide, ammonia, and alkali metals. The sulfur constitunets will further add complications to a gas reforming process by poisoning the catalyst. In addition to hydrocarbons, ammonia is partly converted to nitrogen in the reformer. The alkali metals, in turn, will condense in hot gas filtration taking place at around 500°C (930°F). The final conditioning of gas for synthesis purposes can be realized by conventional commercial technologies for acid gas removal, such as the Rectisol process, and the H2/CO ratio can be adjusted by a water gas shift reactor.

6.4.2.1 Synthesis Gas

Synthesis gas is one of the key products of the gasification of biomass and is a vital source of environmentally fuels (Chapter 13) and chemicals (Chapter 14). Moreover, it is a suitable fuel for the production of electricity. Employing the Fischer-Tropsch (FT) process (Chapter 12), gasoline, diesel and other chemicals can be manufactured (Chadeesingh, 2011; Speight, 2019). Synthesis gas can be readily converted to methanol or DME, which in turn can be transformed to gasoline in the presence of zeolites. Hydrogen is produced from synthesis gas by means of the water gas shift reaction (Chapter 11), which has numerous applications including in fuel cells. Synthetic natural gas (SNG, CH_4) is also one of the significant uses of synthesis gas. Synthesis gas has also found applications in producing medium-BTU gas which is used as turbine fuel and in integrated gasification combined cycle (IGCC).

Contaminants such as particulates, tars, nitrogenous compounds such as ammonia (NH_3) and hydrgoen cyanide (HCN), sulfur-containing inorganic compounds such as hydrgoen sulfide (H_2S), carbonyl sulfide (COS), and carbon disulfide (CS_2). Halogens such as hydrogen chloride and traces of metals such as sodium (Na) and potassium (K) are present in varying quantities in synthesis gas produced from gasification. As compared to other contaminants, tar is present in varying quantities per unit wt of feedstock. The type of feedstock, operational conditions, and the gasifier type are the variables which determine the amount of tar produced. These contaminants in synthesis gas pose numerous technical and working problems and must be removed (Chapter 9). For example, hydrgoen sulfide is responsible for equipment corrosion, tar causes fouling, and catalyst deactivation occurs due to the presence of tar, hydrgoen sulfide, ammonia, hydrogen chloride, and trace metals.

Synthesis gas produced from biomass sources is comparable in combustion efficiency to natural gas (Speight, 2008; Chadeesingh, 2011) which reduces the emissions of sulfur, nitrogen oxides, and mercury, resulting in a much cleaner fuel (Nordstrand *et al.*, 2008; Lee *et al.*, 2006; Sondreal *et al.*, 2004, 2006; Yang *et al.*, 2007; Wang *et al.*, 2008). The resulting hydrogen gas can be used for electricity generation or as a transport fuel or in any aspects of refining that requires hydrogen (Speight, 2016). The gasification process also facilitates capture of carbon dioxide emissions from the combustion effluent.

Although bio-synthesis gas can be used as a stand-alone fuel, the energy density of synthesis gas is approximately half that of natural gas and is therefore mostly suited for the production of transportation fuels and other chemical products. Synthesis gas is mainly used as an intermediary building block for the final production (synthesis) of various fuels such as synthetic natural gas, methanol and synthetic crude oil fuel (dimethyl ether – synthesized gasoline and diesel fuel) (Chadeesingh, 2011; Speight, 2013a). Synthesis gas is a clean and renewable form of energy generated from biomass that could very well substitute for conventional sources of energy. The gas is generally composed of methane (55 to 65%), carbon dioxide (35 to 45%), nitrogen (0 to 3%), hydrogen (0 to 1%), and hydrogen sulfide (0 to 1%).

The use of bio-synthesis gas offers the opportunity to furnish a broad range of environmentally clean fuels and chemicals and there has been steady growth in the traditional uses of synthesis gas. Almost all hydrogen gas is manufactured from synthesis gas and there has been an increase in the demand for this basic chemical. In fact, the major use of synthesis gas is in the manufacture of hydrogen for a growing number of purposes, especially in crude oil refineries (Speight, 2014a, 2016). Methanol not only remains the second-largest consumer of synthesis gas but has shown remarkable growth as part of the methyl ethers used as octane enhancers in automotive fuels.

The Fischer-Tropsch synthesis remains the third-largest consumer of synthesis gas, mostly for transportation fuels but also as a growing feedstock source for the manufacture of chemicals, including polymers. The hydroformylation of olefin derivatives (the Oxo reaction), a completely chemical use of synthesis gas, is the fourth-largest use of carbon monoxide and hydrogen mixtures. A direct application of synthesis gas as fuel (and eventually also for chemicals) that promises to increase is its use for *integrated gasification combined cycle* (IGCC) units for the generation of electricity (and also chemicals), crude oil coke or viscous feedstocks (Holt, 2001). Finally, synthesis gas is the principal source of carbon monoxide, which is used in an expanding list of carbonylation reactions, which are of major industrial interest.

6.4.2.2 Low-Btu Gas

During the production of gas by oxidation with air, the oxygen is not separated from the air and, as a result, the gas product invariably has a low Btu content (low heat-content, 150 to 300 Btu/ft^3). Several important chemical reactions and a host of side reactions are involved in the manufacture of low heat-content gas under the high-temperature conditions employed (Speight, 2013a). Low heat-content gas contains several components, four of which are always major components present at levels of at least several percent; a fifth component, methane, is marginally a major component.

The nitrogen content of low heat-content gas ranges from somewhat less than 33% v/v to slightly more than 50% v/v and cannot be removed by any reasonable means; the presence of nitrogen at these levels makes the product gas *low heat-content* by definition. The nitrogen also strongly limits the applicability of the gas to chemical synthesis. Two other non-combustible components (water, H_2O, and carbon dioxide, CO_2) further lower the heating value of the gas; water can be removed by condensation and carbon dioxide by relatively straightforward chemical means.

The two major combustible components are hydrogen and carbon monoxide; the H_2/CO ratio varies from approximately 2:3 to approximately 3:2. Methane may also make an appreciable contribution to the heat content of the gas. Of the minor components hydrogen sulfide is the most significant and the amount produced is, in fact, proportional to the sulfur content of the feedstock. Any hydrogen sulfide present must be removed by one, or more, of several procedures (Mokhatab *et al.*, 2006; Speight, 2007, 2014).

6.4.2.3 Medium-Btu Gas

Medium-Btu gas (medium heat-content gas) has a heating value in the range 300 to 550 Btu/ft^3) and the composition is much like that of low heat-content gas, except that there is virtually no nitrogen. The primary combustible gases in medium heat-content gas are hydrogen and carbon monoxide. Medium heat-content gas is considerably more versatile than low heat-content gas; like low heat-content gas, medium heat-content gas may be used directly as a fuel to raise steam, or used through a combined power cycle to drive a gas turbine, with the hot exhaust gases employed to raise steam, but medium heat-content gas is especially amenable to synthesize methane (by methanation), higher hydrocarbon derivatives (by Fischer-Tropsch synthesis), methanol, and a variety of synthetic chemicals.

The reactions used to produce medium heat-content gas are the same as those employed for low heat-content gas synthesis, the major difference being the application of a nitrogen barrier (such as the use of pure oxygen) to keep diluent nitrogen out of the system.

In medium heat-content gas, the H_2/CO ratio varies from 2:3 C to 3:1 and the increased heating value correlates with higher methane and hydrogen contents as well as with lower carbon dioxide contents. Furthermore, the very nature of the gasification process used to produce the medium heat-content gas has a marked effect upon the ease of subsequent processing. For example, the CO_2-acceptor product is quite amenable to use for methane production because it has (i) the desired H_2/CO ratio just exceeding 3:1, (ii) an initially high methane content, and (iii) relatively low water and carbon dioxide contents. Other gases may require appreciable shift reaction and removal of large quantities of water and carbon dioxide prior to methanation.

6.4.2.4 High-Btu Gas

High-Btu gas (high heat-content gas) is essentially pure methane and often referred to as *synthetic natural gas* or *substitute natural gas* (SNG) (Speight, 1990, 2013a). In the context of this book, substitute natural gas (SNG) that is produced from biomass can be considered a type of synthesis gas (Chapter 1). Synthetic natural gas contains mainly methane and can be produced from a product gas stream that contains significant amounts of methane. This generally means that gasification should take place non-catalytically at approximately 900°C (1650°F). At these temperatures, methane destruction generally is limited and the fluidized-bed gasification process is one of the obvious options.

However, to qualify as substitute natural gas, a product must contain at least 95% methane, giving an energy content (heat content) of synthetic natural gas on the order of 980 to 1080 Btu/ft^3). The commonly accepted approach to the synthesis of high heat-content gas is the catalytic reaction of hydrogen and carbon monoxide:

$$3H_2 + CO \rightarrow CH_4 + H_2O$$

To avoid catalyst poisoning, the feed gases for this reaction must be quite pure and, therefore, impurities in the product are rare. The large quantities of water produced are removed by condensation and recirculated as very pure water through the gasification system. The hydrogen is usually present in slight excess to ensure that the toxic carbon monoxide is reacted; this small quantity of hydrogen will lower the heat content to a small degree.

The carbon monoxide/hydrogen reaction is somewhat inefficient as a means of producing methane because the reaction liberates large quantities of heat. In addition, the methanation catalyst is troublesome and prone to poisoning by sulfur compounds and the decomposition of metals can destroy the catalyst. Hydrogasification may be thus employed to minimize the need for methanation:

$$[C]_{feedstock} + 2H2 \rightarrow CH_4$$

The product of hydrogasification is far from pure methane and additional methanation is required after hydrogen sulfide and other impurities are removed.

Besides the main components of carbon monoxide, hydrogen, carbon dioxide, methane and nitrogen, several trace substances are also present in the product gas. These include particulate matter) such as entrained ash, and bed material), sulphur compounds (hydrogen sulphide, carbonyl sulphide), alkali compounds (mainly chlorine compounds) and tars or higher hydrocarbons that are prone to condensation at temperatures on the order of 300 to 350°C (570 to 660°F) and can, therefore, foul pipes and equipment. Removal of these substances is of particular importance in the production of synthetic natural gas since the methanation reaction uses catalysts that are highly sensitive to impurities in the gas. A number of different gas cleaning technologies exist (Chapter 7).

6.4.3 Liquid Products

The production of liquid fuels from a carbonaceous feedstock via gasification is often referred to as the *indirect liquefaction* of the feedstock (Speight, 2013a, 2014). In these processes, the feedstock is not converted directly into liquid products but involves a two-stage conversion operation in which the feedstock is first converted (by reaction with steam and oxygen) to produce a gaseous mixture that is composed primarily of carbon monoxide and hydrogen (synthesis gas). The gas stream is subsequently purified (to remove sulfur, nitrogen, and any particulate matter) after which it is catalytically converted to a mixture of liquid hydrocarbon products.

The synthesis of hydrocarbon derivatives from carbon monoxide and hydrogen (synthesis gas) (the Fischer-Tropsch synthesis) is a procedure for the indirect liquefaction of various carbonaceous feedstocks (Speight, 2011a, 2011b). This process is the only liquefaction scheme currently in use on a relatively large commercial scale (for the production of liquid fuels from coal using the Fischer-Tropsch process (Singh, 1981).

Thus, the feedstock is converted to gaseous products at temperatures in excess of 800°C (1470°F), and at moderate pressures, to produce synthesis gas:

$$[C]_{feedstock} + H_2O \rightarrow CO + H_2$$

In practice, the Fischer-Tropsch reaction is carried out at temperatures of 200 to 350°C (390 to 660°F) and at pressures of 75 to 4000 psi. The hydrogen/carbon monoxide ratio is typically on the order of 2/2:1 or 2/5:1, since up to three volumes of hydrogen may be required to achieve the next stage of the liquids production, the synthesis gas must then be converted by means of the water-gas shift reaction) to the desired level of hydrogen:

$$CO + H_2O \rightarrow CO_2 + H_2$$

After this, the gaseous mix is purified and converted to a wide variety of hydrocarbon derivatives:

$$nCO + (2n + 1)H_2 \rightarrow C_nH_{2n+2} + nH_2O$$

These reactions result primarily in low- and medium-boiling aliphatic compounds suitable for gasoline and diesel fuel.

6.4.4 Solid Products

The solid product (solid waste) of a gasification process is typically ash which is the oxides of metals-containing constituents of the feedstock. The amount and type of solid waste produced is very much feedstock dependent. The waste is a significant environmental issue due to the large quantities produced, chiefly fly ash if coal is the feedstock or a co-feedstock, and the potential for leaching of toxic substances (such as heavy metals such as lead and arsenic) into the soil and groundwater at disposal sites.

At the high temperature of the gasifier, most of the mineral matter of the feedstock is transformed and melted into slag, an inert glass-like material and, under such conditions, non-volatile metals and mineral compounds are bound together in molten form until the slag is cooled in a water bath at the bottom of the gasifier, or by natural heat loss at the bottom of an entrained-bed gasifier. Slag production is a function of mineral matter content of the feedstock – coal produces much more slag per unit weight than crude oil coke. Furthermore, as long as the operating temperature is above the fusion temperature of the ash, slag will be produced. The physical structure of the slag is sensitive to changes in operating temperature and pressure of the gasifier and a quick physical examination of the appearance of the slag can often be an indication of the efficiency of the conversion of feedstock carbon to gaseous product in the process.

Slag is comprised of black, glassy, silica-based materials and is also known as *frit*, which is a high-density, vitreous, and abrasive material low in carbon and formed in various shapes from jagged and irregular pieces to rod and needle-like forms. Depending upon the gasifier process parameters and the feedstock properties, there may also be residual carbon char. Vitreous slag is much preferable to ash, because of its habit of encapsulating toxic constituents (such as heavy metals) into a stable, non-leachable material. Leachability data obtained from different gasifiers unequivocally shows that gasifier slag is highly non-leachable, and can be classified as non-hazardous. Because of its particular properties and non-hazardous, non-toxic nature, slag is relatively easily marketed as a byproduct for multiple advantageous uses, which may negate the need for its long-term disposal.

The physical and chemical properties of gasification slag are related to (i) the composition of the feedstock, (ii) the method of recovering the molten ash from the gasifier, and (iii) the proportion of devolatilized carbon particles (char) discharged with the slag. The rapid water-quench method of cooling the molten slag inhibits recrystallization, and results in the formation of a granular, amorphous material. Some of the differences in the properties of the slag may be attributed to the specific design and operating conditions prevailing in the gasifiers.

Char is the finer component of the gasifier solid residuals, composed of unreacted carbon with various amounts of siliceous ash. Char can be recycled back into the gasifier to increase carbon usage and has been used as a supplemental fuel source for use in a combustor. The irregularly shaped particles have a well-defined pore structure and have excellent potential as an adsorbent and precursor to activated carbon. In terms of recycling char to the gasifier, a property that is important to fluidization is the effective particle density. If the char has a large internal void space, the density will be much less than that of the feedstock (especially coal) or char from slow carbonization of a carbonaceous feedstock.

6.5 The Future

Currently, in biomass gasification plants, clean gas is produced at ambient temperature after filtration and scrubbing, limiting its applications. The reduction in gas temperature owing to cleaning followed by conditioning reduces the overall profitability of the plant (although the synthesis gas cooling step generates high-quality steam which can be of use elsewhere in the process or exported depending on the setup). Moreover, if the tar separation is not very effective, the gas quality and yield will suffer, making it unfit for applications where high levels of purity are essential. Therefore, gas conditioning preceded by cleanup at elevated temperatures (i.e., hot gas cleanup) is necessary, to ensure high efficiency in industrial applications, especially in the case of steam gasification.

Progress in catalysts, sorbents and filtration techniques operating at high temperatures have paved a way to integrate gasification and gas cleanup in one reactor. An example of a novel hot gas cleanup process is the use of plasma torches to crack tars; this differs from plasma gasification where the plasma is used for energy generation by gasifying biomass, municipal solid waste, and refuse-derived fuel.

The future of synthesis gas production by gasification of biomass depends very much on the effect of gasification processes on the surrounding environment. It is these environmental effects and issues that will direct the success of gasification. In fact, there is the distinct possibility that within the foreseeable future the gasification process will increase in popularity in crude oil refineries – some refineries may even be known as gasification refineries (Speight, 2011b). A gasification refinery would have, as the centerpiece of the refinery, gasification technology as is the case of the Sasol refinery in South Africa (Couvaras, 1997). The refinery would produce synthesis gas (from the carbonaceous feedstock) from which liquid fuels would be manufactured using the Fischer-Tropsch synthesis technology.

In fact, gasification to produce synthesis gas can proceed from any carbonaceous material, including biomass. Inorganic components of the feedstock, such as metals and minerals, are trapped in an inert and environmentally safe form as char, which may have use as a fertilizer. Biomass gasification is therefore one of the most technically and economically convincing energy possibilities for a potentially carbon neutral economy.

The manufacture of gas mixtures of carbon monoxide and hydrogen has been an important part of chemical technology for about a century. Originally, such mixtures were obtained by the reaction of steam with incandescent coke and were known as *water gas*. Eventually, steam reforming processes, in which steam is reacted with natural gas (methane) or crude oil naphtha over a nickel catalyst, found wide application for the production of synthesis gas.

A modified version of steam reforming known as autothermal reforming, which is a combination of partial oxidation near the reactor inlet with conventional steam reforming further along the reactor, improves the overall reactor efficiency and increases the flexibility of the process. Partial oxidation processes using oxygen instead of steam also found wide application for synthesis gas manufacture, with the special feature that they could utilize low-value feedstocks such as viscous crude oil residues. In recent years, catalytic partial oxidation employing very short reaction times (milliseconds) at high temperatures (850 to 1000°C, 1560 to 1830°F) is providing still another approach to synthesis gas manufacture (Hickman and Schmidt, 1993).

In a gasifier, the carbonaceous material undergoes several different processes: (i) pyrolysis of carbonaceous fuels, (ii) combustion, and (iii) gasification of the remaining char. The process is very dependent on the properties of the carbonaceous material and determines the structure and composition of the char, which will then undergo gasification reactions.

The conversion of the gaseous products of gasification processes to synthesis gas, a mixture of hydrogen (H_2) and carbon monoxide (CO), in a ratio appropriate to the application, needs additional steps, after purification. The product gases – carbon monoxide, carbon dioxide, hydrogen, methane, and nitrogen – can be used as fuels or as raw materials for chemical or fertilizer manufacture.

Gasification by means other than the conventional methods has also received some attention and has provided rationale for future processes (Rabovitser et al., 2010). In the process, a carbonaceous material and at least one oxygen carrier are introduced into a non-thermal plasma reactor at a temperature in the range of approximately 300°C to approximately 700°C (570 to approximately 1290°F) and a pressure in a range from atmospheric pressure to approximate 1030 psi and a non-thermal plasma discharge is generated within the non-thermal plasma reactor. Plasma gasification technology has been shown to be an effective and environmentally friendly method for solid waste treatment and energy utilization.

Plasma gasification uses an external heat source to gasify the waste, resulting in very little combustion. Almost all of the carbon is converted to fuel synthesis gas. The high operating temperatures (above 1800°C, 3270°F) allow for the breaking down of all tars, char and dioxins. The exit gas from the reactor is cleaner and there is no ash at the bottom of the reactor. The waste feed sub-system is used for the treatment of each type of waste in order to meet the inlet requirements of the plasma furnace. For example, for a waste material with high moisture content, a drier will be required. However, a typical feed system consists of a shredder for solid waste size reduction prior to entering the plasma furnace.

In the plasma reactor two torches are used together with a gas (such as oxygen, helium or air) to generate the plasma. The torches which extend into the plasma furnace are fitted with graphite electrodes. An electric current is passed through the electrodes and an electric arc is generated between the tip of the electrodes and the conducting receiver which is the slag at the bottom of the furnace. Because of the electrical resistivity across the system significant heat is required to strip away electrons from the molecules of the gas introduced resulting in an ionized superheated gas stream or plasma. This gas exits the torch at temperatures up to 10000°C (18000°F). Such high temperatures are used to break down the waste primarily into elemental gas and solid waste (slag) as well as reduce dioxins, sulphur oxides and carbon dioxide emissions. Unlike waste incineration plasma gasification waste-to-energy technology has proven to have a benign impact on the environment.

The plasma furnace is the central component of the system where gasification and vitrification takes place. Plasma torches are mounted at the bottom of the reactor; they provide high-jtemperature air (almost three times higher than traditional combustion temperatures) which allow for the gasification of the waste materials. It is a non-incineration thermal process that uses extremely high temperatures in a partial oxygen environment to decompose completely the input waste material into very simple molecules. In the process, the carbonaceous feedstock and the oxygen carrier are exposed to the non-thermal plasma discharge, resulting in the formation of a product gas which comprises substantial amounts of hydrocarbon derivatives, such as methane, hydrogen and/or carbon monoxide. The products of the process are a fuel or gas known as synthesis gas and an inert vitreous material known as slag.

Furthermore, gasification and conversion of carbonaceous solid fuels to synthesis gas for application of power, liquid fuels and chemicals is practiced worldwide. Crude oil coke, coal, biomass, and refinery waste are major feedstocks for an on-site refinery gasification unit. The concept of blending of a variety of carbonaceous feedstocks (such as coal, biomass, or refinery waste) with a viscous feedstock of the coke from the thermal processing of the viscous feedstock is advantageous in order to obtain the highest value of products as compared to gasification of crude oil coke alone. Furthermore, based on gasifier type, co-gasification of carbonaceous feedstocks can be an advantageous and efficient process. In addition, the variety of upgrading and delivery options that are available for application to synthesis gas enable the establishment of an integrated energy supply system whereby synthesis gases can be upgraded, integrated, and delivered to a distributed network of energy conversion facilities, including power, combined heat and power, and combined cooling, heating and power (sometimes referred to as *tri-generation*) as well as used as fuels for transportation applications.

As a final note, the production of chemicals from biomass is based on thermochemical conversion routes which are, in turn, based on biomass gasification (Roddy and Manson-Whitton, 2012). The products are (i) a gas, which is the desired product and (ii) a solid ash residue whose composition depends on the type of biomass. Continuous gasification processes for various feedstocks have been under development since the early 1930s. Ideally, the gas produced would be a mixture of hydrogen and carbon monoxide, but, in practice, it also contains methane, carbon dioxide, and a range of contaminants.

A variety of gasification technologies is available across a range of sizes, from small updraft and downdraft gasifiers through a range of fluidized-bed gasifiers at an intermediate scale and on to larger entrained flow and plasma gasifiers (Bridgwater, 2003; Roddy and Manson-Whitton, 2012). In an updraft gasifier, the oxidant is blown up through the fixed gasifier bed with the synthesis gas exiting at the top whereas in a downdraft gasifier, the oxidant is blown through the reactor in a downward direction with the synthesis gas exiting at the bottom.

Gasification processes tend to operate either above the ash melting temperature (typically 1200°C, 2190°F) or below the ash melting temperature (typically >1000°C, >1830°F). In the higher temperature processes, there is little methane or tar formation. The question of which gasification technology is the most appropriate depends on whether the priority is to (i) produce a very pure bio-synthesis gas, (ii) accommodate a wide range of feedstock types, (iii) avoid preprocessing of biomass, or (iv) operate at a large scale, and produce chemical products from a variety of feedstocks.

The introduction of steam as a reactant influent during gasification has been shown to enhance the production of hydrogen during the gasification of a variety of fuels that include coal, biomass, and municipal solid waste (Minkova *et al.*, 2000; Senneca, 2007). Introduction of carbon dioxide will enhance the production of carbon monoxide during high-temperature steam gasification while depressing the production of hydrogen and methane. Both the steam and carbon dioxide have been shown to increase the char reactivity (Ye *et al.*, 1998; Ochoa *et al.*, 2001; Demirbaş, 2006) through a modification of the char pore structure and surface activity. Biomass gasification using steam can result in an increased concentration of hydrogen in the synthesis gas with higher concentrations corresponding to higher gasification temperatures in the 700 to 1100°C (1290 to 2110°F) range. If a methane fuel stream is the preferred product, careful monitoring of the process to operate below 600°C would permit optimizing the production of methane while minimizing the production of hydrogen. A general characteristic of the fuel stream produced through gasification is a clean stream with a minimum of tar, soot and particulates. Biomass gasification results in a producer gas whose main components are carbon monoxide, hydrogen, and methane. Other volatile components include carbon dioxide, acetaldehyde (CH_3CHO), acetic acid (CH_3CO_2H), phenol (C_6H_5OH), formaldehyde (HCHO), formic acid (HCO_2H), and acetone CH_3COCH_3). If the gasification process is performed under high levels of nitrogen dilution, many of these species may not be detectable since they are present in the parts per million range (Butterman and Castaldi, 2008).

In summary, a refinery that is equipped with a gasifier is a suitable refinery for the complete conversion of heavy feedstocks and (including crude oil coke) to valuable products (including petrochemicals). In fact, integration between bottoms processing units and gasification, presents some unique synergies including the production of feedstocks for a petrochemical complex.

References

Adhikari, U., Eikeland, M.S., and Halvorsen, B.M. 2015. Gasification of Biomass for Production of Syngas for Biofuel. *Proceedings. 56th SIMS, Linköping, Sweden. October 7-9.* Page 255-260. http://www.ep.liu.se/ecp/119/025/ecp15119025.pdf

Ahmad, A.A., Zawawi, N.A., Kasim, F.H., Inayat, A., and Khasri, A. 2016. Assessing the gasification performance of biomass: A Review on Biomass Gasification Process Conditions, Optimization and Economic Evaluation. *Renew. Sustain. Energy Rev.*, 53: 1333-1347.

Arena, U. 2012. Process and Technological Aspects of Municipal Solid Waste Gasification. A Review. *Waste Management,* 32: 625-639.

Asadullah, M. 2014. Barriers of Commercial Power Generation Using Biomass Gasification Gas: A Review. *Renew. Sustain. Energy Rev.*, 29: 201-215.

Baker, R.T.K., and Rodriguez, N.M. 1990. In: *Fuel Science and Technology Handbook.* Marcel Dekker Inc., New York. Chapter 22.

Balat, M. 2011. Fuels from Biomass – An Overview. In: *The Biofuels Handbook.* J.G. Speight (Editor). Royal Society of Chemistry, London, United Kingdom. Part 1, Chapter 3.

Beenackers, A.A.C.M. 1999. Biomass Gasification in Moving Beds. A Review of European Technologies. *Renew. Energy,* 16: 1180-1186.

Bhattacharya, S., Md. Mizanur Rahman Siddique, A.H., and Pham, H-L. 1999. A Study in Wood Gasification on Low Tar Production. *Energy,* 24: 285-296.

Biermann, C.J. 1993. *Essentials of Pulping and Papermaking.* Academic Press Inc., New York.

Boateng, A.A. Walawender, W.P., Fan, L.T., and Chee, C.S. 1992. Fluidized-bed Steam Gasification of Rice Hull. *Bioresource Technology,* 40(3): 235-239.

Brage, C., Yu, Q., Chen, G., and Sjöström, K. 2000. Tar Evolution Profiles Obtained from Gasification of Biomass and Coal. *Biomass and Bioenergy,* 18(1): 87-91.

Brar, J.S., Singh, K., Wang, J., and Kumar, S. 2012. Cogasification of Coal and Biomass: A Review. *International Journal of Forestry Research,* 2012: 1-10.

Bridgwater, A.V. 2003. Renewable Fuels and Chemicals by Thermal Processing of Biomass. *Chem. Eng. Journal,* 91: 87-102.

Butterman, H.C., and Castaldi, M.J. 2008. CO2 Enhanced Steam Gasification of Biomass Fuels. *Proceedings. NAWTEC16. 16th Annual North American Waste-to-Energy Conference. Philadelphia, Pennsylvania. May 19-21.* https://asmedigitalcollection.asme.org/NAWTEC/proceedings-abstract/NAWTEC16/42932/157/328514

Chadeesingh, R. 2011. The Fischer-Tropsch Process. In: *The Biofuels Handbook.* J.G. Speight (Editor). The Royal Society of Chemistry, London, United Kingdom. Part 3, Chapter 5, Page 476-517.

Chen, G., Sjöström, K. and Bjornbom, E. 1992. Pyrolysis/Gasification of Wood in a Pressurized Fluidized Bed Reactor. *Ind. Eng. Chem. Research,* 31(12): 2764-2768.

Collot, A.G., Zhuo, Y., Dugwell, D.R., and Kandiyoti, R. 1999. Co-Pyrolysis and Cogasification of Coal and Biomass in Bench-Scale Fixed-Bed and Fluidized Bed Reactors. *Fuel,* 78: 667-679.

Couvaras, G. 1997. Sasol's Slurry Phase Distillate Process and Future Applications. *Proceedings. Monetizing Stranded Gas Reserves Conference, Houston. December 1997.*

Cusumano, J.A., Dalla Betta, R.A., and Levy, R.B. 1978. *Catalysis in Coal Conversion.* Academic Press Inc., New York.

Davidson, R.M. 1983. *Mineral Effects in Coal Conversion.* Report No. ICTIS/TR22, International Energy Agency, London, United Kingdom.

Demirbaş, A. 2004. Effects of Temperature and Particle Size on Bio-Char Yield from Pyrolysis of Agricultural Residues. *J. Anal. & Appl. Pyrolysis,* 72: 243-248.

Demirbaş, A. 2006. Production and Characterization of Bio-Chars from Biomass via Pyrolysis. *Energy Sources, Part A: Recovery, Utilization and Environmental Effects,* 28: 413-422.

Demirbaş, A. 2011. Production of Fuels from Crops. In: *The Biofuels Handbook.* J.G. Speight (Editor). Royal Society of Chemistry, London, United Kingdom. Part 2, Chapter 1.

Dogru, M., Howarth, C.R., Akay, G., Keskinler, B., and Malik, A.A. 2002. Gasification of Hazelnut Shells in a Downdraft Gasifier. *Energy,* 27: 415-427.

Fabry, F., Rehmet, C., Rohani, V-J., and Fulcheri, L. 2013. Waste Gasification by Thermal Plasma: A Review. *Waste and Biomass Valorization,* 4(3): 421-439.

Fermoso, J., Plaza, M.G., Arias, B., Pevida, C., Rubiera, F., and Pis, J.J. 2009. Co-Gasification of Coal with Biomass and Petcoke in a High-Pressure Gasifier for Syngas Production. *Proceedings. 1st Spanish National Conference on Advances in Materials Recycling and Eco-Energy. Madrid, Spain. November 12-13.*

Gabra, M. Pettersson, E. Backman, R. Kjellström, B. 2001. Evaluation of Cyclone Gasifier Performance for Gasification of Sugar Cane Residue – Part 1: Gasification of Bagasse. *Biomass and Bioenergy,* 21(5): 351-369.

Gray, D., and Tomlinson, G. 2000. Opportunities For Petroleum Coke Gasification Under Tighter Sulfur Limits For Transportation Fuels. *Proceedings. 2000 Gasification Technologies Conference, San Francisco, California. October 8-11.*

Hanaoka, T., Inoue, S., Uno, S., Ogi, T., and Minowa, T. 2005. Effect of Woody Biomass Components on Air-Steam Gasification. *Biomass and Bioenergy,* 28(1): 69-76.

Hickman, D.A., and Schmidt, L.D. 1993. Syngas Formation by Direct Catalytic Oxidation of Methane. *Science.* 259: 343-346.

Higman, C., and Van der Burgt, M. 2008. *Gasification* 2nd Edition. Gulf Professional Publishing, Elsevier, Amsterdam, Netherlands.

Holt, N.A.H. 2001. Integrated Gasification Combined Cycle Power Plants. *Encyclopedia of Physical Science and Technology* 3rd Edition. Academic Press Inc., New York.

Hu, J., Fei Yu, F., and Lu, Y. 2012. Application of Fischer–Tropsch Synthesis in Biomass to Liquid Conversion. *Catalysts,* 2: 303-326.

Irfan, M.F. 2009. Research Report: Pulverized Coal Pyrolysis & Gasification in $N_2/O_2/CO_2$ Mixtures by Thermo-Gravimetric Analysis. *Novel Carbon Resource Sciences Newsletter,* Kyushu University, Fukuoka, Japan. Volume 2: 27-33.

Jenkins, B.M., and Ebeling, J.M. 1985. Thermochemical Properties of Biomass Fuels. *California Agriculture* (May-June), Page 14-18.

John, E., and Singh, K. 2011. Properties of Fuels from Domestic and Industrial Waste. In: *The Biofuels Handbook.* J.G. Speight (Editor). The Royal Society of Chemistry, London, United Kingdom. Part 3, Chapter 2, Page 377-407.

Johnson J.L. 1979. *Kinetics of Coal Gasification.* John Wiley and Sons Inc., Hoboken, New Jersey.

Khosravi, M., and Khadse, A., 2013. Gasification of Petcoke and Coal/Biomass Blend: A Review. *International Journal of Emerging Technology and Advanced Engineering,* 3(12): 167-173.

Ko, M.K., Lee, W.Y., Kim, S.B., Lee, K.W., and Chun, H.S. 2001. Gasification of Food Waste With Steam in Fluidized Bed. *Korean Journal of Chemical Engineering,* 18(6): 961-964.

Kumabe, K., Hanaoka, T., Fujimoto, S., Minowa, T., and Sakanishi, K. 2007. Cogasification of Woody Biomass and Coal with Air and Steam. *Fuel,* 86: 684-689.

Kumar, A., Jones, D.D., and Hanna, M.A. 2009. Thermochemical Biomass Gasification: A Review of the Current Status of the Technology. *Energies,* 2: 556-581.

Kunii, D., and Levenspiel, O. 2013. *Fluidization Engineering* 2nd Edition. Butterworth-Heinemann, Elsevier, Amsterdam, Netherlands.

Lahaye, J., and Ehrburger, P. (Editors). 1991. *Fundamental Issues in Control of Carbon Gasification Reactivity*. Kluwer Academic Publishers, Dordrecht, Netherlands.

Lapuerta M., Hernández J.J., Pazo A., and López, J. 2008. Gasification and co-gasification of biomass wastes: Effect of the biomass origin and the gasifier operating conditions. *Fuel Processing Technology*, 89(9): 828-837.

Lee, S. 2007. Gasification of Coal. In: *Handbook of Alternative Fuel Technologies*. S. Lee, J.G. Speight, and S. Loyalka (Editors). CRC Press, Taylor & Francis Group, Boca Raton, Florida. 2007.

Lee, S., Speight, J.G., and Loyalka, S. 2007. *Handbook of Alternative Fuel Technologies*. CRC-Taylor & Francis Group, Boca Raton, Florida.

Lee, S., and Shah, Y.T. 2013. *Biofuels and Bioenergy*. CRC Press, Taylor & Francis Group, Boca Raton, Florida.

Luque, R., and Speight, J.G. (Editors). 2015. *Gasification for Synthetic Fuel Production: Fundamentals, Processes, and Applications*. Woodhead Publishing, Elsevier, Cambridge, United Kingdom.

Lv, P.M., Xiong, Z.H., Chang, J., Wu, C.Z., Chen, Y., and Zhu, J.X. 2004. An Experimental Study on Biomass Air-Steam Gasification in a Fluidized Bed. *Bioresource Technology*, 95(1): 95-101.

Marano, J.J. 2003. *Refinery Technology Profiles: Gasification and Supporting Technologies*. Report prepared for the United States Department of Energy, National Energy Technology Laboratory. United States Energy Information Administration, Washington, DC. June.

Martinez-Alonso, A., and Tascon, J.M.D. 1991. In: *Fundamental Issues in Control of Carbon Gasification Reactivity*. Lahaye, J., and Ehrburger, P. (Editors). Kluwer Academic Publishers, Dordrecht, Netherlands.

Matsukata, M., Kikuchi, E., and Morita, Y. 1992. A New Classification of Alkali and Alkaline Earth Catalysts for Gasification of Carbon. *Fuel*. 71: 819-823.

McKendry, P. 2002. Energy Production from Biomass Part 3: Gasification Technologies. *Bioresource Technology*, 83(1): 55-63.

McLendon T.R., Lui A.P., Pineault R.L., Beer S.K., and Richardson S.W. 2004. High-Pressure Co-Gasification of Coal and Biomass in a Fluidized Bed. *Biomass and Bioenergy*, 26(4): 377-388.

Mims, C.A. 1991. In *Fundamental Issues in Control of Carbon Gasification Reactivity*. Lahaye, J., and Ehrburger, P. (Editors). Kluwer Academic Publishers, Dordrecht, Netherlands. Page 383.

Minkova, V., Marinov, S.P., Zanzi, R., Bjornbom, E., Budinova, T., Stefanova, M., and Lakov, L. 2000. Thermochemical Treatment of Biomass in a Flow of Steam or in a Mixture of Steam and Carbon Dioxide. *Fuel Processing Technology*, 62: 45-52.

Mokhatab, S., Poe, W.A., and Speight, J.G. 2006. *Handbook of Natural Gas Transmission and Processing*. Elsevier, Amsterdam, Netherlands.

Molino, A., Iovane, P., Donatelli, A., Braccio, G., Chianese, S., Musmarra, D. 2013. Steam Gasification of Refuse-Derived Fuel in a Rotary Kiln Pilot Plant: Experimental Tests. *Chem. Eng. Trans.*, 32: 337-342.

Molino, A., Chianese, S., and Musmarra, D. 2016. Biomass Gasification Technology: The State Of The Art Overview. *J. Energy Chem.*, 25: 10-25.

Nordstrand D., Duong D.N.B., Miller, B.G. 2008. *Combustion Engineering Issues for Solid Fuel Systems*. Chapter 9, Post-combustion Emissions Control. B.G. Miller and D. Tillman (Editors). Elsevier, London, United Kingdom.

Ochoa, J., Cassanello, M.C., Bonelli, P.R., and Cukierman, A.L. 2001. CO2 Gasification of Argentinian Coal Chars: A Kinetic Characterization. *Fuel Processing Technology*, 74: 161-176.

Pakdel, H., and Roy, C. 1991. Hydrocarbon Content of Liquid Products and Tar from Pyrolysis and Gasification of Wood. *Energy & Fuels*, 5: 427-436.

Pan, Y.G., Velo, E., Roca, X., Manyà, J.J., and Puigjaner, L. 2000. Fluidized-Bed Cogasification of Residual Biomass/Poor Coal Blends for Fuel Gas Production. *Fuel*, 79: 1317-1326.

Pandey, A. 2009. *Handbook of Plant-Based Biofuels*. CRC Press, Taylor and Francis Group, Boca Raton, Florida.

Penrose, C.F., Wallace, P.S., Kasbaum, J.L., Anderson, M.K., and Preston, W.E. 1999. Enhancing Refinery Profitability by Gasification, Hydroprocessing and Power Generation. *Proceedings. Gasification Technologies Conference San Francisco, California. October*. https://www.globalsyngas.org/uploads/eventLibrary/GTC99270.pdf

Pepiot, P., Dibble, and Foust, C.G. 2010. Computational Fluid Dynamics Modeling of Biomass Gasification and Pyrolysis. In: *Computational Modeling in Lignocellulosic Biofuel Production*. M.R. Nimlos and M.F. Crowley (Editors). ACS Symposium Series; American Chemical Society, Washington, DC.

Pfeifer C., Puchner B., and Hofbauer H. 2009. Comparison of Dual Fluidized Bed Steam Gasification of Biomass With and Without Selective Transport of CO2. *Chem Eng Sci.*, 64: 5073-5083.

Probstein, R.F., and Hicks, R.E. 1990. *Synthetic Fuels*. pH Press, Cambridge, Massachusetts. Chapter 4.

Rabovitser, I.K., Nester, S., and Bryan, B. 2010. Plasma Assisted Conversion of Carbonaceous Materials into a Gas. United States Patent 7,736,400. June 25.

Radovanović, L., and Speight, 2018. J.G. Global Warming – Truth and Myths. *Petroleum and Chemical Industry International*, 1(1). https://www.opastonline.com/wp-content/uploads/2018/10/Globalwarming-truthandmyths-pcii-18.pdf

Rajvanshi, A.K. 1986. Biomass Gasification. In: *Alternative Energy in Agriculture*, Vol. II. D.Y. Goswami (Editor). CRC Press, Boca Raton, Florida. Page 83-102.

Ramroop Singh, N. 2011. Biofuel. In: *The Biofuels Handbook*. J.G. Speight (Editor). Royal Society of Chemistry, London, United Kingdom. Part 1, Chapter 5.

Rapagnà, S., and Latif, A. 1997. Steam Gasification of Almond Shells in a Fluidized Bed Reactor: The Influence of Temperature and Particle Size on Product Yield and Distribution. *Biomass and Bioenergy*, 12(4): 281-288.

Rapagnà, S., Jand, N., Kiennemann, A., Foscolo, P.U. 2000. Steam Gasification of Biomass in a Fluidized-Bed of Olivine Particles. *Biomass and Bioenergy*, 19(3): 187-197.

Rauch, R., Hrbek, J., Hofbauer, H. 2014. Biomass Gasification for Synthesis Gas Production and Applications of the Syngas. *Wiley Interdiscip. Rev. Energy Environ.*, 3: 343-362.

Ricketts, B., Hotchkiss, R., Livingston, W., and Hall, M. 2002. Technology Status Review of Waste/Biomass Co-Gasification with Coal. *Proceedings. Inst. Chem. Eng. Fifth European Gasification Conference. Noordwijk, Netherlands. April 8-10*.

Roddy, D.J., and Manson-Whitton, C. 2012. Biomass Gasification and Pyrolysis. In: *Comprehensive Renewable Energy*. Volume 5: *Biomass and Biofuels*. D.J. Roddy (Editor). Elsevier, Amsterdam, Netherlands.

Rodríguez-Olalde, N.E., Mendoza-Chávez, E., Castro-Montoya, A.J., Saucedo-Luna, J., Maya-Yescas, R., Rutiaga-Quiñones, J.G., and Ponce Ortega, J.M. 2015. Simulation of Syngas Production from Lignin Using Guaiacol as a Model Compound. *Energies*, 8: 6705-6714.

Senneca, O. 2007. Kinetics of Pyrolysis, Combustion and Gasification of Three Biomass Fuels. *Fuel Processing Technology*, 88: 87-97.

Shabbar, S., and Janajreh, I. 2013. Thermodynamic Equilibrium Analysis of Coal Gasification Using Gibbs Energy Minimization Method. *Energy Conversion and Management*, 65: 755-763.

Shen, C-H., Chen, W-H., Hsu, H-W., Sheu, J-Y., and Hsieh, T-H. 2012. Co-Gasification Performance of Coal and Petroleum Coke Blends in A Pilot-Scale Pressurized Entrained-Flow Gasifier. *Int. J. Energy Res.*, 36: 499-508.

Sikarwar, V.S., Zhao, M., Clough, P., Yao, J., Zhong, X., Memon, M.Z., Shah, N., Anthony. E.J., and Fennell, P.S. 2016. An Overview of Advances in Biomass Gasification. *Energy Environ. Sci.*, 9; 2939-2977.

Sikarwar, V.S., Zhao, M., Fennell, P.S., Shah, N., Anthony, E.J. 2017. Progress in Biofuel Production from Gasification. *Prog. Energy Combust. Sci.*, 61: 189-248.

Singh, S.P., Weil, S.A., and Babu, S.P. 1980. Thermodynamic Analysis of Coal Gasification Processes. *Energy*, 5(8-9): 905–914.

Sjöström, K., Chen, G., Yu, Q., Brage, C., and Rosén, C. 1999. Promoted Reactivity of Char in Cogasification of Biomass and Coal: Synergies in the Thermochemical Process. *Fuel*, 78: 1189-1194.

Sondreal, E. A., Benson, S. A., Pavlish, J. H., and Ralston, N.V.C. 2004. An Overview of Air Quality III: Mercury, Trace Elements, and Particulate Matter. *Fuel Processing Technology*. 85: 425-440.

Speight, J.G. 1990. In: *Fuel Science and Technology Handbook*. J.G. Speight (Editor). Marcel Dekker Inc., New York. Chapter 33.

Speight, J.G. 2007. *Natural Gas: A Basic Handbook*. GPC Books, Gulf Publishing Company, Houston, Texas.

Speight, J.G. 2008. *Synthetic Fuels Handbook: Properties, Processes, and Performance*. McGraw-Hill, New York.

Speight, J.G. (Editor). 2011a. *Biofuels Handbook*. Royal Society of Chemistry, London, United Kingdom.

Speight, J.G. 2011b. *The Refinery of the Future*. Gulf Professional Publishing, Elsevier, Oxford, United Kingdom.

Speight, J.G. 2013a. *The Chemistry and Technology of Coal* 3rd Edition. CRC Press, Taylor & Francis Group, Boca Raton, Florida.

Speight, J.G. 2013b. *Coal-Fired Power Generation Handbook*. Scrivener Publishing, Salem, Massachusetts.

Speight, J.G. 2014a. *The Chemistry and Technology of Petroleum* 5th Edition. CRC Press, Taylor & Francis Group, Boca Raton, Florida.

Speight, J.G. 2014b. *Gasification of Unconventional Feedstocks*. Gulf Professional Publishing, Elsevier, Oxford, United Kingdom.

Speight. J.G. 2016. Hydrogen in Refineries. In: *Hydrogen Science and Engineering – Materials, Processes, Systems, and Technology*. D. Stolten and B. Emonts (Editors). Wiley-VCH Verlag GmbH & Co., Weinheim, Germany. Chapter 1. Page 3-18.

Speight, J.G., and Islam, M.R. 2016. *Peak Energy – Myth or Reality*. Scrivener Publishing, Beverly, Massachusetts.

Speight, J.G. 2017. *Handbook of Petroleum Refining*. CRC Press, Taylor & Francis Group, Boca Raton, Florida.

Speight, J.G. 2019. *Handbook of Petrochemical Processes*. CRC Press, Taylor & Francis Group, Boca Raton, Florida.

Sundaresan, S., and Amundson, N.R. 1978. Studies in Char Gasification – I: A lumped Model. *Chemical Engineering Science*, 34: 345-354.

Sutikno, T., and Turini, K. 2012. Gasifying Coke to Produce Hydrogen in Refineries. *Petroleum Technology Quarterly*, Q3: 105.

Tijmensen, M.J.A., Faaij, A,P,C,, Hamelinck, C.N., and Van Hardeveld, M.R.M. 2002. Exploration of The Possibilities for Production of Fischer Tropsch liquids and Power via Biomass Gasification. *Biomass Bioenergy*, 23: 129-152.

Tinaut, F.V., Melgar, A., Pérez, J.F., and Horillo, A. 2008. Effect of Biomass Particle Size and Air Superficial Velocity on the Gasification Process in a Downdraft Fixed Bed Gasifier. An Experimental And Modelling Study. *Fuel Process. Technol.*, 89: 1076-1089.

Usón, S., Valero, A., Correas, L., and Martínez, Á. 2004. Co-Gasification of Coal and Biomass in an IGCC Power Plant: Gasifier Modeling. *Int. J. Thermodynamics*, 7(4): 165-172.

Vélez, F.F., Chejne, F., Valdés, C.F., Emery, E.J., and Londoño, C.A. 2009. Cogasification of Colombian Coal and Biomass in a Fluidized Bed: An Experimental Study. *Fuel*, 88(3): 424-430.

Wang, Y., Duan Y., Yang L., Jiang Y., Wu C., Wang Q., and Yang X. 2008. Comparison of Mercury Removal Characteristic between Fabric Filter and Electrostatic Precipitators of Coal-Fired Power Plants. *Journal of Fuel Chemistry and Technology*, 36(1): 23-29.

Yang, H., Xua, Z., Fan, M., Bland, A.E., and Judkins, R.R. 2007. Adsorbents for Capturing Mercury in Coal-Fired Boiler Flue Gas. *Journal of Hazardous Materials*. 146:1-11.

Ye, D.P., Agnew, J.B., and Zhang, D.K. 1998. Gasification of a South Australian Low-Rank Coal with Carbon Dioxide and Steam: Kinetics and Reactivity Studies. *Fuel*, 77: 1209-1219.

7
Gasification of Waste

7.1 Introduction

Waste is an unavoidable byproduct of human activity and the improvement in living standards has led to increases in the quantity and complexity of generated waste, whilst industrial diversification and the provision of expanded health-care facilities have added substantial quantities of industrial waste and biomedical waste into the waste. Thus, there is a need for the management and disposal of the growing volume of waste, in addition to waste disposal (landfill) operations that are being stretched to the limit, and suitable disposal areas are in short supply. Moreover, the potential for rain water (and snow melt) to leach chemical constituents from landfills and into the groundwater table is of immediate concern. Landfills have been, and at the time of writing, remain the most common way to dispose of municipal and industrial waste. In fact, landfills are also gas generators through the decompositon of the organic constituents of the landfill over time (Speight, 2019).

Gasification has a high potential for application in waste processing compared to other existing techniques such as landfill and incineration because a wide variety of feedstocks can be processed to produce multiple useful products (Malkow, 2004). As organic materials such as wood and sewage sludge are gasified, the chemical process creates a clean synthesis gas, a fuel that can then be used like natural gas. Compared to other waste-to-energy methods like burning, or incineration, the gasification process allows the synthesis gas to be cleaned of contaminants prior to its use. Incineration plants do attempt to clean their emissions, but it is achieved at the post-combustion stage, making it harder to capture and isolate hazardous emissions. Creating synthesis gas through gasification is almost like closing the circle, or closing the loop, on the lifecycle of the waste and the energy embedded in the waste can be used to power engines that generate electricity.

Gasification is an intricate process involving drying the feedstock followed by pyrolysis, partial combustion of intermediates, and finally gasification of the resulting products using a variety of specifically designed reactors (Chapter 3). The process can be performed in the presence of a gasifying media which can be air, oxygen (O_2), steam (H_2O) or carbon dioxide (CO_2), inside the gasifier, of which there are various types (Chapter 3). The quality of the gaseous product is dependent on the type of gasifier and the gasifying agent. The gasifying medium also plays a vital role of converting the carbonaceous feedstock to low-molecular-weight gases such as carbon monoxide (CO) and hydrogen – the essential components of synthesis gas (Chapter 2). The quality and properties of the product are dependent on (i) the feedstock, (ii) the gasifying agent, (iii) the feedstock dimensions, (iv) the temperature, (v) the pressure, (vi) the reactor design, and (vii) the presence of a catalyst and/or a sorbent (Luque and Speight, 2015; Sikarwar et al., 2016). In fact, gasification is a unique process that transforms any carbon-based material, such as municipal solid waste (MSW), into energy without burning it and converts the carbonaceous waste into gaseous products of which synthesis gas is (in the current context) of prime importance – removal of pollutants and impurities results in clean gas that can be converted into electricity and valuable products (Chapter 1, Chapter 6).

Municipal solid waste is predominantly composed of carbon, hydrogen and oxygen, and as such is a potential energy source and development of the technology necessary for gasification of the waste to synthesis gas and to hydrogen has the potential to address a number of economic, environmental, and resource issues. A path for the conversion of municipal solid waste hydrogen has the potential to be economically advantageous because of the reduced costs of management and disposal of the waste. Thus, with gasification, municipal solid waste and other types of wastes are no longer of environmental concern but are feedstocks for a gasifier. Instead of the associated costs of disposal of and landfill management, using waste as a feedstock for a gasification process reduces disposal costs and required landfill space, and converts the carbonaceous wastes to electricity and fuels.

Initially, the gasification process was applied to coal as a means of producing fuel gases and chemicals as well as electricity but has seen considerable evolution in terms of acceptance of feedstocks other than coal as well as the technologies used for the process and represents *significant* advances over the incineration process (Chapter 1, Chapter 2) (Orr and Maxwell, 2000; Malkow, 2004; Speight, 2008; E4Tech, 2009; 2011a, 2011b, 2013a, 2013b, 2014). In order to understand the advantages of gasification when compared to incineration for waste materials, it is important to understand the differences between the two processes.

Incineration, which does have a place in waste disposal operations, uses municipal solid waste as a fuel. The waste is burned with high volumes of air to form carbon dioxide and heat. In a waste-to-energy plant that uses incineration, the hot gaseous products are used to generate steam, which is then used in a steam turbine to generate electricity. On the other hand, gasification converts municipal solid waste to usable synthesis gas and it is the production of this synthesis gas which makes gasification different from the incineration process. In the gasification process, the municipal solid waste is not a fuel but a feedstock for a high-temperature chemical conversion process. Instead of making just heat and electricity, as is done in a waste-to-energy plant using incineration, the synthesis gas produced by gasification can be turned into higher value commercial products such as transportation fuels, chemicals, fertilizers, and substitute natural gas.

In addition, one of the concerns with incineration of municipal solid waste is the formation and reformation of toxic dioxins and furans, especially from PVC plastics (polyvinyl chloride plastics). These toxins end up in exhaust streams by three pathways: (i) by decomposition, as low-molecular weight volatile constituents, (ii) by re-forming in which lower molecular weight constituents combine to form new products, and/or (iii) by the unusual step of passing through the incinerator without change. Incineration does not always allow adequate control of these processes.

In respect of municipal solid waste, gasification is significantly different from and cleaner than incineration: (i) in the high-temperature environment in gasification, higher molecular weight materials, such as plastics, are effectively decomposed to synthesis gas, which can be cleaned and processed before any further use, (ii) dioxins and furans need sufficient oxygen to form and the oxygen-deficient atmosphere in a gasifier does not provide the environment needed for the formation of dioxins and furans, (iii) when the synthesis gas is primarily used as a fuel for making heat, it can be cleaned as necessary *before* combustion; this cannot occur in incineration.

Thus, waste-to-energy plants based on the gasification principle are high-efficiency power plants that utilize municipal solid waste as fuel rather than conventional sources of energy such as crude oil or coal, although the co-gasification of waste with biomass, coal, and crude oil residua is always an option (Speight, 2011a, 2011b, 2013a, 2013c, 2014). Such plants recover the thermal energy contained in the waste in highly efficient boilers that generate steam that can then be used to drive turbines for electricity production.

This chapter presents descriptions of the various types of waste and the recovery of energy from waste by gasification, including the benefits such as: (i) reduction in the total quantity of waste depending upon the waste composition and the gasification technology employed, (ii) reduction in environmental pollution, and (iii) improved commercial viability of the waste disposal project from the sale of energy/products.

It is for the pertinent reason of variability that gasification – a thermochemical conversion process – that produces synthesis gas and char has risen to prominence. It is, therefore, to be expected that the products will vary according to the composition of the waste and product composition must be monitored in order to apply the necessary gas cleaning operations (Chapter 9).

7.2 Waste Types

Waste (also called in the United States *garbage* or *trash*) is a substance, object, or collection of substances and objects selected for disposal or required to be disposed of by the provisions of local, regional, or national laws. In addition, waste is also substances or objects that are not the prime products of a process (or processes) for which the initial user has no further use in terms of his/her own purposes of production, transformation or consumption, and of which s/he wants to dispose. Wastes may be generated during (i) the extraction of raw materials, (ii) the processing of raw materials into intermediate and final products, (iii) the consumption of final products, and (iv) other human activities (Table 7.1).

Furthermore, the composition of waste is one of the main factors influencing not only the emissions from solid waste treatment, as different waste types contain different amount of degradable organic carbon (DOC) and fossil carbon, but also the manner by which the waste can be treated and disposed in landfills (Speight, 2019). In addition to the organic constituents, waste also contains ash, as well as non-fossil carbon. Some textiles, plastics, rubber and electronic waste contain the bulk part of fossil carbon in municipal solid waste. Paper (with coatings) and leather (synthetic) can also include small amounts of fossil carbon.

Moreover, the composition of waste will vary by city in the same country and will also vary by the day of the week, season and year in the same city. Also, sampling at solid waste disposal sites on rainy days will change moisture content (i.e., wet weight composition) significantly, and needs attention in interpretation when the waste is destined for further treatment or landfill disposal. Thus, representative (or average) composition data should be treated with caution because of the potential difference in the composition of the waste which can have a significant effect on the method of disposal.

Table 7.1 Sources and types of waste.

Source	Waste generators	Types of solid waste
Residential	Single and multifamily dwellings	Food wastes, paper, cardboard, plastics, textiles, leather, yard wastes, wood, glass, metals, ashes, special wastes (e.g., bulky items, consumer electronics, white goods, batteries, oil, tires), and household hazardous wastes).
Industrial	Light and heavy manufacturing, refineries, chemical plants, power plants, mineral extraction and processing	Housekeeping wastes, packaging, food wastes, construction and demolition materials, hazardous wastes, ashes, Industrial process wastes, scrap materials, off-specification products, tailings.
Institutional	Schools, hospitals, prisons, government centers	Paper, cardboard, plastics, wood, food wastes, glass, metals, special wastes, hazardous wastes.
Construction and demolition	New construction sites, road repair, renovation sites, demolition of buildings	Wood, steel, concrete, dirt, etc.
Municipal services	Street cleaning, landscaping, parks, beaches, other recreational areas, water and wastewater treatment plants	Street sweepings; landscape and tree trimmings; general wastes from parks, beaches, and other recreational areas; sludge.
Agriculture	Crops, orchards, vineyards, dairies, feedlots, farms	Spoiled food wastes, agricultural wastes, hazardous wastes (e.g., pesticides).

7.2.1 Solid Waste

Solid waste is a general term that includes garbage, rubbish, refuse, sludge from a wastewater treatment plant, water supply treatment plant, or air pollution control facility, sewage sludge, and other discarded material, including solid, liquid, semisolid, or contained gaseous material resulting from industrial, municipal, commercial, mining, and agricultural operations and from community and institutional activities. *Soil, dirt, rock, sand, and other inert solid materials,* whether natural or of human origin, used to fill land are not classified as waste if the object of the fill is to make the land suitable for the construction of surface improvements. Solid waste does not include waste materials that result from activities associated with the exploration, development, or production of oil or gas or geothermal resources, or other substance or material regulated by the local or federal governments.

Solid waste that is typically excluded from gasification feedstocks includes uncontaminated solid waste resulting from the construction, remodeling, repair and demolition of utilities, structures and roads; and uncontaminated solid waste resulting from land clearing.

Such waste includes, but is not limited to, bricks, concrete and other masonry materials, soil, rock, wood (including painted, treated and coated wood and wood products), land clearing debris, wall coverings, plaster, drywall, plumbing fixtures, non-asbestos insulation, roofing shingles and other roof coverings, asphaltic pavement, glass, plastics that are not sealed in a manner that conceals other wastes, empty buckets 10 gallons or less in size and having no more than one inch of residue remaining on the bottom, electrical wiring and components containing no hazardous liquids, and pipe and metals that are incidental to any of the above.

In summary, solid waste as the definition applies to this text is any unwanted or discarded carbonaceous (containing carbon) or hydrocarbonaceous (containing carbon and hydrogen) material that originates from a variety of sources and is not a liquid or a gas. Furthermore, the disposal of a wide variety of wastes has become an important problem because the traditional means of disposal to a landfill has become environmentally much less acceptable than previously. Newer and stricter regulation of the conventional disposal method(s) has made the economics of waste processing for resource recovery much more favorable.

However, before moving on to the various aspects of the gasification process, it is worthwhile to describe in more detail the types of waste that arise from human activities and which might be suitable for gasification.

7.2.2 Municipal Solid Waste

Municipal solid waste (MSW) is solid waste resulting from, or incidental to, municipal, community, commercial, institutional, and recreational activities; it includes garbage, rubbish, ashes, street cleanings, dead animals, medical waste, and all other nonindustrial solid waste.

Municipal solid waste (MSW) is generated from households, offices, hotels, shops, schools and other institutions. The major components are food waste, paper, plastic, rags, metal and glass, although demolition and construction debris is often included in collected waste, as are small quantities of hazardous waste, such as electric light bulbs, batteries, automotive parts and discarded medicines and chemicals.

Municipal solid waste is a negatively priced, abundant and essentially renewable feedstock. The composition of municipal solid waste (Table 7.2) can vary from one community to the next, but the overall differences are not substantial. In fact, there are several types of waste that might also be classified under the municipal solid waste umbrella (Table 7.1).

The heat content of raw MSW depends on the concentration of combustible organic materials in the waste and its moisture content. Typically, raw municipal solid waste has a heating value on the order of approximately half that of bituminous coal (Speight, 2013a). The moisture content of raw municipal solid waste is usually 20% w/w.

7.2.3 Industrial Solid Waste

Industrial solid waste is solid waste resulting from or incidental to any process of industry, manufacturing, mining, or agricultural operations. Industrial solid waste is classified as either *hazardous* or *non-hazardous*. *Hazardous industrial waste* includes any industrial solid waste or combination of industrial solid wastes identified or listed as a hazardous waste.

Table 7.2 General composition of municipal solid waste.

Component	% w/w
Paper	33.7
Cardboard	5.5
Plastics	9.1
Textiles	3.6
Rubber, Leather, "other"	2.0
Wood	7.2
Horticultural Wastes	14.0
Food Wastes	9.0
Glass and Metals	13.1

Industrial solid waste encompasses a wide range of materials of varying environmental toxicity. Typically this range would include paper, packaging materials, waste from food processing, oils, solvents, resins, paints and sludge, glass, ceramics, stones, metals, plastics, rubber, leather, wood, cloth, straw, abrasives, etc. As with municipal solid waste, the absence of a regularly updated and systematic database on industrial solid waste ensures that the exact rates of generation are largely unknown.

Nonhazardous industrial waste is an industrial solid waste that is not identified or listed as a hazardous waste. A generator of nonhazardous industrial solid waste must further classify the waste as:

- Class 1 waste, including any industrial solid waste or mixture of industrial solid wastes that – because of its concentration, or physical or chemical characteristics – is toxic; corrosive; flammable; a strong sensitizer or irritant; or a generator of sudden pressure by decomposition, heat, or other means; or may pose a substantial present or potential danger to human health or the environment when improperly processed, stored, transported, or disposed of or otherwise.
- Class 2 waste, consisting of any individual solid waste or combination of industrial solid wastes that are not described as Hazardous, Class 1, or Class 3.
- Class 3 waste, consisting of inert and essentially insoluble industrial solid waste, usually including, but not limited to, materials such as rock, brick, glass, dirt, and certain plastics and rubber, etc., that are not readily decomposable.

7.2.4 Bio-Solids

Bio-solids include livestock waste, agricultural crop residues and agro-industrial byproducts. In most traditional, sedentary agricultural systems, farmers use the land application of raw or composted agricultural wastes as a means of recycling valuable nutrients and organics back into the soil and this remains the most widespread means of disposal. Similarly, fish farming communities commonly integrate fish rearing with agricultural activities such as livestock husbandry, vegetable and paddy cultivation and fruit farming.

Many countries with agricultural-based economies use agricultural wastes to produce biogas through anaerobic digestion (Speight, 2008, 2011a). The biogas (approximately 60% v/v methane) is primarily used directly for cooking, heating and lighting, whilst the slurry from the anaerobic digesters is used as liquid fertilizer, a feed supplement for cattle and pigs and as a medium for soaking seeds.

7.2.5 Biomedical Waste

Biomedical waste is the waste materials produced by hospitals and health care institutions, which have been increasing over the past four decades to meet the medical and health care requirements of the growing world population. Until recent years, little attention was paid to the wastes generated from these facilities, which are potentially hazardous to human health and the environment. In fact, serious concern has arisen regarding the potential for spreading pathogens, as well as causing environmental contamination due to the improper handling and management of clinical and biomedical waste.

Regulated medical waste (RMW) is a waste stream that contains potentially infectious material – also called *red bag waste* or *biohazardous waste*. Regulated medical waste is regulated state-by-state, but also falls under the Blood-borne Pathogen Standard as defined by the US Office of Safety and Health Administration (OSHA). Such waste is subject to state and federal regulations and may not be suitable as gasification feedstock and requires higher temperature to assure complete disposal of the constituents.

7.2.6 Sewage Sludge

The thermochemical conversion of the organic content of sewage sludge into high-value gases such as synthesis gas, as well as carbon dioxide, methane, and other hydrocarbon derivatives is the main basis for gasification. This reaction occurs in a partially oxidized reaction atmosphere at high temperature (800 to 1000°C) (1470 to (1830°F). The process can use air, carbon dioxide, oxygen, steam, or mixtures of such gases but, as anticipated, the gasifying agent has significant impact on the calorific value of the synthesis gas with highest heating value from the oxy-gasification (Oladejo *et al.*, 2019; Werle and Dudziak, 2019). The product gas can be used directly for heating or electricity generation via heat engine or can be further processed for chemicals or liquid fuels (Chapter 10).

The gasification reaction can be divided into four sub-stages: (i) drying of sample (70 to 200°C, 158 to 390°F), (ii) devolatilization (350 to 600°C, 660 to 1110°F), (iii) oxidation of the volatiles, and (iv) char gasification. The use of *in-situ* drying is only ideal for samples with low moisture content. After the drying stage, the pyrolysis of the samples is done for generating volatiles and char that can be fully oxidized to drive the other reaction stages. Therefore the oxidation stage (with temperatures on the order of 1200°C, 2190°F) produces heat to run the gasification, pyrolysis and drying stage. Finally, the high temperature reduction of the char produced from pyrolysis generates light hydrocarbon gases in the fourth stage. The position of these sub-stages in the gasifier immensely affects the flow of gasifying agent, reaction process and operating efficiency, thereby becoming a deciding factor in the choice of reactors.

There are three main types of gasifiers that can be employed for the process: (i) the fixed-bed downdraft gasifier, (ii) the fixed-bed updraft gasifier, and (iii) the fluidized-bed gasifier. The fixed-bed alignment involves a flow of gasifying agent and heat up or down the reactor chambers to activate the drying, pyrolysis and gasifying stages consecutively. This leads to efficiency reduction and shorter residence time (particularly for char oxidation which is the rate-limiting step) in comparison to the fluidized bed that allows instantaneous occurrence of all sub-stages which allows completion of the gasification process. The reactor type also affects the tar and pollutant concentration in the product which hinders the efficiency of the process (Oladejo *et al.*, 2019).

7.3 Feedstock Properties

Because of the nature of the constituents in any of the above waste streams that are used in a waste-to-energy gasification plant, it is necessary to consider at this point the feedstocks properties and any safety (and health) issues that might arise from the use of such feedstocks. In fact, such feedstock materials typically comprise biomass waste (or biomass), municipal solid waste (MSW), refuse-derived fuel (RDF), or solid recovered fuel (SRF) and waste materials are not always of such a composition that behavior in the process can be predicted with any degree of accuracy (Speight, 2011a; Al Asfar, 2014).

The individual feedstock constituents will typically have their own hazards including fire, dust explosion and toxic gas formation but when used in conjunction with other constituents, the need for safety and handling of the combined feedstocks may require extra precautions. For example, where feed materials such as biomass wood are stored in large piles, there is potential for self-heating (spontaneous ignition) – which is always an issue when coal is stockpiled (Speight, 2013a, 2013b). Wood fuel is a source of nutrients for microbial activity, which, in the presence of moisture, over extended time periods, can lead to the generation of heat, and self-ignition. Other feed safety considerations include hazards associated with dust, such as (i) explosion hazards requiring protection by, for example, hot particle detection and (ii) explosion venting to mitigate the effects of explosions.

7.4 Fuel Production

The gasification of solid waste gasification includes a number of physical and chemical interactions that occur at temperatures generally higher than 600°C (1110°F), the exact temperature depending on the reactor type and the waste characteristics, in particular the ash softening and melting temperatures (Higman and van der Burgt, 2003; Arena, 2012). The different types of waste gasification processes are generally classified on the basis of oxidation medium: (i) partial oxidation with air, (ii) oxygen-enriched air, (iii) or pure oxygen, (iv) steam gasification, or (v) plasma gasification. Some processes are operated with oxygen-enriched air, i.e., a mixture of nitrogen and oxygen having oxygen content higher than 21% v/v but less than 50% v/v in order to higher heating value gas as a consequence of the reduced nitrogen content, that makes possible to carry out auto-thermal processes at higher temperature, without expensive consumption of oxygen (Mastellone et al., 2010a). The partial oxidation process using pure oxygen generates synthesis gas free (or almost free) of atmospheric nitrogen. The steam gasification option generates a high hydrogen concentration, medium heating value nitrogen-free synthesis gas. In this case, steam is the only gasifying agent and the process does not include exothermic reactions but does need an external source of energy for the endothermic gasification reactions.

As a guide to the gasification chemistry, two main steps have been proposed for the thermal degradation of municipal solid waste have been observed: (i) thermal degradation step at temperatures from 280 to 350°C (535 to 660°F) and consists mainly of the decomposition of any waste biomass component into low-boiling (methane, ethane, and propane) hydrocarbon derivatives and (ii) thermal degradation step at temperatures from 380 to 450°C (715 to 840°F) and is mainly attributed to polymer components, such as plastics and rubber in municipal solid waste. The polymer component can also evolve significant amounts of benzene derivatives, such as styrene (Kwon et al., 2009). However, the complexity of municipal solid waste should warrant more complex thermal decompositon regimes than the two proposed.

In the case of plasma gasification (Moustakas et al., 2005; Lemmens et al., 2007), the heat source of the gasifier is one or more plasma arc torches that create an electric arc and produce a very high-temperature plasma gas (up to 15000°C, 27000°F), which in turn allows control of temperature independently from fluctuations in the feed quality and supply of a gasification agent (air, oxygen or steam). This allows variations in the feeding rate, moisture content and elemental composition of the waste material: plasma gasifiers can therefore accept feedstocks of variable particle size, containing coarse lumps and fine powders, with minimal feed preparation (Gomez et al., 2009).

7.4.1 Preprocessing

Gasification is a thermochemical process that generates a gaseous, fuel-rich product (Chapter 1) and, regardless of how the gasifier is designed (Chapter 2), two processes must take place in order to produce a useable fuel gas. In the first stage, pyrolysis releases the volatile components of the fuel at temperatures below 600°C (1110°F). The byproduct of pyrolysis that is not vaporized is *char* and consists mainly of fixed carbon and ash. In the second gasification stage, the char remaining after pyrolysis is either reacted with steam or hydrogen or combusted with air or pure oxygen. Gasification with air results in a nitrogen-rich, low-Btu fuel gas. Gasification with pure oxygen results in a higher-quality mixture of carbon monoxide and hydrogen and virtually no nitrogen, and gasification with steam (steam reforming) (Chapter 6) also results in a synthesis gas that is rich in hydrogen and carbon dioxide with only minor amounts of impurities (Richardson et al., 1995). Typically, the exothermic reaction between the feedstock carbon and oxygen provides the heat energy required to drive the pyrolysis and char gasification reactions.

Municipal solid waste is not a homogenous waste stream. Since inorganic materials (metals, glass, concrete, and rocks) do not enter into the thermal conversion reactions, part of the energy which could be used to gasify the feedstock is expended in heating the inorganic materials to the pyrolysis reactor temperature. Then the inorganic materials are cooled in cleanup processes, and the heat energy is lost, which reduces the overall efficiency of the system. To make the process more efficient, some preprocessing of waste is typically required and includes the separation of thermally non-degradable material such as metals, glass, and concrete debris. Preprocessing may include sorting, separation, size reduction, and densification (for reducing overall volume of feedstock being fed into the gasifier). Such preprocessing techniques are common in the waste recycling industry for recovery of paper, glass, and metals from the municipal solid waste streams.

Thus, the first function of the front-end (preprocessing) system is to accept solid waste directly from the collection vehicle and to separate the solid waste into two fractions – combustible waste and non-combustible waste. The front-end separation produces the feedstock for the gasification process.

Before the gasification proper and in order to enhance the process, feedstock pregasification systems (preprocessing systems) are necessary to accomplish the extraction of metals, glass and inorganic materials, resulting in the increased recycling and utilization of materials. In addition, a wide range of plastics cannot be recycled or cannot be recycled any further, feedstock for gasification. Thus, the main steps involved in preprocessing of municipal solid waste are analogous to the preprocessing of coal (Speight, 2013a, 2013b) or biomass (Speight, 2008, 2011a) and include (i) sorting – manual and/or mechanical, (ii) shredding, (iii) grinding, (iv) blending with other materials, (v) drying, and (vi) pelletization. The purpose of preprocessing is to produce a feed material with, as best as can be achieved, near-consistent physical characteristics and chemical properties. Preprocessing operations are also designed to produce a material that can be safely handled, transported, and stored prior to the gasification process. In addition, particle size or pellet size affects the product distribution (Luo *et al.*, 2010).

If the municipal solid waste has high moisture content, a dryer may be added to the preprocessing stage to lower the moisture content of the waste stream to 25% w/w, or lower (CH2MHill, 2009). Lower moisture content of the feedstock increases its heating value and the system becomes more efficient. The waste heat or fuel produced by the system can be used to dry the incoming municipal solid waste.

In some case, the preprocessing operation may be used for the production of a combustible fraction (i.e., a solid fuel) from municipal solid waste and from mixed waste and its thermal conversion requires two basic and distinct subsystems – the *front-end* and the *back-end*. The combustible fraction recovered from mixed municipal solid waste has been given the name *refuse-derived fuel* (RDF). The composition of the recovered combustible fraction is a mixture that has higher concentrations of combustible materials (e.g., paper and plastics) than those present in the parent mixed municipal solid waste.

The main components (i.e., unit operations) of a front-end subsystem are usually any combination of size reduction, screening, magnetic separation, and density separation (e.g., air classification). The types and configurations of unit operations selected for the front-end design depend on the types of secondary materials that will be recovered and on the desired quality of the recovered fuel fraction. The fuel quality must be specified by the designer or supplier of the thermal conversion system.

Typically, systems that recover a combustible fraction from mixed municipal solid waste utilize size reduction, screening, and magnetic separation. Some designs and facilities have used screening, followed by size reduction (e.g., pretrommel screening – a trommel is a drum screen), as the fundamental foundation of the system design, while others have reversed the order of these two operations. A number of considerations enter into the determination and the selection of the optimum order of screening and size reduction for a given application. Among others, the considerations include composition of the waste. Other unit operations may also be included in the system design, including manual sorting, magnetic separation, air classification, and pelletization (i.e., densification), as the need dictates for recovery of other materials (e.g., aluminum, etc.) and for achieving the desired specification of the solid fuel product (Diaz and Savage, 1996).

7.4.2 Process Design

After being preprocessed into, for example, suitable particle-size pieces – or fed directly (if a gas or liquid) – the waste is injected into the gasifier, along with a controlled amount of air or oxygen. The high-temperature conditions in the gasifier decompose the feedstock, eventually forming synthesis gas, which consists primarily of hydrogen and carbon monoxide and, depending upon the specific gasification technology, smaller quantities of methane, carbon dioxide, hydrogen sulfide, and water vapor. Typically, 70 to 85% w/w/of the carbon in the feedstock is converted into the synthesis gas.

The ratio of carbon monoxide to hydrogen depends in part upon (i) the hydrogen and carbon content of the feedstock and (ii) the type of gasifier used, but can also be adjusted or *shifted* downstream of the gasifier through use of catalysts. This ratio is important in determining the type of product to be manufactured (electricity, chemicals, fuels, hydrogen) (Chapter 1, Chapter 6). For example, a refinery would use a synthesis gas consisting primarily of hydrogen, important in producing transportation fuels (Speight 2011b, 2014). Conversely, a chemical plant will require synthesis gas with approximately equal proportions of hydrogen and carbon monoxide, both of which are basic building blocks for a broad range of products – including consumer and agricultural products such as fertilizers, plastics, and fine chemicals (i.e., complex, single, pure chemical compounds). Thus, the inherent flexibility of the gasification process to adapt to feedstock requirements and the desired product slate can lead to the production of one or more products from the same process.

As a result of the gasifier section and the prerequisites for use of the selected reactor, the correct design of the front-end system is obviously a necessity for the successful operation of a waste-to-energy facility. The key function of the preprocessing system is the segregation of the combustible components from the non-combustible components. In the production of a refuse-derived fuel, particular attention must be paid to the combustion unit in which the fuel is to be burned. For example, in order to facilitate handling, storage, and transportation, it may be necessary in some cases to produce a densified fuel (i.e., a pelletized fuel) that meets necessary specifications (Pellet Fuels Institute, 2011).

Processing of municipal solid wastes for the production of a fuel is a seemingly straightforward process in terms of design and system operation. The performance and operation of the processing system is strongly and fundamentally determined by (i) the characteristics of the feedstock, (ii) the type of equipment chosen, and (iii) the location of the equipment in the overall processing configuration. Although some of the equipment available for waste processing applications may have been proven to be well suited to the processing tasks of other industries (e.g., mining, forestry, etc.), waste differs substantially from the raw materials that serve as feedstocks for other industries.

The failure to take the difference into account can result in operational problems at waste processing facilities, such as (i) use of equipment that was improperly applied, (ii) use of equipment that was improperly designed, or (iii) use of equipment that was improperly operated. Plant operators and designers now must be aware of the need for a thorough understanding of the operating parameters of each piece of equipment as they pertain specifically to waste preprocessing and gasification. This need for specialized knowledge extends to a detailed familiarization with the physical and chemical characteristics of the waste feedstocks (Savage, 1996).

In summary, gasification technology is selected on the basis of (i) feedstock properties and quality, (ii) gasifier operation, and (iii) desired product slate, and (iv) product quality. The main reactors used for gasification of municipal solid waste are fixed-bed and fluidized-bed units. Larger-capacity gasifiers are preferable for treatment of municipal solid waste because they allow for variable fuel feed, uniform process temperatures due to highly turbulent flow through the bed, good interaction between gases and solids, and high levels of carbon conversion (Chapter 2).

7.5 Process Products

By general definition, the goal of the gasification process is to produce gaseous products, in particular synthesis gas from which hydrogen can be isolated on an as-needed basis (Chapter 6). Furthermore, the product gas resulting from waste gasification contains carbon dioxide, tar, particulate matter, halogens/acid gases, heavy metals and alkaline compounds depending on the feedstock composition and the particular gasification process. The downstream power generating and gas cleaning equipment typically requires removal of these contaminants.

7.5.1 Synthesis Gas

Like many gasification processes, the goal of *waste gasification* is to produce a gas that can either be (i) used as fuel gas or (ii) used for hydrocarbon derivatives and/or chemicals production. In either case, the gas is synthesis gas – mixtures of carbon monoxide and hydrogen – and the yield and composition of the gas and yield of byproducts are dependent upon (i) the properties and character of the feedstock, (ii) the gasifier type, and (iii) the conditions in the gasifier (Chapter 1, Chapter 2) (Orr and Maxwell, 2000).

The raw synthesis gas produced in the gasifier contains trace levels of impurities that must be removed prior to its ultimate use. After the gas is cooled, virtually all the trace minerals, particulates, sulfur, mercury, and unconverted carbon are removed using commercially proven cleaning processes common to the gas processing industry as well as the chemical and refining industries (Mokhatab *et al.*, 2006; Hsu and Robinson, 2006; Gary *et al.*, 2007; Speight, 2007, 2014). For feedstocks containing mercury, more than 90% w/w of the mercury can be removed from the synthesis gas using relatively small and commercially available activated carbon beds.

7.5.2 Carbon Dioxide

Carbon dioxide can also be removed in the synthesis gas clean-up stage using a number of commercial technologies (Mokhatab *et al.*, 2006; Speight, 2007). In fact, carbon dioxide is routinely removed with a commercially proven process in gasification-based ammonia, hydrogen, and chemical manufacturing plants. Gasification-based plants for the production of ammonia are equipped to separate and capture approximately 90% v/v of their carbon dioxide and gasification-based methanol plants separate and capture approximately 70% v/v of the produced carbon dioxide. In fact, the gasification process is considered to offer a cost-effective and efficient means of capturing carbon dioxide during the energy production process.

Carbon dioxide can also be removed in the synthesis gas clean-up stage using a number of commercial technologies. In fact, carbon dioxide is routinely removed with a commercially proven process in gasification-based ammonia, hydrogen, and chemical manufacturing plants. Gasification-based ammonia plants already capture/separate approximately 90% v/v of the carbon dioxide and gasification-based methanol plants capture 70% v/v. The gasification process offers the most cost-effective and efficient means of capturing carbon dioxide during the energy production process.

7.5.3 Tar

By definition for this text, tar is any condensable or non-condensable organic material in the product stream, and is largely intractable and comprised of aromatic compounds.

When municipal solid waste is gasified, significant amounts of tar are produced. If tar is allowed to condense (condensation temperatures range from 200 to 600°C, 390 to 1110°F) it can cause coke to form on fuel reforming catalysts, deactivate sulfur removal systems, erode compressors, heat exchangers, ceramic filters, and damage gas turbines and engines. Non-condensable tar can also cause problems for advanced power conversion devices, such as fuel cell catalysts, and complicate environmental emissions compliance.

The amount and composition of tars are dependent on the fuel, the operating conditions and the secondary gas phase reactions – tar can be sub-divided into three categories based on the reaction temperature ranges in which they form (Table 7.3). This categorization is important for assessing gasification processes, as the effectiveness of conversion and/or removal schemes depend greatly on the specific tar composition and their concentration in the fuel gas.

The primary tars are mixed oxygenates and are a product of pyrolysis. As gasification takes over at higher temperatures, the primary products thermally decompose to lesser

Table 7.3 General properties of classification of tars by formation temperature.

Category	Formation temperature	Constituents
Primary	400-600°C	Mixed oxygenates
	750-1110°F	Phenolic ethers
Secondary	600-800°C	Alkyl phenols
	1110-1470°F	Heterocyclic ethers
Tertiary	800-1000°C	Polynuclear aromatic hydrocarbons
	1470-1830°F	Phenolic ethers

amounts of secondary and tertiary products and a greater quantity of light gases. Tertiary products are the most stable and difficult to crack catalytically. Provided that there is adequate gas mixing, primary and tertiary tars are mutually exclusive in the product gas. Both lignin and cellulose in the fuel result in the formation of tertiary tar compounds. However, lignin rich fuels have been shown to form heavier tertiary aromatics more quickly.

Both physical and chemical treatment processes can reduce the presence of tar in the product gas. The physical processes are classified into wet and dry technologies depending on whether water is used. Various forms of wet or wet/dry scrubbing processes are commercially available, and these are the most commonly practiced techniques for physical removal of tar.

Wet physical processes involve tar condensation, droplet filtration, and/or gas/liquid mixture separation. Cyclones, cooling towers, venturi scrubbers, baghouses, electrostatic precipitators, and wet/dry scrubbers are the primary tools. The main disadvantage to using wet physical processes is that the tar is transferred to wastewater, so the heating value is lost and the water must be disposed of in an environmentally acceptable way. Wastewater that contains tar is classified as *hazardous waste* – treatment and disposal can add significantly to the overall cost of the gasification plant.

Dry tar removal using ceramic, metallic, or fabric filters are alternatives to wet tar removal processes. However, at temperatures above 150°C (300°F), tars can become semi-solid and adhesive and cause operational problems with such barriers – as a result, dry tar removal schemes are rarely implemented. Injection of activated carbon in the product gas stream or in a granular bed may also reduce tars through adsorption and collection with a baghouse. The carbonaceous material containing the tars can be recycled back to the gasifier to encourage further thermal and/or catalytic decomposition – i.e., the tar is recycled to extinction.

Chemical tar treatment processes are the most widely practiced in the gasification industry. They can be divided into four generic categories: (i) thermal, (ii) steam, (iii) partially oxidative, and (iv) catalytic processes. Thermal destruction has been shown to break down aromatics at temperatures above 1000°C (1830°F). However, such high temperatures can have adverse effects on heat exchangers and refractory surfaces due to ash sintering in the gasification vessel. The introduction of steam does encourage reformation of primary and some secondary oxygenated tar compounds, but has a lesser on many nitrogen-containing organic compounds.

The presence of oxygen during gasification has been shown to accelerate both the destruction of primary tar products and the formation of aromatic compounds from phenol cracking increases in the presence of low oxygen environments (less than 10% v/v) of the gas. Only above 10% v/v was a decrease in the amount of tertiary tars observed. A net increase in the carbon monoxide may also be observed as the product from the oxidative cracking of tar. Benzene levels are not usually affected by the presence of oxygen.

The most widely used and studied tar cracking catalyst is dolomite (a mixture of calcium carbonate, $CaCO_3$, and magnesium carbonate, $MgCO_3$). Dolomite has been shown to work more effectively when placed in a vessel downstream from the gasifier and in a low carbon monoxide environment. However, when used within the gasifier, catalytic materials often accumulate a layer of coke that causes rapid loss of catalytic efficiency.

The specific tar conversion and tar destruction schemes depend on the nature and composition of the tars present, as well as the intended end-use equipment. However, the advantages of recycling the tar product for further treatment include (ii) increased waste-to-energy efficiency, (ii) lower emissions, and (iii) lower effluent treatment costs. Although progress in mitigating tar formation and increasing tar remove (if formed), the need for effective and less costly tar removal processes has a barrier to widespread commercialization municipal solid waste integrated gasification combined cycle (IGCC) power generation.

7.5.4 Particulate Matter

The detrimental effect of particulate matter on the atmosphere has been of some concern for several decades. Species such as mercury, selenium, and vanadium which can be ejected into the atmosphere and from fossil fuel combustion are particularly harmful to the flora and fauna. There are many types of particulate collection devices in use and they involve a number of different principles for removal of particles from gasification product streams (Speight, 2013a, 2013b). However, the selection of an appropriate particle removal device must be based upon equipment performance as anticipated/predicted under the process conditions. To enter into a detailed description of the various devices available for particulate removal is well beyond the scope of this text but it is essential for the reader to be aware of the equipment available for particulate removal and the means by which this might be accomplished: (i) cyclones, which are particle collectors that have many potential applications in coal gasification systems; (ii) electrostatic precipitators, which are efficient collectors of fine particulates matter and are capable of reducing the amount of submicron particles by 90%, or more and they also have the capability of collecting liquid mists as well as dust; (iii) granular-bed filters, which comprise a class of filtration equipment that is distinguished by a bed of separate, closely packed granules which serve as the filter medium and collect particulates at high temperature and pressure; (iv) wet scrubbers, which represent a simple method to clean exhaust air or exhaust gas and remove toxic or smelling compounds using the principle of close contact with fine water drops in a co-current or countercurrent flow of the gas stream.

7.5.5 Halogens/Acid Gases

The principal combustion products of halogen-containing organic waste is either hydrogen halides (such as hydrogen chloride, HCl, or hydrogen bromide, HBr) or metal halides (such as mercuric chloride, $HgCl_2$, or mercurous chloride, HgCl) that volatilize out of the reactor along with the other gases. In the gasification of *pure* municipal solid waste (i.e., without coal, biomass, or any other feedstock added), hydrogen chloride is the prevailing chlorine-containing product while bromine constituents can accumulate to a greater extent in the bottom ash but in the presence of hydrogen bromine is transformed to hydrogen bromide (HBr), which with the hydrogen chloride (HCl) is readily removed in a scrubbing system and hence causes no emission problems.

A significant advantage of gasification is that it takes place in a reducing atmosphere, which prevents sulfur and nitrogen compounds from oxidizing. As a result, most elemental nitrogen or sulfur in the waste stream end up as hydrogen sulfide (H_2S), carbonyl sulfide (COS), nitrogen (N_2) or ammonia (NH_3) rather than sulfur oxides (SOx) or nitrogen oxides (NOx), respectively. The reduced sulfur species can then be recovered as elemental sulfur at efficiencies between 95 and 99% w/w, or converted to a sulfuric acid byproduct (Mokhatab *et al.*, 2006; Speight, 2007).

The typical sulfur removal and recovery processes used to treat the raw synthesis gas are the same as commercially available methods used in other industrial applications, such as oil refining and natural gas recovery (Speight, 2007, 2008, 2014). One commonly used process to remove sulfur compounds is the selective-amine (olamine) technology where sulfur species are removed from the synthesis gas using, for example, an amine-based solvent in an absorber tower. The reduced sulfur species removed in the solvent stripper are converted to elemental sulfur in a sulfur recovery process such as the Selectox/Claus process.

When municipal solid waste is gasified, nitrogen in the fuel is converted primarily to ammonia, which when fired in a turbine or other combustion engine forms nitrogen oxides, a harmful pollutant. Removal of ammonia and other nitrogen compounds in the product gas prior to combustion can be accomplished with wet scrubbers or by catalytic destruction. Catalytic destruction of ammonia has been studied with dolomite and iron-based catalysts. This technique is of interest because tars are simultaneously decomposed (cracked) to lower weight gaseous compounds. Destruction of 99% v/v of the ammonia in the gas stream has been demonstrated with these catalysts.

If the product gas is cooled first, wet scrubbing with lime is also an effective ammonia removal technique. Gasification processes that use pure oxygen, steam or hydrogen, will only have nitrogen contents brought in through the fuel stream. Typical municipal solid waste has a nitrogen content of less than 1% w/w.

7.5.6 Heavy Metals

Trace amounts of metals and other volatile materials are also present in MSW. These are typically toxic substances that pose ecological and human health risks when released into the environment.

Mercury that occurs in the fly ash and flue gas is likely to be in the elemental form but when oxidizing conditions are prevalent in the gasifier, the presence of hydrogen chloride (HCl) and chlorine (Cl_2) can cause some of the elemental mercury to form mercuric chloride ($HgCl_2$):

$$Hg + 4HCl + O_2 \leftrightarrow 2HgCl_2 + 2H_2O$$

$$Hg + Cl_2 \leftrightarrow HgCl_2$$

Volatilized heavy metals (or heavy metals that are entrained in the gas stream due to the high gas velocity) that are not collected in the gas cleanup system can bio-accumulate in the environment and can be carcinogenic and damage human nervous systems (Speight and Arjoon, 2012). For this reason, mercury must be removed from the product gas prior to combustion or further use. However, there has been extraordinary success removing heavy metals with activated carbon, baghouses filters and electrostatic precipitators (Mokhatab *et al.*, 2006; Speight, 2007, 2013a).

7.5.7 Alkalis

The primary elements causing alkali slagging are potassium, sodium, chlorine and silica. Sufficient volatile alkali content in a feedstock causes a reduction in the ash fusion temperature and promotes slagging and/or fouling. Alkali compounds in the ash from the gasification of municipal solid waste gasification can cause serious slagging in the boiler or gasification vessel. Sintered or fused deposits can form agglomerates in fluidized beds and on grates. Potassium sulfate (K_2SO_4) and potassium chloride (KCl) have been found to mix with flue dust and deposit/condense on the upper walls of the gasifier.

Alkali deposit formation is a result of particle impaction, condensation, and chemical reaction. Unfortunately, most deposits occur subsequent to gasification and cannot always be predicted solely on the basis of analysis of the feedstock. There are two characteristic temperature intervals for alkali metal emission. A small fraction of the alkali content is released below 500°C (930°F) and is attributed to the decomposition of the organic structure. Another fraction of alkali compounds is released from the char residue at temperatures above 500°C (930°F).

Thus, the presence of alkali metals in gasification processes is known to cause several operational problems. Eutectic systems consisting of alkali salts are formed on the surfaces of fly ash particles or on the fluidized-bed material – the eutectic system is a mixture of chemical compounds or elements that have a single chemical composition that solidifies at a lower temperature than any other composition made up of the same ingredients. The semi-solid or adhesive particle surfaces can lead to the formation of bed material agglomerates, which must be replaced by fresh material. The deposition of fly ash particles and the condensation of vapor-phase alkali compounds on heat exchanging surfaces lower the heat conductivity and may eventually require temporary plant shutdowns for the removal of deposits.

The challenges of removing alkali vapor and particulate matter are closely connected, since alkali metal compounds play an important role in the formation of new particles as well as the chemical degradation of ceramic barrier filters used in some hot gas cleaning systems. The most convenient method is to cool the gas and condense out the alkali compounds.

7.5.8 Slag

Most solid and liquid feed gasifiers produce a hard glass-like byproduct (*slag*, also called *vitreous frit*) that is composed primarily of sand, rock, and any minerals (or thermal derivatives thereof) originally contained in the gasifier feedstock. Slag is the result of gasifier operation at temperatures above the fusion, or melting temperature of the mineral matter. Under these conditions, non-volatile metals are bound together in a molten form until it is cooled in a pool of water at the bottom of a quench gasifier, or by natural heat loss at the bottom of an entrained-bed gasifier. Volatile metals such as mercury, if present in the feedstock, are typically not recovered in the slag, but are removed from the raw synthesis gas during cleanup. Typically, the slag is non-hazardous (depending upon the type of mineral matter in the feedstock) and can be used in roadbed construction, cement manufacturing or in roofing materials.

234 Synthesis Gas

Slag production is a function of the amount of mineral matter present in the gasifier feedstock, so materials such as municipal solid waste (as well as, for example, coal and biomass) produce much more slag than crude oil residua. Regardless of the character of the feedstock, as long as the operating temperature is above the fusion temperature of the ash (true for the modern gasification technologies under discussion), slag will be produced. As well as dependency on the waste feedstock, the physical structure of the slag is sensitive to changes in operating temperature and pressure and, in some cases, physical examination of the appearance of the slag can provide a good indication of carbon conversion in the gasifier.

Furthermore, because the slag is in a fused vitrified state, it rarely fails the toxicity characteristic leaching procedure (TCLP) protocols for metals (Speight and Arjoon, 2012). Slag is not a good substrate for binding organic compounds so it is usually found to be nonhazardous, exhibiting none of the characteristics of a hazardous waste. Consequently, it may be disposed of in a nonhazardous landfill, or sold as an ore to recover the metals concentrated within its structure. The hardness of slag also makes it suitable as an abrasive or road-bed material as well as an aggregate in concrete formulations (Speight, 2013a, 2014).

7.6 Advantages and Limitations

Gasification has several advantages over the traditional disposal of municipal solid waste and other waste materials by combustion. The process takes place in a low-oxygen environment that limits the formation of dioxins and of large quantities of sulfur oxides (SOx) and nitrogen oxides (NOx).

Furthermore, the process requires just a fraction of the stoichiometric amount of oxygen necessary for combustion. As a result, the volume of process gas is low, requiring smaller and less expensive gas cleaning equipment. The lower gas volume is reflected in a higher partial pressure of contaminants in the off-gas, which favors more complete adsorption and particulate capture according to chemical thermodynamics:

$$\Delta G = -RT\ln(P_1/P_2)$$

ΔG is the Gibbs free energy of the system, T is the temperature, $P_1/$ is the initial pressure, and P_2 is the final pressure. The lower gas volume also means a higher partial pressure of contaminants in the off-gas, which favors more complete adsorption and particulate capture.

In fact, one of the important advantages of gasification is that the contaminants can be removed from the synthesis prior to use, thereby eliminating many of the types of after-the-fact (post-combustion) emission control systems required by incineration plants. Whether generated from conventional gasification or from plasma gasification, the synthesis gas can be used in reciprocating engines or turbines to generate electricity or further processed to produce substitute natural gas, chemicals, fertilizers or transportation fuels, such as ethanol. In summary, gasification of waste generates a gas product that can be integrated with combined cycle turbines, reciprocating engines and, potentially, with fuel cells that convert fuel energy to electricity more than twice as efficiently as conventional steam boilers.

Furthermore, the ash produced from gasification is more amenable to use – the ash exits from the gasifier in a molten form where, after quench-cooling, it forms a glassy, *non-leachable* slag that can be used for cement, roofing shingles, asphalt filler, or for sandblasting. Some gasifiers are designed to recover valuable molten metals in a separate stream, taking advantage of the ability of gasification technology to enhance recycling.

On the other hand, during gasification, tars, heavy metals, halogens and alkaline compounds are released within the product gas and can cause environmental and operational problems. Tars are high molecular weight organic gases that ruin reforming catalysts, sulfur removal systems, ceramic filters and increase the occurrence of slagging in boilers and on other metal and refractory surfaces. Alkalis can increase agglomeration in fluidized beds that are used in some gasification systems and also can ruin gas turbines during combustion. Heavy metals are toxic and bio-accumulate if released into the environment. Halogens are corrosive and are a cause of acid rain if emitted to the environment. The key to achieving cost efficient, clean energy recovery from municipal solid waste gasification will be overcoming problems associated with the release and formation of these contaminants.

In terms of power generation, the co-utilization of waste with biomass and/or with coal may provide economies of scale that help achieve the policy objectives identified above at an affordable cost. In some countries, governments propose cogasification processes as being *well suited for community-sized developments*, suggesting that waste should be dealt with in smaller plants serving towns and cities, rather than moved to large, central plants (satisfying the so-called *proximity principle*).

References

Arena, U. 2012. Process and Technological Aspects of Municipal Solid Waste Gasification. A Review. *Waste Management*, 32: 625-639.

Al Asfar, J. 2014. Gasification of Solid Waste Biomass. *Jordan Journal of Mechanical and Industrial Engineering*, 8(1): 13-19.

CH2MHill. 2009. *Waste-to-Energy Review of Alternatives*. Report Prepared for Regional District of North Okanagan, by CH2MHill, Burnaby, British Columbia, Canada.

Diaz, L.F. and Savage, G.M. 1996. Pretreatment Options for Waste-to-Energy Facilities. Solid Waste Management: Thermal Treatment & Waste-to-Energy Technologies, VIP-53. *Proceedings. International Technologies Conference, Air & Waste Management Association, Washington, DC.*

Ducharme, C. 2010. Technical and Economic Analysis of Plasma-assisted Waste-to-Energy processes. Thesis Submitted in partial fulfillment of requirements for M.S. Degree in Earth Resources Engineering. Department of Earth and Environmental Engineering, Fu Foundation of Engineering and Applied Science, Columbia University. September.

E4Tech. 2009. Review of Technologies for Gasification of Biomass and Wastes. NNFCC project 09/008. NNFCC Biocenter, York, United Kingdom. June.

Fabry, F., Rehmet, C., Rohani, V., and Fulcheri, L. 2013. Waste Gasification by Thermal Plasma: A Review. *Waste and Biomass Valorization*. 4: 421-439.

Gary, J.H., Handwerk, G.E., and Kaiser, M.J. 2007. *Petroleum Refining: Technology and Economics* 5th Edition. CRC Press, Taylor & Francis Group, Boca Raton, Florida.

Gomez, E., Amutha, D., Rani, Cheeseman, C., Deegan, D., Wisec, M., Boccaccini, A., 2009. Thermal Plasma Technology for the Treatment of Wastes: A Critical Review. *Journal of Hazardous Materials* 161, 614–626.

Grimshaw A.J. and Lago A. 2010. Small Scale Energos Gasification Technology. *Proceedings. 3rd International Symposium on Energy from Biomass and Waste. Venice, Italy. November 8-11, 2010.* CISA Publishers, Padova, Italy.

Hankalin, V., Helanti, V., and Isaksson, J., 2011. High Efficiency Power Production by Gasification. *Proceedings. 13th International Waste Management and Landfill Symposium. October 3-7, 2011, S. Margherita di Pula, Cagliari, Italy.* CISA Publishers, Padova, Italy.

He, M., Hu, Z., Xiao, B., Li, J., Guo, A., Luo, S., Yang, F., Feng, Y., and Yang, G. 2009. Hydrogen-Rich Gas from Catalytic Steam Gasification of Municipal Solid Waste (MSW): Influence of Catalyst and Temperature on Yield and Product Composition. *International Journal of Hydrogen Energy*, 34(1): 195-203.

Heberlein, J., and Murphy, A.B.. 2008. Thermal Plasma Waste Treatment. *J. Phys. D: Appl. Phys.* 41 (2008) 053001. stacks.iop.org/JPhysD/41/053001; accessed August 31, 2013.

Higman C. and van der Burgt M. 2003. *Gasification*. Gulf Professional Publishing, Elsevier, Amsterdam, Netherlands.

Hsu, C.S., and Robinson, P.R. 2006. *Practical Advances in Petroleum Processing*, Volumes 1 and 2. Springer, New York.

Kalinenko, R.A., Kuznetsov, A.P., Levitsky, A.A., Messerle, V.E., Mirokhin, Yu.A., Polak, L.S., Sakipov, Z.B., Ustimenko, A.B. 1993. Pulverized Coal Plasma Gasification. *Plasma Chemistry and Plasma Processing*, 3(1): 141-167.

Kwon, E., Westby, K.J., and Castaldi, M.J. 2009. An Investigation into Syngas Production from Municipal Solid Waste (MSW) Gasification under Various Pressures and CO2 Concentration Atmospheres. Paper No. NAWTEC17-2351. *Proceedings. 17th Annual North American Waste-to-Energy Conference (NAWTEC17). Chantilly, Virginia. May 18-20.*

Leal-Quirós, E. 2004. Plasma Processing of Municipal Solid Waste. *Braz. J. Phys.*, 34(4b): 1587-1593.

Lemmens, B., Elslander, H., Vanderreydt, I., Peys, K., Diels, L., Osterlinck, M., Joos, M., 2007. Assessment of plasma gasification of high caloric waste streams. *Waste Management*, 27: 1562-1569.

Luo, S., Xiao, B., Hu, Z., Liu, S., Guan, Y., and Cai, L. 2010. Influence of Particle Size on Pyrolysis and Gasification Performance of Municipal Solid Waste in a Fixed Bed Reactor. *Bioresource Technology*, 101(16): 6517-6520.

Luque, R., and Speight, J.G. (Editors). 2015. *Gasification for Synthetic Fuel Production: Fundamentals, Processes, and Applications.* Woodhead Publishing, Elsevier, Cambridge, United Kingdom.

Malkow, T. 2004. Novel and Innovative Pyrolysis and Gasification Technologies for Energy Efficient and Environmentally Sound MSW Disposal. *Waste Management*, 24: 53-79.

Mastellone M.L., Santoro D., Zaccariello L., Arena U. 2010a. The Effect of Oxygen-enriched Air on the Fluidized Bed Co-Gasification of Coal, Plastics and Wood. *Proceedings. 3rd International Symposium on Energy from Biomass and Waste. Venice, Italy. November 8-11, 2010.* CISA Publishers, Padova, Italy.

Mastellone, M.L., Zaccariello, L., Arena, U., 2010b. Co-Gasification of Coal, Plastic Waste and Wood in a Bubbling Fluidized Bed Reactor. *Fuel* 89(10): 2991-3000.

Messerle, V.E., and Ustimenko, A.B. 2007. Solid Fuel Plasma Gasification. *Advanced Combustion and Aerothermal Technologies.* NATO Science for Peace and Security Series C. Environmental Security. Page 141-1256.

Mokhatab, S., Poe, W.A., and Speight, J.G. 2006. *Handbook of Natural Gas Transmission and Processing.* Elsevier, Amsterdam, Netherlands.

Moustakas, K., Fatta, D., Malamis, S., Haralambous, K., Loizidou, M., 2005. Demonstration Plasma Gasification/Vitrification System for Effective Hazardous Waste Treatment. *Journal of Hazardous Materials*, 123, 120-126.

Oladejo, J., Shi, K., Luo, X., Yang, G., and Wu, T. 2019. A Review of Sludge-to-Energy Recovery Methods. *Energies*, 12: 60. www.mdpi.com/journal/energies

Orr, D., and Maxwell, D. 2000. *A Comparison of Gasification and Incineration of Hazardous Wastes.* Report No. DCN 99.803931.02. United States Department of Energy, Morgantown, West Virginia. March 30.

Pellet Fuels Institute. 2011. *Pellet Fuels Institute Standard Specification for Residential/Commercial Densified Fuel.* Pellet Fuels Institute, Arlington, Virginia.

Richardson, J.H., Rogers, R.S., Thorsness, C.B., Wallman, P.H., Leininger, T.F., Richter, G.N., Robin, A.M., Wiese, H.C., and Wolfenbarger, J.K. 1995. Conversion of Municipal Solid Waste to Hydrogen. Report No. UCRL-JC-120142. *Proceedings. DOE Hydrogen Program Review, in Coral Gables, Florida. April 19-21. Available from United states department of Energy, Washington, DC.*

Savage, G.M., and Trezek, G.J. 1976. Screening Shredded Municipal Solid Waste. *Compost Science*, 17(1): 7-11.

Savage, G.M. 1996. The History and Utility of Waste Characterization Studies. *Proceedings. 86th Annual Meeting & Exhibition, Air & Waste Management Association Denver, Colorado.*

Sikarwar, V.S., Zhao, M., Clough, P., Yao, J., Zhong, X., Memon, M.Z., Shah, N., Anthony. E.J., and Fennell, P.S. 2016. An Overview of Advances in Biomass Gasification. *Energy Environ. Sci.*, 9; 2939-2977.

Speight, J.G. 2007. *Natural Gas: A Basic Handbook.* GPC Books, Gulf Publishing Company, Houston, Texas.

Speight, J.G. 2008. *Synthetic Fuels Handbook: Properties, Processes, and Performance.* McGraw-Hill, New York.

Speight, J.G. (Editor). 2011a. *The Biofuels Handbook.* Royal Society of Chemistry, London, United Kingdom.

Speight, J.G. 2011b. *The Refinery of the Future.* Gulf Professional Publishing, Elsevier, Oxford, United Kingdom.

Speight, J.G., and Arjoon, K.K. 2012. *Bioremediation of Petroleum and Petroleum Products.* Scrivener Publishing, Salem, Massachusetts.

Speight, J.G. 2013a. *The Chemistry and Technology of Coal* 3rd Edition. CRC Press, Taylor & Francis Group, Boca Raton, Florida.

Speight, J.G. 2013b. *Coal-Fired Power Generation Handbook.* Scrivener Publishing, Salem, Massachusetts.

Speight, J.G. 2014. *The Chemistry and Technology of Petroleum* 5th Edition. CRC Press, Taylor & Francis Group, Boca Raton, Florida.

Speight, J.G. 2019. *Biogas: Production and Properties.* Nova Science publishers, New York.

Suzuki, A., and Nagayama, S., 2011. High efficiency WtE power plant using high temperature gasifying and direct melting furnace. *Proceedings. 13th International Waste Management and Landfill Symposium. October 3-7, 2011, S. Margherita di Pula, Cagliari, Italy. CISA Publishers, Padova, Italy.*

Werle, S., and Dudziak, M. 2019. Gasification of Sewage Sludge. In: *Industrial and Municipal Sludge: Emerging Concerns and Scope for Resource Recovery.* M.N.V. Prasad, P.J. de Campos Favas, M. Vithanage, and S.V. Mohan (Editors). Butterworth-Heinemann, Elsevier BV, Amsterdam, Netherlands. Chapter 25. Page 575-593.

8

Reforming Processes

8.1 Introduction

Throughout the previous chapters there have been several references and/or acknowledgments of a very important property of crude oil and crude oil products, and that is the hydrogen content or the use of hydrogen during refining in hydrotreating processes, such as desulfurization and in hydroconversion processes, such as hydrocracking (Parkash, 2003; Gary *et al.*, 2007; Speight, 2011a, 2014; Hsu and Robinson, 2017; Speight, 2017). Although the hydrogen recycle gas may contain up to 40% by volume of other gases (usually hydrocarbon derivatives), hydrotreater catalyst life is a strong function of hydrogen partial pressure. Optimum hydrogen purity at the reactor inlet extends catalyst life by maintaining desulphurization kinetics at lower operating temperatures and reducing carbon laydown. Typical purity increases resulting from hydrogen purification equipment and/or increased hydrogen sulfide removal as well as tuning hydrogen circulation and purge rates, may extend catalyst life up to approximately 25%.

A critical issue facing the world's refiners today is the changing landscape in processing crude oil crude into refined transportation fuels under an environment of increasingly stringent clean fuel regulations, decreasing high-boiling fuel oil demand and increasingly higher-density, more sour (higher sulfur) supply of crude oil. Hydrogen network optimization is at the forefront of world refineries' options to address clean fuel trends, to meet growing transportation fuel demands and to continue to make a profit from their crudes. A key element of a hydrogen network analysis in a refinery involves the capture of hydrogen in its fuel streams and extending its flexibility and processing options. Thus, innovative hydrogen network optimization will be a critical factor influencing refiners' future operating flexibility and profitability in a shifting world of crude feedstock supplies and ultra-low-sulfur (ULS) gasoline and diesel fuel.

As hydrogen use has become more widespread in refineries, hydrogen production has moved from the status of a high-tech specialty operation to an integral feature of most refineries (Raissi, 2001; Vauk *et al.*, 2008). This has been made necessary by the increase in hydrotreating and hydrocracking, including the treatment of progressively heavier (more viscous) feedstocks (Parkash, 2003; Gary *et al.*, 2007; Speight, 2011a, 2014; Hsu and Robinson, 2017; Speight, 2017). In fact, the use of hydrogen in thermal processes is perhaps the single most significant advance in refining technology during the 20th century (Scherzer and Gruia, 1996; Bridge, 1997; Dolbear, 1998). The continued increase in hydrogen demand over the last several decades is a result of the conversion of crude oil to match changes in product slate and the supply of high-boiling, high-sulfur oil, and in order to make lower-boiling, cleaner, and more salable products. There are also many reasons other than product quality for using hydrogen in processes adding to the need to add hydrogen at relevant stages of the refining process and, most important, according to the availability of hydrogen (Bezler, 2003; Miller and Penner, 2003; Ranke and Schödel, 2003).

With the increasing need for *clean* fuels, the production of hydrogen for refining purposes requires a major effort by refiners. In fact, the trend to increase the number of hydrogenation (*hydrocracking* and/or *hydrotreating*) processes in refineries coupled with the need to process the heavier oils, which require substantial quantities of hydrogen for upgrading because of the increased use of hydrogen in hydrocracking processes, has resulted in vastly increased demands for this gas. The hydrogen demands can be estimated to a very rough approximation using API gravity and the extent of the reaction, particularly the hydrodesulfurization reaction (Speight, 2000; Speight, 2017). But accurate estimation requires equivalent process parameters and a thorough understanding of the nature of each individual process. Thus, as hydrogen production grows, a better understanding of the capabilities and requirements of a hydrogen plant becomes ever more important to overall refinery operations as a means of making the best use of hydrogen supplies in the refinery.

The chemical nature of the crude oil used as the refinery feedstock has always played the major role in determining the hydrogen requirements of that refinery. For example, the lower-density, more paraffinic crude oils will require somewhat less hydrogen for upgrading to, say, a gasoline product than a heavier more asphaltic crude oil (Speight, 2000). It follows that the hydrodesulfurization of high-boiling oils and residua (which, by definition, is a hydrogen-dependent process) needs substantial amounts of hydrogen as part of the processing requirements.

In general, considerable variation exists from one refinery to another in the balance between hydrogen produced and hydrogen consumed in the refining operations. However, what is more pertinent to the present text is the excessive amounts of hydrogen that are required for hydroprocessing operations, whether these be hydrocracking or the somewhat milder hydrotreating processes. For effective hydroprocessing, a substantial hydrogen partial pressure must be maintained in the reactor and, in order to meet this requirement, an excess of hydrogen above that actually consumed by the process must be fed to the reactor. Part of the hydrogen requirement is met by recycling a stream of hydrogen-rich gas. However, the need still remains to generate hydrogen as makeup material to accommodate the process consumption of 500 to 3000 scf/bbl depending upon whether the high-boiling feedstock is being subjected to a predominantly hydrotreating (hydrodesulfurization) or to a predominantly hydrocracking process.

In some refineries, the hydrogen needs can be satisfied by hydrogen recovery from catalytic reformer product gases, but other external sources are required. However, for the most part, many refineries now require on-site hydrogen production facilities to supply the gas for their own processes. Most of this non-reformer hydrogen is manufactured either by steam-methane reforming or by oxidation processes. However, other processes, such as steam-methanol interaction or ammonia dissociation, may also be used as sources of hydrogen. Electrolysis of water produces high-purity hydrogen, but the power costs may be prohibitive.

An early use of hydrogen in refineries was in naphtha hydrotreating, as feed pretreatment for catalytic reforming (which in turn was producing hydrogen as a byproduct). As environmental regulations tightened, the technology matured and heavier streams were hydrotreated. Thus in the early refineries, the hydrogen for hydroprocesses was provided as a result of catalytic reforming processes in which dehydrogenation is a major chemical reaction and, as a consequence, hydrogen gas is produced (Parkash, 2003; Gary *et al.*, 2007; Speight, 2011a, 2014; Hsu and Robinson, 2017; Speight, 2017). The low-boiling fractions from the catalytic reformer contain a high ratio of hydrogen to methane so the stream is freed from ethane and/or propane to get a high concentration of hydrogen in the stream.

The hydrogen is recycled though the reactors where the reforming takes place to provide the atmosphere necessary for the chemical reactions and also prevents the carbon from being deposited on the catalyst, thus extending its operating life. An excess of hydrogen above whatever is consumed in the process is produced, and, as a result, catalytic reforming processes are unique in that they are the only crude oil refinery processes to produce hydrogen as a byproduct. However, as refineries and refinery feedstocks evolved during the last four decades, the demand for hydrogen has increased and reforming processes are no longer capable of providing the quantities of hydrogen necessary for feedstock hydrogenation. Within the refinery, other processes are used as sources of hydrogen. Thus, the recovery of hydrogen from the byproducts of the coking units, visbreaker units and catalytic cracking units is also practiced in some refineries.

In coking units and visbreaker units, high-boiling feedstocks are converted to crude oil coke, oil, low-boiling hydrocarbon derivatives (benzene, naphtha, liquefied petroleum gas) and gas (Parkash, 2003; Gary *et al.*, 2007; Speight, 2011a, 2014; Hsu and Robinson, 2017; Speight, 2017). Depending on the process, hydrogen is present in a wide range of concentrations. Since coking processes need gas for heating purposes, adsorption processes are best suited to recover the hydrogen because they feature a very clean hydrogen product and an off-gas suitable as fuel.

Catalytic cracking is the most important process step for the production of low-boiling products from gas oil and increasingly from vacuum gas oil and high-boiling viscous feedstocks (Parkash, 2003; Gary *et al.*, 2007; Speight, 2011a, 2014; Hsu and Robinson, 2017; Speight, 2017). In catalytic cracking the molecular mass of the main fraction of the feed is lowered, while another part is converted to coke that is deposited on the hot catalyst. The catalyst is regenerated in one or two stages by burning the coke off with air that also provides the energy for the endothermic cracking process. In the process, paraffins and naphthenes are cracked to olefins and to alkanes with shorter chain length, mono-aromatic compounds are dealkylated without ring cleavage, di-aromatics and poly-aromatics are dealkylated and converted to coke. Hydrogen is formed in the last type of reaction, whereas the first two reactions produce low-boiling hydrocarbon derivatives and therefore require hydrogen. Thus, a catalytic cracker can be operated in such a manner that enough hydrogen for subsequent processes is formed.

In reforming processes, naphtha fractions are reformed to improve the quality of gasoline (Speight, 2000; Speight, 2017). The most important reactions occurring during this process are the dehydrogenation of naphthenes to aromatics. This reaction is endothermic and is favored by low pressures and the reaction temperature lies in the range of 300 to 450°C (570 to 840°F). The reaction is performed on platinum catalysts, with other metals, e.g., rhenium, as promoters.

Hydrogen is generated in a refinery by the catalytic reforming process, but there may not always be the need to have a catalytic reformer as part of the refinery sequence. Nevertheless, assuming that a catalytic reformer is part of the refinery sequence, the hydrogen production from the reformer usually falls well below the amount required for hydroprocessing purposes. For example, in a 100,000-bbl/day hydrocracking refinery, assuming intensive reforming of hydrocracked gasoline, the hydrogen requirements of the refinery may still fall some 500 to 900 scf/bbl of crude charge below that necessary for the hydrocracking sequences. Consequently, an *external* source of hydrogen is necessary to meet the daily hydrogen requirements of any process where the heavier feedstocks are involved.

The trend to increase the number of hydrogenation (*hydrocracking* and/or *hydrotreating*) processes in refineries (Dolbear, 1998) coupled with the need to process the heavier oils, which require substantial quantities of hydrogen for upgrading, has resulted in vastly increased demands for this gas.

Hydrogen has historically been produced during catalytic reforming processes as a byproduct of the production of the aromatic compounds used in gasoline and in solvents. As reforming processes changed from fixed-bed to cyclic to continuous regeneration, process pressures have dropped and hydrogen production per barrel of reformate has tended to increase. However, hydrogen production as a byproduct is not always adequate to the needs of the refinery and other processes are necessary. Thus, hydrogen production by steam reforming or by partial oxidation of residua has also been used, particularly where high-boiling oil is available. Steam reforming is the dominant method for hydrogen production and is usually combined with pressure-swing adsorption (PSA) to purify the hydrogen to greater than 99% by volume (Bandermann and Harder 1982).

The gasification of residua and coke to produce hydrogen and/or power may become an attractive option for refiners (Campbell, 1997; Dickenson *et al.*, 1997; Fleshman, 1997; Gross and Wolff, 2000). The premise that the gasification section of a refinery will be the *garbage can* for deasphalter residues, high-sulfur coke, as well as other refinery wastes is worthy of consideration.

Several other processes are available for the production of the additional hydrogen that is necessary for the various high-boiling feedstock hydroprocessing sequences (Parkash, 2003; Gary *et al.*, 2007; Speight, 2011a, 2014; Hsu and Robinson, 2017; Speight, 2017), and it is the purpose of the present chapter to present a general description of these processes. In general, most of the external hydrogen is manufactured by steam-methane reforming or by oxidation processes. Other processes such as ammonia dissociation, steam-methanol interaction, or electrolysis are also available for hydrogen production, but economic factors and feedstock availability assist in the choice between processing alternatives.

The processes described in this chapter are those gasification processes that are often referred to as the *garbage disposal units* of the refinery. Hydrogen is produced for use in other parts of the refinery as well as for energy and it is often produced from process byproducts that may not be of any use elsewhere. Such byproducts might be the highly aromatic, heteroatom, and metal containing reject from a deasphalting unit or from a mild hydrocracking process. However attractive this may seem, there will be the need to incorporate a gas cleaning operation to remove any environmentally objectionable components from the hydrogen gas.

8.2 Processes Requiring Hydrogen

Crude oil is rarely used in its raw form but must instead be processed into its various products, generally as a means of forming products with hydrogen content different from that of the original feedstock. Thus, the chemistry of the refining process is concerned primarily with the production not only of better products but also of salable materials (Parkash, 2003; Gary et al., 2007; Speight, 2011a, 2014; Hsu and Robinson, 2017; Speight, 2017).

The distinguishing feature of the hydrogenating processes is that, although the composition of the feedstock is relatively unknown and a variety of reactions may occur simultaneously, the final product may actually meet all the required specifications for its particular use (Furimsky, 1983; Speight, 2000). Hydrogenation processes for the conversion of petroleum and petroleum products may be classified as *destructive* and *nondestructive* (Parkash, 2003; Gary et al., 2007; Speight, 2011a, 2014; Hsu and Robinson, 2017; Speight, 2017). The former (*hydrogenolysis* or *hydrocracking*) is characterized by the rupture of carbon-carbon bonds and is accompanied by hydrogen saturation of the fragments to produce lower-boiling products. Such treatment requires rather high temperatures and high hydrogen pressures, the latter to minimize coke formation. Many other reactions, such as isomerization, dehydrogenation, and cyclization, can occur under these conditions.

On the other hand, nondestructive, or simple, hydrogenation is generally used for the purpose of improving product (or even feedstock) quality without appreciable alteration of the boiling range. Treatment under such mild conditions is often referred to as *hydrotreating* or *hydrofining* and is essentially a means of eliminating nitrogen, oxygen, and sulfur as ammonia, water, and hydrogen sulfide, respectively.

8.2.1 Hydrotreating

Catalytic hydrotreating is a hydrogenation process used to remove approximately 90% of contaminants such as nitrogen, sulfur, oxygen, and metals from liquid crude oil fractions (Parkash, 2003; Gary et al., 2007; Speight, 2011a, 2014; Hsu and Robinson, 2017; Speight, 2017). These contaminants, if not removed from the crude oil fractions as they travel through the refinery processing units, can have detrimental effects on the equipment, the catalysts, and the quality of the finished product. Typically, hydrotreating is done prior to processes such as catalytic reforming so that the catalyst is not contaminated by untreated feedstock. Hydrotreating is also used prior to catalytic cracking to reduce sulfur and improve product yields, and to upgrade middle-distillate crude oil fractions into finished kerosene, diesel fuel, and heating fuel oils. In addition, hydrotreating converts olefins and aromatics to saturated compounds.

In a typical catalytic hydrodesulfurization unit, the feedstock is deaerated and mixed with hydrogen, preheated in a fired heater (315 to 425°F; 600 to 800°F) and then charged under pressure (up to 1,000 psi) through a fixed-bed catalytic reactor (Parkash, 2003; Gary et al., 2007; Speight, 2011a, 2014; Hsu and Robinson, 2017; Speight, 2017). In the reactor, the sulfur and nitrogen compounds in the feedstock are converted into hydrogen sulfide and ammonia. The reaction products leave the reactor and after cooling to a low temperature enter a liquid/gas separator. The hydrogen-rich gas from the high-pressure separation is recycled to combine with the feedstock, and the low-pressure gas stream rich in hydrogen sulfide is sent to a gas treating unit where the hydrogen sulfide is removed. The clean gas is then suitable as fuel for the refinery furnaces. The liquid stream is the product from hydrotreating and is normally sent to a stripping column for removal of hydrogen sulfide and other undesirable components. In cases where steam is used for stripping, the product is sent to a vacuum drier for removal of water. Hydrodesulfurized products are blended or used as catalytic reforming feedstock.

Hydrotreating processes differ depending upon the feedstock available and catalysts used. Hydrotreating can be used to improve the burning characteristics of distillates such as kerosene. Hydrotreatment of a kerosene fraction can convert aromatics into naphthenes, which are cleaner-burning compounds. Lube-oil hydrotreating uses catalytic treatment of the oil with hydrogen to improve product quality. The objectives in mild lube hydrotreating include saturation of olefins and improvements in color, odor, and acid nature of the oil. Mild lube hydrotreating also may be used following solvent processing. Operating temperatures are usually below 315°C (600°F) and operating pressures below 800 psi. Severe lube hydrotreating, at temperatures in the 315 to 400°C (600 to 750°F) range and hydrogen pressures up to 3000 psi, is capable of saturating aromatic rings, along with sulfur and nitrogen removal, to impart specific properties not achieved at mild conditions.

Hydrotreating also can be employed to improve the quality of pyrolysis gasoline (*pygas*), a byproduct from the manufacture of ethylene. Traditionally, the outlet for pygas has been motor gasoline blending, a suitable route in view of its high octane number. However, only small portions can be blended untreated owing to the unacceptable odor, color, and gum-forming tendencies of this material. The quality of pygas, which is high in di-olefin content, can be satisfactorily improved by hydrotreating, whereby conversion of di-olefins into mono-olefins provides an acceptable product for motor gas blending.

8.2.2 Hydrocracking

Hydrocracking is a two-stage process combining catalytic cracking and hydrogenation, wherein heavier feedstocks are cracked in the presence of hydrogen to produce more desirable products. Hydrocracking also produces relatively large amounts of iso-butane for alkylation feedstock and the process also performs isomerization for pour-point control and smoke-point control, both of which are important in high-quality jet fuel.

Hydrocracking employs high pressure, high temperature, and a catalyst. Hydrocracking is used for feedstocks that are difficult to process by either catalytic cracking or reforming, since these feedstocks are characterized usually by high polycyclic aromatic content and/or high concentrations of the two principal catalyst poisons, sulfur and nitrogen compounds. The hydrocracking process largely depends on the nature of the feedstock and the relative rates of the two competing reactions, hydrogenation and cracking. High-boiling aromatic feedstock is converted into lower-boiling products under a wide range of very high pressures (1,000 to 2,000 psi) and fairly high temperatures (400 to 815°C; 750 to 1500°F), in the presence of hydrogen and special catalysts. When the feedstock has a high paraffinic content, the primary function of hydrogen is to prevent the formation of polycyclic aromatic compounds. Another important role of hydrogen in the hydrocracking process is to reduce tar formation and prevent buildup of coke on the catalyst. Hydrogenation also serves to convert sulfur and nitrogen compounds present in the feedstock to hydrogen sulfide and ammonia (Parkash, 2003; Gary *et al.*, 2007; Speight, 2011a, 2014; Hsu and Robinson, 2017; Speight, 2017).

In the first stage of the process preheated feedstock is mixed with recycled hydrogen and sent to the first-stage reactor, where catalysts convert sulfur and nitrogen compounds to hydrogen sulfide and ammonia. Limited hydrocracking also occurs. After the hydrocarbon leaves the first stage, it is liquefied by cooling and run through a hydrocarbon separator.

The hydrogen is recycled to the feedstock. The liquid is charged to a fractionator. Depending on the products desired (gasoline components, jet fuel, and gas oil), the fractionator is run to cut out some portion of the first stage reactor out-turn. Kerosene-range material can be taken as a separate side-draw product or included in the fractionator bottoms with the gas oil. The fractionator bottoms are again mixed with a hydrogen stream and charged to the second stage. Since this material has already been subjected to some hydrogenation, cracking, and reforming in the first stage, the operations of the second stage are more severe (higher temperatures and pressures). Like the outturn of the first stage, the second-stage product is separated from the hydrogen and charged to the fractionator.

8.3 Feedstocks

The most common, and perhaps the best, feedstocks for steam reforming are low-boiling saturated hydrocarbon derivatives that have a low sulfur content; including natural gas, refinery gas, liquefied petroleum gas (LPG), and low-boiling naphtha.

Natural gas is the most common feedstock for hydrogen production since it meets all the requirements for reformer feedstock. Natural gas typically contains more than 90% methane and ethane with only a few percent of propane and higher-boiling hydrocarbon derivatives (Mokhatab *et al.*, 2006; Speight, 2014, 2019). Natural gas may (or most likely will) contain traces of carbon dioxide with some nitrogen and other impurities. Purification of natural gas, before reforming, is usually relatively straightforward. Traces of sulfur must be removed to avoid poisoning the reformer catalyst; zinc oxide treatment in combination with hydrogenation is usually adequate.

Refinery gas, containing a substantial amount of hydrogen, can be an attractive steam reformer feedstock since it is produced as a byproduct. Processing of refinery gas will depend on its composition, particularly the levels of olefins and of propane and higher-boiling hydrocarbon derivatives. Olefins, which can cause problems by forming coke in the reformer, are converted to saturated compounds in the hydrogenation unit. Higher-boiling hydrocarbon derivatives in refinery gas can also form coke, either on the primary reformer catalyst or in the preheater. If there is more than a few percent of C_3 and higher compounds, a promoted reformer catalyst should be considered, in order to avoid carbon deposits.

Refinery gas from different sources varies in suitability as hydrogen plant feed. Catalytic reformer off-gas for example, is saturated, very low in sulfur, and often has high hydrogen content (Parkash, 2003; Gary *et al.*, 2007; Speight, 2011a, 2014; Hsu and Robinson, 2017; Speight, 2017). The process gases from a coking unit or from a fluid catalytic cracking unit (Parkash, 2003; Gary *et al.*, 2007; Speight, 2011a, 2014; Hsu and Robinson, 2017; Speight, 2017) are much less desirable because of the content of unsaturated constituents. In addition to olefins, these gases contain substantial amounts of sulfur that must be removed before the gas is used as feedstock. These gases are also generally unsuitable for direct hydrogen recovery, since the hydrogen content is usually too low. Hydrotreater off-gas lies in the middle of the range. It is saturated, so it is readily used as hydrogen plant feed. The content of hydrogen and higher molecular weight hydrocarbon derivatives depends to a large extent on the upstream pressure. Sulfur removal will generally be required.

8.4 Process Chemistry

Before the feedstock is introduced to a process, there is the need for application of a strict feedstock purification protocol. Prolonging catalyst life in hydrogen production processes is attributable to effective feedstock purification, particularly sulfur removal. A typical natural gas or other low-boiling hydrocarbon feedstock contains traces of hydrogen sulfide and organic sulfur.

In order to remove sulfur compounds, it is necessary to hydrogenate the feedstock to convert the organic sulfur to hydrogen that is then reacted with zinc oxide (ZnO) at approximately 370°C (700°F) that results in the optimal use of the zinc oxide as well as ensuring complete hydrogenation. Thus, assuming assiduous feedstock purification and removal of all of the objectionable contaminants, the chemistry of hydrogen production can be defined.

In *steam reforming*, low-boiling hydrocarbon derivatives such as methane are reacted with steam to form hydrogen:

$$CH_4 + H_2O \rightarrow 3H_2 + CO \qquad \Delta H_{298K} = +97,400 \text{ Btu/lb}$$

H is the heat of reaction. A more general form of the equation that shows the chemical balance for higher-boiling hydrocarbon derivatives is:

$$C_nH_m + nH_2O \rightarrow (n + m/2)H_2 + nCO$$

The reaction is typically carried out at approximately 815°C (1500°F) over a nickel catalyst packed into the tubes of a reforming furnace. The high temperature also causes the hydrocarbon feedstock to undergo a series of cracking reactions, plus the reaction of carbon with steam:

$$CH_4 \rightarrow 2H_2 + C$$

$$C + H_2O \rightarrow CO + H_2$$

Carbon is produced on the catalyst at the same time that hydrocarbon is reformed to hydrogen and carbon monoxide. With natural gas or similar feedstock, reforming predominates and the carbon can be removed by reaction with steam as fast as it is formed. When higher-boiling feedstocks are used, the carbon is not removed fast enough and builds up thereby requiring catalyst regeneration or replacement. Carbon buildup on the catalyst (when high-boiling feedstocks are employed) can be avoided by addition of alkali compounds, such as potash, to the catalyst thereby encouraging or promoting the carbon-steam reaction.

However, even with an alkali-promoted catalyst, feedstock cracking limits the process to hydrocarbon derivatives with a boiling point less than of 180°C (350°F). Natural gas, propane, butane, and low-boiling naphtha are most suitable. Pre-reforming, a process that uses an adiabatic catalyst bed operating at a lower temperature, can be used as a pretreatment to allow heavier feedstocks to be used with lower potential for carbon deposition (coke formation) on the catalyst.

After reforming, the carbon monoxide in the gas is reacted with steam to form additional hydrogen (the *water-gas shift* reaction):

$$CO + H_2O \rightarrow CO_2 + H_2 \quad \Delta H_{298K} = -16,500 \text{ Btu/lb}$$

This leaves a mixture consisting primarily of hydrogen and carbon monoxide that is removed by conversion to methane:

$$CO + 3H_2O \rightarrow CH_4 + H_2O$$

$$CO_2 + 4H_2 \rightarrow CH_4 + 2H_2O$$

The critical variables for steam reforming processes are (i) temperature, (ii) pressure, and (iii) the steam/hydrocarbon ratio. Steam reforming is an equilibrium reaction, and conversion of the hydrocarbon feedstock is favored by high temperature, which in turn requires higher fuel use. Because of the volume increase in the reaction, conversion is also favored by low pressure, which conflicts with the need to supply the hydrogen at high pressure. In practice, materials of construction limit temperature and pressure.

On the other hand, and in contrast to reforming, shift conversion is favored by low temperature. The gas from the reformer is reacted over iron oxide catalyst at 315 to 370°C (600 to 700°F) with the lower limit being dictated activity of the catalyst at low temperature.

Hydrogen can also be produced by *partial oxidation* (POX) of hydrocarbon derivatives in which the hydrocarbon is oxidized in a limited or controlled supply of oxygen:

$$2CH_4 + O_2 \rightarrow CO + 4H_2 \quad \Delta H_{298K} = -10,195 \text{ Btu/lb}$$

The shift reaction also occurs and a mixture of carbon monoxide and carbon dioxide is produced in addition to hydrogen. The catalyst tube materials do not limit the reaction temperatures in partial oxidation processes and higher temperatures may be used that enhance the conversion of methane to hydrogen. Indeed, much of the design and operation of hydrogen plants involves protecting the reforming catalyst and the catalyst tubes because of the extreme temperatures and the sensitivity of the catalyst. In fact, minor variations in feedstock composition or operating conditions can have significant effects on the life of the catalyst or the reformer itself. This is particularly true of changes in molecular weight of the feed gas, or poor distribution of heat to the catalyst tubes.

Since the high temperature takes the place of a catalyst, partial oxidation is not limited to the lower-boiling feedstocks that are required for steam reforming. Partial oxidation processes were first considered for hydrogen production because of expected shortages of lower-boiling feedstocks and the need to have available a disposal method for higher-boiling, high-sulfur streams such as asphalt or crude oil coke.

Catalytic partial oxidation, also known as auto-thermal reforming, reacts oxygen with a low-boiling feedstock and by passing the resulting hot mixture over a reforming catalyst. The use of a catalyst allows the use of lower temperatures than in non-catalytic partial oxidation and which causes a reduction in oxygen demand.

The feedstock requirements for catalytic partial oxidation processes are similar to the feedstock requirements for steam reforming and low-boiling hydrocarbon derivatives from refinery gas to naphtha are preferred. The oxygen substitutes for much of the steam in preventing coking and a lower steam/carbon ratio is required. In addition, because a large excess of steam is not required, catalytic partial oxidation produces more carbon monoxide and less hydrogen than steam reforming. Thus, the process is more suited to situations where carbon monoxide is the more desirable product such as, for example, as synthesis gas for chemical feedstocks.

8.5 Commercial Processes

In spite of the use of low-quality hydrogen (that contains up to 40% by volume hydrocarbon gases), a high-purity hydrogen stream (95%-99% by volume hydrogen) is required for hydrodesulfurization, hydrogenation, hydrocracking, and petrochemical processes. Hydrogen, produced as a byproduct of refinery processes (principally hydrogen recovery from catalytic reformer product gases), often is not enough to meet the total refinery requirements, necessitating the manufacturing of additional hydrogen or obtaining supply from external sources.

Catalytic reforming remains an important process used to convert low-octane naphtha into high-octane gasoline blending components called *reformate*. Reforming represents the total effect of numerous reactions such as cracking, polymerization, dehydrogenation, and isomerization taking place simultaneously. Depending on the properties of the naphtha feedstock (as measured by the paraffin, olefin, naphthene, and aromatic content) and catalysts used, reformate can be produced with very high concentrations of toluene, benzene, xylene, and other aromatics useful in gasoline blending and petrochemical processing. Hydrogen, a significant byproduct, is separated from reformate for recycling and use in other processes.

A catalytic reformer comprises a reactor section and a product-recovery section. More or less standard is a feed preparation section in which, by combination of hydrotreatment and distillation, the feedstock is prepared to specification. Most processes use platinum as the active catalyst. Sometimes platinum is combined with a second catalyst (bimetallic catalyst) such as rhenium or another noble metal. There are many different commercial catalytic reforming processes including Platforming, Powerforming, Ultraforming, and Thermofor catalytic reforming (Parkash, 2003; Gary *et al.*, 2007; Speight, 2011a, 2014; Hsu and Robinson, 2017; Speight, 2017). In the Platforming process, the first step is preparation of the naphtha feed to remove impurities from the naphtha and reduce catalyst degradation. The naphtha feedstock is then mixed with hydrogen, vaporized, and passed through a series of alternating furnace and fixed-bed reactors containing a platinum catalyst. The effluent from the last reactor is cooled and sent to a separator to permit removal of the hydrogen-rich gas stream from the top of the separator for recycling. The liquid product from the bottom of the separator is sent to a fractionator called a stabilizer (butanizer) and the bottom product (reformate) is sent to storage and butanes and lighter gases pass overhead and are sent to the saturated gas plant.

Some catalytic reformers operate at low pressure (50-200 psi), and others operate at high pressures (up to 1,000 psi). Some catalytic reforming systems continuously regenerate the catalyst in other systems. One reactor at a time is taken off-stream for catalyst regeneration, and some facilities regenerate all of the reactors during turnarounds. Operating procedures should be developed to ensure control of hot spots during start-up. Safe catalyst handling is very important and care must be taken not to break or crush the catalyst when loading the beds, as the small fines will plug up the reformer screens. Precautions against dust when regenerating or replacing catalyst should also be considered and a water wash should be considered where stabilizer fouling has occurred due to the formation of ammonium chloride and iron salts. Ammonium chloride may form in pretreater exchangers and cause corrosion and fouling. Hydrogen chloride from the hydrogenation of chlorine compounds may form acid or ammonium chloride salt.

8.5.1 Autothermal Reforming

In autothermal reforming process (ATR process) the organic feedstock and steam and sometimes carbon dioxide are mixed directly with oxygen and air in the reformer (Aasberg-Petersen et al., 2002; Hagh, 2004). The reformer itself comprises a refractory lined vessel which contains the catalyst, together with an injector located at the top of the vessel and consists of three zones: (i) the burner – in which the feedstock streams are mixed in a turbulent diffusion flame, (ii) the combustion zone – in which partial oxidation reactions produce a mixture of carbon monoxide and hydrogen, and (iii) the catalytic zone – in which the gases leaving the combustion zone reach thermodynamic equilibrium.

Partial oxidation reactions occur in a region of the reactor referred to as the combustion zone. It is the mixture from this zone which then flows through a catalyst bed where the actual reforming reactions occur. Heat generated in the combustion zone from partial oxidation reactions is utilized in the reforming zone, so that in the ideal case, it is possible that the autothermal reforming process ATR can exhibit excellent heat balance. In addition, the process offers (i) flexible operation, including short start-up periods and fast load changes, as well as (ii) the potential for soot-free operation which, however, is feedstock dependent.

In autothermal (or secondary) reformers, the oxidation of methane supplies the necessary energy and is carried out either simultaneously or in advance of the reforming reaction (Brandmair et al., 2003; Ehwald et al., 2003; Nagaoka et al., 2003). The equilibrium of the methane steam reaction and the water-gas shift reaction determines the conditions for optimum hydrogen yields. The optimum conditions for hydrogen production require (i) high temperature at the exit of the reforming reactor on the order of 800 to 900°C (1470 to 1650°F), (ii) high excess of steam with a molar steam-to-carbon ratio on the order of 2.5 to 3, and (iii) relatively low pressures, typically less than 450 psi). Most commercial plants employ supported nickel catalysts for the process.

8.5.2 Combined Reforming

Combined reforming incorporates the combination of both steam reforming and autothermal reforming (Wang et al., 2004). In such a configuration, the feedstock is first only partially converted, under mild conditions, to synthesis gas in a relatively small steam reformer. The off-gas from the steam reformer is then sent to an oxygen-fired secondary reactor, the autothermal reforming reactor where any hydrocarbon derivatives in the gas stream are converted to synthesis gas by partial oxidation followed by steam reforming.

Partial oxidation:

$$CH_4 + 0.5\ O_2 \rightarrow 2H_2 + CO$$

or (for completeness):

$$2CH_4 + O_2 \rightarrow 4H_2 + 2CO$$

Steam reforming:

$$CH_4 + H_2O \rightarrow 3H_2 + CO$$

Another configuration requires the feedstock to be split into two streams which are then fed in parallel, to the steam reforming and autothermal reactors.

8.5.3 Dry Reforming

Dry reforming is less common than steam reforming, and its main use is for processes that require high proportion of CO in the synthesis gas. The reaction is represented by the equation:

$$CH_4 + CO_2 \leftrightarrows 2CO + 2H_2, \quad \Delta H°(1{,}000°C) = +258.9\ kJ/mol.$$

The thermodynamics of dry reforming is similar to steam reforming and the main operational difference of dry reforming from steam reforming is its tendency for coking, made more severe by the lack of steam to remove carbon according). In some applications, such as in mixed reforming (combination of steam and dry reforming), steam is added for effective containment of coking problems. Since coking quickly deactivates nickel (Ni) catalysts, rhodium (Rh) and ruthenium (Ru) catalysts are used in most dry reforming applications.

However, while nickel-based catalysts are suitable for this reaction and commercially available, the major drawback is that nickel catalysts also catalyze carbon deposition by the decomposition of methane or by the Boudouard reaction, which leads to potential catalyst breakdown and catalyst deactivation as well as reactor clogging.

$$CH_4 \rightarrow C + 2H_2$$

$$2CO \rightarrow C + CO_2$$

A number of methods for reducing the amount of carbon formed on nickel catalysts during reforming reactions have been found, such as the addition of alkali and alkaline-earth metal dopants, such as potassium (K), sodium (Na), magnesium oxide (MgO), calcium oxide (CaO), or lanthanide/actinide dopants, which result to a more carbon resistant catalyst, in most cases due to the promotion of carbon dioxide dissociation and, hence, more surface oxygen for reaction with carbon.

8.5.4 Steam-Methane Reforming

Steam-methane reforming (sometimes referred to as *steam reforming*, SMR), while not truly the subject of this chapter is worthy of note because of the production of synthesis gas. The process is carried out by passing a preheated mixture comprising essentially methane and steam through catalyst filled tubes. Since the reaction is endothermic, heat must be provided in order to effect the conversion and is achieved by the use of burners located adjacent to the tubes. The products of the process are a mixture of hydrogen, carbon monoxide and carbon dioxide.

In order to maximize the conversion of the methane feed, both a primary and secondary reformer are generally utilized. A *primary reformer* is used to effect 90 – 92% conversion of methane. Here, the hydrocarbon feed is partially reacted with steam over a nickel-alumina catalyst to produce a synthesis gas with H_2/CO ratio of approximately 3:1. This is done in a fired tube furnace at 900°C (1650°F) at a pressure of 220 to 450 psi. The unconverted methane is reacted with oxygen at the top of a *secondary autothermal reformer* containing nickel catalyst in the lower region of the vessel.

The deposition of carbon can be an acute problem with the use of nickel-based catalysts in the primary reformer (Rostrup-Nielsen, 1984; Alstrup, 1988, Rostrup-Nielsen, 1993). Considerable research has been done with the aim of finding approaches to prevent carbon formation. A successful technique is to use a steam/carbon ratio in the feed gas that does not allow the formation of carbon. However, this method results in lowering the efficiency of the process. Another approach utilizes sulfur passivation, which led to the development of the SPARG process (Rostrup-Nielsen, 1984; Udengaard *et al.*, 1992). This technique utilizes the principle that the reaction leading to the deposition of carbon requires a larger number of adjacent surface Ni atoms than does steam reforming. When a fraction of the surface atoms are covered by sulfur, the deposition of carbon is thus more greatly inhibited than steam reforming reactions. A third approach is to use Group VIII metals that do not for carbides, e.g., platinum (Pt). However, due to the high cost of such metals they are unable to compare to the economics associated with Ni. A major challenge in development of the steam-methane reforming process is the energy-intensive nature of the process due to the high endothermic character of the reactions. The trend in development thus is one which seeks higher energy efficiency. Improvements in catalysts and metallurgy require adaption to lower steam/carbon ratios and higher heat flux.

The steam-methane reforming process is the benchmark process that has been employed over a period of several decades for hydrogen production. The process involves reforming natural gas in a continuous catalytic process in which the major reaction is the formation of carbon monoxide and hydrogen from methane and steam:

$$CH_4 + H_2O \rightarrow CO + 3H_2 \qquad \Delta H_{298K} = +97,400 \text{ Btu/lb}$$

Higher molecular weight feedstocks can also be reformed to hydrogen:

$$C_3H_8 + 3H_2O \rightarrow 3CO + 7H_2$$

252 SYNTHESIS GAS

That is,

$$C_nH_m + nH_2O \rightarrow nCO + (0.5m + n)H_2$$

In the actual process, the feedstock is first desulfurized by passage through activated carbon, which may be preceded by caustic and water washes. The desulfurized material is then mixed with steam and passed over a nickel-based catalyst (730 to 845°C, 1350 to 1550°F and 400 psi, 2758 kPa). Effluent gases are cooled by the addition of steam or condensate to approximately 370°C (700°F), at which point carbon monoxide reacts with steam in the presence of iron oxide in a shift converter to produce carbon dioxide and hydrogen:

$$CO + H_2O = CO_2 + H_2 \qquad \Delta H_{298K} = -41.16 \text{ kJ/mol}$$

The carbon dioxide is removed by amine washing; the hydrogen is usually a high-purity (>99%) material.

Since the presence of any carbon monoxide or carbon dioxide in the hydrogen stream can interfere with the chemistry of the catalytic application, a third stage is used to convert of these gases to methane:

$$CO + 3H_2 \rightarrow CH_4 + H_2O$$

$$CO_2 + 4H_2 \rightarrow CH_4 + 2H_2O$$

For many refiners, sulfur-free natural gas (CH_4) is not always available to produce hydrogen by this process. In that case, higher-boiling hydrocarbon derivatives (such as propane, butane, or naphtha) may be used as the feedstock to generate hydrogen (*q.v.*).

The net chemical process for steam methane reforming is then given by:

$$CH_4 + 2H_2O \rightarrow CO_2 + 4H_2 \qquad \Delta H_{298K} = +165.2 \text{ kJ/mol}$$

Indirect heating provides the required overall endothermic heat of reaction for the steam-methane reforming.

One way of overcoming the thermodynamic limitation of steam reforming is to remove either hydrogen or carbon dioxide as it is produced, hence shifting the thermodynamic equilibrium towards the product side. The concept for sorption-enhanced methane steam reforming is based on in situ removal of carbon dioxide by a sorbent such as calcium oxide (CaO).

$$CaO + CO_2 \rightarrow CaCO_3$$

Sorption enhancement enables lower reaction temperatures, which may reduce catalyst coking and sintering, while enabling use of less expensive reactor wall materials. In addition, heat release by the exothermic carbonation reaction supplies most of the heat required by the endothermic reforming reactions. However, energy is required to regenerate the sorbent to its oxide form by the energy-intensive calcination reaction:

$$CaCO_3 \rightarrow CaO + CO_2$$

Use of a sorbent requires either that there be parallel reactors operated alternatively and out of phase in reforming and sorbent regeneration modes, or that sorbent be continuously transferred between the reformer/carbonator and regenerator/calciner (Balasubramanian et al., 1999; Hufton et al., 1999).

The steam-methane reforming process described briefly above would be an ideal hydrogen production process if it was not for the fact that large quantities of natural gas, a valuable resource, are required as both feed gas and combustion fuel. For each mole of methane reformed, more than one mole of carbon dioxide is co-produced and must be disposed. This can be a major issue as it results in the same amount of greenhouse gas emission as would be expected from direct combustion of natural gas or methane. In fact, the production of hydrogen as a clean burning fuel by way of steam reforming of methane and other fossil-based hydrocarbon fuels is not in environmental balance if in the process, carbon dioxide and carbon monoxide are generated and released into the atmosphere, although alternate scenarios are available (Gaudernack, 1996). Moreover, as the reforming process is not totally efficient, some of the energy value of the hydrocarbon fuel is lost by conversion to hydrogen but with no tangible environmental benefit, such as a reduction in emission of greenhouse gases. Despite these apparent shortcomings, the process has the following advantages: (i) produces 4 moles of hydrogen for each mole of methane consumed, (ii) feedstocks for the process (methane and water are readily available, (iii) the process is adaptable to a wide range of hydrocarbon feedstocks, (iv) operates at low pressures, less than 450 psi, (v) requires a low steam/carbon ratio (2.5 to 3), (vi) good utilization of input energy (reaching 93%), (vii) can use catalysts that are stable and resist poisoning, and (viii) good process kinetics.

Liquid feedstocks, either liquefied petroleum gas or naphtha (*q.v.*), can also provide backup feed, if there is a risk of natural gas curtailments. The feed handling system needs to include a surge drum, feed pump, vaporizer (usually steam-heated) followed by further heating before desulfurization. The sulfur in liquid feedstocks occurs as mercaptans, thiophene derivatives, or higher-boiling compounds. These compounds are stable and will not be removed by zinc oxide; therefore a hydrogenation unit will be required. In addition, as with refinery gas, olefins must also be hydrogenated if they are present.

The reformer will generally use a potash-promoted catalyst to avoid coke buildup from cracking of the heavier feedstock. If liquefied petroleum gas is to be used only occasionally, it is often possible to use a methane-type catalyst at a higher steam/carbon ratio to avoid coking. Naphtha will require a promoted catalyst unless a pre-former is used.

8.5.5 Steam-Naphtha Reforming

Steam-naphtha reforming is a continuous process for the production of hydrogen from liquid hydrocarbon derivatives and is, in fact, similar to steam-methane reforming that is one of several possible processes for the production of hydrogen from low-boiling hydrocarbon derivatives other than ethane Murata, 1998; Muradov, 1997, 2000; Brandmair et al., 2003; Find et al., 2003). A variety of naphtha-types in the gasoline boiling range may be employed, including feeds containing up to 35% aromatics. Thus, following pretreatment to remove sulfur compounds, the feedstock is mixed with steam and taken to the reforming furnace (675 to 815°C, 1250 to 1500°F, 300 psi, 2068 kPa), where hydrogen is produced.

8.6 Catalysts

Hydrogen plants are one of the most extensive users of catalysts in the refinery. Catalytic operations include hydrogenation, steam reforming, shift conversion, and methanation. This process is quite different from and not to be confused with the catalytic reforming process which is a chemical process used to convert naphtha distilled from crude oil (typically having a low octane rating) into a high-octane liquid products (the reformate), which is a premium blending stock for the manufacture of high-octane gasoline.

In the current context, the reforming process is used to produce products such as hydrogen, ammonia, and methanol from natural gas (methane), naphtha, and/or other crude oil-derived feedstocks. This process is also used to convert methanol or biomass-derived feedstocks to produce hydrogen.

8.6.1 Reforming Catalysts

The reforming catalyst is usually supplied as nickel oxide that, during start-up, is heated in a stream of inert gas, then steam. When the catalyst is near the normal operating temperature, hydrogen or a low-boiling hydrocarbon is added to reduce the nickel oxide to metallic nickel.

The high temperatures (up to 870°C, 1600°F), and the nature of the reforming reaction require that the reforming catalyst be used inside the radiant tubes of a reforming furnace. The active agent in reforming catalyst is nickel, and normally the reaction is controlled both by diffusion and by heat transfer. Catalyst life is limited as much by physical breakdown as by deactivation.

Sulfur is the main catalyst poison and the catalyst poisoning is theoretically reversible with the catalyst being restored to near full activity by steaming. However, in practice the deactivation may cause the catalyst to overheat and coke, to the point that it must be replaced. Reforming catalysts are also sensitive to poisoning by heavy metals, although these are rarely present in low-boiling hydrocarbon feedstocks and in naphtha feedstocks.

While methane-rich streams such as natural gas or refinery gas are the most common feeds to hydrogen plants, there is often a requirement for a variety of reasons to process a variety of higher-boiling feedstocks, such as liquefied petroleum gas and naphtha. Feedstock variations may also be inadvertent due, for example, to changes in refinery off-gas composition from another unit or because of variations in naphtha composition because of feedstock variance to the naphtha unit. Coking deposition on the reforming catalyst and ensuing loss of catalyst activity is the most characteristic issue that must be assessed and mitigated.

Thus, when using higher-boiling feedstocks in a hydrogen plant, coke deposition on the reformer catalyst becomes a major issue. Coking is most likely in the reformer unit at the point where both temperature and hydrocarbon content are high enough. In this region, hydrocarbon derivatives crack and form coke faster than the coke is removed by reaction with steam or hydrogen and when catalyst deactivation occurs, there is a simultaneous temperature increases with a concomitant increase in coke formation and deposition. In other zones, where the hydrocarbon-to-hydrogen ratio is lower, there is less risk of coking.

Coking depends to a large extent on the balance between catalyst activity and heat input with the more active catalysts producing higher yields of hydrogen at lower temperature thereby reducing the risk of coking. A uniform input of heat is important in this region of the reformer since any catalyst voids or variations in catalyst activity can produce localized hot spots leading to coke formation and/or reformer failure.

Coke formation results in hotspots in the reformer that increases pressure drop, reduces feedstock (methane) conversion, leading eventually to reformer failure. Coking may be partially mitigated by increasing the steam/feedstock ratio to change the reaction conditions but the most effective solution may be to replace the reformer catalyst with one designed for higher-boiling feedstocks.

A *standard* steam-methane reforming catalyst uses nickel on an alpha-alumina ceramic carrier that is acidic in nature. Promotion of hydrocarbon cracking with such a catalyst leads to coke formation from higher-boiling feedstocks. Some catalyst formulations use a magnesia/alumina (MgO/Al_2O_3) support that is less acidic than α-alumina that reduces cracking on the support and allows higher-boiling feedstocks (such as liquefied petroleum gas) to be used.

Further resistance to coking can be achieved by adding an alkali promoter, typically some form of potash (KOH) to the catalyst. Besides reducing the acidity of the carrier, the promoter catalyzes the reaction of steam and carbon. While carbon continues to be formed, it is removed faster than it can build up. This approach can be used with naphtha feedstocks boiling point up to approximately 180°C (350°F). Under the conditions in a reformer, potash is volatile and it is incorporated into the catalyst as a more complex compound that slowly hydrolyzes to release potassium hydroxide (KOH). Alkali-promoted catalyst allows the use of a wide range of feedstocks but, in addition to possible potash migration, which can be minimized by proper design and operation, the catalyst is also somewhat less active than conventional catalyst.

Another option to reduce coking in steam reformers is to use a *pre-reformer* in which a fixed-bed of catalyst, operating at a lower temperature, upstream of the fired reformer is used. In a pre-reformer, adiabatic steam-hydrocarbon reforming is performed outside the fired reformer in a vessel containing high nickel catalyst. The heat required for the endothermic reaction is provided by hot flue gas from the reformer convection section. Since the feed to the fired reformer is now partially reformed, the steam methane reformer can operate at an increased feed rate and produce 8 to 10% additional hydrogen at the same reformer load. An additional advantage of the pre-reformer is that it facilitates higher mixed feed preheat temperatures and maintains relatively constant operating conditions within the fired reformer regardless of variable refinery off-gas feed conditions. Inlet temperatures are selected so that there is minimal risk of coking and the gas leaving the pre-reformer contains only steam, hydrogen, carbon monoxide, carbon dioxide, and methane. This allows a standard methane catalyst to be used in the fired reformer and this approach has been used with feedstocks up to low-boiling kerosene. Since the gas leaving the pre-reformer poses reduced risk of coking, it can compensate to some extent for variations in catalyst activity and heat flux in the primary reformer.

8.6.2 Shift Conversion Catalysts

The second important reaction in a steam reforming plant is the shift conversion reaction:

$$CO + H_2O \rightarrow CO_2 + H_2$$

Two basic types of shift catalyst are used in steam reforming plants: iron/chrome high temperature shift catalysts, and copper/zinc low temperature shift catalysts.

High-temperature shift catalysts operate in the range of 315 to 430°C (600 to 800°F) and consist primarily of magnetite (Fe_3O_4) with three-valent chromium oxide (Cr_2O_3) added as a stabilizer. The catalyst is usually supplied in the form of ferric oxide (Fe_2O_3) and six-valent chromium oxide (CrO_3) and is reduced by the hydrogen and carbon monoxide in the shift feed gas as part of the start-up procedure to produce the catalyst in the desired form. However, caution is necessary since if the steam/carbon ratio of the feedstock is too low and the reducing environment too strong, the catalyst can be reduced further, to metallic iron. Metallic iron is a catalyst for Fischer-Tropsch reactions and hydrocarbon derivatives will be produced.

Low-temperature shift catalysts operate at temperatures on the order of 205 to 230°C (400 to 450°F). Because of the lower temperature, the reaction equilibrium is more controllable and lower amounts of carbon monoxide are produced. The low-temperature shift catalyst is primarily used in wet scrubbing plants that use a methanation for final purification. Pressure-swing adsorption plants do not generally use a low-temperature because since any unconverted carbon monoxide is recovered as reformer fuel. Low-temperature shift catalysts are sensitive to poisoning by sulfur and are sensitive to water (liquid) that can cause softening of the catalyst followed by crusting or plugging.

The catalyst is supplied as copper oxide (CuO) on a zinc oxide (ZnO) carrier and the copper must he reduced by heating it in a stream of inert gas with measured quantities of hydrogen. The reduction of the copper oxide is strongly exothermic and must be closely monitored.

8.6.3 Methanation Catalysts

In wet scrubbing plants, the final hydrogen purification procedure involves methanation in which the carbon monoxide and carbon dioxide are converted to methane:

$$CO + 3H_2O \rightarrow CH_4 + H_2O$$

$$CO_2 + 4H_2 \rightarrow CH_4 + 2H_2O$$

The active agent is nickel, on an alumina carrier.

The catalyst has a long life, as it operates under ideal conditions and is not exposed to poisons. The main source of deactivation is plugging from carryover of carbon dioxide from removal solutions.

The most severe hazard arises from high levels of carbon monoxide or carbon dioxide that can result from breakdown of the carbon dioxide removal equipment or from exchanger tube leaks that quench the shift reaction. The results of breakthrough can be severe, since the methanation reaction produces a temperature rise of 70°C (125°F) per 1% of carbon monoxide or a temperature rise of 33°C (60°F) per 1% of carbon dioxide. While the normal operating temperature during methanation is approximately 315°C (600°F), it is possible to reach 700°C (1300°F) in cases of major breakthrough.

8.7 Hydrogen Purification

When the hydrogen content of the refinery gas is greater than 50% by volume, the gas should first be considered for hydrogen recovery, using a membrane (Brüschke, 1995, 2003)) or pressure-swing adsorption unit (Mokhatab *et al.*, 2006; Speight, 2014, 2019). The tail gas or reject gas that will still contain a substantial amount of hydrogen can then be used as steam reformer feedstock. Generally, the feedstock purification process uses three different refinery gas streams to produce hydrogen. First, high-pressure hydrocracker purge gas is purified in a membrane unit that produces hydrogen at medium pressure and is combined with medium pressure off-gas that is first purified in a pressure-swing adsorption unit. Finally, low-pressure off-gas is compressed, mixed with reject gases from the membrane and pressure-swing adsorption units, and used as steam reformer feed.

Various processes are available to purify the hydrogen stream but since the product streams are available as a wide variety of composition, flows, and pressures, the best method of purification will vary. And there are several factors that must also be taken into consideration in the selection of a purification method. These are: (i) hydrogen recovery, (ii) product purity, (iii) pressure profile, (iv) reliability, and (v) cost, an equally important parameter is not considered here since the emphasis is on the technical aspects of the purification process.

8.7.1 Wet Scrubbing

Wet scrubbing systems, particularly amine or potassium carbonate systems, are used for removal of acid gases such as hydrogen sulfide or carbon dioxide (Speight, 1993; Dalrymple *et al.*, 1994). Most systems depend on chemical reaction and can be designed for a wide range of pressures and capacities. They were once widely used to remove carbon dioxide in steam reforming plants, but have generally been replaced by pressure swing adsorption units except where carbon monoxide is to be recovered. Wet scrubbing is still used to remove hydrogen sulfide and carbon dioxide in partial oxidation plants.

Wet scrubbing systems remove only acid gases or high-boiling hydrocarbon derivatives but they do not methane or other hydrocarbon gases, hence have little influence on product purity. Therefore, wet scrubbing systems are most often used as a pretreatment step, or where a hydrogen-rich stream is to be desulfurized for use as fuel gas.

8.7.2 Pressure-Swing Adsorption Units

Pressure-swing adsorption units use beds of solid adsorbent to separate impurities from hydrogen streams leading to high-purity high-pressure hydrogen and a low-pressure tail gas stream containing the impurities and some of the hydrogen. The beds are then regenerated by depressuring and purging. Part of the hydrogen (up to 20%) may be lost in the tail gas.

Pressure-swing adsorption is generally the purification method of choice for steam reforming units because of its production of high-purity hydrogen and is also used for purification of refinery off-gases, where it competes with membrane systems.

Many hydrogen plants that formerly used a *wet scrubbing* process for hydrogen purification are now using the *pressure-swing adsorption* (PSA) for purification (Mokhatab *et al.*, 2006; Speight, 2014, 2019). The pressure-swing adsorption process is a cyclic process that uses beds of solid adsorbent to remove impurities from the gas and generally produces higher-purity hydrogen (99.9 volume percent purity compared to less than 97 volume percent purity). The purified hydrogen passes through the adsorbent beds with only a tiny fraction absorbed and the beds are regenerated by depressurization followed by purging at low pressure.

When the beds are depressurized, a waste gas (or *tail gas*) stream is produced and consists of the impurities from the feed (carbon monoxide, carbon dioxide, methane, and nitrogen) plus some hydrogen. This stream is burned in the reformer as fuel and reformer operating conditions in a pressure-swing adsorption plant are set so that the tail gas provides no more than approximately 85% of the reformer fuel. This gives good burner control because the tail gas is more difficult to burn than regular fuel gas and the high content of carbon monoxide can interfere with the stability of the flame. As the reformer operating temperature is increased, the reforming equilibrium shifts, resulting in more hydrogen and less methane in the reformer outlet and hence less methane in the tail gas.

8.7.3 Membrane Systems

Membrane systems separate gases by taking advantage of the difference in rates of diffusion through membranes (Brüschke, 1995, 2003). Gases that diffuse faster (including hydrogen) become the permeate stream and are available at low pressure whereas the slower-diffusing gases become the non-permeate and leave the unit at a pressure close to the pressure of the feedstock at entry point. Membrane systems contain no moving parts or switch valves and have potentially very high reliability. The major threat is from components in the gas (such as aromatics) that attack the membranes, or from liquids, which plug them.

Membranes arc fabricated in relatively small modules; for larger capacity more modules are added. Cost is therefore virtually linear with capacity, making them more competitive at lower capacities. The design of membrane systems involves a tradeoff between pressure drop (or diffusion rate) and surface area, as well as between product purity and recovery. As the surface area is increased, the recovery of fast components increases; however, more of the slow components are recovered, which lowers the purity.

8.7.4 Cryogenic Separation

Cryogenic separation units operate by cooling the gas and condensing some, or all, of the constituents for the gas stream. Depending on the product purity required, separation may involve flashing or distillation. Cryogenic units offer the advantage of being able to separate a variety of products from a single feed stream. One specific example is the separation of low-boiling olefins from a hydrogen stream.

Hydrogen recovery is in the range of 95%, with purity above 98% obtainable.

8.8 Hydrogen Management

Many existing refinery hydrogen plants use a conventional process, which produces a medium-purity (94% to 97%) hydrogen product by removing the carbon dioxide in an absorption system and methanation of any remaining carbon oxides. Since the 1980s, most hydrogen plants are built with pressure-swing adsorption (PSA) technology to recover and purify the hydrogen to purities above 99.9%. Since many refinery hydrogen plants utilize refinery off-gas feeds containing hydrogen, the actual maximum hydrogen capacity that can be synthesized via steam reforming is not certain since the hydrogen content of the off-gas can change due to operational changes in the hydrotreaters.

Hydrogen management has become a priority in current refinery operations and when planning to produce lower sulfur gasoline and diesel fuels (Zagoria *et al.*, 2003; Davis and Patel, 2004; Luckwal and Mandal, 2009). Along with increased hydrogen consumption for deeper hydrotreating, additional hydrogen is needed for processing heavier and higher sulfur crude slates. In many refineries, hydroprocessing capacity and the associated hydrogen network is limiting refinery throughput and operating margins. Furthermore, higher hydrogen purities within the refinery network are becoming more important to boost hydrotreater capacity, achieve product value improvements and lengthen catalyst life cycles.

Improved hydrogen utilization and expanded or new sources for refinery hydrogen and hydrogen purity optimization are now required to meet the needs of the future transportation fuel market and the drive towards higher refinery profitability. Many refineries developing hydrogen management programs fit into the two general categories of either a catalytic reformer supplied network or an on-purpose hydrogen supply.

Some refineries depend solely on catalytic reformer(s) as their source of hydrogen for hydrotreating. Often, they are semi-regenerative reformers where off-gas hydrogen quantity, purity, and availability change with feed naphtha quality, as octane requirements change seasonally, and when the reformer catalyst progresses from start-of-run to end-of-run conditions and then goes offline for regeneration. Typically, during some portions of the year, refinery margins are reduced as a result of hydrogen shortages.

Multiple hydrotreating units compete for hydrogen – either by selectively reducing throughput, managing intermediate tankage logistics, or running the catalytic reformer sub-optimally just to satisfy downstream hydrogen requirements. Part of the operating year still runs in hydrogen surplus, and the network may be operated with relatively low hydrogen utilization (consumption/production) at 70 to 80%. Catalytic reformer off-gas hydrogen supply may swing from 75 to 85% hydrogen purity. Hydrogen purity upgrade can be achieved through some hydrotreaters by absorbing high-boiling hydrocarbon derivatives. But without supplemental hydrogen purification, critical control of hydrogen partial pressure in hydroprocessing reactors is difficult, which can affect catalyst life, charge rates, and/or gasoline yields.

More complex refineries, especially those with hydrocracking units, also have on-purpose hydrogen production, typically with a steam methane reformer steam methane reformer that utilizes refinery off-gas and supplemental natural gas as feedstock. The steam methane reformer plant provides the swing hydrogen requirements at higher purities (92% to more than 99% hydrogen) and serves a hydrogen network configured with several purity and pressure levels. Multiple purities and existing purification units allow for more optimized hydroprocessing operation by controlling hydrogen partial pressure for maximum benefit. Typical hydrogen utilization is 85% to 95%.

References

Aasberg-Petersen, K., Christensen, T.S., Nielsen, C.S., and Dybkjær, I. 2002. Recent Developments in Autothermal Reforming and Pre-reforming for Synthesis Gas Production in GTL Applications. *Preprints. Am. Chem. Soc. Div. Fuel Chem.*, 47(1): 96-97.

Balasubramanian, B., Ortiz, A.L., Kaytakoglu, S. and Harrison, D.P. 1999. Hydrogen from Methane in a Single-Step Process. *Chemical Engineering Science*. 54: 3543-3552.

Bandermann, F., Harder, K.B. 1982. Production of Hydrogen via Thermal Decomposition of Hydrogen Sulfide and Separation of Hydrogen and Hydrogen Sulfide by Pressure Swing Adsorption. *Int. J. Hydrogen Energy*. 7(6): 471-475.

Bezler, J. 2003. Optimized Hydrogen Production – A Key Process Becoming Increasingly Important in Refineries. *Proceedings. DGMK Conference on Innovation in the Manufacture and Use of Hydrogen. Dresden, Germany. October 15-17*. Page 65.

Bishara, A., Salman, O.S., Khraishi, N., and Marafi, A. 1987. Thermochemical Decomposition of Hydrogen Sulfide by Solar Energy. *Int. J. Hydrogen Energy*, 12(10): 679-685.

Brandmair, M., Find, J., and Lercher, J.A. 2003. Combined Autothermal Reforming and Hydrogenolysis of Alkanes. *Proceedings. DGMK Conference on Innovation in the Manufacture and Use of Hydrogen. Dresden, Germany. October 15-17*. Page 273-280.

Bridge, A.G. 1997. In *Handbook of Petroleum Refining Processes*. 2nd Edition. R.A. Meyers (Editor). McGraw-Hill, New York. Chapter 14.1.

Brüschke, H. 1995. Industrial Application of Membrane Separation Processes. *Pure Appl. Chem.*, 345 67(6): 993-1002.

Brüschke, H. 2003. Separation of Hydrogen from Dilute Streams (e.g. Using Membranes). *Proceedings. DGMK Conference on Innovation in the Manufacture and Use of Hydrogen. Dresden, Germany. October 15-17*. Page 47.

Campbell, W.M. 1997. In *Handbook of Petroleum Refining Processes*. 2nd Edition. R.A. Meyers (Editor). McGraw-Hill, New York. Chapter 6.1.

Clark, P.D., Wassink B. 1990. A Review of Methods for the Conversion of Hydrogen Sulfide to Sulfur and Hydrogen. *Alberta Sulfur Res. Quart. Bull.*, 26(2/3/4): 1.

Clark, P.D., N.I. Dowling, J.B. Hyne, D.L. Moon. 1995. Production of Hydrogen and Sulfur from Hydrogen Sulfide in Refineries and Gas Processing Plants. *Quarterly Bulletin (Alberta Sulfur Research Ltd.)*, 32(1): 11-28.

Dahl, J., and Weimer, A.W.. 2001. *Preprints. Div. Fuel Chem. Am. Chem. Soc.* 221.

Dalrymple, D.A., Trofe, T.W., and Leppin, D. Gas Industry Assesses New Ways to Remove Small Amounts of Hydrogen Sulfide. *Oil & Gas Journal*, May 23.

Davis, R.A., and Patel, N.M. 2004. *Petroleum Technology Quarterly*. 2004 Spring. Page 28-35.

Dickenson, R.L., Biasca, F.E., Schulman, B.L., and Johnson, H.E. 1997. Refiner Options for Converting and Utilizing Heavy Fuel Oil. *Hydrocarbon Processing*. 76(2): 57.

Dolbear, G.E. 1998. *Hydrocracking: Reactions, Catalysts, and Processes. In Petroleum Chemistry and Refining*. J.G. Speight (Editor). Taylor & Francis, Washington, DC. Chapter 7.

Donini, J. C. 1996. Separation and Processing of Hydrogen Sulfide in the Fossil Fuel Industry. *Minimum Effluent Mills Symposium*. 357-363.

Dragomir, R., Drnevich, R.F., Morrow, J., Papavassiliou, V., Panuccio, G., and Watwe, R. 2010. Technologies for Enhancing Refinery Gas Value. *Proceedings. AIChE 2010 SPRING Meeting.* San Antonio, Texas. November 7-12.

Ehwald, H., Kürschner, U., Smejkal, Q., and Lieske, H. 2003. Investigation of Different Catalysts for Autothermal Reforming of i-Octane. *Proceedings. DGMK Conference on Innovation in the Manufacture and Use of Hydrogen.* Dresden, Germany. October 15-17. Page 345.

Find, J., Nagaoka, K., and Lercher, J.A.. 2003. Steam Reforming of Light Alkanes in Micro-Structured Reactors. *Proceedings. DGMK Conference on Innovation in the Manufacture and Use of Hydrogen. Dresden, Germany. October 15-17.* Page 257.

Fleshman, J.D. 1997. In *Handbook of Petroleum Refining Processes.* 2nd Edition. R.A. Meyers (Editor). McGraw-Hill, New York. Chapter 6.2.

Gary, J.G., Handwerk, G.E., and Kaiser, M.J. 2007. *Petroleum Refining: Technology and Economics.* 5th Edition. CRC Press, Taylor & Francis Group, Boca Raton, Florida.

Gaudernack, B. 1996. Hydrogen from Natural Gas without Release of Carbon Dioxide into the Atmosphere. Hydrogen Energy Prog. *Proceedings. 11th World Hydrogen Energy Conference.* 1: 511-523.

Gross, M., and Wolff, J. 2000. Gasification of Residue as a Source of Hydrogen for the Refining Industry in India. *Proceedings. Gasification Technologies Conference.* San Francisco, California. October 8-11.

Hagh, B.F. 2004. Comparison of Autothermal Reforming For Hydrocarbon Fuels. *Preprints. Am. Chem. Soc. Div. Fuel Chem.*, 49(1): 144-147.

Hsu, C.S., and Robinson, P.R. (Editors). 2017. *Handbook of Petroleum Technology.* Springer International Publishing AG, Cham, Switzerland.

Hufton, J.R., Mayorga, S. and Sircar, S., 1999. Sorption-Enhanced Reaction Process for Hydrogen Production. *AIChE. Journal*, 45: 248-256.

Kiuchi, H. 1982. Recovery of Hydrogen from Hydrogen Sulfide with Metals and Metal Sulfides. *Int. J. Hydrogen Energy*, 7(6).

Kotera, Y., Todo, N., and Fukuda, K. 1976. Process for Production of Hydrogen and Sulfur from Hydrogen Sulfide as Raw Material. U.S. Patent No. 3,962,409. June 8.

Lacroix, M., H. Marrakchi, C. Calais, M. Breysse, C. Forquy.1991. Catalytic Properties of Transition Metal Sulfides for the Dehydrogenation of Sulfur Containing Molecules. *Studies in Surface Science and Catalysis*, 59: 277-285.

Luckwal, K., and Mandal, K.K. 2009. Improve Hydrogen Management of Your Refinery. *Hydrocarbon Processing*, 88(2): 55-61.

Luinstra, E. 1996. Hydrogen from Hydrogen Sulfide – A Review of the Leading Processes. *Proceedings. 7th Sulfur Recovery Conference.* Gas Research Institute, Chicago. Page 149-165.

Megalofonos, S.K., and Papayannakos, N.G. 1997. Kinetics of Catalytic Reaction of Methane and Hydrogen Sulfide over MoS_2. *Journal of Applied Catalysis A: General.* 65(1-2): 249-258.

Miller G.Q., and Penner, D.W. 2003. Meeting Future Needs for Hydrogen – Possibilities and Challenges. *Proceedings. DGMK Conference on Innovation in the Manufacture and Use of Hydrogen.* Dresden, Germany. October 15-17. Page 7.

Mokhatab, S., Poe, W.A., and Speight, J.G. 2006. *Handbook of Natural Gas Transmission and Processing.* Elsevier, Amsterdam, Netherlands.

Muradov, N.Z. 1998. CO2-Free Production of Hydrogen by Catalytic Pyrolysis of Hydrocarbon Fuel. *Energy & Fuels.* 12(1): 41-48.

Muradov, N.Z. 2000. Thermocatalytic Carbon Dioxide-Free Production of Hydrogen from Hydrocarbon Fuels. *Proceedings. Hydrogen Program Review*, NREL/CP-570-28890.

Murata, K., H. Ushijima, K. Fujita. 1997. Process for Producing Hydrogen from Hydrocarbon. United States Patent 5,650,132.

Nagaoka, K., Jentys, A., and Lecher, J.A. 2003. Autothermal Reforming of Methane over Mono- and Bi-metal Catalysts Prepared from Hydrotalcite-like Precursors. Proceedings. *DGMK Conference on Innovation in the Manufacture and Use of Hydrogen. Dresden, Germany. October 15-17.* Page 171.

Parkash, S. 2003. *Refining Processes Handbook.* Gulf Professional Publishing, Elsevier, Amsterdam, Netherlands.

Raissi, A.T. 2001. Technoeconomic Analysis of Area II Hydrogen Production. Part 1. *Proceedings. US DOE Hydrogen Program Review Meeting, Baltimore, Maryland.*

Ranke, H., and Schödel, N 2003. Hydrogen Production Technology – Status and New Developments. *Proceedings. DGMK Conference on Innovation in the Manufacture and Use of Hydrogen. Dresden, Germany. October 15-17.* Page 19.

Scherzer, J., and Gruia, A.J. *Hydrocracking Science and Technology.* Marcel Dekker Inc., New York.

Speight, J.G. 2000. *The Desulfurization of Heavy Oils and Residua.* 2nd Edition. Marcel Dekker Inc., New York.

Speight, J.G. 2014. *The Chemistry and Technology of Petroleum* 5th Edition. CRC Press, Taylor and Francis Group, Boca Raton, Florida.

Speight, J.G. 2017. *Handbook of Petroleum Refining.* CRC Press, Taylor and Francis Group, Boca Raton, Florida.

Speight, J.G. 2019. *Natural Gas: A Basic Handbook* 2nd Edition. GPC Books, Gulf Publishing Company, Elsevier, Cambridge, Massachusetts.

Uemura, Y., Ohe, H., Ohzuno, Y., and Hatate, Y. 1999. Carbon and Hydrogen from Hydrocarbon Pyrolysis. *Proceedings. Int. Conf. Solid Waste Technology Management.* 15: 5E/25-5E/30.

Vauk, D., Di Zanno, P., Neri, B., Allevi, C., Visconti, A., and Rosanio, L. 2008. What Are Possible Hydrogen Sources for Refinery Expansion? *Hydrocarbon Processing,* 87(2): 69-76.

Wang, W., Stagg-Williams, S.M., Noronha, F.B., Mattos, L.V., and Passos, F.B. 2004. *Preprints. Am. Chem. Soc., Div. Fuel Chem.,* 49(1): 133

Zagoria, A., Huychke, R., and Boulter, P. H. 2003. Refinery Hydrogen Management – The Big Picture. *Proceedings. DGMK Conference on Innovation in the Manufacture and Use of Hydrogen. Dresden, Germany. October 15-17.* Page 95.

9
Gas Conditioning and Cleaning

9.1 Introduction

Synthesis gas is experiencing a period of rapid development and synthesis gas upgrading is attracting increasing attention (Ryckebosch *et al.*, 2011; Niesner *et al.*, 2013; Andriani *et al.*, 2014). Currently, synthesis gas is mostly used as a fuel in power generators and boilers. For these uses, the hydrogen sulfide content in synthesis gas should be less than 200 parts per million (ppm v/v) to ensure a long life for the power and heat generators. Consequently, the market for synthesis gas upgrading is facing significant challenges in terms of energy consumption and operating costs. Selection of upgrading technology is site-specific, case-sensitive and dependent on the synthesis gas utilization requirements and local circumstances (Ramírez *et al.*, 2015). But first it is important to recognize the composition of the gas so that the appropriate technologies can be selected for use to remove the specific contaminants. Thus, when any biomass-derived gas (either synthesis gas or bio-synthesis gas) is used for biofuel production, the cleaning of the raw gas is needed strictly in order to remove contaminants and potential catalyst poisons as well as to achieve the qualitative composition required before the gas is sold for use.

Synthesis gas, as produced by any of the methods described in earlier chapters (i.e., by anaerobic digestion, pyrolysis, or gasification) is typically saturated with water vapor and contains, in addition to methane (CH_4) and carbon dioxide (CO_2), various amounts of hydrogen sulfide (H_2S) as well as other contaminants that vary in amounts depending upon the composition of the biomass feedstocks and the process parameters employed for production of the gas. The properties of the contaminants are such that failure to remove them will make the gas unusable and even poisonous to the user. For example, hydrogen sulfide is a toxic gas, with a specific, unpleasant odor, similar to rotten eggs, forming acidic products in combination with the water vapors in the synthesis gas which can also result in corrosion of equipment. To prevent this, and similar example of the adverse effects of other contaminants, the synthesis gas must be dried and any contaminants removed before the synthesis gas is sent to one or more conversion units or for sales.

Typically, in the gas processing industry (also called the *gas cleaning industry* or the *gas refining industry*), the feedstock is not used directly as fuel because of the complex nature of the feedstock and the presence of one of more of the aforementioned impurities that are corrosive or poisonous. It is, therefore, essential that any gas should be contaminant free when it enters any one of the various reactors (Parkash, 2003; Gary *et al.*, 2007; Speight, 2014; Hsu and Robinson, 2017; Speight, 2017, 2019). Typically, the primary raw synthesis gas has been subjected to chemical and/or physical changes (refining) after being produced. Thus for many applications (Chapter 8) the quality of synthesis gas has to be improved. In addtion, landfill gas (Chapter 1, Chapter 4) often contains significant amounts of halogenated compounds which need to be removed prior to use. Occasionally the oxygen content is high when air is allowed to invade the gas producer during collection of the landfill gas or when the biomass from which the gas is produced contains an abundance of oxygen-containing organic constituents.

The main contaminants (but not necessarily the only contaminants) that may require removal in gas cleaning systems are hydrogen sulfide, water, carbon dioxide, and halogenated compounds. Desulfurization is to prevent corrosion and avoid concentrations of the toxic hydrogen sulfide for safety in use and in the workplace. Also, if hydrogen sulfide and other sulfur-containing species such as thiol derivatives (RSH, also called mercaptan derivative) and carbonyl sulfide (COS) are not removed, combustion of the gas produces sulfur dioxide and sulfur trioxide which is even more poisonous than hydrogen sulfide. At the same time, the presence of sulfur dioxide in the gas stream lowers the dew point (the temperature to which the gas must be cooled to become saturated with water vapor) of in the stack gas. The sulfurous acid formed (H_2SO_3) by reaction of the sulfur dioxide and the water vapor as well as reaction of carbon dioxide and the water vapor forms highly corrosive acidic species:

$$SO_2 + H_2O \rightarrow H_2SO_3$$

$$CO_2 + H_2O \rightarrow H_2CO_3$$

Thus, water removal from the synthesis gas is also essential, not only because of the potential for the reaction of water with contaminants in the gas stream but also because of potential accumulation of water vapor condensing in the gas line, the formation of a corrosive acidic solution when hydrogen sulfide is dissolved or to achieve low dew points when synthesis gas is stored under elevated pressures in order to avoid condensation and freezing. Also, if the non-combustible carbon dioxide is not removed the energy content of the synthesis gas is diluted and there may also be an environmental impact due to the presence and emission of this contaminant.

Generally, the contaminants are categorized as (i) particulate matter, (ii) condensable hydrocarbon derivatives, including tar products, (iii) alkali metals, such as sodium and potassium, (iv) nitrogen-containing derivatives, including ammonia and hydrogen cyanide, (v) sulfur-containing derivatives, such as hydrogen sulfide, carbonyl sulfide, and carbon disulfide, and halogen containing derivatives, such as hydrogen chloride and hydrogen bromide, and hydrogen fluoride.

In fact, regarding the constituents that make up the tar products that are formed during gasification when biomass decomposes in pyrolysis and gasification reactions, primary tar compounds are mostly oxygenated compounds that are decomposition products of biomass. These compounds react further and form secondary and tertiary tar compounds, which consist of compounds that do not exist in the source biomass. The secondary tar typically consists of alkylated one-ring and two-ring aromatic compounds (including heterocyclic compounds) whereas tertiary tar consists of aromatic hydrocarbon derivatives such as benzene, naphthalene, and various polycyclic aromatic hydrocarbon derivatives (PAHs, also called polynuclear aromatic compounds, PNAs). Tar constituents are typically classified on the basis of the number of rings in the constituents or by boiling point distribution or physical properties. More generally, tar is defined as aromatic compounds that are higher-boiling than benzene and, in addtion, an operational definition for tar depending on the end-use application has been used.

Process selectivity indicates the preference with which the process removes one acid gas component relative to (or in preference to) another. For example, some processes remove both hydrogen sulfide and carbon dioxide; other processes are designed to remove hydrogen sulfide only. It is very important to consider the process selectivity for, say, hydrogen sulfide removal compared to carbon dioxide removal that ensures minimal concentrations of these components in the product, thus the need for consideration of the carbon dioxide to hydrogen sulfide in the gas stream.

To include a description of all of the possible process variations for gas cleaning is beyond the scope of this book. Therefore, the focus of this chapter is a selection of the processes that are an integral part within the concept of production of a specification-grade product (methane) for sale to the consumer.

Furthermore, it is the purpose of this chapter to present the methods by which various gaseous and liquid feedstocks can be processed and prepared for petrochemical production. This requires removal of impurities that would otherwise by deleterious to petrochemical production and analytical assurance that those feedstocks are, indeed, free of deleterious contaminants (Speight, 2015, 2018).

9.2 Gas Streams

The actual practice of processing gas streams to pipeline dry gas quality levels can be quite complex, but usually involves four main processes to remove the various impurities. Gas streams produced during natural gas and crude oil refining, while ostensibly being hydrocarbon in nature, may contain substantial proportions of acid gases such as hydrogen sulfide and carbon dioxide. Most commercial plants employ hydrogenation to convert organic sulfur compounds into hydrogen sulfide. Hydrogenation is effected by means of recycled hydrogen-containing gases or external hydrogen over a nickel molybdate or cobalt molybdate catalyst.

In summary, refinery process gas, in addition to hydrocarbons, may contain other contaminants, such as carbon oxides (CO_x, where x = 1 and/or 2), sulfur oxides (So_x, where x = 2 and/or 3), as well as ammonia (NH_3), mercaptans (R-SH), and carbonyl sulfide (COS).

The presence of these impurities may eliminate some of the sweetening processes, since some processes remove large amounts of acid gas but not to a sufficiently low concentration. On the other hand, there are those processes not designed to remove (or incapable of removing) large amounts of acid gases whereas they are capable of removing the acid gas impurities to very low levels when the acid gases are present only in low-to-medium concentration in the gas (Katz, 1959).

The processes that have been developed to accomplish gas purification vary from a simple once-through wash operation to complex multi-step recycling systems (Speight, 1993). In many cases, the process complexities arise because of the need for recovery of the materials used to remove the contaminants or even recovery of the contaminants in the original, or altered, form (Katz, 1959; Kohl and Riesenfeld, 1985; Newman, 1985).

There are many variables in treating refinery gas or natural gas. The precise area of application of a given process is difficult to define and it is necessary to consider several factors:

(i) the types of contaminants in the gas stream, (ii) the concentration of each contaminant in the gas stream, (iii) the degree of contaminant removal desired (iv) the selectivity of acid gas removal required, (v) the temperature of the gas stream to be processed, (vi) the pressure of the gas stream to be processed, (vii) the volume of the gas stream to be processed, (viii) the composition of the gas to be processed, (ix) the carbon dioxide-hydrogen sulfide ratio in the gas, and (x) the desirability of sulfur recovery due to process economics or environmental issues.

In addition to hydrogen sulfide (H_2S) and carbon dioxide (CO_2), the gas stream may contain other contaminants, such as mercaptans (RSH) and carbonyl sulfide (COS). The presence of these impurities may eliminate some of the sweetening processes since some processes remove large amounts of acid gas but not to a sufficiently low concentration. On the other hand, there are those processes that are not designed to remove (or are incapable of removing) large amounts of acid gases. However, these processes are also capable of removing the acid gas impurities to very low levels when the acid gases are there in low to medium concentrations in the gas.

Process selectivity indicates the preference with which the process removes one acid gas component relative to (or in preference to) another. For example, some processes remove both hydrogen sulfide and carbon dioxide; other processes are designed to remove hydrogen sulfide only. It is important to consider the process selectivity for, say, hydrogen sulfide removal compared to carbon dioxide removal that ensures minimal concentrations of these components in the product, thus the need for consideration of the carbon dioxide to hydrogen sulfide in the gas stream. Gas processing involves the use of several different types of processes but there is always overlap between the various processing concepts. In addition, the terminology used for gas processing can often be confusing and/or misleading because of the overlap (Nonhebel, 1964; Curry, 1981; Maddox, 1982). There are four general processes used for emission control (often referred to in another, more specific context as flue gas desulfurization): (i) adsorption. (ii) absorption, (iii) catalytic oxidation, and (iv) thermal oxidation (Soud and Takeshita, 1994).

Adsorption is a physical-chemical phenomenon in which the gas is concentrated on the surface of a solid or liquid to remove impurities. Usually, carbon is the adsorbing medium, which can be regenerated upon *desorption* (Fulker, 1972; Mokhatab *et al.*, 2006; Speight, 2007). The quantity of material adsorbed is proportional to the surface area of the solid and, consequently, adsorbents are usually granular solids with a large surface area per unit mass. Subsequently, the captured gas can be desorbed with hot air or steam either for recovery or for thermal destruction.

Adsorbing units (adsorbers) are widely used to increase a low gas concentration prior to incineration unless the gas concentration is very high in the inlet air stream. Adsorption also is employed to reduce problem odors from gases. There are several limitations to the use of adsorption systems, but it is generally felt that the major one is the requirement for minimization of particulate matter and/or condensation of liquids (e.g., water vapor) that could mask the adsorption surface and drastically reduce its efficiency (Mokhatab *et al.*, 2006; Speight, 2014, 2019).

Absorption differs from *adsorption*, in that it is not a physical-chemical surface phenomenon, but an approach in which the absorbed gas is ultimately distributed throughout the absorbent (liquid). The process depends only on physical solubility and may include chemical reactions in the liquid phase (*chemisorption*). Common absorbing media used are water, aqueous amine solutions, caustic, sodium carbonate, and nonvolatile hydrocarbon oils, depending on the type of gas to be absorbed. Usually, the gas-liquid contactor designs which are employed are plate columns or packed beds (Mokhatab et al., 2006; Speight, 2014, 2019).

Absorption is achieved by dissolution (a physical phenomenon) or by reaction (a chemical phenomenon) (Barbouteau and Galaud, 1972; Ward, 1972; Mokhatab et al., 2006; Speight, 2007). Chemical adsorption processes adsorb sulfur dioxide onto a carbon surface where it is oxidized (by oxygen in the flue gas) and absorbs moisture to give sulfuric acid impregnated into and on the adsorbent.

Liquid absorption processes (which usually employ temperatures below 50°C (120°F) are classified either as *physical solvent processes* or *chemical solvent processes*. The former processes employ an organic solvent, and absorption is enhanced by low temperatures, or high pressure, or both. Regeneration of the solvent is often accomplished readily (Staton et al., 1985). In chemical solvent processes, absorption of the acid gases is achieved mainly by use of alkaline solutions such as amines or carbonates (Kohl and Riesenfeld, 1985). Regeneration (desorption) can be brought about by use of reduced pressures and/or high temperatures, whereby the acid gases are stripped from the solvent.

Solvents used for emission control processes should have (Speight, Mokhatab et al., 2006; Speight, 2007):

1. A high capacity for acid gas.
2. A low tendency to dissolve hydrogen.
3. A low tendency to dissolve low-molecular weight hydrocarbons.
4. Low vapor pressure at operating temperatures to minimize solvent losses;
5. Low viscosity.
6. Low thermal stability.
7. Absence of reactivity toward gas components.
8. Low tendency for fouling.
9. Low tendency for corrosion.
10. Economically acceptable.

Amine washing of gas emissions involves chemical reaction of the amine with any acid gases with the liberation of an appreciable amount of heat and it is necessary to compensate for the absorption of heat. Amine derivatives such as ethanolamine (monoethanolamine, MEA), diethanolamine (DEA), triethanolamine (TEA), methyldiethanolamine (MDEA), diisopropanolamine (DIPA), and diglycolamine (DGA) have been used in commercial applications (Katz, 1959; Kohl and Riesenfeld, 1985; Maddox et al., 1985; Polasek and Bullin, 1985; Jou et al., 1985; Pitsinigos and Lygeros, 1989; Speight, 1993; Mokhatab et al., 2006; Speight, 2014, 2019).

The chemistry can be represented by simple equations for low partial pressures of the acid gases:

$$2RNH_2 + H_2S \rightarrow (RNH_3)_2S$$

$$2RHN_2 + CO_2 + H_2O \rightarrow (RNH_3)_2CO_3$$

At high acid gas partial pressure, the reactions will lead to the formation of other products:

$$(RNH_3)_2S + H_2S \rightarrow 2RNH_3HS$$

$$(RNH_3)_2CO_3 + H_2O \rightarrow 2RNH_3HCO_3$$

The reaction is extremely fast, the absorption of hydrogen sulfide being limited only by mass transfer; this is not so for carbon dioxide.

Regeneration of the solution leads to near complete desorption of carbon dioxide and hydrogen sulfide. A comparison between monoethanolamine, diethanolamine, and diisopropanolamine shows that monoethanolamine is the cheapest of the three but shows the highest heat of reaction and corrosion; the reverse is true for diisopropanolamine.

Carbonate washing is a mild alkali process for emission control by the removal of acid gases (such as carbon dioxide and hydrogen sulfide) from gas streams (Speight, 1993) and uses the principle that the rate of absorption of carbon dioxide by potassium carbonate increases with temperature. It has been demonstrated that the process works best near the temperature of reversibility of the reactions:

$$K_2CO_3 + CO_2 + H_2O \rightarrow 2KHCO_3$$

$$K_2CO_3 + H_2S \rightarrow KHS + KHCO_3$$

Water washing, in terms of the outcome, is analogous to washing with potassium carbonate (Kohl and Riesenfeld, 1985), and it is also possible to carry out the desorption step by pressure reduction. The absorption is purely physical and there is also a relatively high absorption of hydrocarbons, which are liberated at the same time as the acid gases.

In *chemical conversion processes*, contaminants in gas emissions are converted to compounds that are not objectionable or that can be removed from the stream with greater ease than the original constituents. For example, a number of processes have been developed that remove hydrogen sulfide and sulfur dioxide from gas streams by absorption in an alkaline solution.

Catalytic oxidation is a chemical conversion process that is used predominantly for destruction of volatile organic compounds and carbon monoxide. These systems operate in a temperature regime of 205 to 595°C (400 to 1100°F) in the presence of a catalyst. Without the catalyst, the system would require higher temperatures. Typically, the catalysts used are a combination of noble metals deposited on a ceramic base in a variety of configurations (e.g., honeycomb-shaped) to enhance good surface contact.

Catalytic systems are usually classified on the basis of bed types such as *fixed bed* (or *packed bed*) and *fluid bed* (*fluidized bed*). These systems generally have very high destruction efficiencies for most volatile organic compounds, resulting in the formation of carbon dioxide, water, and varying amounts of hydrogen chloride (from halogenated hydrocarbons). The presence in emissions of chemicals such as heavy metals, phosphorus, sulfur, chlorine, and most halogens in the incoming air stream act as poison to the system and can foul up the catalyst.

Thermal oxidation systems, without the use of catalysts, also involve chemical conversion (more correctly, chemical destruction) and operate at temperatures in excess of 815°C (1500°F), or 220 to 610°C (395 to 1100°F) higher than catalytic systems.

Historically, *particulate matter control* (*dust control*) (Mody and Jakhete, 1988) has been one of the primary concerns of industries, since the emission of particulate matter is readily observed through the deposition of fly ash and soot as well as in impairment of visibility. Differing ranges of control can be achieved by use of various types of equipment. Upon proper characterization of the particulate matter emitted by a specific process, the appropriate piece of equipment can be selected, sized, installed, and performance tested. The general classes of control devices for particulate matter are as follows:

Cyclone collectors are the most common of the inertial collector class. Cyclones are effective in removing coarser fractions of particulate matter. The particle-laden gas stream enters an upper cylindrical section tangentially and proceeds downward through a conical section. Particles migrate by centrifugal force generated by providing a path for the carrier gas to be subjected to a vortex-like spin. The particles are forced to the wall and are removed through a seal at the apex of the inverted cone. A reverse-direction vortex moves upward through the cyclone and discharges through a top center opening. Cyclones are often used as primary collectors because of their relatively low efficiency (50-90% is usual). Some small-diameter high-efficiency cyclones are utilized. The equipment can be arranged either in parallel or in series to both increase efficiency and decrease pressure drop. However, there are disadvantages that must be recognized (Mokhatab *et al.*, 2006; Speight, 2014, 2019). These units for particulate matter operate by contacting the particles in the gas stream with a liquid. In principle the particles are incorporated in a liquid bath or in liquid particles which are much larger and therefore more easily collected.

Fabric filters are typically designed with non-disposable filter bags. As the dusty emissions flow through the filter media (typically cotton, polypropylene, Teflon, or fiberglass), particulate matter is collected on the bag surface as a dust cake. Fabric filters are generally classified on the basis of the filter bag cleaning mechanism employed. Fabric filters operate with collection efficiencies up to 99.9% although other advantages are evident. There are several issues that arise during use of such equipment (Mokhatab *et al.*, 2006; Speight, 2014, 2019).

Wet scrubbers are devices in which a countercurrent spray liquid is used to remove particles from an air stream. Device configurations include plate scrubbers, packed beds, orifice scrubbers, venturi scrubbers, and spray towers, individually or in various combinations. Wet scrubbers can achieve high collection efficiencies at the expense of prohibitive pressure drops (Mokhatab *et al.*, 2006; Speight, 2014, 2019).

Other methods include use of high-energy input *venturi scrubbers* or electrostatic scrubbers where particles or water droplets are charged, and flux force/condensation scrubbers where a hot humid gas is contacted with cooled liquid or where steam is injected into saturated gas. In the latter scrubber the movement of water vapor toward the cold water surface carries the particles with it (*diffusiophoresis*), while the condensation of water vapor on the particles causes the particle size to increase, thus facilitating collection of fine particles.

The *foam scrubber* is a modification of a wet scrubber in which the particle-laden gas is passed through a foam generator, where the gas and particles are enclosed by small bubbles of foam.

Electrostatic precipitators (Mokhatab *et al.*, 2006; Speight, 2014, 2019) operate on the principle of imparting an electric charge to particles in the incoming air stream, which are then collected on an oppositely charged plate across a high-voltage field. Particles of high resistivity create the most difficulty in collection. Conditioning agents such as sulfur trioxide (SO_3) have been used to lower resistivity.

Important parameters include design of electrodes, spacing of collection plates, minimization of air channeling, and collection-electrode rapping techniques (used to dislodge particles). Techniques under study include the use of high-voltage pulse energy to enhance particle charging, electron-beam ionization, and wide plate spacing. Electrical precipitators are capable of efficiencies >99% under optimum conditions, but performance is still difficult to predict in new situations.

9.3 Synthesis Gas Cleaning

Whatever the source of the synthesis gas, methane (in varying amounts) is the predominant hydrocarbon derivative in the gas stream methane. However, there are other hydrocarbon derivatives in addition to water vapor, hydrogen sulfide (H_2S), carbon dioxide, nitrogen, and other compounds.

Treated synthesis gas consists mainly of methane (often referred to as biomethane). However, synthesis gas is not pure methane, and its properties are modified by the presence of impurities, such as nitrogen, carbon dioxide, and small amounts of unrecovered higher boiling (non-gaseous at STP) hydrocarbon derivatives. An important property of any gas stream is its heating value – relatively high amounts of nitrogen and/or carbon dioxide reduce the heating value of the gas. Synthesis gas, if it was pure methane, would have a heating value of 1,010 Btu/ft^3. This value is reduced to approximately 900 Btu/ft^3 if the gas contains approximately 10% v/v nitrogen and carbon dioxide – the heating value of carbon dioxide is zero.

In order to remove the unwanted byproducts and contaminants in the product gas, primary treatments, such as optimization of the properties of biomass feedstock, design of the gasifier, and the operational parameters of the gasifier are first implemented. Secondary treatment, such as a downstream cleaning system based on physical (scrubbers, filters) or catalytic strategies, have to be incorporated for the hot gas cleaning to achieve a more satisfactory reduction for end-applications and also to meet sales specification, which include meeting environmental regulations.

Thus, to reach the condition of acceptability, the synthesis gas must be the end-product of being treated by a series of processes that have successfully removed the contaminants that are present in the raw (untreated) gas. Gas processing consists of separating all of the non-methane derivatives (both the combustible and non-combustible environmentally unfriendly and performance-unfriendly constituents) and fluids from the gas (Kidnay and Parrish, 2006; Mokhatab *et al.*, 2006; Speight, 2014, 2019). While often assumed to be hydrocarbon derivatives in nature, there are also components of the gaseous products that must be removed prior to release of the gases to the atmosphere or prior to use of the gas, i.e., as a fuel gas or as a process feedstock.

9.3.1 Composition

Trace quantities of contaminants in hydrocarbon products can be harmful to many catalytic chemical processes in which these products are used. In general, the raw product gas of biomass gasification contains a range of minor species and contaminants, including particles, tar, alkali metals, chlorine compounds such as hydrogen chloride

(HCl), nitrogen compounds such as ammonia (NH_3) and hydrogen cyanide (HCN), sulfur compounds such as hydrogen sulfide (H_2S) and carbonyl sulfide (COS), as well other species.

The maximum permissible levels of total contaminants are normally included in specifications for such hydrocarbon derivatives. It is recommended that this test method be used to provide a basis for agreement between two laboratories when the determination of sulfur in hydrocarbon gases is important. In the case of liquefied petroleum gas (LPG), total volatile sulfur is measured on an injected gas sample. One test method (ASTM D3246) describes a procedure for the determination of sulfur in the range from 1.5 to 100 mg/kg (ppm w/w) in hydrocarbon products that are gaseous at normal room temperature and pressure. There is also a variety of other standard test methods that can be applied to the determination of the properties of synthesis gas and, hence the suitability of the gas for use (ASTM, 2018; Speight, 2018).

Acidic constituents such as carbon dioxide and hydrogen sulfide as well as thiol derivatives (RSH, also called mercaptan derivatives, RSH) can contribute to corrosion of refining equipment, harm catalysts, pollute the atmosphere, and prevent the use of hydrocarbon components in petrochemical manufacture (Mokhatab *et al.*, 2006; Speight, 2014, 2019). When the amount of hydrogen sulfide is high, it may be removed from a gas stream and converted to sulfur or sulfuric acid; a recent option for hydrogen sulfide removal is the use of chemical scavengers (Kenreck, 2014). Some gases contain sufficient carbon dioxide to warrant recovery as dry ice (Bartoo, 1985).

Synthesis gas is not always truly hydrocarbon in nature and may contain contaminants, such as carbon oxides (CO_x, where x = 1 and/or 2), sulfur oxides (SO_x, where x = 2 and/or 3), as well as ammonia (NH_3), mercaptan derivatives (RSH), carbonyl sulfide (COS), and mercaptan derivatives (RSH). The presence of these impurities may eliminate some of the sweetening processes from use since some of these processes remove considerable amounts of acid gas but not to a sufficiently low concentration. On the other hand, there are those processes not designed to remove (or incapable of removing) large amounts of acid gases whereas they are capable of removing the acid gas impurities to very low levels when the acid gases are present only in low-to-medium concentration in the gas (Katz, 1959; Mokhatab *et al.*, 2006; Speight, 2014, 2019).

The sources of the various synthesis gas stream are varied but, in terms of gas cleaning (i.e., removal of the contaminants before petrochemical production), the processes are largely the same but it is a question of degree. For example, gas streams for some sources may produce gases that may contain higher amounts of carbon dioxide and/or hydrogen sulfide and the processes will have to be selected accordingly.

In addition to its primary importance as a fuel, synthesis gas is also a source of hydrocarbon derivatives for petrochemical feedstocks. While the major hydrocarbon constituent of synthesis gas is methane, there are components such as carbon dioxide (CO_2), hydrogen sulfide (H_2S), and mercaptan derivatives (thiols, also called mercaptans, RSH), as well as trace amounts of sundry other emissions such as carbonyl sulfide (COS). The fact that methane has a foreseen and valuable end-use makes it a desirable product, but in several other situations it is considered a pollutant, having been identified as a greenhouse gas.

In practice, heaters and scrubbers are usually installed at or near to the source. The scrubbers serve primarily to remove sand and other large-particle impurities and the heaters ensure that the temperature of the gas does not drop too low. With gas that contains even low quantities of water, gas hydrates ($C_nH_{2n+2} \cdot xH_2O$) tend to form when temperatures drop. These hydrates are solid or semi-solid compounds, resembling ice like crystals. If the hydrates accumulate, they can impede the passage of gas through valves and gathering systems (Zhang *et al.*, 2007). To reduce the occurrence of hydrates small gas-fired heating units are typically installed along the gathering pipe wherever it is likely that hydrates may form.

9.3.2 Process Types

Synthesis gas can also be upgraded to pipeline-quality gas and this upgraded gas may be used for residential heating and as vehicle fuel. When distributing the synthesis gas using pipelines, gas pipeline standards will, more than likely, become applicable. Removing water vapor is easier than removing carbon dioxide and hydrogen sulfide from synthesis gas (Table 9.1). A condensate trap at a proper location on the gas pipeline can remove water vapor as warm synthesis gas cools by itself after leaving the digester or the gas-generating equipment.

Gas processing involves the use of several different types of processes to remove contaminants from gas streams but there is always overlap between the various processing concepts. In addition, the terminology used for gas processing can often be confusing and/or misleading because of the overlap (Curry, 1981; Maddox, 1982). Gas processing is necessary to ensure that the gas prepared for transportation (usually by pipeline) and for sales must be as clean and pure as the specifications dictate. Thus, synthesis gas, as it is used by consumers, is much different from the gas that is brought from, for example, the anaerobic digester (Chapter 3). Moreover, although gas produced in the digester may be composed primarily of methane, is by no means as pure.

The processes that have been developed to accomplish gas purification vary from a simple once-through wash operation to complex multi-step recycling systems (Abatzoglou and Boivin, 2009; Mokhatab *et al.*, 2006; Speight, 2014, 2019). In many cases, the process complexities arise because of the need for recovery of the materials used to remove the contaminants or even recovery of the contaminants in the original, or altered, form (Katz, 1959; Kohl and Riesenfeld, 1985; Newman, 1985; Mokhatab *et al.*, 2006; Speight, 2014, 2015, 2019). In addition to the corrosion of equipment by acid gases (Speight, 2014) the escape into the atmosphere of sulfur-containing gases can eventually lead to the formation of the constituents of acid rain, i.e., the oxides of sulfur (sulfur dioxide, SO_2, and sulfur trioxide, SO_3). Similarly, the nitrogen-containing gases can also lead to nitrous and nitric acids (through the formation of the oxides NO_x, where x = 1 or 2) which are the other major contributors to acid rain. The release of carbon dioxide and hydrocarbon derivatives as constituents of refinery effluents can also influence the behavior and integrity of the ozone layer.

Finally, throughout the various cleaning processes, it is important to minimize the loss of methane in order to achieve an economically viable cleaning operation. However, it is also important to minimize the methane slip since methane is a strong greenhouse gas. Thus, the release of methane to the atmosphere should be minimized by treating the off-gas, air or water streams leaving the plant even though the methane cannot be

Table 9.1 Common methods for the removal of carbon dioxide and hydrogen sulfide from gas streams.

Carbon dioxide removal	
Water scrubbing	Uses principle of the higher solubility of carbon dioxide in water to separate the carbon dioxide from the gas.
	The process uses high pressure and removes hydrogen sulfide as well as carbon dioxide.
	The main disadvantage of this process is that it requires a large volume of water that must be purified and recycled.
Polyethylene glycol scrubbing	This process is more efficient than water scrubbing.
	It also requires the regeneration of a large volume of polyethylene glycol.
Chemical absorption	Chemical reaction between carbon dioxide and amine-based solvents (olamines)
Carbon molecular sieves	Uses differential adsorption characteristics to separate methane and carbon dioxide, which is carried out at high pressure (pressure-swing adsorption).
	Hydrogen sulfide should be removed before the adsorption process.
Membrane separation	Selectively separates hydrogen sulfide from carbon dioxide from methane.
	The carbon dioxide and the hydrogen sulfide dissolve while the methane is collected for use.
Cryogenic separation	Cooling until condensation or sublimation of the carbon dioxide.
Hydrogen sulfide removal	
Biological desulfurization	Natural bacteria can convert hydrogen sulfide into elemental sulfur in the presence of oxygen and iron. This can be done by introducing a small amount (2 to 5%) of air into the head space of the digester. As a result, deposits of elemental sulfur will be formed in the digester.
	This process may be optimized by a more sophisticated design where air is bubbled through the digester feed material. It is critical that the introduction of the air be carefully controlled to avoid reducing the amount of gas that is produced.
Iron/iron oxide reaction	Hydrogen sulfide reacts readily with either iron oxide or iron chloride to form insoluble iron sulfide.

(Continued)

Table 9.1 Common methods for the removal of carbon dioxide and hydrogen sulfide from gas streams. (*Continued*)

	The reaction can be exploited by adding the iron chloride to the digester feed material or passing the gas through a bed of iron oxide-containing material.
	The iron oxide media needs to be replaced periodically.
	The regeneration process is highly exothermic and must be controlled to avoid problems.
Activated carbon	Activated carbon impregnated with potassium iodide can catalytically react with oxygen and hydrogen sulfide to form water and sulfur.
	The activated carbon beds need regeneration or replacement when saturated.
Scrubbing/membrane separation	The carbon dioxide and hydrogen sulfide can be removed by washing with water, glycol solutions, or separated using the membrane technique.

utilized. Methane can be present in the off-gas leaving a pressure-swing adsorption-column, in air from a water scrubber with water recirculation or in water in a water scrubber without water recirculation. The off-gas from an upgrading plant seldom contains a high enough methane concentration to maintain a flame without addition of natural gas or synthesis gas. One way of limiting the methane slip is to mix the off-gas with air that is used for combustion. Alternatively the methane can be oxidized by thermal or catalytic oxidation. Thus:

$$CH_4 + O_2 \rightarrow CO_2 + H_2O$$

In the catalytic process, the oxidation takes place at the surface of the catalyst which lowers the energy needed to oxidize the methane, thus enabling oxidation at a lower temperature. The active component of the catalyst is platinum, palladium or cobalt.

9.4 Water Removal

The relative humidity of synthesis gas inside the digester is 100%, so the gas is saturated with water vapors. To protect the energy conversion equipment from wear and from eventual damage, water must be removed from the produced synthesis gas. The quantity of water contained by synthesis gas depends on temperature. A part of the water vapor can be condensed by cooling of the gas. This is frequently done in the gas pipelines transporting synthesis gas from digester to CHP unit. The water condensates on the walls of the sloping pipes and can be collected in a condensation separator, at the lowest point of the pipeline.

Water in the liquid phase causes corrosion or erosion problems in pipelines and equipment, particularly when carbon dioxide and hydrogen sulfide are present in the gas. The simplest method of water removal (refrigeration or cryogenic separation) is to cool the gas to a temperature at least equal to or (preferentially) below the dew point (Mokhatab *et al.*, 2006; Speight, 2014, 2019).

In addition to separating any condensate from the wet gas stream, it is necessary to remove most of the associated water. Most of the liquid, free water associated with the extracted gas stream is removed by simple separation methods but, however, the removal of the water vapor that exists in solution in a gas stream requires a more complex treatment. This treatment consists of dehydrating the gas stream, which usually involves one of two processes: either absorption, or adsorption.

Moisture may be removed from hydrocarbon gases at the same time as hydrogen sulfide is removed. Moisture removal is necessary to prevent harm to anhydrous catalysts and to prevent the formation of hydrocarbon hydrates (e.g., $C_3H_8.18H_2O$) at low temperatures. A widely used dehydration and desulfurization process is the glycolamine process, in which the treatment solution is a mixture of ethanolamine and a large amount of glycol. The mixture is circulated through an absorber and a reactivator in the same way as ethanolamine is circulated in the Girbotol process. The glycol absorbs moisture from the hydrocarbon gas passing up the absorber; the ethanolamine absorbs hydrogen sulfide and carbon dioxide. The treated gas leaves the top of the absorber; the spent ethanolamine-glycol mixture enters the reactivator tower, where heat drives off the absorbed acid gases and water.

Absorption occurs when the water vapor is taken out by a dehydrating agent. Adsorption occurs when the water vapor is condensed and collected on the surface. In a majority of cases, cooling alone is insufficient and, for the most part, impractical for use in field operations. Other more convenient water removal options use (i) *hygroscopic* liquids (e.g., diethylene glycol or triethylene glycol) and (ii) solid adsorbents or desiccants (e.g., alumina, silica gel, and molecular sieves). Ethylene glycol can be directly injected into the gas stream in refrigeration plants.

9.4.1 Absorption

An example of absorption dehydration is known as *glycol dehydration* – the principal agent in this process is diethylene glycol which has a chemical affinity for water (Mokhatab *et al.*, 2006; Abdel-Aal *et al.*, 2016; Speight, 2019). Glycol dehydration involves using a solution of a glycol such as diethylene glycol (DEG) or triethylene glycol (TEG), which is brought into contact with the wet gas stream in a *contactor*. In practice, absorption systems recover 90 to 99% by volume of methane that would otherwise be flared into the atmosphere.

In the process, a liquid desiccant dehydrator serves to absorb water vapor from the gas stream. The glycol solution absorbs water from the wet gas and, once absorbed, the glycol particles become heavier and sink to the bottom of the contactor where they are removed. The dry gas stream is then transported out of the dehydrator. The glycol solution, bearing all of the water stripped from the gas stream, is recycled through a specialized boiler designed to vaporize only the water out of the solution. The boiling point differential between water (100°C, 212°F) and glycol (204°C, 400°F) makes it relatively easy to remove water from the glycol solution.

As well as absorbing water from the wet gas stream, the glycol solution occasionally carries with it small amounts of methane and other compounds found in the wet gas. In order to decrease the amount of methane and other compounds that would otherwise be lost, flash tank separator-condensers are employed to remove these compounds before the glycol solution reaches the boiler. The flash tank separator consists of a device that reduces the pressure of the glycol solution stream, allowing the methane and other hydrocarbon derivatives to vaporize (*flash*). The glycol solution then travels to the boiler, which may also be fitted with air or water cooled condensers, which serve to capture any remaining organic compounds that may remain in the glycol solution. The regeneration (stripping) of the glycol is limited by temperature: diethylene glycol and triethylene glycol decompose at or even before their respective boiling points. Such techniques as stripping of hot triethylene glycol with dry gas (e.g., heavy hydrocarbon vapors, the *Drizo process*) or vacuum distillation are recommended.

Another absorption process, the Rectisol process, is a physical acid gas removal process using an organic solvent (typically methanol) at subzero temperatures, and characteristic of physical acid gas removal processes, it can purify synthesis gas down to 0.1 ppm v/v total sulfur, including hydrogen sulfide (H_2S) and carbonyl sulfide (COS), and carbon dioxide (CO_2) in the ppm v/v range (Mokhatab *et al.*, 2006; Liu *et al.*, 2010; Abdel-Aal *et al.*, 2016). The process uses methanol as a wash solvent and the wash unit operates under favorable at temperatures below 0°C (32°F). To lower the temperature of the feed gas temperatures, it is cooled against the cold-product streams, before entering the absorber tower. At the absorber tower, carbon dioxide and hydrogen sulfide (with carbonyl sulfide) are removed. By use of an intermediate flash, co-absorbed products such as hydrogen and carbon monoxide are recovered, thus increasing the product recovery rate. To reduce the required energy demand for the carbon dioxide compressor, the carbon dioxide product is recovered in two different pressure steps (medium pressure and lower pressure). The carbon dioxide product is essentially sulfur-free (hydrogen sulfide-free, carbonyl sulfide-free) and water free. The carbon dioxide products can be used for enhanced oil recovery (EOR) and/or sequestration or as pure carbon dioxide for other processes.

9.4.2 Adsorption

Adsorption is a physical-chemical phenomenon in which the gas is concentrated on the surface of a solid or liquid to remove impurities. It must be emphasized that *adsorption* differs from *absorption* in that absorption is not a physical-chemical surface phenomenon but a process in which the absorbed gas is ultimately distributed throughout the absorbent (liquid). Dehydration using a solid adsorbent or solid-desiccant is the primary form of dehydrating gas stream using adsorption, and usually consists of two or more adsorption towers, which are filled with a solid desiccant (Mokhatab *et al.*, 2006; Abdel-Aal *et al.*, 2016; Speight, 2019). Typical desiccants include activated alumina or a granular silica gel material. A wet gas stream is passed through these towers, from top to bottom. As the wet gas passes around the particles of desiccant material, water is retained on the surface of these desiccant particles. Passing through the entire desiccant bed, almost all of the water is adsorbed onto the desiccant material, leaving the dry gas to exit the bottom of the tower. There are several solid desiccants which possess the physical characteristic to adsorb water from the gas stream. These desiccants are generally used in dehydration systems consisting of two or more towers and associated regeneration equipment.

Molecular sieves – a class of aluminosilicates which produce the lowest water dew points and which can be used to simultaneously sweeten dry gases and liquids (Mokhatab *et al.*, 2006; Maple, and Williams 2008; Abdel-Aal *et al.*, 2016; Speight, 2019) – are commonly used in dehydrators ahead of plants designed to recover ethane and other higher-boiling hydrocarbon derivatives. These plants operate at very cold temperatures and require very dry feed gas to prevent formation of hydrates. Dehydration to -100°C (-148°F) dew point is possible with molecular sieves. Water dew points less than -100°C (-148°F) can be accomplished with special design and definitive operating parameters (Mokhatab *et al.*, 2006).

Molecular sieves are commonly used to selectively adsorb water and sulfur compounds from light hydrocarbon streams such as liquefied petroleum gas, propane, butane, pentane, light olefin derivatives, and alkylation feed. Sulfur compounds that can be removed are hydrogen sulfide, mercaptan derivatives, sulfide derivatives, and disulfide derivatives. In the process, the sulfur-containing feedstock is passed through a bed of sieves at ambient temperature. The operating pressure must be high enough to keep the feed in the liquid phase. The operation is cyclic in that the adsorption step is stopped at a predetermined time before sulfur breakthrough occurs. Sulfur and water are removed from the sieves by purging with fuel gas at 205 to 315°C (400 to 600°F).

Solid-adsorbent dehydrators are typically more effective than liquid absorption dehydrators (e.g., glycol dehydrators) and are usually installed as a type of straddle system along gas pipelines. These types of dehydration systems are best suited for large volumes of gas under very high pressure, and are thus usually located on a pipeline downstream of a compressor station. Two or more towers are required due to the fact that after a certain period of use, the desiccant in a particular tower becomes saturated with water. To regenerate and recycle the desiccant, a high-temperature heater is used to heat gas to a very high temperature and passage of the heated gas stream through a saturated desiccant bed vaporizes the water in the desiccant tower, leaving it dry and allowing for further gas stream dehydration.

Although two-bed adsorbent treaters have become more common (while one bed is removing water from the gas, the other undergoes alternate heating and cooling), on occasion, a three-bed system is used: one bed adsorbs, one is being heated, and one is being cooled. An additional advantage of the three-bed system is the facile conversion of a two-bed system so that the third bed can be maintained or replaced, thereby ensuring continuity of the operations and reducing the risk of a costly plant shutdown.

Silica gel (SiO_2) and alumina (Al_2O_3) have good capacities for water adsorption (up to 8% by weight). Bauxite (crude alumina, Al_2O_3) adsorbs up to 6% by weight water, and molecular sieves adsorb up to 15% by weight water. Silica is usually selected for dehydration of sour gas because of its high tolerance to hydrogen sulfide and to protect molecular sieve beds from plugging by sulfur. Alumina *guard beds* serve as protectors by the act of attrition and may be referred to as an *attrition reactor* containing an *attrition catalyst* (Speight, 2000, 2014, 2017) may be placed ahead of the molecular sieves to remove the sulfur compounds. Downflow reactors are commonly used for adsorption processes, with an upward flow regeneration of the adsorbent and cooling using gas flow in the same direction as adsorption flow.

278 Synthesis Gas

Solid desiccant units generally cost more to buy and operate than glycol units. Therefore, their use is typically limited to applications such as gases having a high hydrogen sulfide content, very low water dew point requirements, simultaneous control of water, and hydrocarbon dew points. In processes where cryogenic temperatures are encountered, solid desiccant dehydration is usually preferred over conventional methanol injection to prevent hydrate and ice formation (Kidnay and Parrish, 2006).

9.4.3 Cryogenics

Another possibility of synthesis gas drying is by cooling the gas in electrically powered gas coolers, at temperatures below 10°C (50°F), which allows a lot of humidity to be removed. In order to minimize the relative humidity, but not the absolute humidity, the gas can be warmed up again after cooling, in order to prevent condensation along the gas pipelines.

Thus, cryogenic upgrading makes use of the distinct boiling/sublimation points of the different gases particularly for the separation of carbon dioxide and methane. In the process, the raw (untreated) synthesis gas is cooled down to the temperatures where the carbon dioxide in the gas condenses or sublimates and can be separated as a liquid or a solid fraction, while methane accumulates in the gas phase. Water and siloxane derivatives are also removed during cooling of the gas. However, the content of methane in the synthesis gas affects the characteristics of the gas, i.e., higher pressures and/or lower temperatures are needed to condense or sublime carbon dioxide when it is in a mixture with methane. Cooling usually takes place in several steps in order to remove the different gases in the synthesis gas individually and to optimize the energy recovery.

As an example, a hot synthesis gas stream containing water and carbon dioxide can be upgraded by cooling the gas to 40°C (104°F); most of the water is condensed. The remaining water is removed in the carbon dioxide removal step – the carbon dioxide to be removed from the gas stream to meet the specifications. The final concentration of carbon dioxide in the gas stream is determined by the specification of the Wobbe index. For carbon dioxide removal, considering the high partial pressure of carbon dioxide both membranes and physical solvents can be chosen, where membranes are at their maximum scale and physical solvents are at their minimum scale.

The main advantage of the cryogenic process is that it is possible to obtain synthesis gas with high methane content of up to 99% v/v. The main disadvantage is that for the upgrading process it is necessary to use much technological equipment, especially compressors, turbines and heat exchangers. This significant demand for additional equipment can make the cryogenic separation process extremely expensive.

9.5 Acid Gas Removal

In addition to water removal, one of the most important parts of gas processing involves the removal of hydrogen sulfide and carbon dioxide, which are generally referred to as contaminants. Synthesis gas from some sources contains significant amounts of hydrogen sulfide and carbon dioxide and, analogous to natural gas that contains these same impurities, is usually referred to as *sour gas*. Sour gas is undesirable because the sulfur compounds it contains can be extremely harmful, even lethal, to breathe and the gas can also be extremely corrosive. The process for removing hydrogen sulfide from sour gas is commonly referred to as *sweetening* the gas. Thus, sulfur compounds, mainly hydrogen sulfide, must be removed from the syngas to the best possible extent since they can poison catalysts. The methanation catalyst is particularly prone to sulfur poisoning. Regenerative sorbents (so-called sulfur guards) can be used to reduce sulfur concentrations to the necessary limits, well below 1 ppm v/v. Washing techniques (based on physical or chemical adsorption) can also be implemented, making sulfur recovery via a Claus process possible if economically viable.

There are four general processes used for emission control (often referred to in another, more specific context as flue gas desulfurization): (i) physical adsorption in which a solid adsorbent is used, (ii) physical absorption in which a selective absorption solvent is used, (iii) chemical absorption is which a selective absorption solvent is used, (iv) and catalytic oxidation thermal oxidation (Soud and Takeshita, 1994; Mokhatab *et al.*, 2006; Speight, 2014, 2019).

9.5.1 Adsorption

Adsorption is a physical-chemical phenomenon in which the gas is concentrated on the surface of a solid or liquid to remove impurities. Activated carbon and molecular sieves are also capable of adsorbing water in addition to the acid gases.

In order to avoid any confusion, it must be emphasized here that *absorption* differs from *adsorption* in that absorption is not a physical-chemical surface phenomenon but a process in which the absorbed gas is ultimately distributed throughout the absorbent (liquid). The process depends only on physical solubility and may include chemical reactions in the liquid phase (*chemisorption*). Common absorbing media used are water, aqueous amine solutions, caustic, sodium carbonate, and nonvolatile hydrocarbon oils, depending on the type of gas to be absorbed. In these processes, a solid with a high surface area is used. Molecular sieves (zeolites) are widely used and are capable of adsorbing large amounts of gas. In practice, more than one adsorption bed is used for continuous operation – one bed is in use while the other is being regenerated.

On the other hand, adsorption is usually a gas-solid interaction in which an adsorbent such as activated carbon (the *adsorbent* or *adsorbing medium*) which can be regenerated upon *desorption* (Mokhatab *et al.*, 2006; Boulinguiez and Le Cloirec, 2009; Speight, 2014, 2019). The quantity of material adsorbed is proportional to the surface area of the solid and, consequently, adsorbents are usually granular solids with a large surface area per unit mass. Activated carbon impregnated with potassium iodide can catalytically react with oxygen and hydrogen sulfide to form water and sulfur. The reaction is best achieved at 50 to 70°C (122 to 158°F) and 100 to 120 psi. The activated carbon adsorption beds must be regenerated or replaced before they become ineffective due to over-saturation.

Regeneration is accomplished by passing hot dry fuel gas through the bed. Molecular sieves are competitive only when the quantities of hydrogen sulfide and carbon disulfide are low. The captured (adsorbed) gas can be desorbed with hot air or steam either for recovery or for thermal destruction. Adsorber units are widely used to increase a low gas concentration prior to incineration unless the gas concentration is very high in the inlet air stream and the process is also used to reduce problem odors (or obnoxious odors) from gases. There are several limitations to the use of adsorption systems, but it is generally the case that the major limitation is the requirement for minimization of particulate matter and/or condensation of liquids (e.g., water vapor) that could mask the adsorption surface and drastically reduce its efficiency.

9.5.2 Absorption

Absorption is achieved by dissolution (a physical phenomenon) or by reaction (a chemical phenomenon) (Barbouteau and Galaud, 1972; Mokhatab *et al.*, 2006; Speight, 2014, 2019). In addition to economic issues or constraints, the solvents used for gas processing should have: (i) a high capacity for acid gas, (ii) a low tendency to dissolve hydrogen, (iii) a low tendency to dissolve low-molecular weight hydrocarbon derivatives, (iv) low vapor pressure at operating temperatures to minimize solvent losses, (v) low viscosity, (vi) low thermal stability, (vii) absence of reactivity toward gas components, (viii) low tendency for fouling, and (ix) a low tendency for corrosion, and (x) economically acceptable (Mokhatab *et al.*, 2006; Speight, 2014, 2019).

Noteworthy commercial processes used are the Selexol process, the Sulfinol process, and the Rectisol process (Mokhatab *et al.*, 2006; Speight, 2014, 2019). In these processes, no chemical reaction occurs between the acid gas and the solvent. The solvent, or absorbent, is a liquid that selectively absorbs the acid gases and leaves out the hydrocarbons.

The Selexol process uses a mixture of the dimethyl ether of propylene glycol as a solvent. It is nontoxic and its boiling point is not high enough for amine formulation. The selectivity of the solvent for hydrogen sulfide (H_2S) is much higher than that for carbon dioxide (CO_2), so it can be used to selectively remove these different acid gases, minimizing carbon dioxide content in the hydrogen sulfide stream sent to the sulfur recovery unit (SRU) and enabling regeneration of solvent for carbon dioxide recovery by economical flashing. In the process, a synthesis gas stream is injected in the bottom of the absorption tower operated at 1000 psi. The rich solvent is flashed in a flash drum (flash reactor) at 200 psi where methane is flashed and recycled back to the absorber and joins the sweet (low-sulfur or no-sulfur) gas stream. The solvent is then flashed at atmospheric pressure and acid gases are flashed off. The solvent is then stripped by steam to completely regenerate the solvent, which is recycled back to the absorber. Any hydrocarbon derivatives are condensed and any remaining acid gases are flashed from the condenser drum. This process is used when there is a high acid gas partial pressure and no heavy hydrocarbon derivatives. Diisopropanolamine (DIPA) can be added to this solvent to remove carbon dioxide to a level suitable for pipeline transportation.

Noteworthy commercial processes commercially used are the Selexol, the Sulfinol, and the Rectisol processes. In these processes, no chemical reaction occurs between the acid gas and the solvent. The solvent, or absorbent, is a liquid that selectively absorbs the acid gases and leaves out the hydrocarbon derivatives. In the Selexol process for example, the solvent is dimethyl ether or polyethylene glycol. Raw synthesis gas passes countercurrently to the descending solvent. When the solvent becomes saturated with the acid gases, the pressure is reduced, and hydrogen sulfide and carbon dioxide are desorbed. The solvent is then recycled to the absorption tower.

The Sulfinol process uses a solvent that is a composite solvent, consisting of a mixture of diisopropanolamine (30 to 45% v/v) or methyl diethanolamine (MDEA), sulfolane (tetrahydrothiophene dioxide) (40 to 60% v/v), and water (5 to 15% v/v). The acid gas loading of the Sulfinol solvent is higher and the energy required for its regeneration is lower than those of purely chemical solvents. At the same time, it has the advantage over purely physical solvents that severe product specifications can be met more easily and co-absorption of hydrocarbon derivatives is relatively low. Aromatic compounds, higher molecular weight hydrocarbon derivatives, and carbon dioxide are soluble to a lesser extent. The process is typically used when the hydrogen sulfide-carbon dioxide ratio is greater than 1:1 or where carbon dioxide removal is not required to the same extent as hydrogen sulfide removal. The process uses a conventional solvent absorption and regeneration cycle in which the sour gas components are removed from the feed gas by countercurrent contact with a lean solvent stream under pressure. The absorbed impurities are then removed from the rich solvent by stripping with steam in a heated regenerator column. The hot lean solvent is then cooled for reuse in the absorber. Part of the cooling may be by heat exchange with the rich solvent for partial recovery of heat energy. The solvent reclaimer is used in a small ancillary facility for recovering solvent components from higher boiling products of alkanolamine degradation or from other high-boiling or solid impurities.

The big difference between water scrubbing and the Selexol process is that carbon dioxide and hydrogen sulfide are more soluble in Selexol which results in a lower solvent demand and reduced pumping. In addition, water and halogenated hydrocarbon derivatives (contaminants in synthesis gas from landfills) are removed when scrubbing synthesis gas with Selexol. Selexol scrubbing is always designed with recirculation. Due to formation of elementary sulfur stripping the Selexol solvent with air is not recommended but with steam or inert gas (upgraded synthesis gas or natural gas). Removing hydrogen sulfide on beforehand is an alternative.

9.5.3 Chemisorption

Chemisorption (chemical absorption) processes are characterized by a high capability of absorbing large amounts of acid gases. They use a solution of a relatively weak base, such as monoethanolamine. The acid gas forms a weak bond with the base which can be regenerated easily. Mono- and diethanolamine derivatives are frequently used for this purpose. The amine concentration normally ranges between 15 and 30%. The gas stream is passed through the amine solution where sulfides, carbonates, and bicarbonates are formed. Diethanolamine is a favored absorbent due to its lower corrosion rate, smaller amine loss potential, fewer utility requirements, and minimal reclaiming needs. Diethanolamine also reacts reversibly with 75% of carbonyl sulfides (COS), while the mono- reacts irreversibly with 95% of the carbonyl sulfide and forms a degradation product that must be disposed in an environmentally acceptable manner.

The ethanolamine process, known as the *Girbotol* process, removes acid gases (hydrogen sulfide and carbon dioxide) from gases. The Girbotol process uses an aqueous solution of ethanolamine ($H_2NCH_2CH_2OH$) that reacts with hydrogen sulfide at low temperatures and releases hydrogen sulfide at high temperatures. The ethanolamine solution fills a tower (the absorber) through which the sour gas is bubbled. Purified gas leaves the top of the tower, and the ethanolamine solution leaves the bottom of the tower with the absorbed acid gases. The ethanolamine solution enters a reactivator tower where heat drives the acid gases from the solution. Ethanolamine solution, restored to its original condition, leaves the bottom of the reactivator tower to go to the top of the absorber tower, and acid gases are released from the top of the reactivator.

Alkanolamine scrubbers can be used to remove hydrogen sulfide or hydrogen sulfide and carbon dioxide simultaneously (Katz, 1959; Kohl and Riesenfeld, 1985; Maddox et al., 1985; Polasek and Bullin, 1985; Jou et al., 1985; Pitsinigos and Lygeros, 1989; Kohl and Nielsen, 1997; Mokhatab et al., 2006; Speight, 2014; Abdel-Aal et al., 2016; Speight, 2019). For instance, monoethanolamine (MEA), diethanolamine (DEA), methyldiethanolamine (MDEA), diisopropanolamine (DIPA) and triethanolamine (TEA) can be used. The alkanolamine scrubbing process is also carried out in two steps: in the first step involves the amine contactor and the second step involves the regenerator. In a process with diethanolamine and methyldiethanolamine the main reactions in equilibrium in the system are:

$$2H_2O \rightarrow H_3O^+ + OH^-$$

$$CO_2 + 2H_2O \rightarrow HCO_3^- + H_3O^+$$

$$HCO_3^- + H_2O \rightarrow H_3O^+ + CO_3^{2-}$$

$$H_2S + H_2O \rightarrow H_3O^+ + HS^-$$

$$HS^- + H_2O \rightarrow H_3O^+ + S^{2-}$$

$$DEA^+ + H_2O \rightarrow DEA + H_3O^+$$

$$MDEA^+ + H_2O \rightarrow MDEA + H_3O^+$$

$$DEA^+COO^- + H_2O \rightarrow DEA + HCO_3^-$$

The pH is an important parameter because it affects the concentration of the ionic species and therefore the reactions with the amines. Moreover, the temperature and pressure also have a significant effect on the process.

The reactions that involve carbon dioxide and hydrogen sulfide with amines are as follows:

$$DEA + CO_2 + H_2O \rightarrow DEA^+COO^- + H_3O^+$$

$$MDEA + CO_2 + H_2O \rightarrow MDEA^+ + HCO_3^- \quad \text{(Eq. 14)}$$

$$MDEA + H_2S \rightarrow MDEA^+ + HS^-$$

$$DEA + H_2S \rightarrow DEA^+ + HS^-$$

This system is very complex due to the number of reactions involved and other reactions may also occur in the system. Tertiary amines such as DMEA do not react directly with carbon dioxide and hydrogen sulfide with amines is almost instantaneous by proton transfer. Thus, tertiary amines are used for selective hydrogen sulfide removal and for selective hydrogen sulfide and carbon dioxide removal a mixed amine system (tertiary and primary or secondary) is employed. Moreover, other compounds can be mixed with amine solutions.

Thus, treatment of a gas stream to remove the acid gas constituents (hydrogen sulfide and carbon dioxide) is most often accomplished by contact of the gas stream with an alkaline solution. The most commonly used treating solutions are aqueous solutions of the ethanolamine or alkali carbonates, although a considerable number of other treating agents have been developed in recent years (Mokhatab et al., 2006; Speight, 2014, 2019). Most of these newer treating agents rely upon physical absorption and chemical reaction. When only carbon dioxide is to be removed in large quantities or when only partial removal is necessary, a hot carbonate solution or one of the physical solvents is the most economical selection.

The primary process for sweetening sour gas uses an amine (*olamine*) solution to remove the hydrogen sulfide (the *amine process*). The sour gas is run through a tower, which contains the olamine solution. There are two principle amine solutions used, monoethanolamine (MEA) and diethanolamine (DEA). Either of these compounds, in liquid form, will absorb sulfur compounds from the gas stream as it passes through. The effluent gas is virtually free of sulfur compounds, and thus loses its sour gas status. The amine solution used can be regenerated for reuse and, although most sour gas sweetening involves the amine absorption process, it is also possible to use solid desiccants like iron sponge to remove hydrogen sulfide and carbon dioxide (Mokhatab et al., 2006; Abdel-Aal et al., 2016; Speight, 2019), including bio-based iron sponge (Cherosky and Li, 2013).

Diglycolamine (DGA), is another amine solvent used in the Econamine process in which absorption of acid gases occurs in an absorber containing an aqueous solution of diglycolamine, and the heated rich solution (saturated with acid gases) is pumped to the regenerator (Reddy and Gilmartin, 2008). Diglycolamine solutions are characterized by low freezing points, which make them suitable for use in cold climates.

The most well-known hydrogen sulfide removal process is based on the reaction of hydrogen sulfide with iron oxide (often also called the iron sponge process or the dry box method) in which the gas is passed through a bed of wood chips impregnated with iron oxide. The iron oxide is converted to the corresponding sulfur which is reaerated by oxidation (Mokhatab et al., 2006; Speight, 2014, 2019):

$$Fe_2O_3 + 3H_2S \rightarrow Fe_2S_3 + 3H_2O$$

$$2Fe_2S_3 + 3O_2 \rightarrow 2Fe_2O_3 + 6S$$

The iron oxide process (which was implemented during the 19th century and also referred to as the iron sponge process) is the oldest and still the most widely used batch process for sweetening the synthesis gas (Zapffe, 1963; Anerousis and Whitman, 1984; Mokhatab et al., 2006; Speight, 2006, 2014). The reaction can be exploited by adding the iron chloride to the digester feed material or passing the synthesis gas through a bed of iron oxide-containing material. The iron oxide comes in different forms such as rusty steel wool, iron oxide pellets or wood pellets coated with iron oxide. The iron oxide media needs to be replaced periodically. The regeneration process is highly exothermic and must be controlled to avoid problems.

In the process, the sour gas is passed down through the bed. In the case where continuous regeneration is to be utilized a small concentration of air is added to the sour gas before it is processed. This air serves to continuously regenerate the iron oxide, which has reacted with hydrogen sulfide, which serves to extend the on-stream life of a given tower but probably serves to decrease the total amount of sulfur that a given weight of bed will remove. Ferric hydroxide [Fe(OH)$_3$] can also be used in this process.

The process is usually best applied to gases containing low to medium concentrations (300 ppm v/v) of hydrogen sulfide or mercaptan derivatives. This process tends to be highly selective and does not normally remove significant quantities of carbon dioxide. As a result, the hydrogen sulfide stream from the process is usually high purity. The use of iron oxide process for sweetening sour gas is based on adsorption of the acid gases on the surface of the solid sweetening agent followed by chemical reaction of ferric oxide (Fe$_2$O$_3$) with hydrogen sulfide:

$$2Fe_2O_3 + 6H_2S \rightarrow 2Fe_2S_3 + 6H_2O$$

The reaction requires the presence of slightly alkaline water and a temperature below 43°C (110°F) and bed alkalinity (pH: 8 to 10) should be checked regularly, usually on a daily basis. The pH level is to be maintained through the injection of caustic soda with the water. If the gas does not contain sufficient water vapor, water may need to be injected into the inlet gas stream.

The ferric sulfide produced by the reaction of hydrogen sulfide with ferric oxide can be oxidized with air to produce sulfur and regenerate the ferric oxide:

$$2Fe_2S_3 + 3O_2 \rightarrow 2Fe_2O_3 + 6S$$

$$2S + 2O_2 \rightarrow 2SO_2$$

The regeneration step is exothermic and air must be introduced slowly so the heat of reaction can be dissipated. If air is introduced quickly the heat of reaction may ignite the bed. Some of the elemental sulfur produced in the regeneration step remains in the bed. After several cycles this sulfur will form a cake over the ferric oxide, decreasing the reactivity of the bed. Typically, after 10 cycles the bed must be removed and a new bed introduced into the vessel.

The iron oxide process is one of several metal oxide-based processes that scavenge hydrogen sulfide and organic sulfur compounds (mercaptan derivatives) from gas streams through reactions with the solid based chemical adsorbent (Kohl and Riesenfeld, 1985). They are typically non-regenerable, although some are partially regenerable, losing activity upon each regeneration cycle. Most of the processes are governed by the reaction of a metal oxide with hydrogen sulfide to form the metal sulfide. For regeneration, the metal oxide is reacted with oxygen to produce elemental sulfur and the regenerated metal oxide. In addition, to iron oxide, the primary metal oxide used for dry sorption processes is zinc oxide.

In the zinc oxide process, the zinc oxide media particles are extruded cylinders 3-4 mm in diameter and 4 to 8 mm in length (Kohl and Nielsen, 1997; Mokhatab et al., 2006; Abdel-Aal et al., 2016; Speight, 2019) and react readily with the hydrogen sulfide:

$$ZnO + H_2S \rightarrow ZnS + H_2O$$

At increased temperatures (205 to 370°C, 400 to 700°F), zinc oxide has a rapid reaction rate, therefore providing a short mass transfer zone, resulting in a short length of unused bed and improved efficiency.

Removal of larger amounts of hydrogen sulfide from gas streams requires a continuous process, such as the *Ferrox* process or the Stretford process. The *Ferrox process* is based on the same chemistry as the iron oxide process except that it is fluid and continuous. The *Stretford* process employs a solution containing vanadium salts and anthraquinone disulfonic acid (Maddox, 1974; Mokhatab *et al.*, 2006; Abdel-Aal *et al.*, 2016).

Most hydrogen sulfide removal processes return the hydrogen sulfide unchanged, but if the quantity involved does not justify installation of a sulfur recovery plant (usually a Claus plant) it is necessary to select a process that directly produces elemental sulfur. In the *Beavon-Stretford* process, a hydrotreating reactor converts sulfur dioxide in the off-gas to hydrogen sulfide which is contacted with Stretford solution (a mixture of vanadium salt, anthraquinone disulfonic acid, sodium carbonate, and sodium hydroxide) in a liquid-gas absorber. The hydrogen sulfide reacts stepwise with sodium carbonate and anthraquinone disulfonic acid to produce elemental sulfur, with vanadium serving as a catalyst. The solution proceeds to a tank where oxygen is added to regenerate the reactants. One or more froth or slurry tanks are used to skim the product sulfur from the solution, which is recirculated to the absorber.

Even though the removal of hydrogen sulfide can be achieved by chemical or physical absorption, chemical scrubbers are the most commonly used systems for hydrogen sulfide removal (Mokhatab *et al.*, 2006; Speight, 2014, 2019). The most important processes are iron-based chelation methods and absorption by alkanolamines. In iron-based chelation processes there are two main reactions:

$$H_2S + 2Fe^{3+} \text{ Chelant}^{n-} \rightarrow S + 2H^+ + 2\text{ Fe}^{2+} \text{ Chelant}^{n-}$$

$$O_2 + 4Fe^{2+} \text{ Chelant-} + 2H_2O \rightarrow 4Fe^{3+} \text{ Chelant-} + 4\text{ OH}^-$$

The first step involves the chemical reaction to produce elemental sulfur and in the second step iron(III) is regenerated. Several commercial processes are based on iron chelation, such as the Lo-cat and Sulferox processes. Ethylenediaminetetraacetic acid (EDTA), hydroxyethylenediaminetriacetic acid (HEDTA), diethylenetriaminepentaacetic acid (DTPA), and nitrilotriacetic acid (NTA) are the most conventional ligands used in this process. The operating conditions are typically: 4 to 8°C (39.5 to 46.5°F), pH 4 to pH 8, iron concentration on the order of 1,000 to 10,000 ppm v/v with a chelant/iron ratio in the range 1.1 to 2.0 (Deshmukh and Shete, 2013; Ramírez *et al.*, 2015).

The process using potassium phosphate is known as phosphate desulfurization, and it is used in the same way as the Girbotol process to remove acid gases from liquid hydrocarbon derivatives as well as from gas streams. The treatment solution is a water solution of potassium phosphate (K_3PO_4), which is circulated through an absorber tower and a reactivator tower in much the same way as the ethanolamine is circulated in the Girbotol process; the solution is regenerated thermally. Processes using ethanolamine and potassium phosphate are now widely used.

9.5.4 Other Processes

There is a series of alternate processes that involve (i) the use of chemical reactions to remove contaminants from gas streams or (ii) the use of specialized equipment to physically remove contaminants from gas streams.

As example of the first category, i.e., the use of chemical reactions to remove contaminants from gas streams, strong basic solutions are effective solvents for acid gases. However, these solutions are not normally used for treating large volumes of synthesis gas because the acid gases form stable salts, which are not easily regenerated. For example, carbon dioxide and hydrogen sulfide react with aqueous sodium hydroxide to yield sodium carbonate and sodium sulfide, respectively.

$$CO_2 + 2NaOH \rightarrow Na_2CO_3 + H_2O$$

$$H_2S + 2NaOH \rightarrow Na_2S + 2H_2O$$

However, a strong caustic solution is used to remove mercaptans from gas and liquid streams. In the Merox Process, for example, a caustic solvent containing a catalyst such as cobalt, which is capable of converting mercaptans (RSH) to caustic insoluble disulfides (RSSR), is used for streams rich in mercaptans after removal of hydrogen sulfide. Air is used to oxidize the mercaptan derivatives to disulfide derivatives. The caustic solution is then recycled for regeneration. The Merox process is mainly used for treatment of refinery gas streams.

As one of the major contaminants in synthesis gas streams, carbon dioxide must optimally be removed as it reduces the energy content of the gas and affects the selling price of the gas. Moreover, it becomes acidic and corrosive in the presence of water that has a potential to damage the pipeline and the equipment system. Hence, the presence of carbon dioxide in synthesis gas remains one of the challenging gas separation problems in process engineering for carbon dioxide/methane systems. Therefore, the removal of carbon dioxide from the synthesis gas through the purification processes is vital for an improvement in the quality of the product (Mokhatab *et al.*, 2006; Speight, 2014, 2019).

Carbonate washing is a mild alkali process (typically the alkali is potassium carbonate, K_2CO_3) for gas processing for the removal of acid gases (such as carbon dioxide and hydrogen sulfide) from gas streams and uses the principle that the rate of absorption of carbon dioxide by potassium carbonate increases with temperature (Mokhatab *et al.*, 2006; Speight, 2014, 2019). It has been demonstrated that the process works best near the temperature of reversibility of the reactions:

$$K_2CO_3 + CO_2 + H_2O \rightarrow 2KHCO_3$$

$$K_2CO_3 + H_2S \rightarrow KHS + KHCO_3$$

The Fluor process uses propylene carbonate to remove carbon dioxide, hydrogen sulfide, carbonyl sulfide, water and higher boiling hydrocarbon derivatives $\left(C_2^+\right)$ from gas stream (Abdel-Aal *et al.*, 2016).

Water washing, in terms of the outcome, is almost analogous to (but often less effective than) washing with potassium carbonate (Kohl and Riesenfeld, 1985; Kohl and Nielsen, 1997), and it is also possible to carry out the desorption step by pressure reduction. The absorption is purely physical and there is also a relatively high absorption of hydrocarbon derivatives, which are liberated at the same time as the acid gases. The water scrubbing processes uses the higher solubility of carbon dioxide in water to separate the carbon dioxide from synthesis gas. This process is done under high pressure and removes hydrogen sulfide as well as carbon dioxide. The main disadvantage of this process is that it requires a large volume of water that must be purified and recycled. An analogous process, polyethylene glycol scrubbing, is similar to water scrubbing but it is more efficient. However, the process, does require the regeneration of a large volume of polyethylene glycol.

In *chemical conversion processes*, contaminants in gas emissions are converted to compounds that are not objectionable or that can be removed from the stream with greater ease than the original constituents. For example, a number of processes have been developed that remove hydrogen sulfide and sulfur dioxide from gas streams by absorption in an alkaline solution.

Catalytic oxidation is a chemical conversion process that is used predominantly for destruction of volatile organic compounds and carbon monoxide. These systems operate in a temperature regime on the order of 205 to 595°C (400 to 1100°F) in the presence of a catalyst – in the absence of the catalyst, the system would require a higher operating temperature. The catalysts used are typically a combination of noble metals deposited on a ceramic base in a variety of configurations (e.g., honeycomb-shaped) to enhance good surface contact. Catalytic systems are usually classified on the basis of bed types such as *fixed bed* (or *packed bed*) and *fluid bed* (*fluidized bed*). These systems generally have very high destruction efficiencies for most volatile organic compounds, resulting in the formation of carbon dioxide, water, and varying amounts of hydrogen chloride (from halogenated hydrocarbon derivatives). The presence in emissions of chemicals such as heavy metals, phosphorus, sulfur, chlorine, and most halogens in the incoming air stream act as poison to the system and can foul up the catalyst. Thermal oxidation systems, without the use of catalysts, also involve chemical conversion (more correctly, chemical destruction) and operate at temperatures in excess of 815°C (1500°F), or 220 to 610°C (395 to 1100°F) higher than catalytic systems.

Other processes include the *Alkazid process* for removal of hydrogen sulfide and carbon dioxide using concentrated aqueous solutions of amino acids. The hot potassium carbonate process decreases the acid content of natural and refinery gas from as much as 50% to as low as 0.5% and operates in a unit similar to that used for amine treating. The *Giammarco-Vetrocoke* process is used for hydrogen sulfide and/or carbon dioxide removal. In the hydrogen sulfide removal section, the reagent consists of sodium carbonate (Na_2CO_3) or potassium carbonate (K_2CO_3) or a mixture of the carbonates which contains a mixture of arsenite derivatives and arsenate derivatives; the carbon dioxide removal section utilizes hot aqueous alkali carbonate solution activated by arsenic trioxide (As_2O_3) or selenous acid (H_2SeO_3) or tellurous acid (H_2TeO_3). A word of caution might be added about the last three chemicals which are toxic and can involve stringent environmental-related disposal protocols.

Molecular sieves are highly selective for the removal of hydrogen sulfide (as well as other sulfur compounds) from gas streams and over continuously high absorption efficiency. They are also an effective means of water removal and thus offer a process for the simultaneous dehydration and desulfurization of gas. Gas that has excessively high water content may require upstream dehydration, however (Mokhatab *et al.*, 2006; Speight, 2014; Abdel-Aal *et al.*, 2016; Speight, 2019). The carbon molecular sieve method uses differential adsorption characteristics to separate methane and The carbon molecular sieve method uses differential adsorption characteristics to separate methane and carbon dioxide.

In addition, the *molecular sieve process* is similar to the iron oxide process. Regeneration of the bed is achieved by passing heated clean gas over the bed. As the temperature of the bed increases, it releases the adsorbed hydrogen sulfide into the regeneration gas stream. The sour effluent regeneration gas is sent to a flare stack, and up to 2% v/v of the gas seated can be lost in the regeneration process. A portion of the gas stream may also be lost by the adsorption of hydrocarbon components by the sieve (Mokhatab *et al.*, 2006; Speight, 2014, 2019).

In this process, unsaturated hydrocarbon components, such as olefin derivatives and aromatic derivatives, tend to be strongly adsorbed by the molecular sieve. Molecular sieves are susceptible to poisoning by such chemicals as glycols and require thorough gas cleaning methods before the adsorption step. Alternatively, the sieve can be offered some degree of protection by the use of *guard beds* in which a less expensive catalyst is placed in the gas stream before contact of the gas with the sieve, thereby protecting the catalyst from poisoning. This concept is analogous to the use of guard beds or attrition catalysts in the crude oil industry (Speight, 2000, 2014, 2017).

Carbon molecular sieves are excellent products to separate specifically a number of different gaseous compounds in synthesis gas. Thereby the molecules are usually loosely adsorbed in the cavities of the carbon sieve but not irreversibly bound. The selectivity of adsorption is achieved by different mesh sizes and/or application of different gas pressures. When the pressure is released the compounds extracted from the synthesis gas are desorbed. The process – pressure-swing adsorption (PSA) technology – can be used to enrich methane from synthesis gas; the molecular sieve is applied which is produced from coke rich in pores in the micrometer range. The pores are then further reduced by cracking of the hydrocarbon derivatives.

In order to reduce the energy consumption for gas compression, a series of vessels are linked together. The gas pressure released from one vessel is subsequently used by the others. Usually four vessels in series are used filled with the molecular sieve which removes at the same time carbon dioxide and any water vapor. After removal of hydrogen sulfide, i.e., using activated carbon and water condensation in a cooler at 4°C, the synthesis gas flows at a pressure of 90 psi into the adsorption unit. The first column cleans the raw gas at 90 psi to an upgraded synthesis gas with a vapor pressure of less than 10 ppm water and a methane content of 96% or more.

In the second column the pressure of 90 psi is first released to approximately 45 psi by pressure communication with column 4, which was previously degassed by a slight vacuum. In a second step the pressure is then reduced to atmospheric pressure. The released gas flows back to the digester in order to recover the methane. The third column is evacuated from 15 psi to 0.15 psi. The desorbed gas consists predominantly of carbon dioxide but also some methane and is therefore normally released to the environment. In order to reduce methane losses the system can be designed with recirculation of the desorbed gases.

A number of other possible impurities, such as organic acids (e.g., formic and acetic acid) and unsaturated and higher-boiling hydrocarbon derivatives, are normally not included in purity specifications for synthesis processes because they are already undesirable in upstream process stages of biosyngas production (e.g., compression steps). However, organic compounds present must be below their dewpoint at pressure of the gas application to prevent condensation and fouling in the system. For organic compounds with sulfur or nitrogen heteroatoms (such as thiophene derivatives and pyridine derivatives) the additional specification applies that they need to be removed below ppm v/v level, as they are intrinsically poisonous for the catalyst.

Thiophene

Pyridine

9.6 Removal of Condensable Hydrocarbons

Hydrocarbon derivatives that are higher molecular weight than methane that are present in gas streams are valuable raw materials and important fuels. They can be recovered by lean oil extraction. The first step in this process is to cool the treated gas by exchange with liquid propane after which the cooled gas is then washed with a cold hydrocarbon liquid, which dissolves most of the condensable hydrocarbon derivatives. The uncondensed gas is dry synthesis gas and is composed mainly of methane. Dry synthesis gas may then be used either as a fuel or as a chemical feedstock. Another way to recover any higher molecular weight hydrocarbon constituents from synthesis gas is by using cryogenic cooling (cooling to very low temperatures on the order of -100 to -115°C, -150 to -175°F), which is achieved primarily through lowering the temperatures to below the dew point.

To prevent hydrate formation, synthesis gas streams may be treated with glycols, which dissolve water efficiently. Ethylene glycol (EG), diethylene glycol (DEG), and triethylene glycol (TEG) are typical solvents for water removal. Triethylene glycol is preferable in vapor phase processes because of its low vapor pressure, which results in less glycol loss. The triethylene glycol absorber unit typically contains 6 to 12 bubble-cap trays to accomplish the water absorption. However, more contact stages may be required to reach dew points below -40°C (-40°F). Calculations to determine the number of trays or feet of packing, the required glycol concentration, or the glycol circulation rate require vapor-liquid equilibrium data. In addition, predicting the interaction between triethylene glycol and water vapor in synthesis gas over a broad range allows the designs for ultra-low dew point applications to be made.

One alternative to using bubble-cap trays is the use of the adiabatic expansion of the inlet gas. In the process, the inlet gas is first treated to remove water and acid gases, then cooled via heat exchange and refrigeration. Further cooling of the gas is accomplished through turbo expanders, and the gas is sent to a demethanizer to separate methane from the higher-boiling hydrocarbon derivatives. Improved recovery of the higher-boiling hydrocarbon derivatives could be achieved through better control strategies and by use of on-line gas chromatographic analysis.

Membrane separation process are very versatile and are designed to process a wide range of feedstocks and offer a simple solution for removal and recovery of higher-boiling hydrocarbon derivatives from gas streams (Foglietta, 2004; May-Britt, 2008; Basu et al., 2010; Rongwong et al., 2012; Ozturk and Demirciyeva, 2013; Abdel-Aal et al., 2016). There are two membrane separation techniques: (i) high-pressure gas separation and (ii) gas-liquid adsorption. The high-pressure separation process selectively separates hydrogen sulfide and carbon dioxide from methane. Usually, this separation is performed in three stages and produces methane with a high degree of purity (on the order of 96% v/v methane).

The separation process is based on high-flux membranes that selectively permeates higher-boiling hydrocarbon derivatives (compared to methane) and are recovered as a liquid after recompression and condensation. The residue stream from the membrane is partially depleted of higher-boiling hydrocarbon derivatives, and is then sent to sales gas stream. Gas permeation membranes are usually made with vitreous polymers that exhibit good selectivity but, to be effective, the membrane must be very permeable with respect to the separation process (Rojey et al., 1997).

Membrane separation is based on the selectivity properties of the membrane material. The driving force is the difference in chemical potential, which includes the effect of temperature and partial pressure régimes. Membranes can be operated at high pressure (>300 psi) or low pressure (120 to 150 psi). Several membranes can be used for synthesis gas upgrading and these can be classified by the type of material: non-polymeric membranes (such as alumina, zeolites, carbon), ceramic membranes, palladium membranes, and carbon molecular sieve membranes. Nevertheless, polymeric membranes are the most widely used because they are cheaper than inorganic membranes. Moreover, polymeric materials can be easily fabricated into flat sheets or asymmetric hollow fibres, both shapes that are usually used in industrial applications.

The material properties that define the performance are molecular structure, glass transition temperature, crystallinity and degree of crosslinking. However, commercially available polymeric membranes are usually degraded by hydrogen sulfide, siloxanes, and/or volatile organic compounds (Accettola et al., 2008; Ajhar et al., 2010; Gislon et al., 2013). For instance, hydrogen sulfide can cause plasticization of glassy polymeric membranes and this alters the polymeric structure and weakens the mechanical strength of the membrane.

In the high-pressure separation option, pressurized gas (36 bar) is first cleaned over for example an activated carbon bed to remove (halogenated) hydrocarbon derivatives and hydrogen sulfide from the raw gas as well as oil vapor from the compressors. The carbon bed is followed by a particle filter and a heater. The membranes made of acetate-cellulose separate small polar molecules such as carbon dioxide, moisture and the remaining hydrogen sulfide. These membranes are not effective in separating nitrogen from methane.

The raw gas is upgraded in 3 stages to a clean gas with 96% methane or more. The waste gas from the first two stages is recycled and the methane can be recovered. The waste gas from stage 3 (and in part of stage 2) is flared or used in a steam boiler as it still contains 10 to 20% v/v methane. First experiences have shown that the membranes can last up to 3 years which is comparable to the lifetime of membranes for natural gas purification – a primary market for membrane technology – which last typically two to five years.

The clean gas is further compressed up to 3600 psi and stored in steel cylinders in capacities of 276 m3 divided in high-, medium- and low-pressure banks. The membranes are very specific for given molecules, i.e., hydrogen sulfide and carbon dioxide are upgraded in different modules. The utilization of hollow-fiber membranes allows the construction of very compact modules working in cross flow.

The gas-liquid absorption membrane method is a separation technique which was developed for synthesis gas upgrading only recently. The essential element is a microporous hydrophobic membrane separating the gaseous from the liquid phase. The molecules from the gas stream, flowing in one direction, which are able to diffuse through the membrane will be absorbed on the other side by the liquid flowing in countercurrent.

The absorption membranes work at approx. atmospheric pressure (1 bar) which allows low-cost construction. The removal of gaseous components is very efficient. At a temperature of 25 to 35°C (77 to 95°F) the hydrogen sulfide concentration in the raw gas of 2% v/v is reduced to less than 0.025% v/v (250 ppm v/v) – the absorbent is sodium hydroxide (NaOH) or another alkaline material. Sodium hydroxide saturated with hydrogen sulfide can be used in water treatment to remove heavy metals. The concentrated hydrogen sulfide solution is fed into a Claus reaction or oxidized to elementary sulfur. The synthesis gas is upgraded very efficiently from 55% v/v methane (43% v/v carbon dioxide) to more than 96% methane. The amine solution is regenerated by heating and the carbon dioxide released is pure and can be sold for industrial applications.

The main advantages of membranes are (i) safety and simplicity of operation, (ii) relative ease of maintenance and scale-up, (iii) excellent reliability, (iv) small footprint, and (v) operation without hazardous chemicals. Several operation modes can be employed – as an example, synthesis gas upgrading can be achieved by a two-stage cascade process with recycling and a single stage provided good flexibility for integration into synthesis gas plants.

9.6.1 Extraction

There are two principle techniques for removing higher-boiling hydrocarbon constituents from gas stream liquids: (i) the absorption method and (ii) the cryogenic expander process. In the latter process (the cryogenic expander process), a turboexpander is used to produce the necessary refrigeration and very low temperatures and high recovery of light components, such as ethane and propane, can be attained. The gas stream is first dehydrated using a molecular sieve followed by cooling of the dry stream. The separated liquid containing most of the heavy fractions is then demethanized, and the cold gases are expanded through a turbine that produces the desired cooling for the process. The expander outlet is a two-phase stream that is fed to the top of the demethanizer column. This serves as a separator in which: (i) the liquid is used as the column reflux and the separator vapors combined with vapors stripped in the demethanizer are exchanged with the feed gas, and (ii) the heated gas, which is partially recompressed by the expander compressor, is further recompressed to the desired distribution pressure in a separate compressor.

The extraction of higher-boiling hydrocarbon constituents from the gas stream produces both cleaner, purer synthesis gas, as well as the valuable hydrocarbon derivatives for use in, say, a petrochemical plant (Speight, 2019). This process allows for the recovery of approximately 90 to 95% v/v of the ethane originally in the gas stream. In addition, the expansion turbine is able to convert some of the energy released when the gas stream is expanded into recompressing the gaseous methane effluent, thus saving energy costs associated with extracting ethane.

9.6.2 Absorption

The absorption method of high molecular weight recovery of hydrocarbon derivatives is very similar to using absorption for dehydration (Mokhatab *et al.*, 2006; Speight, 2014, 2019). The main difference is that, in the absorption of higher-boiling hydrocarbon constituents from the gas stream, absorbing oil is used as opposed to glycol. This absorbing oil has an affinity for the higher-boiling hydrocarbon constituents in much the same manner as glycol has an affinity for water. Before the oil has picked up any higher-boiling hydrocarbon constituents, it is termed *lean* absorption oil.

The *oil absorption process* involves the countercurrent contact of the lean (or stripped) oil with the incoming wet gas with the temperature and pressure conditions programmed to maximize the dissolution of the liquefiable components in the oil. The *rich* absorption oil (sometimes referred to as *fat* oil), containing higher-boiling hydrocarbon constituents, exits the absorption tower through the bottom. It is now a mixture of absorption oil, propane, butanes, pentanes, and other higher boiling hydrocarbon derivatives. The rich oil is fed into lean oil stills, where the mixture is heated to a temperature above the boiling point of the higher-boiling hydrocarbon constituents but below that of the oil. This process allows for the recovery of higher boiling constituents from the synthesis gas stream.

The basic absorption process is subject to modifications that improve process effectiveness and even to target the extraction of specific higher-boiling hydrocarbon constituents. In the refrigerated oil absorption method, where the lean oil is cooled through refrigeration, propane recovery can be on the order of 90%+ v/v and approximately 40% v/v of the ethane can be extracted from the gas stream. Extraction of the other, higher-boiling hydrocarbon constituents is typically near-quantitative using this process.

9.6.3 Fractionation

Fractionation processes are very similar to those processes classed as *liquids removal* processes but often appear to be more specific in terms of the objectives: hence the need to place the fractionation processes into a separate category. The fractionation processes are those processes that are used (i) to remove the more significant product stream first, or (ii) to remove any unwanted light ends from the higher-boiling liquid products.

In the general practice of gas processing, the first unit is a de-ethanizer followed by a depropanizer then by a debutanizer and, finally, a butane fractionator. Thus each column can operate at a successively lower pressure, thereby allowing the different gas streams to flow from column to column by virtue of the pressure gradient, without necessarily the use of pumps. The purification of hydrocarbon gases by any of these processes is an important part of gas processing operations.

Thus, after any higher-boiling hydrocarbon constituents have been removed from the gas stream, they must be separated (fractionated) into the individual constituents prior to sales. The process of fractionation occurs in stages with each stage involving separation of the hydrocarbon derivatives as individual products. The process commences with the removal of the lower-boiling hydrocarbon derivatives from the feedstock. The particular fractionators are used in the following order: (i) the de-ethanizer, which is used to separate the ethane from the stream from the gas stream, (ii) the depropanizer, which is used to separate the propane from the de-ethanized gas stream, (iii) the debutanizer, which is used to separate the butane isomers, leaving the pentane isomers and higher-boiling hydrocarbon derivatives in the gas stream, and (iv) the butane splitter or de-isobutanizer, which is used to separate n-butane and iso-butane.

After the recovery of the higher-boiling hydrocarbon constituents, sulfur-free dry gas (methane) may be liquefied for transportation through cryogenic tankers. Further treatment may be required to reduce the water vapor below 10 ppm and carbon dioxide and hydrogen sulfide to less than 100 and 50 ppm v/v, respectively. Two methods are generally used to liquefy the gas: (i) the expander cycle and (ii) mechanical refrigeration. In the expander cycle, part of the gas is expanded from a high transmission pressure to a lower pressure. This lowers the temperature of the gas. Through heat exchange, the cold gas cools the incoming gas, which in a similar way cools more incoming gas until the liquefaction temperature of methane is reached.

In mechanical refrigeration, a multicomponent refrigerant consisting of nitrogen, methane, ethane, and propane is used through a cascade cycle. When these liquids evaporate, the heat required is obtained from the gas, which loses energy/temperature till it is liquefied. The refrigerant gases are recompressed and recycled.

9.6.4 Enrichment

The gas product must meet specific quality measures in order for the pipeline grid to operate properly. Consequently, synthesis gas which, in most cases contains contaminants, must be processed, i.e., cleaned, before it can be safely delivered to the high-pressure, long-distance pipelines that transport the product to the consuming public. A gas stream that is not within certain specific gravities, pressures, Btu content range, or water content levels will cause operational problems, pipeline deterioration, or can even cause pipeline rupture. Thus, the purpose of *enrichment* is to produce a gas stream for sale. Therefore, the process concept is essentially the separation of higher-boiling hydrocarbon constituents from the methane to produce a lean, dry gas.

As an example, carbon dioxide is to some extent soluble in water and therefore some carbon dioxide will be dissolved in the liquid phase of the digester tank. In upgrading with the in situ methane enrichment process, sludge from the digester is circulated to a desorption column and then back to the digester. In the desorption column carbon dioxide is desorbed by pumping air through the sludge. The constant removal of carbon dioxide from the sludge leads to an increased concentration of methane in the synthesis gas phase leaving the digester.

The gas stream received and transported must (especially in the United States and many other countries) meet the quality standards specified by pipeline. These quality standards vary from pipeline to pipeline and are usually a function of (i) the design of the pipeline system, (ii) the design of any downstream interconnecting pipelines, and (iii) the requirements of the customer. In general, these standards specify that the gas stream should (i) be within a specific Btu content range, typically 1,035 Btu ft^3 ± 50 Btu ft^3, (ii) be delivered at a specified hydrocarbon dew point temperature level to prevent any vaporized gas liquid in the mix from condensing at pipeline pressure, (iii) contain no more than trace amounts of elements such as hydrogen sulfide, carbon dioxide, nitrogen, water vapor, and oxygen, (iv) be free of particulate solids and liquid water that could be detrimental to the pipeline or its ancillary operating equipment. Gas processing equipment, whether in the field or at processing/treatment plants, assures that these specifications can be met.

In most cases processing facilities extract contaminants and higher-boiling hydrocarbon derivatives from the gas stream but, in some cases, the gas processors blend some higher-boiling hydrocarbon derivatives into the gas stream in order to bring it within acceptable Btu levels. For instance, in some areas if the produced gas (including coalbed methane) does not meet (is below) the Btu requirements of the pipeline operator, in which case a blend of higher Btu-content gas stream or a propane-air mixture is injected to enrich the heat content (Btu value) prior for delivery to the pipeline.

The number of steps and the type of techniques used in the process of creating a pipeline-quality gas stream most often depends upon the source and makeup of the wellhead production stream. Among the several stages of gas processing are: (i) gas-oil separation, (ii) water removal, (iii) liquids removal, (iv) nitrogen removal, (v) acid gas removal, and (vi) fractionation.

In many instances pressure relief at the wellhead will cause a natural separation of gas from oil (using a conventional closed tank, where gravity separates the gas hydrocarbon derivatives from the heavier oil). In some cases, however, a multi-stage gas-oil separation process is needed to separate the gas stream from the crude oil. These gas-oil separators are commonly closed cylindrical shells, horizontally mounted with inlets at one end, an outlet at the top for removal of gas, and an outlet at the bottom for removal of oil. Separation is accomplished by alternately heating and cooling (by compression) the flow stream through multiple steps. However, the number of steps and the type of techniques used in the process of creating a pipeline-quality gas stream most often depends upon the source and makeup of the gas stream. In some cases, several of the steps may be integrated into one unit or operation, performed in a different order or at alternative locations (lease/plant), or not required at all.

9.7 Tar Removal

Tar is a highly viscous liquid product (typically produced in thermal processes) that condenses in the low temperature zones of the equipment (the pyrolizer or the gasifier) thereby interfering with the gas flow and leading to system disruption. It is, perhaps, the most undesirable product of the process and high residual tar concentrations in the can gas prevent utilization (Devi *et al.*, 2003; Torres *et al.*, 2007; Anis and Zainal, 2011; Huang *et al.*, 2011). Methods for reduction or elimination of tar can be divided into two broad groups: (i) *in situ* tar reduction also called primary tar reduction, which focus on reducing or avoiding tar formation in the reactor, and (ii) post-gasification tar reduction also called secondary tar reduction, which strips the tar from the product gas. A combination of in situ and post-gasification tar reductions can prove to be more effective than either of the single stages alone. The two basic post-gasification methods are physical removal and cracking (catalytic or thermal).

9.7.1 Physical Methods

In a process for the physical removal of tar, the tar is treated tar as dust particles or mist and the tar is condensed before separation. Physical tar removal can be accomplished by cyclones, barrier filters, wet electrostatic precipitators (ESPs) or wet scrubbers. The selection of the equipment application depends on the load concentrations of particulate matter (PM) and tar, particle size distribution and particulate tolerance of downstream users.

Cyclone collectors are the most common of the inertial collector class and are effective in removing coarser fractions of particulate matter and operate by contacting the particles in the gas stream with a liquid. In principle the particles are incorporated in a liquid bath or in liquid particles which are much larger and therefore more easily collected. In the process, the particle-laden gas stream enters an upper cylindrical section tangentially and proceeds downward through a conical section. Particles migrate by centrifugal force generated by providing a path for the carrier gas to be subjected to a vortex-like spin. The particles are forced to the wall and are removed through a seal at the apex of the inverted cone. A reverse-direction vortex moves upward through the cyclone and discharges through a top center opening. Cyclones are often used as primary collectors because of their relatively low efficiency (50 to 90% is usual).

Cyclones may not be efficient when removing small tar droplets and barrier filters are porous material which can capture a certain amount of tar when the product gas passes through the filters. As an aid, catalyst grains can be integrated as a fixed bed inside the filter to promote the simultaneous removal of particulate matter and tar (Rapagna *et al.*, 2010).

Fabric filters are typically designed with non-disposable filter bags. As the gaseous (dust-containing) emissions flow through the filter media (typically cotton, polypropylene, fiberglass, or Teflon), particulate matter is collected on the bag surface as a dust cake. Fabric filters operate with collection efficiencies up to 99.9% although other advantages are evident but there are several issues that arise during use of such equipment.

Wet scrubbers are devices in which a countercurrent spray liquid is used to remove particles from an air stream. Device configurations include plate scrubbers, packed bed scrubbers, orifice scrubbers, venturi scrubbers, and spray towers, individually or in various combinations. Wet scrubbers can achieve high collection efficiencies at the expense of prohibitive pressure drops. The *foam scrubber* is a modification of a wet scrubber in which the particle-laden gas is passed through a foam generator, where the gas and particles are enclosed by small bubbles of foam.

Other methods include use of high-energy input *venturi scrubbers* or electrostatic scrubbers where particles or water droplets are charged, and flux force/condensation scrubbers where a hot humid gas is contacted with cooled liquid or where steam is injected into saturated gas. In the latter scrubber the movement of water vapor toward the cold water surface carries the particles with it (*diffusiophoresis*), while the condensation of water vapor on the particles causes the particle size to increase, thus facilitating collection of fine particles.

Electrostatic precipitators operate on the principle of imparting an electric charge to particles in the incoming air stream, which are then collected on an oppositely charged plate across a high-voltage field. Particles of high resistivity create the most difficulty in collection. Conditioning agents such as sulfur trioxide (SO_3) have been used to lower resistivity. Important parameters include design of electrodes, spacing of collection plates, minimization of air channeling, and collection-electrode rapping techniques (used to dislodge particles). Techniques under study include the use of high-voltage pulse energy to enhance particle charging, electron-beam ionization, and wide plate spacing. Electrical precipitators are capable of efficiencies >99% under optimum conditions, but performance is still difficult to predict in new situations.

Wet electrostatic precipitator units have a high collection efficiency (on the order of 90%) over the entire range of particle size down to 0.5 mm with a very low-pressure drop. Wet scrubbers can achieve a high collection efficiency (also on the order of 90%) as well.

9.7.2 Thermal Methods

Cracking methods are more advantageous in terms of recovering the energy content in the tar by converting the high molecular weight constituents of the tar into lower molecular weight (gaseous) products such as hydrocarbon derivatives and hydrogen at high temperature (up to 1200°C, 2190°F) or catalytic reactions (up to 800°C, 1470°F). Catalytic cracking is commercially used in many refineries plants and has been demonstrated to be one of the most effective processes for the conversion of high molecular weight products (such as crude oil residua, heavy oi, extra heavy oil, and tar sand bitumen) into gases and lower molecular weight distillable and liquids (Parkash, 2003; Gary *et al.*, 2007; Speight, 2014; Hsu and Robinson, 2017; Speight, 2017).

Non-metallic catalysts such as dolomite ($CaCO_3.MgCO_3$), calcite ($CaCO_3$), zeolite and metallic catalysts such as nickel-based catalysts, nickel/molybdenum-based catalysts, nickel/cobalt-based catalysts, molybdenum-based catalysts, platinum-based catalysts, and ruthenium-based catalysts Mo, NiO, Pt and Ru have been applied to various tar conversion processes.

Tar removal can also be achieved by catalytic reforming which has the advantage of keeping the carbon contained in the tars available for further conversion to fuel. Another option to reduce the tar content of the syngas is the use of catalytic bed material in the gasification reactor. Olivine sand has been shown to effectively reduce the tar content in synthesis gas from steam gasification.

9.8 Other Contaminant Removal

The types and the amounts of the contaminants that occur in synthesis gas are, like those contaminants presented above, dependent on the composition of the biomass feedstock and the type of process used to produce the gas as well as the individual process parameters. In addition to the gas from anaerobic digesters, the contaminants found in landfill gas vary widely from landfill site to landfill site as the items disposed of also vary widely (i.e., there is site specificity). Moreover, there is variation in how frequently gas is extracted and in what volumes, and the stage of decomposition of the waste. This section presents some of the contaminants that are not always presented in other works related to contaminant presence and removal. For example, siloxanes derivatives generally occur to a greater extent in synthesis gas from landfill sites compared to the presence of these chemicals in synthesis gas from other sources.

9.8.1 Nitrogen Removal

Nitrogen may often occur in sufficient quantities in gas streams and, consequently, lower the heating value of the gas. Thus several plants for *nitrogen removal* from gas streams have been built, but it must be recognized that nitrogen removal requires liquefaction and fractionation of the entire gas stream, which may affect process economics. In some cases, the nitrogen-containing gas stream is blended with a gas having a higher heating value and sold at a reduced price depending upon the thermal value (Btu/ft^3).

For high flow-rate gas streams, a cryogenic process is typical and involves the use of the different volatility of methane (b.p.-161.6°C/-258.9°F) and nitrogen (b.p. -195.7°C/-320.3°F) to achieve separation. In the process, a system of compression and distillation columns drastically reduces the temperature of the gas mixture to a point where methane is liquefied and the nitrogen is not. On the other hand, for smaller volumes of gas, a system utilizing pressure swing adsorption (PSA) is a more typical method of separation.

The pressure-swing adsorption process is a fixed-bed adsorption process that is carried out at constant temperature and variable pressure. In the process, carbon dioxide is separated from the synthesis gas by adsorption on a surface under elevated pressure. The adsorbing material, usually activated carbon or zeolites, is regenerated by a sequential decrease in pressure before the column is reloaded again. An upgrading plant, using this technique, typically has four, six or nine vessels working in parallel. When the adsorbing material in one vessel becomes saturated the raw gas flow is switched to another vessel in which the adsorbing material has been regenerated. During regeneration the pressure is decreased in several steps. The gas that is desorbed during the first and eventually the second pressure drop may be returned to the inlet of the raw gas, since it will contain some methane that was adsorbed together with carbon dioxide. The gas desorbed in the following pressure reduction step is either led to the next column or if it is almost entirely methane free it is released to the atmosphere. If hydrogen sulphide is present in the raw gas, it will be irreversibly adsorbed on the adsorbing material. In addition, water present in the raw gas can destroy the structure of the material. Therefore hydrogen sulphide and water needs to be removed before the PSA-column.

Thus, the pressure-swing adsorption process enables the separation of carbon dioxide from synthesis gas. Nevertheless, pressure-swing adsorption can be used for the separation of other compounds such as nitrogen, oxygen, and carbon monoxide. Several packing materials can be used for the removal of carbon dioxide and these include zeolite, silica gel and activate carbon. In order to be effective continuously the pressure-swing adsorption process is composed of various fixed-bed columns running in alternative cycles of adsorption, regeneration, and pressure build-up (Nikolić *et al.*, 2009; Alonso-Vicario *et al.*, 2010; Santos *et al.*, 2013). In some cases, the regeneration step is carried out under vacuum and this technology is known as vacuum-swing adsorption (VSA), which requires a vacuum pump at the outlet.

Moreover, the regeneration step can be carried out by increasing the temperature to enhance the desorption of compounds such as hydrogen sulfide (which is strongly adsorbed on the packing material). On increasing the temperature at constant pressure the process is also known as temperature-swing adsorption (TSA). The temperature-swing adsorption process is used for gas sweetening. Synthesis gas must be fed dry into pressure-swing adsorption, vacuum-swing adsorption, and temperature-swing adsorption processes and the water content must therefore be reduced prior to these treatments.

The pressure-swing adsorption operation can be based on equilibrium or kinetic separation. The more strongly adsorbed and faster diffusing compounds are retained on the packing material in equilibrium and kinetic separation, respectively. Natural zeolites have been used for hydrogen sulfide and carbon dioxide removal by pressure-swing adsorption with thermal desorption working at 25°C (77°F) in the pressure range 45 to 100 psi. However, variations in the concentration of carbon dioxide can affect the performance of the pressure-swing adsorption process. For example, an increase of 5% in the flow rate of the carbon dioxide can lead to an increase in the temperature at the top of the packed bed – thus, the use of a thermocouple could enhance the control strategy.

Also, in the pressure-swing adsorption method, methane and nitrogen can be separated by using an adsorbent with an aperture size very close to the molecular diameter of the larger species (the methane) which allows nitrogen to diffuse through the adsorbent. This results in a purified gas stream that is suitable for pipeline specifications. The adsorbent can then be regenerated, leaving a highly pure nitrogen stream. The pressure-swing adsorption method is a flexible method for nitrogen rejection, being applied to both small and large flow rates.

9.8.2 Ammonia Removal

Biomass with a relatively high nitrogen content will generate a product gas that contains ammonia (NH_3) and hydrogen cyanide (HCN) (Yuan *et al.*, 2010) which, in turn, will generate nitrogen oxides (NOx) during combustion. In the case of the gasification of biomass (to produce synthesis gas), the concentration of ammonia in the product gas depends not only on the nature of the biomass feedstock used but also on the gasifier design parameters and operating conditions.

Ammonia can be removed by wet scrubbing technology which has been widely adopted in the existing biomass gasification processes (Dou *et al.*, 2002; Proell *et al.*, 2005) and the ammonia can be removed when the gas is dried. As a result, a separate cleaning step is therefore usually not necessary.

Compared with wet scrubbing technology, hot-gas cleanup technology, preferably employing catalysts, is more advantageous with respect to energy efficiencies as it eliminates the needs of cooling the product gas and reheating again for the syngas applications (Torres *et al.*, 2007). Catalytic processes effectively remove ammonia by converting it to nitrogen, hydrogen, and water. Nickel-Ni-based catalysts have higher activity, while other catalysts do not have good potentials for ammonia.

9.8.3 Particulate Matter Removal

Particulate matter that is present in the product gas can also be a serious problem for some end-users and catalysts used for cleaning product gases have been demonstrated to be negatively affected by particulate matter (Gustafsson *et al.*, 2011).

Particulate matter can be removed using standard technologies such as cyclones, filters and separators, which also reduces the tar content of the gas stream, the extent of tar removal depending on the particle separation technology used. The presence of alkali is of particular importance, because alkali can form silicates with low melting temperatures that may negatively affect the filter operation.

9.8.4 Siloxane Removal

Siloxane compounds are a subgroup of silicone derivatives containing silicon-oxygen (Si-O) bonds with organic radicals which are widely used for a variety of industrial processes.

Siloxanes are used in products such as deodorants and shampoos, and can therefore be found in synthesis gas from sewage sludge treatment plants and in landfill gas. When siloxanes are burned, silicon oxide, a white powder, is formed which can create a problem in gas engines. Although most siloxanes disperse into the atmosphere where they are decomposed, some end up in wastewater, but more generally, siloxane derivatives generally end up as a significant component in the sludge.

As sludge undergoes anaerobic digestion, it may be subjected to temperatures up to 60°C (140°F). At this point the siloxanes contained in the sludge will volatize and become an unwanted constituent of the resulting synthesis gas. This problem can be exacerbated by the fact that silicone-based anti-foaming agents are frequently added to the anaerobic digesters and these silicones sometimes biodegrade into siloxanes. Unfortunately, when siloxanes gases are burned, they are usually converted into silicon dioxide particles, which are chemically and physically similar to sand.

Currently, there are six primary technologies for removing siloxanes from synthesis gas and include the following process options: (i) activated carbon is widely used to remove organic substances from gases and liquids due to its superior adsorbent properties, (ii) activated alumina (Al_2O_3) absorbs siloxanes from synthesis gas; when the alumina becomes saturated, the absorption capability can be recovered by passing a regeneration gas through it, (iii) refrigeration with condensation in combination can be used to selectively remove specific compounds by lowering the temperature or pressure of the gas, and then allowing the compound to precipitate out to a liquid, and then settle out, (iv) synthetic resins remove volatile materials through adsorption. They can be specially formulated to remove specific classes of compounds, (v) liquid absorbents are used by a small number of landfill operators to treat synthesis gas prior to use in combustion devices such as gas turbines, and (vi) membrane technology is a relatively recent development for siloxane removal; however, membranes are subject to acid deterioration from the acidic content usually found in raw synthesis gas. Of these technologies, activated carbon appears to be among the most dominant in the industry.

9.8.5 Alkali Metal Salt Removal

Compared with fossil fuels, biomass is rich in alkali salts that typically vaporize at high gasifier temperatures but condense downstream below 600°C (1110°F). Methods are available to strip the alkali contents since condensation of alkali salts can cause serious corrosion problems. The alkali will condense onto fine solid particles and can be subsequently captured in a cyclone, electrostatic precipitators, or filters when the gas temperature is below 600°C (1110°F). The hot gas can also be passed through a bed of active bauxite to remove alkali when cooling of gas is not permitted.

9.8.6 Biological Methods

Biological processes have mainly been used to remove hydrogen sulfide (Syed *et al.*, 2006). Hydrogen sulfide can be oxidized by microorganisms of the species Thiobacillus and Sulfolobus. The degradation requires oxygen and therefore a small amount of air (or pure oxygen if levels of nitrogen should be minimized) is added for biological desulfurization to take place. The degradation can occur inside the digester and can be facilitated by immobilizing the microorganisms occurring naturally in the digestate. An alternative is to use a trickling filter which the synthesis gas passes through when leaving the digester. In the trickling filter the microorganisms grow on a packing material. Synthesis gas with added air meets a counter flow of water containing nutrients. The sulfur-containing solution is removed and replaced when the pH drops below a certain level. Both methods are widely applied; however, they are not suitable when the synthesis gas is used as vehicle fuel or for grid injection due to the remaining traces of oxygen. An alternative system is available in which the absorption of the hydrogen sulfide is separated from the biological oxidation to sulfur – hence, the synthesis gas flow remains free of oxygen. There has also been focus on the removal of siloxane derivatives and carbon dioxide (Jensen and Webb, 1995; Soreanu, et al., 2011).

As another example, natural bacteria can convert hydrogen sulfide into elemental sulfur in the presence of oxygen and iron. This can be done by introducing a small amount (2 to 5%) of air into the head space of the digester. As a result, deposits of elemental sulfur will be formed in the digester. Even though this situation will reduce the hydrogen level, it will not lower it below that recommended for pipeline-quality gas. This process may be optimized by a more sophisticated design where air is bubbled through the digester feed material. It is critical that the introduction of the air be carefully controlled to avoid reducing the amount of synthesis gas that is produced.

In this section, the main biological processes for the removal of hydrogen sulfide are presented.

9.8.6.1 Biofiltration

Traditional gas cleaning and air pollution control technologies for pollutant gases, such as adsorption, absorption and combustion, were developed to treat high concentration waste gas streams associated with process emissions from stationary point sources (Mokhatab et al., 2006). Although these technologies rely on established physico-chemical principles to achieve effective control of gaseous pollutants, in many cases the control technique yields products which require further treatment before disposal or recycling of treatment materials. In the case of treatment of dilute waste gas streams; however, these traditional methods are relatively less effective, more expensive and wasteful in terms of energy consumption and identification of alternative control measures is warranted. A suitable alternate air pollution control technology is biofiltration, which utilizes naturally occurring microorganisms supported on a stationary bed (filter) to continuously treat contaminants in a flowing waste gas stream (Allen and Uang, 1992; Accettola et al., 2008).

Biofiltration by definition is the aerobic degradation of pollutants from (in the current context) synthesis gas in the presence of a carrier media. The early development work on biofiltration technology concentrated on organic media, such as, peat, compost, and wood bark. In general terms, organic compounds are degraded to carbon dioxide and water, while inorganic compounds, such as sulfur compounds, are oxidized to form oxygenated derivatives. The formation of these acidic compounds can lead to a lowering of pH of the filtration media, which in turn impacts on the performance of the system. Removal of the oxidized compound from the media is an important consideration in the design of biofiltration systems. Biofiltration, like many processes based on bio-treatment in the crude oil refining and gas processing industries (El-Gendy and Speight, 2016; Speight and El-Gendy, 2018), biofiltration is successfully emerging as a reliable, low-cost option for a broad range of air treatment applications. It is now becoming apparent that biological treatment will play a far more significant role in achieving environmental control on air emissions.

In the biofiltration process, three types of bioreactor designs are usually considered: the biofilter, (ii) the biotrickling filter, and (iii) the bioscrubber. The main differences between these systems concern their design and mode of operation: microorganism conditioning, the nature of the fluid phase (gas or liquid), and the presence or absence of stationary solid phases. Nevertheless, synthesis gas desulfurization has been carried out by the biotrickling filter and the bioscrubber.

The biofilter is a pollution control technique which involves using a bioreactor that contains living material to capture and biologically degrade pollutants such as hydrogen sulfide. Common uses include processing wastewater, capturing harmful chemicals or silt from surface runoff, and the macrobiotic oxidation of contaminants in gas streams. The technology finds greatest application in treating malodorous compounds and water-soluble volatile organic compounds (VOCs). Compounds treated are typically mixed VOCs and various sulfur compounds, including hydrogen sulfide. Very large airflows may be treated and although a large area (footprint) has typically been required. Engineered biofilters, designed and built since the early 1990s, have provided significant footprint reductions over the conventional flat-bed, organic media type.

In the process, the polluted gas flow is purified with biofiltration by conducting the gas flow upward through a filter bed that consists from biological material, e.g., compost, tree bark or peat. The filter material is a carrier of a thin water film in which micro-organisms live. The pollution in the gas flow is held back by ad- and absorption on the filter material, and then broken by present micro-organisms. The filter material serves as a supplier of necessary nutrients. The products of the conversion are carbon-dioxide, sulfate, nitrate. The dry weight of the filter varies typically from 40 to 60%. To reduce desiccation of the bed the gas flow must has to saturated with water. For this reason polluted gas flow is moistened before it goes through the biofilter, which is achieved by using a pre-scrubber. The relative humidity of the gas must be 95%. In practice it is always better to apply a moistener to protect the biofilter against dehydration.

A biotrickling filter is a packed-bed bioreactor with immobilized biomass (Montebello *et al.*, 2012; Fernández *et al.*, 2013a, 2013b; Montebello *et al.*, 2013). The gas flows through a fixed bed co-currently or countercurrently to a mobile liquid phase. Synthetic carriers are usually used and these include plastic, ceramic, lava rocks, polyurethane foam, etc. The synthetic carrier does not provide any nutrients so the liquid mobile phase must contain nutrients for the growth and maintenance of the biomass. Programmed or continuous discharge of recirculation medium help to remove the oxidation products. The hydrogen sulfide must be transferred from the gas to liquid phase and the degradation is finally carried out in the biofilm.

According to the microbial cycling of sulfur, the biological oxidation of hydrogen sulfide to sulfate is one of the major reactions involved. There are numerous sulfur-oxidizing microorganisms, but hydrogen sulfide is exclusively oxidized by prokaryotes. In any case, the biotrickling filter is typically inoculated with aerobic active sludge and during the process a specific biomass population is developed.

A biotrickling filter can be operated under aerobic or anoxic conditions. The removal of hydrogen sulfide from synthesis gas has been mainly studied under aerobic conditions and the overall reaction is:

$$HS^- + 2O_2 \rightarrow SO_4^{2-} + 2H^+$$

In summary, two compounds can be produced; sulfate and elemental sulfur, and the production ratio of sulfate (SO_4^{2-}) will be dependent on the oxygen-hydrogen sulfide ratio and the trickling liquid velocity. Air is used to supply oxygen and must, therefore, be supplied in excess. However, control of the air supply is an important issue for safety and operational reasons – the lower and upper explosive limits for methane in air are 5% v/v and 15% v/v, respectively.

On the other hand, synthesis gas can be diluted (mainly by the nitrogen content in air) and undesired residual oxygen can be found in the outlet stream. Therefore, a high oxygen supply is a drawback because the calorific power is also decreased and, moreover, a lack of oxygen increases clogging problems caused by the formation of elemental sulfur. The synthesis gas dilution depends on the efficiency in the mass transfer and the concentration of hydrogen sulfide. An alternative to the direct supply of air mixed with the synthesis gas stream involves the use of venture-based devices to increase the oxygen mass transfer. However, the installation of an aerated liquid recirculation system could produce hydrogen sulfide stripping of the dissolved sulfide.

The anoxic biotrickling filter is based on dissimilatory nitrate reduction. Dissimilatory nitrate reduction is carried out by certain bacteria that can use nitrate and/or nitrite as electron acceptors instead of oxygen:

- Complete denitrification *vs.* partial hydrogen sulfide oxidation

$$5H_2S + NO_3^- \rightarrow 5S + N_2 + 4H_2O + 2OH^-$$

- Complete denitrification *vs.* complete hydrogen sulfide oxidation

$$5H_2S + 8NO_3^- \rightarrow 5SO_4^{2-} + 4N_2 + 4H_2O + 2H_+$$

- Partial denitrification *vs.* partial hydrogen sulfide oxidation

$$H_2S + 2NO_3^- \rightarrow S + 4NO_2^- + H_2O$$

- Partial denitrification *vs.* complete hydrogen sulfide oxidation

$$H_2S + 4NO_3^- \rightarrow SO_4^{2-} + 4NO_2^- + 2H_+$$

9.8.6.2 Bioscrubbing

Bioscrubber systems have been used for hydrogen sulfide removal from synthesis gas (Nishimura and Yoda, 1997). The bioscrubber involves a two-stage process with an absorption tower and a bioreactor, in which the sulfide is oxidized to sulfur and/or sulfate. For example, the Shell-Paques THIOPAQ® process employs alkaline conditions to produce elemental sulfur. In the first step of this process, the hydrogen sulfide is absorbed into an alkaline solution by reaction with hydroxyl and bicarbonate ions. In the second step the hydrosulfide is oxidized to elemental sulfur under oxygen-limiting conditions. Thus:

$$H_2S + OH^- \rightarrow HS^- + H_2O$$

$$H_2S + HCO_3^- \rightarrow HS^- + CO_2 + H_2O$$

$$H_2S + CO_3^{2-} \rightarrow HS^- + HCO_3^-$$

9.8.6.3 Bio-Oxidation

Bio-oxidation (biological oxidation) is one of the most used methods of desulfurization, based on injection of a small amount of air (2 to 8 % v/v) into the raw synthesis gas. This way, the hydrogen sulfide is biologically oxidized either to solid free sulfur or to liquid sulfurous acid (H_2SO_3) (Mokhatab *et al.*, 2006; Speight, 2014, 2019):

$$2H_2S + O_2 \rightarrow 2H_2O + 2S$$

$$2H_2S + 3O_2 \rightarrow 2H_2SO_3$$

In practice, the precipitated sulfur is collected and added to the storage tanks where it is mixed with digestate, in order to improve the fertilizer properties of digestate. Biological desulfurization is frequently carried out inside the digester and, for this kind of desulfurization, oxygen and *Sulfobacter oxydans* bacteria must be present, to convert hydrogen sulfide into elementary sulfur, in the presence of oxygen. Typically, *Sulfobacter oxydans* is present inside the digester (does not have to be added) as the anaerobic digester substrate contains the necessary nutrients for their metabolism. In the process, the air is injected directly in the headspace of the digester and the reactions occur in the reactor headspace, on the floating layer (if existing) and on reactor walls. Due to the acidic nature of the products there is the risk of corrosion. The process is dependent of the existence of a stable floating layer inside the digester and the process often takes place in a separate reactor.

Chemical synthesis gas desulfurization can take place outside of digester, using, for example, a base (usually sodium hydroxide). Another chemical method to reduce the content of hydrogen sulfide is to add commercial ferrous solution (Fe^{2+}) to the feedstock. Ferrous compounds bind sulfur in an insoluble compound in the liquid phase, thereby preventing the production of gaseous hydrogen sulfide.

9.9 Tail Gas Cleaning

Tail gas cleaning (also called tail gas treating) involves the removal of the remaining sulfur compounds from gases remaining after sulfur recovery. Tail gas from a typical Claus process, whether a conventional Claus or one of the extended versions of the process (Parkash, 2003; Mokhatab *et al.*, 2006; Gary *et al.*, 2007; Speight, 2014; Hsu and Robinson, 2017; Speight, 2017, 2019) usually contains small but varying quantities of carbonyl sulfide, carbon disulfide, hydrogen sulfide, and sulfur dioxide as well as sulfur vapor. In addition, there may be hydrogen, carbon monoxide, and carbon dioxide in the tail gas. In order to remove the rest of the sulfur compounds from the tail gas, all of the sulfur-bearing species must first be converted to hydrogen sulfide which is then absorbed into a solvent and the clean gas vented or recycled for further processing.

It is necessary to develop and implement reliable and cost-effective technologies to cope with the various sales specifications and environmental requirements. In response to this trend, several new technologies are now emerging to comply with the most stringent regulations. The typical sulfur recovery efficiencies for a Claus plant is on the order of 90 to 96% w/w for a two-stage plant and 95 to 98% w/w for a three-stage plant. Most environmental agencies require the sulfur recovery efficiency to be in the range of 98.5 to 99.9+% w/w and, as a result, there is the need to reduce the sulfur content of the Claus plant tail gas need to be reduced further.

Moreover, a sulfur removal process must be very precise, since some synthesis gas streams may contain only a small quantity of sulfur-containing compounds that must be reduced several orders of magnitude. Most consumers of gas require less than 4 ppm v/v in the gas – a characteristic feature of a gas stream that contains hydrogen sulfide is the presence of carbon dioxide (generally in the range of 1 to 4% v/v). In cases where the gas stream does not contain hydrogen sulfide, there may also be a relative lack of carbon dioxide.

9.9.1 Claus Process

The Claus process is not so much a gas cleaning process but a process for the disposal of hydrogen sulfide, a toxic gas that originates in most gas streams. Burning hydrogen sulfide as a fuel gas component or as a flare gas component is precluded by safety and environmental considerations since one of the combustion products is the highly toxic sulfur dioxide (SO_2), which is also toxic. As described above, hydrogen sulfide is typically removed from the refinery light ends gas streams through an olamine process after which application of heat regenerates the olamine and forms an acid gas stream. Following from this, the acid gas stream is treated to convert the hydrogen sulfide elemental sulfur and water. The conversion process utilized in most modern refineries is the Claus process, or a variant thereof.

The Claus process involves combustion of approximately one-third of the hydrogen sulfide to sulfur dioxide and then reaction of the sulfur dioxide with the remaining hydrogen sulfide in the presence of a fixed bed of activated alumina, cobalt molybdenum catalyst resulting in the formation of elemental sulfur:

$$2H_2S + 3O_2 \rightarrow 2SO_2 + 2H_2O$$

$$2H_2S + SO_2 \rightarrow 3S + 2H_2O$$

Different process flow configurations are in use to achieve the correct hydrogen sulfide/sulfur dioxide ratio in the conversion reactors.

In a split-flow configuration, one-third split of the acid gas stream is completely combusted and the combustion products are then combined with the non-combusted acid gas upstream of the conversion reactors. In a once-through configuration, the acid gas stream is partially combusted by only providing sufficient oxygen in the combustion chamber to combust one-third of the acid gas.

Two or three conversion reactors may be required depending on the level of hydrogen sulfide conversion required. Each additional stage provides incrementally less conversion than the previous stage. Overall, conversion of 96 to 97% v/v of the hydrogen sulfide to elemental sulfur is achievable in a Claus process. If this is insufficient to meet air quality regulations, a Claus process tail gas treater is utilized to remove essentially the entire remaining hydrogen sulfide in the tail gas from the Claus unit. The tail gas treater may employ a proprietary solution to absorb the hydrogen sulfide followed by conversion to elemental sulfur.

9.9.2 SCOT Process

The SCOT (Shell Claus Off-gas Treating) unit is a most common type of tail gas unit and uses a hydrotreating reactor followed by amine scrubbing to recover and recycle sulfur, in the form of hydrogen, to the Claus unit (Nederland, 2004).

Early SCOT units used diisopropanolamine in the Sulfinol (Sulfinol-D) solution. Methyl diethanolamine-based Sulfinol (Sulfinol-M) was used later to enhance hydrogen sulfide removal and to allow for selective rejection of carbon dioxide in the absorber. To achieve the lowest possible hydrogen sulfide content in the treated gas, the Super-SCOT configuration was introduced. In this version, the loaded Sulfinol-M solution is regenerated in two stages. The partially stripped solvent goes to the middle of the absorber, while the fully stripped solvent goes to the top of the absorber. The solvent going to the top of the absorber is cooled below that used in the conventional SCOT process.

In the process, tail gas (containing hydrogen sulfide and sulfur dioxide) is contacted with hydrogen and reduced in a hydrotreating reactor to form hydrogen sulfide and water. The catalyst is typically cobalt/molybdenum on alumina. The gas is then cooled in a water contractor. The hydrogen sulfide-containing gas enters an amine absorber which is typically in a system segregated from the other refinery amine systems. The purpose of segregation is twofold: (i) the tail gas treater frequently uses a different amine than the rest of the plant, and (ii) the tail gas is frequently cleaner than the refinery fuel gas (in regard to contaminants) and segregation of the systems reduces maintenance requirements for the SCOT unit. Amines chosen for use in the tail gas system tend to be more selective for hydrogen sulfide and are not affected by the high levels of carbon dioxide in the off-gas.

However, the SCOT process can be configured in various ways. For example, it can be integrated with the upstream acid gas removal unit if the same solvent is used in both units. Another configuration has been used to cascade the upstream gas cleanup with the SCOT unit. A hydrogen sulfide-lean acid gas from the upstream gas treating unit is sent to a SCOT process with two absorbers. In the first absorber, the hydrogen sulfide-lean acid gas is enriched, while the second absorber treats the Claus tail gas. A common stripper is used for both SCOT absorbers and, in this latter configuration, different solvents could be used in both the upstream and the SCOT units.

A hydrotreating reactor converts sulfur dioxide in the off-gas to hydrogen sulfide that is then contacted with a Stretford solution (a mixture of a vanadium salt, anthraquinone disulfonic acid, sodium carbonate, and sodium hydroxide) in a liquid-gas absorber (Abdel-Aal *et al.*, 2016). The hydrogen sulfide reacts stepwise with sodium carbonate and the anthraquinone sulfonic acid to produce elemental sulfur, with vanadium serving as a catalyst. The solution proceeds to a tank where oxygen is added to regenerate the reactants. One or more froth or slurry tanks are used to skim the product sulfur from the solution, which is recirculated to the absorber. Other tail gas treating processes include: (i) caustic scrubbing, (ii) polyethylene glycol treatment, (iii) Selectox process, and (iv) a sulfite/bisulfite tail gas treating (Mokhatab *et al.*, 2006; Speight, 2014, 2019).

References

Abatzoglou, N. and Boivin, S. 2009. A Review of Biogas Purification Processes. *Biofuels Bioprod. Biorefin.*, 3(1): 42-71.

Abdel-Aal, H.K., Aggour, M.A., and Fahim, M.A. 2016. *Petroleum and Gas Field Processing*. CRC Press, Taylor & Francis Publishers, Boca Raton Florida.

Abraham, H. 1945. *Asphalts and Allied Substances*. Van Nostrand Scientific Publishers, New York.

Accettola, F., Guebitz, G.M. and Schoeftner, R. 2008. Siloxane Removal from Biogas by Biofiltration: Biodegradation Studies. *Clean Technol. Environ. Policy*, 10(2): 211-218.

Ajhar, M., Travesset, M., Yüce, S. and Melin, T. 2010. Siloxane Removal from Landfill and Digester Gas – A Technology Overview. *Bioresour. Technol.*, 101(9): 2913-2923.

Allen, R., and Yang, Y. 1992. Biofiltration: An Air Pollution Control Technology for Hydrogen Sulfide Emissions. In: *Industrial Environmetnal Chemistry: Waste minimization in Industrial Processes and Remediation of Hazardous Waste*. D.T. Sawyer and A.E. Martell Editors). Springer, Boston, Massachusetts. Page 273-287.

Alonso-Vicario, A., Ochoa-Gómez, J.R., Gil-Río, S., Gómez-Jiménez-Aberasturi, O., Ramírez-López, C.A. and Torrecilla-Soria, J. 2010. Purification and Upgrading of Biogas by Pressure Swing Adsorption on Synthetic and Natural Zeolites. *Microporous Mesoporous Mater.*, 134(1–3): 100-107.

Andriani, D., Wresta, A., Atmaja, T.D. and Saepudin, A. 2014. A Review on Optimization Production and Upgrading Biogas Through CO Removal Using Various Techniques. Appl. Biochem. Biotechnol., 172(4): 1909-1928.

Anis, S., and Zainal, Z.A. 2011. Tar reduction in biomass producer gas via mechanical, catalytic and thermal methods: a review. *Renew Sustain Energy Rev.*, 15: 2355-2377.

ASTM. 2018. *Annual Book of Standards,* ASTM International, West Conshohocken, Pennsylvania.

ASTM D3246. 2018. Standard Test Method for Sulfur in Petroleum Gas by Oxidative Microcoulometry. *Annual Book of Standards*, ASTM International, West Conshohocken, Pennsylvania.

Barbouteau, L., and Dalaud, R. 1972. In *Gas Purification Processes for Air Pollution Control*. G. Nonhebel (Editor). Butterworth and Co., London, United Kingdom. Chapter 7.

Basu, S., Khan, A.L., Cano-Odena, A., Liu, C. and Vankelecom, I.F. 2010. Membrane-based Technologies for Biogas Separations. *Chem. Soc. Rev.*, 39(2): 750-768.

Boulinguiez, B. and Le Cloirec, P. 2009. Biogas Pre-upgrading by Adsorption of Trace Compounds onto Granular Activated Carbons and an Activated Carbon Fiber Cloth. *Water Sci. Technol.*, 59(5): 935-944.

Cherosky, P. and Li, Y. 2013. Hydrogen Sulfide Removal from Biogas by Bio-based Iron Sponge. *Biosyst. Eng.*, 114(1): 55-59.

Curry, R.N. 1981. *Fundamentals of Natural Gas Conditioning*. PennWell Publishing Co., Tulsa, Oklahoma.

Deshmukh, M.G. and Shete, A. 2013. Oxidative Absorption of Hydrogen Sulfide Using Iron-Chelate Based Process: Chelate Degradation. *J. Anal. Bioanal. Tech.*, 88(3): 432-36.

Devi, L., Ptasinski, K.J., and Jenssen, F.J.J.G. 2003. A Review of the Primary Measures for Tar Elimination in Biomass Gasification Processes. *Biomass Bioenergy*, 24: 125-140.

Dou, B., Zhang, M., and Gao, J. 2002. High-temperature removal of NH3, Organic Sulfur, HCl, and Tar Component from Coal-Derived Gas. *Ind. Eng. Chem. Res.*, 41: 4195-4200.

El-Gendy, N.Sh., and Speight, J.G. 2016. *Handbook of Refinery Desulfurization*. CRC Press, Taylor & Francis Group, Boca Raton, Florida.

Fernández, M., Ramírez, M., Gómez, J.M. and Cantero, D. 2013a. Biogas Biodesulfurization in an anoxic Biotrickling Filter Packed with Open-Pore Polyurethane Foam. J. Hazard. Mater., 264: 529-535.

Fernández, M., Ramírez, M., Pérez, R.M., Gómez, J.M. and Cantero, D. 2013b. Hydrogen Sulfide Removal From Biogas by an Anoxic Biotrickling Filter Packed with Pall Rings. *Chem. Eng. J.*, 225: 456-463.

Forbes, R. J. 1958a. *A History of Technology*. Oxford University Press, Oxford, United Kingdom.

Forbes, R. J. 1958b. *Studies in Early Petroleum Chemistry*. E. J. Brill, Leiden, The Netherlands.

Forbes, R.J. 1959. *More Studies in Early Petroleum Chemistry*. E.J. Brill, Leiden, The Netherlands:

Forbes, R. J. 1964. *Studies in Ancient Technology*. E. J. Brill, Leiden, The Netherlands.

Gary, J.G., Handwerk, G.E., and Kaiser, M.J. 2007. *Petroleum Refining: Technology and Economics,* 5th Edition. CRC Press, Taylor & Francis Group, Boca Raton, Florida.

Gislon, P., Galli, S. and Monteleone, G. 2013. Siloxanes Removal from Biogas by High Surface Area Adsorbents. *Waste Manag.*, 33(12): 2687-2693.

Gustafsson, E., Lin, L., and Seemann, M. 2011 Characterization of Particulate Matter in the Hot Product Gas from Indirect Steam Bubbling Fluidized Bed Gasification of Wood Pellets. *Energy Fuels*, 25: 1781–1789.

Hoiberg A.J. 1960. *Bituminous Materials: Asphalts, Tars and Pitches, I & II*. Interscience, New York.

Hsu, C.S., and Robinson, P.R. (Editors). 2017. *Handbook of Petroleum Technology*. Springer International Publishing AG, Cham, Switzerland.

Huang, J., Schmidt, K.G., Bian, Z. 2011. Removal and Conversion of Tar in Syngas from Woody Biomass Gasification for Power Utilization Using Catalytic Hydrocracking. *Energies* 4: 1163-1177.

Jensen, A.B. and Webb, C. 1995. Treatment of H_2S-containing Gases – A Review of Microbiological Alternatives. *Enzyme Microb. Tech.*, 17(1): 2-10.

Jou, F.Y., Otto, F.D., and Mather, A.E. 1985. In *Acid and Sour Gas Treating Processes*. S.A. Newman (Editor). Gulf Publishing Company, Houston, Texas. Chapter 10.

Katz, D.K. 1959. *Handbook of Natural Gas Engineering*. McGraw-Hill Book Company, New York.

Kidnay, A.J., and Parrish, W.R. 2006 *Fundamentals of Natural Gas Processing*. CRC Press, Taylor & Francis Group, Boca Raton, Florida.

Kohl, A.L., and Nielsen, R.B., 1997. *Gas Purification*. Gulf Publishing Company, Houston, Texas.

Kohl, A. L., and Riesenfeld, F.C. 1985. *Gas Purification*. 4th Edition, Gulf Publishing Company, Houston, Texas.

Maddox, R.N. 1982. *Gas Conditioning and Processing*. Volume 4. *Gas and Liquid Sweetening*. Campbell Publishing Co., Norman, Oklahoma.

Maddox, R.N., Bhairi, A., Mains, G.J., and Shariat, A. 1985. In *Acid and Sour Gas Treating Processes*. S.A. Newman (Editor). Gulf Publishing Company, Houston, Texas. Chapter 8.

May-Britt, H. 2008. Membranes in Gas Separation. In: *Handbook of Membrane Separations*. CRC Press, Taylor & Francis Group, Boca Raton, Florida. Page 65-106.

Mody, V., and Jakhete, R. 1988. *Dust Control Handbook*. Noyes Data Corp., Park Ridge, New Jersey.

Mokhatab, S., Poe, W.A., and Speight, J.G. 2006. *Handbook of Natural Gas Transmission and Processing*. Elsevier, Amsterdam, Netherlands.

Montebello, A.M., Fernández, M., Almenglo, F., Ramírez, M., Cantero, D. and Baeza, M. 2012. Simultaneous Methyl Mercaptan and Hydrogen Sulfide Removal in the Desulfurization of Biogas in Aerobic and Anoxic Biotrickling Filters. *Chem. Eng. J.*, 200/202(0): 237-246.

Montebello, A.M., Bezerra, T., Rovira, R., Rago, L., Lafuente, J. and Gamisans, X. 2013. Operational Aspects, pH Transition and Microbial Shifts of a H2S Desulfurizing Biotrickling Filter with Random Packing Material. *Chemosphere*, 93(11): 2675-2682.

Newman, S.A. 1985. *Acid and Sour Gas Treating Processes*. Gulf Publishing, Houston, Texas.

Nikolić, D., Kikkinides, E.S. and Georgiadis, M.C. 2009. Optimization of Multibed Pressure Swing Adsorption Processes. *Ind. Eng. Chem. Res.*, 48(11): 5388-5398.

Niesner, J., Jecha, D. and Stehlík, P. 2013. Biogas Upgrading Technologies: State of the Art Review in the European Region. *Chem. Eng. Trans.*, 35: 517-522.

Nishimura, S. and Yoda, M. 1997. Removal of Hydrogen Sulfide from an Anaerobic Biogas Using a Bio-Scrubber. *Water Sci. Technol.*, 36(6-7): 349-356.

Ozturk, B. and Demirciyeva, F. 2013. Comparison of Biogas Upgrading Performances of Different Mixed Matrix Membranes. *Chem. Eng. J.*, 222: 209-217.

Parkash, S. 2003. *Refining Processes Handbook*. Gulf Professional Publishing, Elsevier, Amsterdam, Netherlands.

Pitsinigos, V.D., and Lygeros, A.I. 1989. Predicting H_2S-MEA Equilibria. *Hydrocarbon Processing* 58(4): 43-44.

Polasek, J., and Bullin, J. 1985. In *Acid and Sour Gas Treating Processes*. S.A. Newman (Editor). Gulf Publishing Company, Houston, Texas. Chapter 7.

Proell, T., Siefert, I.G., and Friedl, A. 2005. Removal of NH3 from Biomass Gasification Producer Gas by Water Condensing in an Organic Solvent Scrubber. *Ind. Eng. Chem. Res.*, 44: 1576-1584.

Ramírez M., Gómez J.M. and Cantero D. 2015. Biogas: Sources, Purification and Uses. In: *Hydrogen and Other Technologies*. U.C. Sharma, S. Kumar, R. Prasad (Editors). Studium Press LLC, New Delhi, India. Chapter: 13, page 296-323.

Rapagna, S., Gallucci, K., and Marcello, M.D. 2010. Gas Cleaning, Gas Conditioning and Tar Abatement by Means of a Catalytic Filter Candle in a Biomass Fluidized-Bed Gasifier. *Bioresour. Technol.*, 101: 7123-7130.

Reddy, S., and Gilmartin, J. 2008. Fluor's Econamine Plus Technology for Post-Combustion CO2 Capture. *Proceedings. GPA Gas Treatment Confernece, Amsterdam, Netherlands. February 20-22.* https://www.fluor.com/SiteCollectionDocuments/FluorEFG-forPost-CombustionCO2CaptureGPAConf-Feb2008.pdf

Rongwong, W., Boributh, S., Assabumrungrat, S., Laosiripojana, N. and Jiraratananon, R. 2012. Simultaneous Absorption of CO2 and H2S from Biogas by Capillary Membrane Contactor. *J. Membrane Sci.*, 392/393: 38-47.

Ryckebosch, E., Drouillon, M. and Vervaeren, H. 2011. Techniques for Transformation of Biogas to Biomethane. *Biomass Bioenergy*, 35(5): 1633-1645.

Santos, M.P.S., Grande, C.A. and Rodrigues, A.E. 2013. Dynamic Study of the Pressure Swing Adsorption Process for Biogas Upgrading and Its Responses to Feed Disturbances. *Ind. Eng. Chem. Res.*, 52(15): 5445-5454.

Soreanu, G., Béland, M., Falletta, P., Edmonson, K., Svoboda, L. and Al-Jamal, M. 2011. Approaches Concerning Siloxane Removal from Biogas – A Review. *Canadian Biosys. Eng.*, 53: 8.1-8.18.

Soud, H., and Takeshita, M. 1994. *FGD Handbook*. No. IEACR/65. International Energy Agency Coal Research, London, United Kingdom.

Speight, J.G. 1978. *Personal Observations at Archeological Digs at The Cities of Babylon, Calah, Nineveh, and Ur*. College of Science, University of Mosul, Iraq.

Speight, J.G. 2008. *Synthetic Fuels Handbook: Properties, Processes, and Performance.* McGraw-Hill, New York.
Speight, J.G. (Editor), 2011. *The Biofuels Handbook.* Royal Society of Chemistry, London, United Kingdom.
Speight, J.G. 2012. *Shale Oil Production Processes.* Gulf Professional Publishing, Elsevier, Oxford, United Kingdom.
Speight, J.G. 2014. *The Chemistry and Technology of Petroleum* 5th Edition. CRC-Taylor and Francis Group, Boca Raton, Florida.
Speight, J.G. 2015. *Handbook of Petroleum Product Analysis* 2nd Edition. John Wiley & Sons Inc., Hoboken, New Jersey.
Speight, J.G. 2017. *Handbook of Petroleum Refining.* CRC Press, Taylor & Francis Group, Boca Raton, Florida.
Speight, J.G. 2018. *Handbook of Natural Gas Analysis.* John Wiley & Sons Inc., Hoboken, New Jersey.
Speight, J.G., and El-Gendy, N.Sh. 2018. *Introduction to Petroleum Biotechnology.* Gulf Professional Publishing Company, Elsevier, Cambridge, Massachusetts.
Speight, J.G. 2019. Natural Gas: *A Basic Handbook* 2nd Edition. Gulf Publishing Company, Elsevier, Cambridge, Massachusetts.
Syed, M., Soreanu, G., Falletta, P. and Béland, M. 2006. Removal of Hydrogen Sulfide from Gas Streams Using Biological Processes – A Review. *Canadian Biosys. Eng.,* 48: 2.1-2.14.
Torres, W., Pansare, S.S., and Goodwin, J.G. 2007. Hot Gas Removal of Tars, Ammonia, and Hydrogen Sulfide From Biomass Gasification Gas. *Catal. Rev.,* 49: 407-56.
Yuan, S., Zhou, Z., and Li, J. 2010. HCN and NH3 Released from Biomass and Soybean Cake Under Rapid Pyrolysis. *Energy Fuels,* 24: 6166-6171.

Part 2
FUELS AND CHEMICALS FROM SYNTHESIS GAS

10

The Fischer-Tropsch Process

10.1 Introduction

In a previous chapter (Chapter 5), there have been references to the use of the gasification process to convert carbonaceous feedstocks, such as crude oil residua, tar sand bitumen, coal, oil shale, and biomass, into the starting chemicals for the production of petrochemicals. The chemistry of the gasification process is based on the thermal decomposition of the feedstock and the reaction of the feedstock carbon and other pyrolysis products with oxygen, water, and fuel gases such as methane and is represented by a sequence of simple chemical reactions (Table 10.1). However, the gasification process is often considered to involve two distinct chemical stages: (i) devolatilization of the feedstock to produce volatile matter and char, (ii) followed by char gasification, which is complex and specific to the conditions of the reaction – both processes contribute to the complex kinetics of the gasification process (Sundaresan and Amundson, 1978).

The Fischer-Tropsch process is a catalytic chemical reaction in which carbon monoxide (CO) and hydrogen (H_2) in the synthesis are converted into hydrocarbon derivatives of various molecular weights. The synthesis gas that is required as the feedstock for the Fischer-Tropsch process is typically produced by gasification of coal (Chapter 4), crude oil resid (Chapter 5), biomass (Chapter 6), or organic waste (Chapter 7), or by various reforming processes (Chapter 8) (Higman and Van der Burgt, 2008; Liu *et al.*, 2010; Rostrup-Nielsen and Christiansen, 2011; Hu *et al.*, 2012). The process can be represented by the simple equation:

$$(2n+1)H_2 + nCO \rightarrow C_nH_{(2n+2)} + nH_2O$$

In this equation, n is an integer. Thus, for n = 1, the reaction represents the formation of methane, which in most gas-to-liquids (GTL) applications is considered an undesirable byproduct.

The Fischer-Tropsch process conditions are usually chosen to maximize the formation of higher molecular weight hydrocarbon liquid fuels which are higher value products. There are other side reactions taking place in the process, among which the water-gas-shift reaction is predominant:

$$CO + H_2O \rightarrow H_2 + CO_2$$

Depending on the catalyst, temperature, and type of process employed, hydrocarbon derivatives ranging from methane to higher molecular paraffin derivatives and olefin derivatives can be obtained. Small amounts of low molecular weight oxygenated derivatives (such as alcohol derivatives and organic acid derivatives) are also formed. Typically,

James G. Speight. *Synthesis Gas: Production and Properties*, (313–348) © 2020 Scrivener Publishing LLC

Table 10.1 Reactions that occur during gasification of a carbonaceous feedstock.

$2C + O_2 \rightarrow 2CO$
$C + O_2 \rightarrow CO_2$
$C + CO_2 \rightarrow 2CO$
$CO + H_2O \rightarrow CO_2 + H_2$ (shift reaction)
$C + H_2O \rightarrow CO + H_2$ (water gas reaction)
$C + 2H_2 \rightarrow CH_4$
$2H_2 + O_2 \rightarrow 2H_2O$
$CO + 2H_2 \rightarrow CH_3OH$
$CO + 3H_2 \rightarrow CH_4 + H_2O$ (methanation reaction)
$CO_2 + 4H_2 \rightarrow CH_4 + 2H_2O$
$C + 2H_2O \rightarrow 2H_2 + CO_2$
$2C + H_2 \rightarrow C_2H_2$
$CH_4 + 2H_2O \rightarrow CO_2 + 4H_2$

Fischer-Tropsch liquids are (unless the process is designed for the production of other products) hydrocarbon products (which vary from naphtha-type liquids to wax) and non-hydrocarbon products. The production of non-hydrocarbon products requires adjustment of the feedstock composition and the process parameters.

Briefly, synthesis gas is the name given to a gas mixture that contains varying amounts of carbon monoxide (CO) and hydrogen (H_2) generated by the gasification of a carbonaceous material. Examples include steam reforming of natural gas, crude oil residua, coal, and biomass. Synthesis gas is used as an intermediate in producing hydrocarbon derivatives via the Fischer-Tropsch process for use as gaseous and liquids fuels.

The synthesis gas is produced by the gasification conversion of carbonaceous feedstock such as crude oil residua, coal, and biomass and production of hydrocarbon products can be represented simply as:

$$CH_{feedstock.} + O_2 \rightarrow CO + H_2$$

$$nCO + nH_2 \rightarrow C_nH_{2n+2}$$

However, before conversion of the carbon monoxide and hydrogen to hydrocarbon products, several reactions are employed to adjust the hydrogen/carbon monoxide ratio. Most important is the water gas shift reaction in which additional hydrogen is produced (at the expense of carbon monoxide) to satisfy the hydrogen/carbon monoxide ratio necessary for the production of hydrocarbon derivatives:

$$H_2O + CO \rightarrow H_2 + CO_2$$

The boiling range of Fischer-Tropsch typically spans the naphtha and kerogen boiling ranges and is suitable for analysis by application of the standard test methods. With the suitable choice of a catalyst the preference for products boiling in the naphtha range (<200°C, <390°F) or for product boiling in the diesel range (approximately 150 to 300°C, 300 to 570°F) can be realized.

The other product that is worthy of consideration is *bio-oil* (*pyrolysis oil, bio-crude*), which is the liquid product produced by the thermal decompositon (destructive distillation) of biomass (Chapter 3) at temperatures on the order of 500°C (930°F). The product is a synthetic crude oil and is of interest as a possible complement (eventually a substitute) to crude oil. The product can vary from a light tarry material to a free-flowing liquid – both require further refining to produce specification grade fuels (Parkash, 2003; Gary *et al.*, 2007; Speight, 2011, 2014a; Hsu and Robinson, 2017; Speight, 2017).

Hydrocarbon moieties are predominant in the product but the presence of varying levels of oxygen (depending upon the character of the feedstock) requires testament (using for example, hydrotreating) during refining. On the other hand, the bio-oil can be used as a feedstock to the Fischer-Tropsch process for the production of lower-boiling products, as is the case when naphtha and gas oil are used as feedstocks for the Fischer-Tropsch process. In summary, the Fischer-Tropsch process produces hydrocarbon products of different molecular weight from a gas mixture of carbon monoxide and hydrogen (synthesis gas) all of which can find use in various energy scenarios.

In the current context, the most valuable product is *synthesis gas* – the mixture of carbon monoxide (CO) and hydrogen (H_2) that is the beginning of a wide range of chemicals (Figure 10.1). The production of synthesis gas, i.e., mixtures of carbon monoxide and hydrogen has been known for several centuries. But it is only with the commercialization of the Fischer-Tropsch reaction that the importance of synthesis gas has been realized. The thermal cracking (pyrolysis) of crude oil or fractions thereof was an important

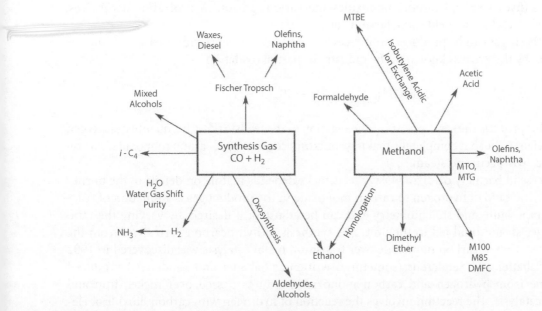

Figure 10.1 Routes to chemicals from synthesis gas and methanol.

method for producing gas in the years following its use for increasing the heat content of water gas.

In addition to the gases obtained by distillation of crude oil, further gaseous products are produced during the processing of naphtha and middle distillate to produce gasoline. Hydrodesulfurization processes involving treatment of naphtha, distillates and residual fuels and from the coking or similar thermal treatment of vacuum gas oils and residual fuel oils also produce gaseous products.

The chemistry of the oil-to-gas conversion has been established for several decades and can be described in general terms although the primary and secondary reactions can be truly complex. The composition of the gases produced from a wide variety of feedstocks depends not only on the severity of cracking but often to an equal or lesser extent on the feedstock type. In general terms, gas heating values are on the order of 950 to 1350 Btu/ft^3.

A second group of refining operations which contribute to gas production are the catalytic cracking processes, such as fluid-bed catalytic cracking, and other variants, in which heavy gas oils are converted into gas, naphtha, fuel oil and coke (Parkash, 2003; Gary *et al.*, 2007; Speight, 2011, 2014a; Hsu and Robinson, 2017; Speight, 2017). The catalysts will promote steam-reforming reactions that lead to a product gas containing more hydrogen and carbon monoxide and fewer unsaturated hydrocarbon products than the gas product from a non-catalytic process. The resulting gas is more suitable for use as a medium heat-value gas than the rich gas produced by straight thermal cracking. The catalyst also influences the reaction rates in the thermal cracking reactions, which can lead to higher gas yields and lower tar and carbon yields.

Almost all crude oil fractions can be converted into gaseous fuels, although conversion processes for the heavier fractions require more elaborate technology to achieve the necessary purity and uniformity of the manufactured gas stream. In addition, the thermal yield from the gasification of heavier feedstocks is invariably lower than that of gasifying light naphtha or liquefied crude oil gas since, in addition to the production of synthesis gas components (hydrogen and carbon monoxide) and various gaseous hydrocarbon derivatives, heavy feedstocks also yield some tar and coke.

Synthesis gas can be produced from heavy oil and other heavy crude feedstocks (such as residua) by the process known in the industry as partial oxidation:

$$[2CH]_{crude\ oil} + O_2 \rightarrow 2CO + H_2$$

In this process, the step consists of the reaction of the feedstock with a quantity of oxygen insufficient to burn it completely, making a mixture consisting of carbon monoxide, carbon dioxide, hydrogen, and steam.

Success in partially oxidizing heavy feedstocks depends mainly on details of the burner design. The ratio of hydrogen to carbon monoxide in the product gas is a function of reaction temperature and stoichiometry and can be adjusted, if desired, by varying the ratio of carrier steam to oil fed to the unit. The synthesis of hydrocarbon derivatives from the hydrogenation of carbon monoxide over transition metal catalysts was discovered in 1902 when Sabatier and Senderens (sometimes written as Sabatier and Sanderens) produced methane from hydrogen and carbon monoxide mixtures passed over nickel, iron and cobalt catalysts. The reaction involves the reaction of hydrogen with carbon dioxide at elevated temperatures (typically 300 to 400°C, 570 to 750°F) and pressures in the presence of a nickel catalyst to produce methane and water. In 1923, Fischer and Tropsch reported the use of alkalized iron catalysts to produce liquid hydrocarbon derivatives rich in oxygenated compounds.

At about the same time, the first commercial hydrogen from synthesis gas produced from steam methane reforming was commercialized. Also, in the early years of the 20th century (i.e., prior to 1905), the first commercial production of hydrogen (from synthesis gas produced from steam methane reforming) became a reality. Following this trend of discovery, Haber and Bosch discovered the synthesis of ammonia from hydrogen and nitrogen in 1910 and the first industrial ammonia synthesis plant was commissioned in 1913. The production of liquid hydrocarbon derivatives and oxygenated derivatives from synthesis gas conversion over iron catalysts was discovered in 1923 by Fischer and Tropsch. Variations on this synthesis pathway were soon to follow for the selective production of methanol, mixed alcohols, and iso-hydrocarbon products. Another outgrowth of Fischer-Tropsch synthesis was the hydroformylation of olefin derivatives discovered in 1938, which is based on the reaction of synthesis gas with olefin derivatives for the production of Oxo aldehyde derivatives and alcohol derivatives.

The Fischer-Tropsch process (*Fischer–Tropsch synthesis*) is a series of catalyzed chemical reactions that convert a mixture of carbon monoxide and hydrogen and into hydrocarbon derivatives. The process is a key component of gas-to-liquids (GTL) technology that produces liquid and solid hydrocarbon derivatives from coal, natural gas, biomass, or other carbonaceous feedstocks. Typical catalysts used are based on iron and cobalt and the hydrocarbon derivatives synthesized in the process are primarily liquid alkanes along with byproducts such as olefin derivatives, alcohols, and solid paraffin derivatives (waxes).

10.2 History and Development of the Process

As originally conceived, the function of the Fischer-Tropsch process was to produce liquid transportation hydrocarbon fuels and various other chemical products (Schulz, 1999). Since the original conception, many refinements and adjustments to the technology have been made, including catalyst development and reactor design. Depending on the source of the synthesis gas, the technology is often referred to as coal-to-liquids (CTL) and/or gas-to-liquids (GTL).

In the simplest terms, the Fischer-Tropsch process is a catalytic chemical reaction in which carbon monoxide (CO) and hydrogen (H_2) in the synthesis gas are converted into hydrocarbon derivatives of various molecular weights according to the following equation:

$$(2n+1)\ H_2 + nCO \rightarrow C_nH_{2n+2} + nH_2O \qquad (n \text{ is an integer})$$

For n=1, the reaction represents the formation of methane, which in most coal-to-liquids or gas-to-liquids applications is considered an undesirable byproduct. The process conditions are usually chosen to maximize the formation of higher molecular weight hydrocarbon liquid fuels which are higher value products. There are other side reactions taking place in the process, among which the water-gas-shift reaction is predominant. Thus:

$$CO + H_2O \rightarrow H_2 + CO_2$$

318 Synthesis Gas

Depending on the catalyst, temperature, and type of process employed, hydrocarbon derivatives ranging from methane to higher molecular paraffin derivatives and olefin derivatives can be obtained. Small amounts of low molecular weight oxygenates (e.g., alcohol and organic acids) are also formed.

The Fischer-Tropsch technology has found industrial application since 1938 in Germany where a total of nine plants produced synthetic hydrocarbon derivatives. However, the history of the commercial Fischer-Tropsch technology dates back to the early years of the 20th century (Table 10.2). Hence, since the turn of the 21st century, as indicated in the above summary of the Fischer-Tropsch history, there has been significantly renewed interest in Fischer-Tropsch technology. In great part this renaissance has been due to the exploitation of cheaper remote, or, "stranded" gas, which has the effect of making the economics of Fischer-Tropsch projects increasingly attractive.

A Fischer-Tropsch plant incorporates three major process sections: (i) production of synthesis gas which is a mixture of carbon monoxide and hydrogen (*steam reforming*), (ii) conversion of the synthesis gas to aliphatic hydrocarbon derivatives and water (*Fischer-Tropsch synthesis process*), and (iii) hydrocracking the longer-chain, waxy synthetic hydrocarbon derivatives to fuel grade fractions. Of the above three steps, the production of synthesis gas is the most energy intensive as well as expensive.

A major attraction of the use of synthesis gas is the very wide range of potential uses by converting synthesis gas into useful downstream products: (i) the Fischer–Tropsch

Table 10.2 History and evolution of the fischer-tropsch process.

1902	Methane formed from mixtures of hydrogen and carbon monoxide over a nickel catalyst (Sabatier and Senderens)
1923	Fischer and Tropsch report work with cobalt, iron and rubidium catalysts at pressure to produce hydrocarbon derivatives
1936	The first four Fischer-Tropsch production plants in Germany began operation
1950	A 5,000 b/d plant was commissioned and began operation in Brownsville, Texas
1950-53	The slurry phase reactor pilot unit developed
1950s	Decline in construction of new Fischer-Tropsch plants due to sudden availability of cheaper crude oil
1955	First Sasol Plant commissioned in South Africa. Iron catalyst was used; two further plants were commissioned in 1980 and 1983.
1970-1980s	Energy crisis initiated renewed interest in Fischer-Tropsch technology as the price of crude oil increased
1990s	Discovery of "stranded" gas reservoirs; renewed interest in Fischer-Tropsch as a viable *gas-to-liquids* (GTL) technology
1992	MossGas plant used Sasol Technology and natural gas as the carbon feedstock
1992-93	Shell used Cobalt based catalyst and natural gas as the feedstock
1993	Sasol slurry phase reactor commissioned using Fe-based catalyst

synthesis of hydrocarbon derivatives, (ii) methanol synthesis, (iii) mixed alcohol synthesis, and (iv) synthesis gas fermentation. By choosing an appropriate catalyst (usually based on iron or cobalt) and appropriate reaction conditions (usually 200 to 350°C, 390 to 650°F, and pressures on the order of 300 to 600 psi), the process with its associated cracking and separation stages can be optimized to produce high molecular weight wax for low molecular weight olefin derivatives and naphtha for petrochemicals production. The ideal feedstock for the Fischer-Tropsch process is synthesis gas consisting of a mixture of hydrogen and carbon monoxide with a molar ratio of 2:1 (Chadeesingh, 2011).

Methanol synthesis is another attractive conversion route because methanol is one of the top 10 petrochemical commodities insofar as, like synthesis gas, it can also be a source of chemicals (Figure 10.1). Synthesis gas can be converted into methanol over a copper-zinc oxide catalyst at 220 to 300°C (430 to 570°F) and 750 to 1500 psi.

$$CO + 2H_2 \rightarrow CH_3OH$$

Methanol can, in turn, be used to make acetic acid, formaldehyde for resins, gasoline additives and petrochemical building blocks such as ethylene and propylene.

Under slightly more severe process conditions (up to 425°C, 800°F, and 4500 psi), a wider range of mixed alcohols can be produced (Chadeesingh, 2011). The processes use catalysts modified from either Fischer-Tropsch synthesis or methanol synthesis, by addition of alkali metals. Finally, the fermentation route for the conversion of synthesis gas uses biochemical processes and reaction conditions that are close to ambient temperature and pressure to make ethanol or other alcohols. Biochemical processes are addressed below.

There are many options for converting synthesis gas into petrochemical feedstocks. For example, the olefin derivatives conversion chain in which ethylene and propylene are converted into polymers (polyethylene, polypropylene, PVC), glycol derivatives (ethylene glycol, $HOCH_2CH_2OH$, propylene glycol, $CH_3CHOHCH_2OH$) and a range of familiar materials such as acetone (CH_3COCH_3), acetic acid (CH_3CO_2H), gasoline additives and surfactants. The olefin derivatives can be produced by synthesizing naphtha in a Fischer-Tropsch process and then cracking it in a conventional naphtha cracker to make ethylene and propylene.

Depending on the source of biomass feedstock and the choice of gasifier technology, the raw synthesis gas can contain varying amounts of particulates (e.g., ash or char, which can lead to erosion, plugging or fouling), alkali metals (which can cause hot corrosion and catalyst poisoning), water-soluble trace components (e.g., halides, ammonia), light oils or tars (e.g., benzene, toluene, xylene or naphthalene, which can lead to catalyst carbonization and fouling), polyaromatic compounds, sulfur components, phosphorus components as well as methane and carbon dioxide. Many of these can be removed (if required) either using standard chemical-industry equipment such as cyclones, filters, electrostatic precipitators, water scrubbers, oil scrubbers, activated carbon and adsorbents, or via clean-up processes such as hydrolysis and various carbon dioxide capture processes (Chapter 4).

Another important factor is the ratio of hydrogen to carbon dioxide in the synthesis gas. Different conversion routes require different ratios, e.g. 1.7:1 and 2.15:1 for producing Fischer-Tropsch naphtha/gasoline and diesel, respectively, or 3:1 for methanol synthesis. Because biomass molecules contain oxygen within their structure, biomass-derived synthesis gas often needs to have the hydrogen to carbon monoxide ratio boosted. One option for achieving this is to react some of the synthesis gas with steam over a catalyst to produce hydrogen and carbon dioxide in the water-gas shift reaction (Chadeesingh, 2011; Speight, 2013a, 2013b) as well as accepting a cost for the removal of carbon dioxide unless there is a byproduct hydrogen source readily available.

The extent to which gas cleanup is required depends on the choice of synthesis gas conversion route. Generally, the level of particulates will need to be reduced considerably for any chemical synthesis process, but the precise extent to which (say) sulfur or halide levels need to be reduced depends on the catalysts that are going to be used. For the methanol synthesis process, for example, the sulfur content of the synthesis gas has to be below 100 ppb v/v. For ammonia synthesis process, there is a similar sulfur constraint, and the carbon dioxide content must be below 10 ppm v/v.

10.3 Synthesis Gas

Synthesis gas, a mixture composed primarily of carbon monoxide and hydrogen but also water, carbon dioxide, nitrogen and methane, has been produced on a commercial scale since the early part of the 20th century. This section provides a general description of the emerging technologies and their potential economic benefits. Recent developments in the technology for synthesis gas production via membrane reactors are also reported (Bayat *et al.*, 2010). During World War II, the Germans obtained synthesis gas by gasifying the carbonaceous feedstock. The mixture was used for producing a liquid hydrocarbon mixture in the gasoline range using Fischer-Tropsch technology. Although this route was abandoned after the war due to the high production cost of these hydrocarbon derivatives, it is currently being used in South Africa, where the carbonaceous feedstock (coal) is relatively inexpensive (SASOL II, and SASOL III).

Almost all carbonaceous materials can be converted into gaseous fuels, although conversion processes for the heavier fractions require more elaborate technology to achieve the necessary purity and uniformity of the manufactured gas stream. In addition, the thermal yield from the gasification of heavier feedstocks is invariably lower than that of gasifying light naphtha or liquefied petroleum gas (LPG) since, in addition to the production of synthesis gas components (hydrogen and carbon monoxide) and various gaseous hydrocarbon derivatives, heavy feedstocks also yield some tar and coke.

Gasification to produce synthesis gas can proceed from just about any organic material, including biomass and plastic waste. The resulting synthesis gas burns cleanly into water vapor and carbon dioxide. Alternatively, synthesis gas may be converted efficiently to methane via the Sabatier reaction, or to a diesel-like synthetic fuel via the Fischer-Tropsch process. Inorganic components of the feedstock, such as metals and minerals, are trapped in an inert and environmentally safe form as char, which may have use as a fertilizer.

In principle, synthesis gas can be produced from any hydrocarbon feedstock. These include natural gas, naphtha, residual oil, crude oil coke, coal, and biomass. The lowest cost routes for synthesis gas production, however, are based on natural gas. The cheapest option is remote or stranded reserves. Current economic considerations dictate that the production of liquid fuels from synthesis gas translates into using natural gas as the hydrocarbon source. Nevertheless, the synthesis gas production operation in a gas-to-liquids plant amounts to greater than half of the capital cost of the plant. The choice of technology for synthesis gas production also depends on the scale of the synthesis operation. Synthesis gas production from solid fuels can require an even greater capital investment with the addition of feedstock handling and more complex synthesis gas purification operations. The greatest impact on improving gas-to-liquids plant economics is to decrease capital costs associated with synthesis gas production and improve thermal efficiency through better heat integration and utilization. Improved thermal efficiency can be obtained by combining the gas-to-liquids plant with a power generation plant to take advantage of the availability of low-pressure steam.

Regardless of the final fuel form, gasification itself and subsequent processing neither emits nor traps greenhouse gases such as carbon dioxide. Combustion of synthesis gas or derived fuels does of course emit carbon dioxide. However, biomass gasification could play a significant role in a renewable energy economy, because biomass production removes carbon dioxide from the atmosphere. While other biofuel technologies such as biogas and biodiesel are also reputed to be carbon neutral, gasification runs on a wider variety of input materials, can be used to produce a wider variety of output fuels, and is an extremely efficient method of extracting energy from biomass. Biomass gasification is therefore one of the most technically and economically convincing energy possibilities for a carbon neutral economy.

Synthesis gas consists primarily of carbon monoxide, carbon dioxide and hydrogen, and has less than half the energy density of natural gas. Synthesis gas is combustible and often used as a fuel source or as an intermediate for the production of other chemicals. Synthesis gas for use as a fuel is most often produced by gasification of the carbonaceous feedstock or municipal waste mainly by the following paths:

$$C + O_2 \rightarrow CO_2$$

$$CO_2 + C \rightarrow 2CO$$

$$C + H_2O \rightarrow CO + H_2$$

When used as an intermediate in the large-scale, industrial synthesis of hydrogen and ammonia, it is also produced from natural gas (via the steam reforming reaction) as follows:

$$CH_4 + H_2O \rightarrow CO + 3H_2$$

The synthesis gas produced in large waste-to-energy gasification facilities is used as fuel to generate electricity.

The manufacture of gas mixtures of carbon monoxide and hydrogen has been an important part of chemical technology for about a century. Originally, such mixtures were obtained by the reaction of steam with incandescent coke and were known as *water gas*. Used first as a fuel, water gas soon attracted attention as a source of hydrogen and carbon monoxide for the production of chemicals, at which time it gradually became known as synthesis gas. Eventually, steam reforming processes, in which steam is reacted with natural gas (methane) or crude oil naphtha over a nickel catalyst, found wide application for the production of synthesis gas.

A modified version of steam reforming known as autothermal reforming, which is a combination of partial oxidation near the reactor inlet with conventional steam reforming further along the reactor, improves the overall reactor efficiency and increases the flexibility of the process. Partial oxidation processes using oxygen instead of steam also found wide application for synthesis gas manufacture, with the special feature that they could utilize low-value feedstocks such as heavy crude oil residua. In recent years, catalytic partial oxidation employing very short reaction times (milliseconds) at high temperatures (850 to 1000°C; 1560 to 1830°F) is providing still another approach to synthesis gas manufacture (Hickman and Schmidt, 1993). Nearly complete conversion of methane, with close to 100% selectivity to hydrogen and carbon dioxide, can be obtained with a rhenium monolith under well-controlled conditions. Experiments on the catalytic partial oxidation of *n*-hexane conducted with added steam give much higher yields of hydrogen than can be obtained in experiments without steam, a result of much interest in obtaining hydrogen-rich streams for fuel cell applications.

The route for a carbonaceous feedstock to synthetic automotive fuels, as practiced by SASOL is technically proven and a series of products with favorable environmental characteristics are produced (Luque and Speight, 2015). As is the case in essentially all conversion processes for carbonaceous feedstocks where air or oxygen is used for the utilization or partial conversion of the energy in the coal, the carbon dioxide burden is a drawback as compared to crude oil.

The uses of synthesis gas include use as a chemical feedstock and in gas-to-liquid processes, which use Fisher-Tropsch chemistry to make liquid fuels as feedstock for chemical synthesis. As well as being used in the production of fuel additives, including diethyl ether and methyl t-butyl ether (MTBE), acetic acid and its anhydride, synthesis gas could also make an important contribution to chemical synthesis through conversion to methanol. There is also the option in which stranded natural gas is converted to synthesis gas production followed by conversion to liquid fuels.

The chemical train for producing synthesis gas (carbon monoxide + hydrogen), from which a variety of products can be produced, can be represented simply as:

$$\text{Carbonaceous feedstock} \rightarrow (\text{partial oxidation}) \rightarrow \text{synthesis gas}$$

$$\text{Synthesis gas} \rightarrow \text{synthetic fuels and petrochemicals}$$

The products designated as synthesis fuels include low-to-high-boiling hydrocarbon derivatives and methanol. Also the high-boiling products (including wax products) can also be used as feedstocks for gas production.

In addition, the actual process is described as comprising three components: (i) synthesis gas generation, (ii) waste heat recovery, and (iii) gas processing. Within each of the above three listed systems are several options. For example, synthesis gas can be generated to yield a range of compositions ranging from high-purity hydrogen to high-purity carbon monoxide. Two major routes can be utilized for high-purity gas production: (i) pressure-swing adsorption and (ii) utilization of a cold box, where separation is achieved by distillation at low temperatures. In fact, both processes can also be used in combination as well. Unfortunately, both processes require high capital expenditure. However, to address these concerns, research and development is ongoing and successes can be measured by the demonstration and commercialization of technologies such as permeable membrane for the generation of high-purity hydrogen, which in itself can be used to adjust the hydrogen/carbon monoxide ratio of the synthesis gas produced (Bayat et al., 2010).

10.4 Production of Synthesis Gas

Gasification processes are used to convert a carbon-containing (carbonaceous) material into a synthesis gas, a combustible gas mixture which typically contains carbon monoxide, hydrogen, nitrogen, carbon dioxide and methane (Chapter 5). The impure synthesis gas has a relatively low calorific value, ranging from 100 to 300 Btu/ft^3. The gasification process can accommodate a wide variety of gaseous, liquid, and solid feedstocks and it has been widely used in commercial applications for the production of fuels and chemicals (Luque and Speight, 2015).

10.4.1 Feedstocks

In principle, synthesis gas can be produced from any hydrocarbon feedstock, which include natural gas, naphtha, residual oil, crude oil coke, coal, biomass and municipal or industrial waste (Chapter 1). The product gas stream is subsequently purified (to remove sulfur, nitrogen, and any particulate matter) after which it is catalytically converted to a mixture of liquid hydrocarbon products. In addition, synthesis gas may also be used to produce a variety of products, including ammonia, and methanol.

Of all of the carbonaceous materials used as feedstocks for gasification process, coal represents the most widely used feedstock and, accordingly, the feedstock about which most is known. In fact, gasification of coal has been a commercially available proven technology (2013a, 2013b). The modern gasification processes have evolved from three first-generation process technologies: (i) Lurgi fixed-bed reactor (ii) high-temperature Winkler fluidized-bed reactor and (iii) Koppers-Totzek entrained-flow reactor. In each case steam/air/oxygen are passed through heated coal which may either be a fixed bed, fluidized bed or entrained in the gas. Exit gas temperatures from the reactor are 500ºC (930ºF), 900 to 1100ºC (1650 to 2010ºF), and 1300 to 1600ºC (2370 to 2910ºF), respectively. In addition to the steam/air/oxygen mixture being used as the feed gases, steam/oxygen mixtures can also be used in which membrane technology and a compressed oxygen-containing gas is employed.

In addition, low-value or negative-value materials and wastes such as crude oil coke, refinery residua, refinery waste, municipal sewage sludge, biomass, hydrocarbon contaminated soils, and chlorinated hydrocarbon byproducts have all been used successfully in gasification operations (Speight, 2008, 2013a, 2013b). In addition, synthesis gas is used as a source of hydrogen or as an intermediate in producing a variety of hydrocarbon products by means of the Fischer-Tropsch synthesis (Chadeesingh, 2011). In fact, gasification to produce synthesis gas can proceed from any carbonaceous material, including biomass and waste.

There are different sources for obtaining synthesis gas. It can be produced by steam reforming or partial oxidation of any hydrocarbon ranging from natural gas (methane) to crude oil residua. It can also be obtained by gasifying the carbonaceous feedstock to a medium Btu gas. (Medium Btu gas consists of variable amounts of carbon monoxide, carbon dioxide, and hydrogen and is used principally as a fuel gas).

As already stated (Chapter 2), the first step in the process is feedstock preparation which may include unit operations such as (i) size reduction, (ii) screening, and (iii) slurrying. The design of the feedstock preparation section depends on the properties of the raw material feed and on the requirements of the gasification technology that is selected. For example, preparing any form of crude oil resid, biomass, or waste for gasification is very different from the preparation of bituminous coal for the gasifier, and the type of gasifier suitable for the gasification may also (or more than likely) be feedstock dependent (Chapter 3).

Once the feedstock is prepared for the gasifier, the major route for producing synthesis gas is the steam reforming of natural gas over a promoted nickel catalyst at temperatures on the order of 800°C (1470°F):

$$CH_4 + H_2O \rightarrow CO + 3H_2$$

In some countries, synthesis gas is mainly produced by steam reforming naphtha. Because naphtha is a mixture of hydrocarbon derivatives ranging approximately from C5 to C10, the steam reforming reaction may be represented using n-heptane:

$$CH_3(CH_2)_5CH_3 + 7H_2O \rightarrow 7CO + 15H_2$$

As the molecular weight of the hydrocarbon increases (lower H/C feed ratio), the hydrogen/carbon monoxide (H_2/CO) product ratio decreases. The hydrogen/carbon monoxide product ratio is approximately 3 for methane, 2.5 for ethane, 2.1 for heptane, and less than 2 for heavier hydrocarbon derivatives. The non-catalytic partial oxidation of hydrocarbon derivatives is also used to produce synthesis gas, but the hydrogen/carbon monoxide ratio is lower than from steam reforming. In practice, this ratio is even lower than what is shown by the stoichiometric equation because part of the methane is oxidized to carbon dioxide and water. When resids are partially oxidized by oxygen and steam at 1400 to 1450°C (2550 to 2640°F) and pressures on the order of 11,500 to 13,000 psi, the gas consists of equal parts of hydrogen and carbon monoxide.

Synthesis gas is an important intermediate. The mixture of carbon monoxide and hydrogen is used for producing methanol. It is also used to synthesize a wide variety of hydrocarbon derivatives ranging from gases to naphtha to gas oil using Fischer-Tropsch technology. This process may offer an alternative future route for obtaining olefin derivatives and chemicals.

Synthesis gas is a major source of hydrogen, which is used for producing ammonia. Ammonia is the host of many chemicals such as urea, ammonium nitrate, and hydrazine. Carbon dioxide, a byproduct from synthesis gas, reacts with ammonia to produce urea (H_2NCONH_2).

$$H_2N-\underset{\underset{\displaystyle}{\|}}{\overset{\overset{\displaystyle O}{\|}}{C}}-NH_2$$

Urea

Urea (also known as carbamide) serves an important role in the metabolism of nitrogen-containing compounds by animals and is the main nitrogen-containing substance in the urine of mamas. It is a colorless, odorless solid, highly soluble in water, and, dissolved in water, it exhibits neither an acid nor an alkali. It is formed in the liver by the combination of two ammonia molecules (NH_3) with a carbon dioxide (CO_2) molecule. It is widely used in fertilizers as a source of nitrogen and is an important raw material for the chemical industry.

Most of the production of hydrocarbon derivatives by Fischer Tropsch method uses synthesis gas produced from sources that yield a relatively low hydrogen/carbon monoxide ratio, such as typical in coal gasifiers. This, however, does not limit this process to low hydrogen/carbon monoxide gas feeds. The only large-scale commercial process using this technology is in South Africa, where coal is an abundant energy source. The process of obtaining liquid hydrocarbon derivatives from coal through the Fischer-Tropsch process is termed indirect coal liquefaction which was originally intended for obtaining liquid hydrocarbon derivatives from solid fuels. However, this method may well be applied in the future to the manufacture of chemicals through cracking the liquid products or by directing the reaction to produce more olefin derivatives.

The reactants in Fischer-Tropsch processes are carbon monoxide and hydrogen. The reaction may be considered a hydrogenative oligomerization of carbon monoxide in presence of a heterogeneous catalyst. The main reactions occurring in Fischer-Tropsch processes are:

(i) Olefin derivatives:

$$2nH_2 + nCO \rightarrow C_nH_{2n} + nH_2O$$

(ii) Paraffin derivatives:

$$(2n+1)H_2 + nCO \rightarrow C_nH_{2n+2} + nH_2O$$

(iii) Alcohol derivatives:

$$2nH_2 + nCO \rightarrow C_nH_{2n+2}O + (n-1)H_2O$$

The coproduct water reacts with carbon monoxide (the shift reaction), yielding hydrogen and carbon dioxide:

$$CO + H_2O \rightarrow H_2 + CO_2$$

The gained hydrogen from the water shift reaction reduces the hydrogen demand for Fischer-Tropsch processes. Water gas shift proceeds at about the same rate as the Fischer-Tropsch reaction. Another side reaction also occurring in Fischer-Tropsch process reactors is the disproportionation of carbon monoxide to carbon dioxide and carbon:

$$2CO \rightarrow CO_2 + C$$

This reaction is responsible for the deposition of carbon in the reactor tubes in fixed-bed reactors and reducing heat transfer efficiency.

Fischer-Tropsch technology is best exemplified by the SASOL projects in South Africa. After the carbonaceous feedstock is gasified to a synthesis gas mixture, it is purified in a Rectisol unit. The purified gas mixture is reacted in a Synthol unit over an iron-based catalyst. The main products are gasoline, diesel fuel, and jet fuels. Byproducts are ethylene, propylene, alpha olefin derivatives, sulfur, phenol, and ammonia which are used for the production of downstream chemicals. However, the exact mechanism is not fully established. One approach assumes a first-step adsorption of carbon monoxide on the catalyst surface followed by a transfer of an adsorbed hydrogen atom from an adjacent site to the metal carbonyl (M-CO). The polymerization continues (as in the last three steps shown above) until termination occurs and the hydrocarbon is desorbed. The last two steps shown above explain the presence of oxygenated derivatives in Fischer-Tropsch products.

Alternatively, an intermediate formation of an adsorbed methylene on the catalyst surface through the dissociative adsorption of carbon monoxide has been considered. The formed metal carbide (M-C) is then hydrogenated to a reactive methylene metal species. The methylene intermediate abstracts a hydrogen and is converted to an adsorbed methyl. Reaction of the methyl with the methylene produces an ethyl-metal species. Successive reactions of the methylene with the formed ethyl produces a long-chain adsorbed alkyl. The adsorbed alkyl species can either terminate to a paraffin by a hydrogenation step or to an olefin by a dehydrogenation step. The carbide mechanism, however, does not explain the formation of oxygenate derivatives in Fischer-Tropsch products.

10.4.2 Product Distribution

The product distribution of hydrocarbon derivatives formed during the Fischer–Tropsch process follows an Anderson-Schulz-Flory distribution:

$$W_n/n = (1-\alpha)^2 \alpha^{n-1}$$

W_n is the weight fraction of hydrocarbon molecules containing n carbon atoms, α is the chain growth probability or the probability that a molecule will continue reacting to form a longer chain. In general, α is largely determined by the catalyst and the specific process conditions.

According to the above equation, methane will always be the largest single product; however, by increasing α close to one, the total amount of methane formed can be minimized compared to the sum of all of the various long-chain products. Increasing α increases the formation of long-chain hydrocarbon derivatives – waxes – which are solid at room temperature. Therefore, for production of liquid transportation fuels it may be necessary to crack the Fischer-Tropsch longer chain products.

The very long-chain hydrocarbon derivatives are waxes, which are solid at room temperature. Therefore, for production of liquid transportation fuels it may be necessary to crack some of the Fischer-Tropsch products. In order to avoid this, some researchers have proposed using zeolites or other catalyst substrates with fixed sized pores that can restrict the formation of hydrocarbon derivatives longer than some characteristic size (usually n<10). This way they can drive the reaction so as to minimize methane formation without producing lots of long-chain hydrocarbon derivatives.

It has been proposed that zeolites or other catalyst substrates with fixed sized pores that can restrict the formation of hydrocarbon derivatives longer than some characteristic size (usually n<10). This would tend to drive the reaction to minimum methane formation without producing the waxy products.

10.5 Process Parameters

The Fischer-Tropsch synthesis is based on the conversion of a mixture of carbon monoxide and hydrogen into liquid hydrocarbon derivatives. The Fischer-Tropsch synthesis can be represented by following chemical reactions (Fischer and Tropsch, 1923):

$$(2n + 1)H_2 + nCO \rightarrow C_nH_{2n+2} + nH_2O$$

$$2nH_2 + nCO \rightarrow C_nH_{2n} + nH_2O$$

The process is a key route for the gas-to-liquids (GTL) technology, produces synthetic fuels and chemicals from mainly natural gas, crude oil resids, coal, biomass, and carbonaceous wastes (Klerk and Furimsky, 2010; Chadeesingh, 2011; Luque and Speight, 2015).

Depending on the reaction temperature, the process can be divided into low-temperature Fischer-Tropsch synthesis (LTFT, at temperatures on the order of 200 to 260°C, 390 to 500°F) and high-temperature Fischer-Tropsch synthesis (HTFT, at temperatures on the order of 300 to 350°C, 570 to 660°F). During the low-temperature Fischer-Tropsch process, higher-boiling hydrocarbon derivatives (boiling above 360°C, 680°F) are produced. Also the total distillate yield is higher than the yield produced by the high-temperature Fischer-Tropsch process. Higher temperature leads to reactions that are more rapid than the reactions in the low-temperature Fischer-Tropsch process and higher conversion rates, and also tends to favor shorter chains as well as the production of methane, olefin derivatives, and aromatic derivative. Typical pressures range from 15 psi to 750 psi, or higher. Increasing the pressure leads to (i) higher conversion rates and (ii) the formation of the much-desired long-chained alkane derivatives.

A variety of catalysts can be used for the Fischer-Tropsch process, but the most common are the transition metals such as cobalt, iron, and ruthenium. Nickel can also be used, but tends to favor the production of methane. In the low-temperature Fischer-Tropsch process, cobalt-based catalyst is the more common catalyst whereas in the high-temperature Fischer-Tropsch process an iron-based catalyst is more common. The advantage of the cobalt-based catalysts is a much longer lifetime and the yields of unsaturated hydrocarbon derivatives and alcohol derivatives are produced compared with the products from the use of iron-based catalysts. On the other hand, iron-based catalysts are more tolerant to sulfur and can be also used (by means of the water gas shift reaction) to adjust the hydrogen-to-carbon monoxide if it is lower than 2.

For large-scale commercial Fischer-Tropsch reactors heat removal and temperature control are the most important design features to obtain optimum product selectivity and long catalyst lifetimes. Over the years, basically four Fischer-Tropsch reactor designs have been used commercially. These are the multi-tubular fixed bed, the slurry reactor, or the fluidized-bed reactor (with either a fixed bed or a circulating bed). The fixed-bed reactor consists of thousands of small tubes with the catalyst as surface-active agent in the tubes. Water surrounds the tubes and regulates the temperature by settling the pressure of evaporation. The slurry reactor is widely used and consists of fluid and solid elements, where the catalyst has no particular position, but flows around as small pieces of catalyst together with the reaction components. The slurry and fixed-bed reactors are used in the low-temperature Fischer-Tropsch process. The fluidized-bed reactors are diverse, but characterized by the fluid behavior of the catalyst.

The multi-tubular fixed-bed reactors (often referred to as Arge reactors) were developed jointly by Lurgi and Ruhrchemie and commissioned in 1955. They were used by Sasol to produce high-boiling Fischer-Tropsch liquid hydrocarbon derivatives and waxes in Sasolburg, in what Sasol called the low-temperature Fischer-Tropsch synthesis process, aiming for liquid fuels production. Most, if not all, of these types of Arge reactors are now being replaced by slurry-bed reactors, which is considered the state-of-the-art technology for low-temperature Fischer-Tropsch synthesis. Slurry-bed Fischer-Tropsch reactors offer better temperature control and higher conversion. Fluidized-bed Fischer-Tropsch reactors were developed for the high-temperature Fischer-Tropsch synthesis to produce low molecular gaseous hydrocarbon derivatives and naphtha. This type of reactor was originally developed in a circulating mode (such as the Sasol synthol reactors) but have been replaced by a fixed fluidized-bed type of reactor (advanced synthol reactors) which is capable of a high throughput.

Sasol in South Africa uses coal and natural gas as a feedstock, and produces a variety of synthetic crude oil products. The process was used in South Africa to meet its energy needs during its isolation under Apartheid. This process has received renewed attention in the quest to produce low-sulfur diesel fuel in order to minimize the environmental impact from the use of diesel engines. The Fischer-Tropsch process as applied at Sasol can be divided into two operating regimes: (i) the high-temperature Fischer-Tropsch process and (ii) the low-temperature Fischer-Tropsch process (Chadeesingh, 2011).

The high-temperature Fischer Tropsch technology uses a fluidized catalyst at 300 to 330°C (570 to 635°F). Originally circulating fluidized-bed units were used (Synthol reactors). Since 1989 a commercial-scale classical fluidized-bed unit has been implemented and improved upon.

The low-temperature Fischer-Tropsch technology has originally been used in tubular fixed-bed reactors at 200 to 230°C (390 to 260°F). This produces more paraffin derivatives and waxy product spectrum than the high-temperature technology. A new type of reactor (the Sasol slurry phase distillate reactor has been developed and is in commercial operation. This reactor uses a slurry phase system rather than a tubular fixed-bed configuration and is currently the favored technology for the commercial production of synfuels.

The commercial Sasol Fischer-Tropsch reactors all use iron-based catalysts on the basis of the desired product spectrum and operating costs. Cobalt-based catalysts have also been known since the early days of this technology and have the advantage of higher conversion for low-temperature cases. Cobalt is not suitable for high-temperature use due to excessive methane formation at such temperatures. For once-through maximum diesel production, cobalt has, despite its high cost, advantages and Sasol has also developed cobalt catalysts which perform very well in the slurry phase process.

However, both the iron and cobalt Fischer-Tropsch catalysts are sensitive to the presence of sulfur compounds in the synthesis gas and can be poisoned by the sulfur compounds. In addition, the sensitivity of the catalyst to sulfur is higher for cobalt-based catalysts than for the iron-based catalysts. This is one reason why cobalt-based catalysts are preferred for Fischer-Tropsch synthesis with synthesis gas derived from natural gas, where the synthesis gas has a higher hydrogen/carbon monoxide ratio and is relatively lower in sulfur content. On the other hand, iron-based catalysts are preferred for lower-quality feedstocks such as coal.

The kerosene (often referred to as diesel although the product could not be sold as diesel fuel without any further treatment to meet specifications) produced by the slurry phase reactor has a highly paraffin derivatives nature, giving a cetane number in excess of 70. The aromatic content of the diesel is typically below 3% and it is also sulfur-free and nitrogen-free. This makes it an exceptional diesel as such or it can be used to sweeten or to upgrade conventional diesels.

The Fischer-Tropsch process is an established technology and already applied on a large scale, although its popularity is hampered by high capital costs, high operation and maintenance costs, and the uncertain and volatile price of crude oil. In particular, the use of natural gas as a feedstock only becomes practical when using stranded gas, i.e., sources of natural gas far from major cities which are impractical to exploit with conventional gas pipelines and liquefied natural gas technology; otherwise, the direct sale of natural gas to consumers would become much more profitable. It is suggested by geologists that supplies of natural gas will peak 5 to 15 years after oil does, although such predictions are difficult to make and often highly uncertain. Hence the increasing interest in a variety of carbonaceous feedstocks as a source of synthesis gas.

Under most circumstances the production of synthesis gas by reforming natural gas will be more economical than from coal gasification, but site-specific factors need to be considered. In fact, any technological advance in this field (such as better energy integration or the oxygen transfer ceramic membrane reformer concept) will speed up the rate at which the synfuels technology will become common practice.

There are large coal reserves which may increasingly be used as a fuel source during oil depletion. Since there are large coal reserves in the world, this technology could be used as an interim transportation fuel if conventional oil were to become more expensive. Furthermore, combination of biomass gasification and Fischer-Tropsch synthesis is a very promising route to produce transportation fuels from renewable or green resources.

Often a higher concentration of some sorts of hydrocarbon derivatives is wanted, which might be achieved by changed reaction conditions. Nevertheless, the product range is wide and infected with uncertainties, due to lack of knowledge of the details of the process and of the kinetics of the reaction. Since the different products have quite different characteristics such as boiling point, physical state at ambient temperature and thereby different use and ways of distribution, often only a few of the carbon chains is wanted. As an example, the low-temperature Fischer-Tropsch is used when longer carbon chains are wanted, because lower temperature increases the portion of longer chains. But too low temperature is not wanted, because of reduced activity. When the desired products are shorter carbon chains, e.g., crude oil, the longer ones might be cracked into shorter chains.

The yield of kerosene (diesel) is therefore highly dependent on the chain growth probability, which again is dependent on (i) pressure, temperature, (ii) the composition of the feedstock gas, (iii) the catalyst type, (iv) the catalyst composition, and (v) the reactor design. The desire to increase the selectivity of some favorable products leads to a need of understanding the relation between reaction conditions and chain growth probability, which in turn request a mathematical expression for the growth probability in order to make a suitable model of the process. The different attempts to model the growth probability have resulted in some models that are regarded in literature as appropriate to describe the product distribution. Two will be presented here to show the influence of temperature and partial pressure.

10.6 Reactors and Catalysts

Since its discovery the Fischer-Tropsch synthesis has undergone periods of rapid development and periods of inaction. Within 10 years of the discovery, German companies were building commercial plants. The construction of these plants stopped about 1940 but existing plants continued to operate during World War II.

10.6.1 Reactors

Currently, two reactor types are used commercially in the Fischer-Tropsch process, a fixed-bed reactor, a fluid-bed reactor, and a slurry-bed reactor. The fixed-bed reactors usually run at lower temperatures to avoid carbon deposition on the reactor tubes. Products from fixed-bed reactors are characterized by low olefin content, and they are generally heavier than products from fluid-beds. Heat distribution in fluid-beds, however, is better than fixed-bed reactors, and fluid-beds are generally operated at higher temperatures. Products are characterized by (i) having more olefin derivatives, (ii) a high proportion of low-boiling hydrocarbon derivatives (gases), and (iii) lower molecular weight product slate than from fixed-bed types.

Originally, the Fischer-Tropsch synthesis was carried out in packed-bed reactors. Gas-agitated multiphase reactors, sometimes called "slurry reactors" or "slurry bubble columns," gained favor, however, because the circulation of the slurry makes it much easier to control the reaction temperature in a slurry-bed reactor than in a fixed-bed reactor. Gas-agitated multiphase reactors operate by suspending catalytic particles in a liquid and feeding gas reactants into the bottom of the reactor through a gas distributor, which produces small gas bubbles. As the gas bubbles rise through the reactor, the reactants are absorbed into the liquid and diffuse to the catalyst particles where, depending on the catalyst system, they are typically converted to gaseous and liquid products.

A slurry-bed reactor is characterized by having the catalyst in the form of a slurry. The feed gas mixture is bubbled through the catalyst suspension. Temperature control is easier than the other two reactor types. An added advantage to a slurry-bed reactor is that it can accept a synthesis gas with a lower hydrogen/carbon monoxide ratio than either the fixed-bed or the fluid-bed reactors.

In the Sasol slurry-phase reactor, preheated synthesis gas is fed into the bottom of the reactor, where it is distributed into slurry consisting of liquid hydrocarbon and catalyst particles. As the gas bubbles upward through the slurry, it diffuses into the slurry and is converted into a range of hydrocarbon derivatives by Fischer-Tropsch reaction. The heat generated from this reaction is removed through the reactor's cooling coils, which generate steam. The heavier (wax) fraction is separated from the slurry containing the catalyst particles in a proprietary process developed by Sasol. The lighter, more volatile fraction is extracted in a gas stream from the top of the reactor. The gas stream is cooled to recover the lighter hydrocarbon derivatives and water. The intermediate hydrocarbon streams are sent to the product upgrading unit, while the water stream is treated in a water recovery unit. The third step upgrades reactor products to diesel and naphtha. The reactor products are mainly paraffin derivatives, but the lighter products contain some olefin derivatives and oxygenated derivatives that need to be removed for product stabilization. Hydrogen is added to hydrotreat the olefin derivatives and oxygenated derivatives, converting them to paraffin derivatives. Hydrogen is also added to the mild hydrocracker, which breaks the long-chain hydrocarbon derivatives into naphtha and diesel. The products are separated out in a fractionation section, which involves hydrocracking and hydroisomerization.

As the process evolved, other types of reactors have been used and include: (i) the parallel plate reactors and (ii) a variety of fixed-bed tubular reactors, and (iii) gas-agitated multiphase reactors. For the parallel plate type of reactor, the catalyst bed is located in tubes fixed between the plates which was cooled by steam/water that passed around the tubes within the catalyst bed. In another version, the reactor may be regarded as finned-tube in which large fins are penetrated by a large number of parallel or connected catalyst-filled tubes. Various designs were utilized for the tubular fixed-bed reactor with the concentrically paced tubes being the preferred one. This type of reactor contained catalyst in the area between the two tubes with cooling water-steam flowing through the inner tube and on the exterior of the outer tube. The gaseous products formed enter the gas bubbles and are collected at the top of the reactor. Liquid products are recovered from the suspending liquid using different techniques, including filtration, settling, and hydrocyclones.

Because the Fischer-Tropsch reaction is exothermic, temperature control is an important aspect of Fischer-Tropsch reactor operation. Gas-agitated multiphase reactors or slurry bubble column reactors have very high heat transfer rates and therefore allow good thermal control of the reaction. On the other hand, because the desired liquid products are mixed with the suspending liquid, recovery of the liquid products can be relatively difficult. This difficulty is compounded by the tendency of the catalyst particles to erode in the slurry, forming fine catalyst particles that are also relatively difficult to separate from the liquid products. Fixed-bed reactors generally avoid the issues that arise from liquid separation and catalyst separation but they (the fixed-bed reactors) do not provide the mixing of phases that allows good thermal control in slurry bubble column reactors.

Furthermore, Fischer-Tropsch reactors are typically sized to achieve a desired volume of production. When a fixed-bed reactor is planned, economies of scale tend to result in the use of long (tall) reactors. Because the Fischer-Tropsch reaction is exothermic, however, a thermal gradient tends to form along the length of the reactor, with the temperature increasing with distance from the reactor inlet. In addition, for most Fischer-Tropsch catalyst systems each 10-degree rise in temperature increases the reaction rate approximately 60%, which in turn results in the generation of still more heat. To absorb the heat generated by the reaction and offset the rise in temperature, a cooling liquid is typically circulated through the reactor.

Thus, for a given reactor system having a known amount of catalyst with a certain specific activity and known coolant temperature, the maximum flow rate of reactants through the reactor is limited by the need to maintain the catalyst below a predetermined maximum catalyst temperature at all points along the length of the catalyst bed and the need to avoid thermal runaway which can result in catalyst deactivation and possible damage to the physical integrity of the reactor system. The net result is that it is unavoidable to operate most of the reactor at temperatures below the maximum temperature, with the corresponding low volumetric productivities over most of the reactor volume.

An innovative technology for combining air separation and natural gas reforming processes is being pursued by Sasol, BP, Praxair and Statoil (Dyer and Chen, 1999). If successful commercialized, this innovation can reduce the cost of synthesis gas generation by as much as 30%. The technology is referred to as *oxygen transport membranes* (OTM) and should combine five unit operations currently in use, viz.: oxygen separation, oxygen compression, partial oxidation (POX), steam methane reforming, and heat exchange. This technology incorporates the use of catalytic components with the membrane to accelerate the reforming reactions.

Air products have also developed and patented a two-step process for synthesis gas generation (Nataraj *et al.*, 2000). This technology can be utilized to generate synthesis gas from several feedstocks, including natural gas, associated gas (from crude oil production), light hydrocarbon gases from refineries, and medium molecular weight (medium boiling) hydrocarbon fractions like naphtha (Parkash, 2003; Gary *et al.*, 2007; Speight, 2011, 2014a; Hsu and Robinson, 2017; Speight, 2017). The first stage comprises conventional steam reforming with partial conversion to synthesis gas. This is followed by complete conversion in an ion transport ceramic membrane (ITM) reactor. This combination solves the problem associated with steam reforming for feedstocks with hydrocarbon derivatives higher-boiling than methane, since higher molecular weight (C_{2+}) hydrocarbon derivatives tend to crack and degrade both the catalyst and membrane.

By shifting the equilibrium in the steam reforming process through removal of hydrogen from the reaction zone, membrane reactors can also be used to increase the equilibrium-limited methane conversion. Using palladium-silver (Pd-Ag) alloy membrane reactors methane conversion can reach as close to 100% (Shu *et al.*, 1995).

10.6.2 Catalysts

Of great importance to the Fischer-Tropsch process is the catalyst. The catalysts used in the latest generation of Fischer-Tropsch technologies are cobalt based, usually carried on alumina supports, sometimes with precious metal promoters. Coal-based processes, including a gasification step, use iron catalysts, which are better suited to high-temperature processes based on feedstocks containing impurities. But iron produces significant quantities of non-paraffin derivatives as byproducts, while cobalt catalysts feature high selectivity and are more efficient for making paraffin derivatives from cleaner feedstocks.

Catalysts play a pivotal role in synthesis gas conversion reactions. In fact, fuels and chemicals synthesis from synthesis gas does not occur in the absence of appropriate catalysts. The basic concept of a catalytic reaction is that reactants adsorb onto the catalyst surface and rearrange and combine into products that desorb from the surface. One of the fundamental functional differences between synthesis gas synthesis catalysts is whether or not the adsorbed carbon monoxide molecule dissociates on the catalyst surface. For the Fischer-Tropsch process and higher alcohol synthesis, carbon monoxide dissociation is a necessary reaction condition. For methanol synthesis, the carbon-oxygen bond remains intact. Hydrogen has two roles in catalytic synthesis gas synthesis reactions. In addition to serving as a reactant needed for carbon monoxide hydrogenation, it is commonly used to reduce the metalized synthesis catalysts and activate the metal surface.

Generally, the Fischer-Tropsch synthesis is catalyzed by a variety of transition metals such as iron, nickel, and cobalt. Iron-based catalysts are relatively low cost and have a higher water-gas-shift activity, and are therefore more suitable for a lower hydrogen/carbon monoxide ratio (H_2/CO) synthesis gas such as those derived from coal gasification. On the other hand, nickel-based catalysts tend to promote methane formation, as in a methanation process. Cobalt-based catalysts are more active and are generally preferred over ruthenium (Ru) and, in comparison to iron, cobalt (Co) has much less water-gas-shift activity.

Thus, in many cases, an iron-containing catalyst is the preferred catalyst due to the higher activity but a nickel-containing catalyst produces large amounts of methane, while a cobalt-containing catalyst has a lower reaction rate and a lower selectivity than the iron-containing catalyst. By comparing cobalt and iron catalysts, it was found that cobalt promotes more middle-distillate products. In the Fischer-Tropsch process, a cobalt-containing catalyst produces hydrocarbon derivatives plus water while iron catalyst produces hydrocarbon derivatives and carbon dioxide. It appears that the iron catalyst promotes the shift reaction more than the cobalt catalyst.

Various metals, including but not limited to iron, cobalt, nickel, and ruthenium, alone and in conjunction with other metals, can serve as Fischer-Tropsch catalysts. Cobalt is particularly useful as a catalyst for converting natural gas to heavy hydrocarbon derivatives suitable for the production of diesel fuel. Iron has the advantage of being readily available and relatively inexpensive but also has the disadvantage of greater water-gas shift activity. Ruthenium is highly active but quite expensive. Consequently, although ruthenium is not the economically preferred catalyst for commercial Fischer-Tropsch production, it is often used in low concentrations as a promoter with one of the other catalytic metals.

The most common catalysts are the transition metals cobalt, iron, and ruthenium. Nickel can also be used, but tends to favor methane formation (methanation). Cobalt seems to be the most active catalyst, although iron may be more suitable for low-hydrogen-content synthesis gases such as those derived from coal due to its promotion of the water-gas-shift reaction. In addition to the active metal the catalysts typically contain a number of promoters, including potassium and copper. Catalysts are supported on high-surface-area binders/supports such as silica (SiO_2), alumina (Al_2O_3), or the more complex zeolites. Cobalt catalysts are more active for Fischer-Tropsch synthesis when the feedstock is natural gas. Natural gas has a high hydrogen to carbon ratio, so the water-gas-shift is not needed for cobalt catalysts. Iron catalysts are preferred for lower-quality feedstocks such as crude oil residua, coal, or biomass.

Unlike the other metals used for this process (Co, Ni, Ru) which remain in the metallic state during synthesis, iron catalysts tend to form a number of chemical phases, including various oxides and carbides during the reaction. Control of these phase transformations can be important in maintaining catalytic activity and preventing breakdown of the catalyst particles. For synthesis of higher molecular weight alcohols, dissociation of carbon monoxide is a necessary reaction condition. For methanol synthesis, the carbon monoxide molecule remains intact. Hydrogen has two roles in catalytic synthesis gas synthesis reactions. In addition to serving as a reactant needed for hydrogenation of carbon monoxide, it is commonly used to reduce the metalized synthesis catalysts and activate the metal surface.

Group 1 alkali metals (including potassium) are poisons for cobalt catalysts but are promoters for iron catalysts. Catalysts are supported on high-surface-area binders/supports such as silica, alumina, and zeolites. Cobalt catalysts are more active for Fischer-Tropsch synthesis when the feedstock is natural gas. Natural gas has a high hydrogen to carbon ratio, so the water-gas-shift is not needed for cobalt catalysts. Iron catalysts are preferred for lower-quality feedstocks such as coal or biomass.

Unlike the other metals used for this process (Co, Ni, Ru), which remain in the metallic state during synthesis, iron catalysts tend to form a number of phases, including various oxides and carbides during the reaction. Control of these phase transformations can be important in maintaining catalytic activity and preventing breakdown of the catalyst particles.

Fischer-Tropsch catalysts are sensitive to poisoning by sulfur-containing compounds. The sensitivity of the catalyst to sulfur is greater for cobalt-based catalysts than for their iron counterparts. Promoters also have an important influence on activity. Alkali metal oxides and copper are common promoters, but the formulation depends on the primary metal, iron or cobalt. Alkali oxides on cobalt catalysts generally cause activity to drop severely even with very low alkali loadings. C_{5+} and carbon dioxide selectivity increase while methane and C_2 to C_4 selectivity decrease. In addition, the olefin to paraffin ratio increases.

Fischer-Tropsch catalysts can lose activity as a result of (i) conversion of the active metal site to an inactive oxide site, (ii) sintering, (iii) loss of active area by carbon deposition, and (iv) chemical poisoning. For example, Fischer-Tropsch catalysts are notoriously sensitive to poisoning by sulfur-containing compounds. The sensitivity of the catalyst to sulfur is greater for cobalt-based catalysts than for their iron counterparts. Some of these mechanisms are unavoidable and others can be prevented or minimized by controlling the impurity levels in the synthesis gas. By far the most abundant, important, and most studied Fischer-Tropsch process catalyst poison is sulfur. Other catalyst poisons include halides and nitrogen compounds (e.g., NH_3, NOx and HCN).

The hydrocarbon derivatives formed are mainly aliphatic, and on a molar basis methane is the most abundant; the amount of higher hydrocarbon derivatives usually decreases gradually with increase molecular weight. *Iso*-paraffin formation is more extensive over zinc oxide (ZnO) or thoria (ThO_2) at 400 to 500°C (750 to 930°F) and at higher pressure. Paraffin waxes are formed over ruthenium catalysts at relatively low temperatures (170 to 200°C, 340 to 390°F), high pressures (1500 psi), and with a carbon monoxide-hydrogen ratio. The more highly branched product made over the iron catalyst is an important factor in a choice for the manufacture of automotive fuels. On the other hand, a high-quality diesel fuel (paraffin character) can be prepared over cobalt.

Secondary reactions play an important part in determining the final structure of the product. The olefin derivatives produced are subjected to both hydrogenation and double-bond shifting toward the center of the molecule; *cis* and *trans* isomers are formed in about equal amounts. The proportion of straight-chain molecules decreases with rise in molecular weight, but even so they are still more abundant than branched-chain compounds up to about C_{10}.

The small amount of aromatic hydrocarbon derivatives found in the product covers a wide range of isomer possibilities. In the C_6 to C_9 range, benzene, toluene, ethylbenzene, xylene, *n*-propyl- and *iso*-propylbenzene, methyl ethyl benzene derivatives and trimethylbenzene derivatives have been identified; naphthalene derivatives and anthracene derivatives are also present.

Alternatively, a methanol-to-olefins (MTO) option is available (Tian *et al.*, 2015). The methanol-to-olefin derivatives reaction is one of the most important reactions in C1 chemistry, which provides a viable option for producing basic petrochemicals from non-oil resources such as coal and natural gas. As olefin-based petrochemicals and relevant downstream processes have been well developed for many years, the methanol-to-olefins provides a link between gasification chemistry and the modern petrochemical industry. In the process, olefin derivatives are produced from methanol using a zeolite catalyst. The methanol feedstocks vaporized, mixed with recovered methanol, superheated and sent to the fluidized-bed reactor. In the reactor, methanol is first converted to a dimethyl ether (DME, CH_3OCH_3) intermediate and then converted to olefin derivatives with a very high selectivity for ethylene and propylene.

Dimethyl ether can be produced by any one of several routes but the most common route is using methanol produced from synthesis gas. In the process, water-free methanol is vaporized and sent to a reactor with an inlet temperature on the order of 220 to 250°C (430 to 480°F), and an outlet temperature on the order of 300 to 350°C (570 to 660°F). Thus:

$$2CH_3OH \rightarrow CH_3OCH_3 + H_2O$$

The reactor effluents are sent to a distillation column where the dimethyl ether is separated from the top and condensed after which the dimethyl ether sent to storage. Water and methanol are discharged from the bottom and fed to a methanol column for methanol recovery. The purified methanol from this column is recycled to the reactor after mixing with feedstock methanol.

Catalysts considered for Fischer-Tropsch synthesis are based on transition metals of iron, cobalt, nickel and ruthenium. Fischer-Tropsch catalyst development has largely been focused on the preference for high molecular weight linear alkanes and diesel fuels production. Among these catalysts, it is generally known that: (i) nickel (Ni) tends to promote methane formation, as in a methanation process; thus generally it is not desirable, (ii) iron (Fe) is relatively low cost and has a higher water-gas-shift activity, and is therefore more suitable for a lower hydrogen/carbon monoxide ratio (H_2/CO) synthesis gas such as those derived from coal gasification, (iii) cobalt (Co) is more active, and generally preferred over ruthenium (Ru) because of the prohibitively high cost of Ruthenium, and (iv) in comparison to iron, Co has much less water-gas-shift activity, and is much more costly. Thus, it is not surprising that commercially available Fischer-Tropsch catalysts are either cobalt or iron based. In addition to the active metal, the iron-containing catalysts at least typically contain a number of promoters, including potassium and copper, as well as high surface area binders/supports such as silica (SiO_2) and/or alumina (Al_2O_3).

Iron-based Fischer-Tropsch catalysts are currently the most popular catalyst for the Fischer-Tropsch process for converting synthesis gas into Fischer-Tropsch liquids, given Fe catalyst's inherent water gas shift capability to increase the hydrogen/carbon monoxide ratio of coal-derived synthesis gas, thereby improving hydrocarbon product yields in the Fischer-Tropsch synthesis. Fe catalysts may be operated in both high-temperature regime (300 to 350°C, 570 to 650°F) and low-temperature regime (220 to 270°C, 430 to 520°F), whereas Co catalysts are only used in the low-temperature range. This is a consequence of higher temperatures causing more methane formation, which is worse for Co compared to Fe.

Cobalt-containing catalysts are a useful alternative to iron-containing catalysts for Fischer-Tropsch synthesis because they demonstrate activity at lower synthesis pressures, so higher catalyst costs can be offset by lower operating costs. Also, coke deposition rate is higher for Fe catalyst than Co catalyst; consequently, Co catalysts have longer lifetimes. Co catalysts have a long lifetime/greater activity; i.e., Co catalysts are replaced less frequently.

Although there are differences in the product distribution of cobalt-containing and iron-containing catalysts at similar temperatures and pressures (for example, 240°C, 465°F, and 450 psi), a cobalt-containing catalyst has somewhat higher selectivity for heavier hydrocarbon derivatives than an iron-containing catalyst) the product distribution is primarily driven by the choice of operating temperature: high temperature results in a naphtha/kerosene ratio of 2:1; low temperature results in naphtha/kerosene ratio on the order of 1:2 more or less no matter if the catalyst is an iron-containing catalyst or a cobalt-containing catalyst. Higher temperatures shift selectivity towards lower carbon number products and more hydrogenated products; branching increases and secondary products such as ketones and aromatic derivatives also increase.

Thus, generally, a low temperature favors yield high molecular mass linear wax derivative while a high temperature favors the production of naphtha and low molecular weight olefin derivatives. In order to maximize production of the naphtha fraction, it is best to use an iron-containing catalyst at a high temperature in a fixed fluid-bed reactor. On the other hand, in order to maximize production of the kerosene fraction, a slurry reactor with a cobalt-containing catalyst is the more appropriate choice.

Both iron-containing catalyst and cobalt-containing are sensitive to the presence of sulfur compounds in the synthesis gas and can be poisoned by them, hence the need for rigorous feedstock preparation (Chapter 4). However, the sensitivity of the catalyst to sulfur is higher for cobalt-containing catalysts than for iron-catalysts and is often the reason why cobalt-containing catalysts are preferred for Fischer-Tropsch synthesis with natural gas-derived synthesis gas, where the synthesis gas has a higher hydrogen/carbon monoxide ratio and is relatively lower in sulfur content.

10.7 Products and Product Quality

The composition of the products from the synthesis gas production processes is varied insofar as the gas composition varies with the type of feedstock and the gasification system employed (Speight, 2013; Luque and Speight, 2015; Speight, 2014a, 2014b). Furthermore, the quality of gaseous product(s) must be improved by removal of any pollutants such as particulate matter and sulfur compounds before further use (Chapter 4), particularly when the intended use is a water gas shift or methanation reaction.

10.7.1 Products

Low Btu gas (low heat-content gas) is the product when the oxygen is not separated from the air and, as a result, the gas product invariably has a low heat-content (150 to 300 Btu/

ft^3). In *medium Btu gas* (medium heat-content gas), the heating value is in the range 300 to 550 Btu/ft^3 and the composition is much like that of low heat-content gas, except that there is virtually no nitrogen and the H$_2$/CO ratio varies from 2:3 to approximately 3:1 and the increased heating value correlates with higher methane and hydrogen contents as well as with lower carbon dioxide content. *High Btu gas* (high heat-content gas) is essentially pure methane and often referred to as synthetic natural gas or substitute natural gas. However, to qualify as substitute natural gas, a product must contain at least 95% methane; the energy content of synthetic natural gas is 980 to 1080 Btu/ft^3. The commonly accepted approach to the synthesis of high heat-content gas is the catalytic reaction of hydrogen and carbon monoxide.

Hydrogen is also produced during gasification of carbonaceous feedstocks. Although several gasifier types exist (Chapter 2), entrained-flow gasifiers are considered most appropriate for producing both hydrogen and electricity from coal since they operate at temperatures high enough (approximately 1500°C, 2730°F) to enable high carbon conversion and prevent downstream fouling from tars and other residuals.

There is also a series of products that are called by older (even archaic) names that evolved from older coal gasification technologies and warrant mention: (i) producer gas, (ii) water gas, (iii) town gas, and (iv) synthetic natural gas. These products are typically low-to-medium Btu gases (Speight, 2013a, 2013b).

10.7.2 Product Quality

Gas processing, although generally simple in chemical and/or physical principles, is often confusing because of the frequent changes in terminology and, often, lack of cross-referencing (Mokhatab *et al.*, 2006; Speight, 2007, 2008, 2013a, 2014a) (Chapter 4). Although gas processing employs different process types there is always overlap between the various concepts. And, with the variety of possible constituents and process operating conditions, a universal purification system cannot be specified for economic application in all cases.

The processes that have been developed for gas cleaning (Mokhatab *et al.*, 2006; Speight, 2007, 2008) vary from a simple once-through wash operation to complex multi-step systems with options for recycle of the gases (Mokhatab *et al.*, 2006). In some cases, process complexities arise because of the need for recovery of the materials used to remove the contaminants or even recovery of the contaminants in the original, or altered, form.

In more general terms, gas cleaning is divided into removal of particulate impurities and removal of gaseous impurities. For the purposes of this chapter, the latter operation includes the removal of hydrogen sulfide, carbon dioxide, sulfur dioxide, and products that are not related to synthesis gas and hydrogen production. However, there is also need for subdivision of these two categories as dictated by needs and process capabilities: (i) coarse cleaning whereby substantial amounts of unwanted impurities are removed in the simplest, most convenient, manner, (ii) fine cleaning for the removal of residual impurities to a degree sufficient for the majority of normal chemical plant operations, such as catalysis or preparation of normal commercial products; or cleaning to a degree sufficient to discharge an effluent gas to atmosphere through a chimney, (iii) ultra-fine cleaning where the extra step (as well as the extra expense) is justified by the nature of the subsequent operations or the need to produce a particularly pure product.

The products can range from (i) high-purity hydrogen (ii) high-purity carbon monoxide (iii) high-purity carbon dioxide and (iv) a range of hydrogen/carbon monoxide mixtures. The plant is often times referred to as a HYCO if it is designed to produce both carbon monoxide and hydrogen at high purity, else it is referred to as a synthesis gas, or, synthesis gas plant. In fact, the hydrogen/carbon monoxide ratio can be selected at will and the plant's process scheme chosen, in part, by the product composition required. The hydrogen/carbon monoxide ratio will likely vary between 1 and 3 for HYCO and synthesis gas plants. However, at one end of the scale, i.e., if hydrogen is the desired product, then the ratio can approach infinity by shifting all of the carbon monoxide to carbon dioxide. By contrast, on the other end, the ratio cannot be adjusted to zero because hydrogen and water is always produced. An interesting general rule of thumb exists in terms of the hydrogen/carbon monoxide ratio produced by the different gasification processes:

Gasification Process	H_2/CO ratio
Steam Methane Reformer (SMR)	3.0 – 5.0
SMR + Oxygen Secondary Reforming (O2R)	2.5 – 4.0
Autothermal Reforming (ATR)	1.6 – 2.65
Partial Oxidation (POx)	1.6 – 1.9

It should be noted that in practice however, the options are not limited to the ranges shown but rather even greater hydrogen/carbon monoxide ratios, if adjustments are made like the inclusion of a Shift converter to effect near-equilibrium water-gas-shift conversion, or, adjusting the amount of steam.

Throughout the previous section there has, of necessity, been frequent reference to the production of hydrogen as an integral part of the production of carbon monoxide, since the two gases make up the mixture known as synthesis gas. Hydrogen is indeed an important commodity in the refining industry because of its use in hydrotreating processes, such as desulfurization, and in hydroconversion processes, such as hydrocracking. Part of the hydrogen is produced during reforming processes but that source, once sufficient, is now insufficient for the hydrogen needs of a modern refinery (Parkash, 2003; Gary *et al.*, 2007; Speight, 2011, 2014a; Hsu and Robinson, 2017; Speight, 2017).

In addition, optimum hydrogen purity at the reactor inlet extends catalyst life by maintaining desulphurization kinetics at lower operating temperatures and reducing carbon laydown. Typical purity increases resulting from hydrogen purification equipment and/or increased hydrogen sulfide removal as well as tuning hydrogen circulation and purge rates, may extend catalyst life up to about 25%. Indeed, since hydrogen use has become more widespread in refineries, hydrogen production has moved from the status of a high-tech specialty operation to an integral feature of most refineries (Raissi, 2001; Vauk *et al.*, 2008).

While the gasification of residua and coke to produce hydrogen and/or power may increase in use in refineries over the next two decades (Speight, 2011b), several other processes are available for the production of the additional hydrogen that is necessary for the various heavy feedstock hydroprocessing sequences (Speight, 2014a) and it is the purpose of this section to present a general description of these processes.

Purities in excess of 99.5% of either the hydrogen or carbon monoxide produced from synthesis gas, can be achieved if desired. Four of the major process technologies available are (i) cryogenics plus methanation, (ii) cryogenics plus pressure-swing adsorption, (iii) methane-wash cryogenic process, and (iv) the Cosorb process, Thus:

(i) *Cryogenics + Methanation*: This process utilizes a cryogenic process (occurring in a cold box) whereby carbon monoxide is liquefied in a number of steps until hydrogen with a purity of ~ 98% is produced. The condensed carbon monoxide product, which would contain methane, is then distilled to produce essentially pure carbon monoxide and a mixture of carbon monoxide/methane. The latter stream can be used as fuel. The hydrogen stream from the cold box is taken to a Shift's Converter where the remaining carbon monoxide is converted to carbon dioxide and hydrogen. The carbon dioxide is then removed and any further carbon monoxide or carbon dioxide can be removed by methanation. The resulting hydrogen stream can be of purities ~ 99.7%.

(ii) *Cryogenics plus Pressure-Swing Adsorption (PSA)*: This process utilizes the similar sequential liquefaction of carbon monoxide in a cold box until hydrogen of ~ 98% purity is achieved. Again, the carbon monoxide stream can be further distilled to remove methane until it is essentially pure. The hydrogen stream is then allowed to go through multiple swings of PSA cycles until the hydrogen purity of even as high as 99.999% is produced.

(iii) *Methane-wash Cryogenic Process*: In this scheme liquid carbon monoxide is absorbed into a liquid methane stream so that the hydrogen stream produced contains only ppm levels of carbon monoxide but about 5 – 8% methane. Hence an hydrogen stream purity of only about 95% is possible. The liquid carbon monoxide/methane stream however, can be distilled to produce an essentially pure carbon monoxide stream and a carbon monoxide/methane stream which can be used as fuel.

(iv) *Cosorb Process*: This process utilizes copper ions (cuprous aluminum chloride, $CuAlCl_4$) in toluene to form a chemical complex with the carbon monoxide, and in effect separating it from the hydrogen, nitrogen, carbon dioxide, and methane. This process can capture about 96% of the carbon monoxide to produce a stream of greater than 99% purity. The downside of this process is that water, hydrogen sulfide and other trace chemicals which can poison the copper catalyst, must be removed prior to the reactor. Further, a hydrogen stream of only up to 97% purity is obtained. However, while the efficiency of cryogenic separation decreases with low carbon monoxide content of the feed, the Cosorb process is able to process gases with a low carbon monoxide content more efficiently.

10.8 Fischer-Tropsch Chemistry

Synthesis gas (synthesis gas) is the name given to a gas mixture that contains varying amounts of carbon monoxide (CO) and hydrogen (H2) generated by the gasification of a

340 SYNTHESIS GAS

carbonaceous material. Examples include steam reforming of natural gas, crude oil residua, coal, and biomass.

10.8.1 Chemical Principles

Synthesis gas consists primarily of carbon monoxide, carbon dioxide and hydrogen, and has less than half the energy density of natural gas. Synthesis gas is combustible and often used as a fuel source or as an intermediate for the production of other chemicals. Synthesis gas for use as a fuel is most often produced by gasification of the carbonaceous feedstock or municipal waste mainly by the following paths:

$$C + O_2 \rightarrow CO_2$$

$$CO_2 + C \rightarrow 2CO$$

$$C + H_2O \rightarrow CO + H_2$$

When used as an intermediate in the large-scale, industrial synthesis of hydrogen and ammonia, it is also produced from natural gas (via the steam reforming reaction) as follows:

$$CH_4 + H_2O \rightarrow CO + 3H_2$$

The synthesis gas produced in large waste-to-energy gasification facilities is used as fuel to generate electricity.

Although the focus of this section is the production of hydrocarbon derivatives from synthesis gas, it is worthy of note that all or part of the clean synthesis gas can also be used (i) as chemical *building blocks* to produce a broad range of chemicals using processes well established in the chemical and petrochemical industry, (ii) as a fuel producer for highly efficient fuel cells (which run off the hydrogen made in a gasifier) or perhaps in the future, hydrogen turbines and fuel cell-turbine hybrid systems, and (iii) as a source of hydrogen that can be separated from the gas stream and used as a fuel or as a feedstock for refineries (which use the hydrogen to upgrade crude oil products) (Parkash, 2003; Gary *et al.*, 2007; Speight, 2011, 2014a; Hsu and Robinson, 2017; Speight, 2017). However, the decreasing availability and increased price of crude oil has renewed the worldwide interest in the production of liquid hydrocarbon derivatives from carbon monoxide and hydrogen using metal catalysts, also known as Fischer-Tropsch synthesis or Fischer-Tropsch process.

Gasification to produce synthesis gas can proceed from just about any organic material, including biomass and plastic waste. The resulting synthesis gas burns cleanly into water vapor and carbon dioxide. Alternatively, synthesis gas may be converted efficiently to methane via the Sabatier reaction, or to a diesel-like synthetic fuel via the Fischer-Tropsch process. Inorganic components of the feedstock, such as metals and minerals, are trapped in an inert and environmentally safe form as char, which may have use as a fertilizer.

In principle, synthesis gas can be produced from any hydrocarbon feedstock. These include natural gas, naphtha, residual oil, crude oil coke, coal, and biomass. The lowest cost routes for synthesis gas production, however, are based on natural gas. The cheapest option is remote or stranded reserves. Current economic considerations dictate that the production of liquid fuels from synthesis gas translates into using natural gas as the hydrocarbon source. Nevertheless, the synthesis gas production operation in a gas-to-liquids plant amounts to greater than half of the capital cost of the plant. The choice of technology for synthesis gas production also depends on the scale of the synthesis operation. Synthesis gas production from solid fuels can require an even greater capital investment with the addition of feedstock handling and more complex synthesis gas purification operations. The greatest impact on improving gas-to-liquids plant economics is to decrease capital costs associated with synthesis gas production and improve thermal efficiency through better heat integration and utilization. Improved thermal efficiency can be obtained by combining the gas-to-liquids plant with a power generation plant to take advantage of the availability of low-pressure steam.

A modified version of steam reforming known as autothermal reforming, which is a combination of partial oxidation near the reactor inlet with conventional steam reforming further along the reactor, improves the overall reactor efficiency and increases the flexibility of the process. Partial oxidation processes using oxygen instead of steam also found wide application for synthesis gas manufacture, with the special feature that they could utilize low-value feedstocks such as heavy crude oil residua. In recent years, catalytic partial oxidation employing very short reaction times (milliseconds) at high temperatures (850 to 1000°C; 1560 to 1830°F) is providing still another approach to synthesis gas manufacture (Hickman and Schmidt, 1993). Nearly complete conversion of methane, with close to 100% selectivity to hydrogen and carbon monoxide, can be obtained with a rhenium monolith under well-controlled conditions. Experiments on the catalytic partial oxidation of *n*-hexane conducted with added steam give much higher yields of hydrogen than can be obtained in experiments without steam, a result of much interest in obtaining hydrogen-rich streams for fuel cell applications.

The route of coal to synthetic automotive fuels, as practiced by SASOL, is technically proven and products with favorable environmental characteristics are produced. As is the case in essentially all the carbonaceous feedstock conversion processes where air or oxygen is used for the utilization or partial conversion of the energy in the carbonaceous feedstock, the carbon dioxide burden is a drawback as compared to crude oil.

The uses of synthesis gas include use as a chemical feedstock and in gas-to-liquid processes which use Fisher-Tropsch chemistry to make liquid fuels as feedstock for chemical synthesis, as well as being used in the production of fuel additives, including diethyl ether and methyl t-butyl ether (MTBE), acetic acid and its anhydride, synthesis gas could also make an important contribution to chemical synthesis through conversion to methanol. There is also the option in which stranded natural gas is converted to synthesis gas production followed by conversion to liquid fuels.

The hydroformylation synthesis (also known as the oxo synthesis or the oxo process) is an industrial process for the production of aldehyde derivatives from alkene derivatives. This chemical reaction entails the addition of a formyl group (CHO) and a hydrogen atom to a carbon-carbon double bond. A key consideration of hydroformylation is the production of normal-isomers or the production of iso-isomers in the product(s). For example, the hydroformylation of propylene can yield two isomeric products, butyraldehyde or iso-butyraldehyde:

$$H_2 + CO + CH_3CH=CH_2 \rightarrow CH_3CH_2CH_2CHO \text{ (n-butyraldehyde)}$$

$$H_2 + CO + CH_3CH=CH_2 \rightarrow (CH_3)_2CHCHO \text{ (iso-butyraldehyde)}$$

These isomers reflect the regio-chemistry (the preference of one direction of chemical bond formation or chemical bond scission over all other possible directions) of the insertion of the alkene into the M–H bond. Generally, both products are not equally desirable.

As an example of the hydroformylation process, the Exxon process, also Kuhlmann-oxo process or the PCUK-oxo process, is used for the hydroformylation of C_6 to C_{12} olefin derivatives using cobalt-based catalysts. In order to recover the catalyst, an aqueous sodium hydroxide solution or sodium carbonate is added to the organic phase. By extraction with olefin and neutralization by addition of sulfuric acid solution under carbon monoxide pressure the metal-carbonyl hydride can recovered. The recovered hydride is stripped out with synthesis gas, absorbed by the olefin, and returned to the reactor. The Exxon process, similar to the BASF process, process is carried out at a temperature on the order of 160 to 180°C (320 to 355°F) and a pressure suitable to the reactants and products.

The Fischer-Tropsch reaction can be described as the synthesis of hydrocarbon derivatives via the hydrogenation of carbon monoxide using transition metal catalysts. The major catalysts used industrially are Fe and Co, but can also be Ru and Ni. From a mechanism perspective the reactions can be regarded as a carbon chain building process where methylene (CH_2) groups are attached sequentially in a carbon chain. (Table 10.3). Thus:

$$nCO + [n + m/2]H_2 \rightarrow C_nH_m + nH_2O$$

For example:

$$CO + 2H_2 \rightarrow \text{-}[CH_{2]}\text{-} + H_2O$$

A common and salient feature of the reactions is the exothermicity of the reactions. As a general rule of thumb, the reactions which produce water and carbon dioxide as a product

Table 10.3 Carbon chain groups of the range of fischer-tropsch products which can be produced.

Carbon number	Group name
C1-C2	Synthetic natural gas (SNG)
C3-C4	Liquefied petroleum gas (LPG)
C5-C7	Low boiling naphtha
C8-C10	High boiling naphtha
C11-C16	Middle distillate
C17-C30*	Low melting wax
C31-C60	High melting wax

*The C17 n-alkane (n-heptadecane) is the first member of the series that is not fully liquid under ambient conditions (melting point: 21°C (70°F).

tend to be more exothermic on account of the very high heat of formation of these species. Some of the reactions proposed are as follows (Rauch, 2001):

$$CO_2 + 3H2 \rightarrow -[CH_2]- + 2H_2O \qquad \Delta H = -125 \text{ kJ/mol}$$

$$CO + 2H_2 \rightarrow -[CH_2]- + H2O \qquad \Delta H = -165 \text{ kJ/mol}$$

$$2 CO + H_2 \rightarrow -[CH_2]- + CO_2 \qquad \Delta H = -204 \text{ kJ/mol}$$

$$3 CO H_2 \rightarrow -[CH_2]- + 2CO_2 \qquad \Delta H = -244 \text{ kJ/mol}$$

There is also the water-gas-shift reaction:

$$CO + H_2O \rightarrow H_2 + CO_2 \qquad \Delta H = -39 \text{ kJ/mol}$$

Due to the very high exothermic nature of the Fischer-Tropsch reactions, as illustrated in the reactions above, an important issue is, not surprisingly, the need to avoid an increase in temperature (Furnsinn et al., 2005). The need for cooling is thus of critical importance in order to: (i) maintain stable reaction conditions, (ii) avoid the tendency to produce lighter hydrocarbon derivatives, and (iii) prevent catalyst sintering and hence reduction in activity. Since the total heat of reaction is in the order of approximately 25% of the heat of combustion of the synthesis gas (i.e., reactants of the Fischer-Tropsch process), a theoretical limit on the maximum efficiency of the Fischer-Tropsch process is imposed (Rauch, 2001).

Two main chemical characteristics of Fischer-Tropsch synthesis are the unavoidable production of a wide range of hydrocarbon products (olefin derivatives, paraffin derivatives, and oxygenated products) and the liberation of a large amount of heat from the highly exothermic synthesis reactions. Product distributions are influenced by temperature, feed gas composition (hydrogen/carbon monoxide), pressure, catalyst type, and catalyst composition. Fischer-Tropsch products are produced in four main steps: synthesis gas generation, gas purification, Fischer-Tropsch synthesis, and product upgrading. Depending on the types and quantities of Fischer-Tropsch products desired, either low (200 to 240°C; 390 to 465°F) or high temperature (300 to 350°C; 570 to 660°F) synthesis is used with either an iron (Fe) or cobalt catalyst (Co) (Van Berge and Everson, 1997).

The required gas mixture of carbon monoxide and hydrogen (synthesis gas) is created through a reaction of coke or the carbonaceous feedstock with water steam and oxygen, at temperatures over 900°C. In the past, town gas and gas for lamps were a carbon monoxide-hydrogen mixture, made by gasifying coke in gas works. In the 1970s it was replaced with imported natural gas (methane). Gasification of carbonaceous feedstocks and Fischer-Tropsch hydrocarbon synthesis together bring about a two-stage sequence of reactions which allows the production of liquid fuels like gasoline and diesel out of the solid combustible the carbonaceous feedstock.

The Fischer-Tropsch synthesis is, in principle, a carbon chain-building process, where methylene groups are attached to the carbon chain. The actual reactions that occur have been, and remain, a matter of controversy, as they have been the last century since the 1930s.

$$(2n+1)H_2 + nCO \rightarrow C_nH_{2n+2} + nH_2O$$

Even though the overall Fischer-Tropsch process is described by the following chemical equation:

$$(2n+1)H2 + nCO \rightarrow CnH_{2n+2} + nH2O$$

The initial reactants in the above reaction (i.e., carbon monoxide and H2) can be produced by other reactions such as the partial combustion of a hydrocarbon:

$$C_nH_{2n+2} + \tfrac{1}{2} nO_2 \rightarrow (n+1)H_2 + nCO$$

For example (when n=1), methane (in the case of gas to liquids applications):

$$2CH_4 + O_2 \rightarrow 4H_2 + 2CO$$

Or by the gasification of any carbonaceous source, such as biomass:

$$C + H_2O \rightarrow H_2 + CO$$

The energy needed for this endothermic reaction is usually provided by (exothermic) combustion with air or oxygen:

$$2C + O_2 \rightarrow 2CO$$

These reactions are highly exothermic, and to avoid an increase in temperature, which results in lighter hydrocarbon derivatives, it is important to have sufficient cooling, to secure stable reaction conditions. The total heat of reaction amounts to approximately 25% of the heat of combustion of the synthesis gas, and lays thereby a theoretical limit on the maximal efficiency of the Fischer-Tropsch process.

The reaction is dependent of a catalyst, mostly an iron or cobalt catalyst where the reaction takes place. There is either a low- or high-temperature process (low-temperature Fischer-Tropsch, high-temperature Fischer-Tropsch), with temperatures ranging between 200 to 240°C (390 to 465°F) for low-temperature Fischer-Tropsch and 300 to 350°C (570 to 660°F) for the high-temperature Fischer-Tropsch process. The high-temperature Fischer-Tropsch process uses an iron catalyst, and the low-temperature Fischer-Tropsch either an iron or a cobalt catalyst. The different catalysts include also nickel-based and ruthenium-based catalysts, which also have enough activity for commercial use in the process. But the availability of ruthenium is limited and the nickel-based catalyst has high activity but produces too much methane, and additionally the performance at high pressure is poor, due to production of volatile carbonyls. This leaves only cobalt and iron as practical catalysts, and this study will only consider these two. Iron is cheap, but cobalt has the advantage of higher activity and longer life, though it is on a metal basis 1,000 times more expensive than iron catalyst.

10.8.2 Refining Fischer-Tropsch Products

The Fischer-Tropsch product stream typically contains hydrocarbon derivatives having a range of numbers of carbon atoms, including gases, liquids and waxes. Depending on the molecular weight product distribution, different Fischer-Tropsch product mixtures are ideally suited to different uses. For example, Fischer-Tropsch product mixtures containing liquids may be processed to yield gasoline, as well as heavier middle distillates. Hydrocarbon waxes may be subjected to additional processing steps for conversion to liquid and/or gaseous hydrocarbon derivatives. Thus, in the production of a Fischer-Tropsch product stream for processing to a fuel it is desirable to obtain primarily hydrocarbon derivatives that are liquids and waxes (e.g., C_{5+} hydrocarbon derivatives). Initially, the light gases in raw product are separated and sent to a gas cleaning operation. The higher-boiling product is distilled to produce separate streams of naphtha, distillate, and wax.

The naphtha stream is first hydrotreated which produces a hydrogen-saturated liquid product (primarily paraffin derivatives), a portion of which are converted by isomerization from normal paraffin derivatives to iso-paraffin derivatives to boost their octane value. Another fraction of the hydrotreated naphtha is catalytically reformed to provide some aromatic content to (and further boost the octane value of) the final gasoline blend stock. The distillate stream is also hydrotreated, resulting directly in a finished diesel blend stock. The wax fraction is hydrocracked into a finished distillate stream and naphtha streams that augment the hydrotreated naphtha streams sent for isomerization and for catalytic cracking.

Thus, conventional refinery processes (Parkash, 2003; Gary *et al.*, 2007; Speight, 2011, 2014a; Hsu and Robinson, 2017; Speight, 2017) can be used for upgrading of Fischer-Tropsch liquid and wax products. A number of possible processes for Fischer-Tropsch products are wax hydrocracking, distillate hydrotreating, catalytic reforming, naphtha hydrotreating, alkylation and isomerization. Fuels produced with the Fischer-Tropsch synthesis are of a high quality due to a very low aromaticity and zero sulfur content.

The diesel fraction has a high cetane number resulting in superior combustion properties and reduced emissions. New and stringent regulations may promote replacement or blending of conventional fuels by sulfur and aromatic-free products. Also, other products besides fuels can be manufactured with Fischer-Tropsch in combination with upgrading processes, for example, ethylene, propylene, α-olefin derivatives, alcohols, ketones, solvents, and waxes. These valuable byproducts of the process have higher added values, resulting in an economically more attractive process economy (Parkash, 2003; Gary *et al.*, 2007; Speight, 2011, 2014a; Hsu and Robinson, 2017; Speight, 2017).

At this point, it is necessary to deal once again with the production of chemicals from the carbonaceous feedstock by gasification followed by conversion of the synthesis gas mixture (carbon monoxide, carbon monoxide, and hydrogen, H2) to higher molecular weight liquid fuels and other chemicals (Penner, 1987; Speight, 2014a, 2019).

The production of synthesis gas involves reaction of the carbonaceous feedstock with steam and oxygen. The gas stream is subsequently purified (to remove sulfur, nitrogen, and any particulate matter) after which it is catalytically converted to a mixture of liquid hydrocarbon products. In addition, synthesis gas may also be used to produce a variety of products, including ammonia, and methanol.

In principle, synthesis gas can be produced from any hydrocarbon (or hydrocarbonaceous) feedstock (Chapter 4, Chapter 5, Chapter 6, Chapyter 7). These include natural gas, naphtha, residual oil, crude oil coke, coal, and biomass. The lowest cost routes for synthesis gas production, however, are based on natural gas. The cheapest option is remote or stranded reserves.

The choice of technology for synthesis gas production also depends on the scale of the synthesis operation. Synthesis gas production from solid fuels can require an even greater capital investment with the addition of feedstock handling and more complex synthesis gas purification operations. The greatest impact on improving gas-to-liquids plant economics is to decrease capital costs associated with synthesis gas production and improve thermal efficiency through better heat integration and utilization. Improved thermal efficiency can be obtained by combining the gas-to-liquids plant with a power generation plant to take advantage of the availability of low-pressure steam.

The synthesis of hydrocarbon derivatives from carbon monoxide and hydrogen (synthesis gas) (the Fischer-Tropsch synthesis) is a procedure for the indirect liquefaction of coal (Storch et al., 1951; Batchelder, 1962). This process is the only coal liquefaction scheme currently in use on a relatively large commercial scale; South Africa is currently using the Fischer-Tropsch process on a commercial scale in its SASOL complex (Singh, 1981), although Germany produced roughly 156 million barrels of synthetic crude oil annually using the Fischer-Tropsch process during World War II.

Briefly, in the gasification process, the carbonaceous feedstock is converted to gaseous products at temperatures in excess of 800°C (1470°F), and at moderate pressures, to produce synthesis gas.

$$C + H_2O \rightarrow CO + H_2$$

The gasification may be attained by means of any one of several processes (Speight, 2013a; 2013b, 2014a, 2014b; Luque and Speight, 2015). The exothermic nature of the process and the decrease in the total gas volume in going from reactants to products suggest the most suitable experimental conditions to use in order to maximize product yields. The process should be favored by high pressure and relatively low reaction temperature.

In practice, the Fischer-Tropsch reaction is generally carried out at temperatures in the range 200 to 350°C (390 to 660°F) and at pressures of 75 to 4000 psi; the hydrogen/carbon monoxide ratio is usually at ca. 2.2:1 or 2.5:1. Since up to three volumes of hydrogen may be required to achieve the next stage of the liquids production, the synthesis gas must then be converted by means of the water-gas shift reaction) to the desired level of hydrogen after which the gaseous mix is purified (acid gas removal, etc.) and converted to a wide variety of hydrocarbon derivatives.

$$CO + H_2O \rightarrow CO_2 + H_2$$

$$CO + (2n+1)H_2 \rightarrow C_nH_{2n+2} + H_2O$$

These reactions result primarily in low- and medium-boiling aliphatic compounds; present commercial objectives are focused on the conditions that result in the production of n-hydrocarbon derivatives as well as olefin derivatives and oxygenated materials.

References

Alstrup, I. 1988. A New Model Explaining Carbon Filament Growth on Nickel, Iron, and Ni-Cu Alloy Catalysts. *J. Catal.* 109: 241-251.

Aasberg-Petersen, K., Bak Hansen, J.-H., Christiansen, T.S., Dybkjær, I., Seier, Christensen, P., Stub Nielsen, C., Winter Madsen, S.E.L., and Rostrup-Nielsen, J.R. 2001. *Technologies for Large-Scale Gas Conversion.* Appl. Cat. A: General, 221: 379-387.

Aasberg-Petersen, K., Christensen, T.S., Stub Nielsen, C., and Dybkjær, I. 2002. Recent Developments in Autothermal Reforming and Pre-reforming for Synthesis Gas Production in GTL Applications. *Preprints, Div. Fuel Chem., Am. Chem. Soc.,* 47(1), 96-97.

Balasubramanian, B., Ortiz, A.L., Kaytakoglu, S. and Harrison, D.P. 1999. Hydrogen from Methane in a Single-Step Process. *Chemical Engineering Science.* 54: 3543-3552.

Batchelder, H.R. 1962. In *Advances in Petroleum Chemistry and Refining.* J.J. McKetta Jr. (Editor). Interscience Publishers Inc., New York. Volume V. Chapter 1.

Bayat, M., Rahmani, F., Mazinani, S., and Rahimpour, M.R. 2010. Enhancement of Gasoline Production in a Novel Optimized Hydrogen-Permselective Membrane Fischer-Tropsch Reactor in GTL Technology. *Chemical Engineering Transactions,* 21: 1477-1482.

Chadeesingh, R. 2011. The Fischer-Tropsch Process. In: *The Biofuels Handbook.* J.G. Speight (Editor). The Royal Society of Chemistry, London, United Kingdom. Part 3, Chapter 5, Page 476-517.

Dyer, P.N. and Chen, C.M. 1999. Engineering Development of Ceramic Membrane Reactor Systems for Converting Natural Gas to Hydrogen and Synthesis Gas for Transportation Fuels. *Proceedings. Energy Products for the 21st Century Conf., Sept 22.*

Fischer, F., and Tropsch, H.U. 1923. Über die Herstellung synthetischer Ölgemische (syn-thol) durch Aufbau aus Kohlenoxyd und Wasserstoff. *Brennstoff Chemie,* 4: 276-285.

Gary, J.G., Handwerk, G.E., and Kaiser, M.J. 2007. *Petroleum Refining: Technology and Economics,* 5th Edition. CRC Press, Taylor & Francis Group, Boca Raton, Florida.

Gunardson, H.H., and Abrardo, J.M. 1999. *Hydrocarbon Processing,* April. Page 87-93.

Hagh, B.F. 2004. Comparison of Autothermal Reforming for Hydrocarbon Fuels. *Preprints. Div. Fuel Chem., Am. Chem. Soc.,* 49 (1): 144-147.

Higman, C., and Van der Burgt, M. 2008. *Gasification* 2nd Edition, Elsevier, Amsterdam, Netherlands.

Hsu, C.S., and Robinson, P.R. (Editors). 2017. *Handbook of Petroleum Technology.* Springer International Publishing AG, Cham, Switzerland.

Hu, J., Fei Yu, F., and Lu, Y. 2012. Application of Fischer–Tropsch Synthesis in Biomass to Liquid Conversion. *Catalysts,* 2: 303-326.

Hufton, J.R., Mayorga, S. and Sircar, S., 1999. Sorption-Enhanced Reaction Process for Hydrogen Production. *AIChE. Journal,* 45: 248-256.

Klerk A, and Furimsky E. 2010. *Catalysis in the Refining of Fischer-Tropsch Syncrude.* RSC Catalysis Series, No. 4. Royal Society of Chemistry, London, United Kingdom. Page 1-294.

Liu, K. Song, C., and Subramani, V. (Editors). 2010. *Hydrogen and Syngas Production and Purification Technologies.* John Wiley & Sons Inc., Hoboken, New Jersey.

Luque, R., and Speight, J.G. (Editors), 2015. *Gasification for Synthetic Fuel Production: Fundamentals, Processes, and Applications.* Woodhead Publishing, Elsevier, Cambridge, United Kingdom.

Mokhatab, S., Poe, W.A., and Speight, J.G. 2006. *Handbook of Natural Gas Transmission and Processing.* Elsevier, Amsterdam, Netherlands.

Nataraj, S., Moore, R.B., Russek, S,L., US 6,048,472; 2000; assigned to Air Products and Chemicals, Inc.

Penner, S.S. 1987. *Proceedings. Fourth Annual Pittsburgh Coal Conference.* University of Pittsburgh, Pittsburgh, Pennsylvania. Page 493.

Rauch, R. 2001. Biomass Gasification to produce Synthesis Gas for Fuel Cells, Liquid Fuels and Chemicals, IEA Bioenergy Agreement, Task 33: Thermal Gasification of Biomass.

Rostrup-Nielsen, J.R. 1984. Sulfur-Passivated Nickel Catalysts for Carbon-Free Steam Reforming of Methane. *J. Catal.,* 85: 31-43.

Rostrup-Nielsen, J.R. 1993. Production of Synthesis Gas. *Catalysis Today,* 19: 305-324.

Rostrup-Nielsen, J., and Christiansen, L.J. 2011. *Concepts in Syngas Manufacture. Catalysis Science Series No. 10.* Imperial College Press, London, United Kingdom.

Schulz, H. 1999. Short History and Present Trends of Fischer–Tropsch Synthesis. *Applied Catalysis A: General,* 86(1-2): 3-12.

Shu, J., Grandjean, B.P.A., and Kaliaguine, S. 1995. *Catalysis Today,* 25: 327-332.

Speight, J.G. 2007. *Natural Gas: A Basic Handbook.* GPC Books, Gulf Publishing Company, Houston, Texas.

Speight, J.G. 2008. *Synthetic Fuels Handbook: Properties, Processes, and Performance.* McGraw-Hill, New York.

Speight, J.G. 2011. *The Refinery of the Future.* Gulf Professional Publishing, Elsevier, Oxford, United Kingdom.

Speight, J.G. 2013a. *The Chemistry and Technology of Coal* 3rd Edition. CRC Press, Taylor & Francis Group, Boca Raton, Florida.

Speight, J.G. 2013b. *Coal-Fired Power Generation Handbook.* Scrivener Publishing, Beverly, Massachusetts.

Speight, J.G. 2014a. *The Chemistry and Technology of Petroleum* 5th Edition. CRC Press, Taylor & Francis Group, Boca Raton, Florida.

Speight, J.G. 2014b. *Gasification of Unconventional Feedstocks.* Gulf Professional Publishing, Elsevier, Oxford, United Kingdom.

Speight, J.G. 2017. *Handbook of Petroleum Refining Processes.* CRC Press, Taylor & Francis Group, Boca Raton, Florida.

Speight, J.G. 2019. *Handbook of Petrochemical Processes.* CRC Press, Taylor & Francis Group, Boca Raton, Florida.

Storch, H.H., Golumbic, N., and Anderson, R.B. 1951. *The Fischer Tropsch and Related Syntheses.* John Wiley & Sons Inc., New York.

Tian, P., Wei, Y., Ye, M., and Liu, Z., 2015. Methanol to Olefins (MTO): From Fundamentals to Commercialization. *ACS Catal.,* 5(3): 1922-1938.

Udengaard, N.R., Hansen, J.H.B., Hanson, D.C., and Stal, J.A. 1992. Sulfur Passivated Reforming Process Lowers Syngas H_2/CO Ratio. *Oil Gas J.,* 90: 62-67.

Van Berge, P.J. and Everson, R.C. 1997. Cobalt as an Alternative Fischer-Tropsch Catalyst to Iron for the Production of Middle Distillates. In: *Stud. Surf. Sci. Catal. 107 Natural Gas Conversion IV.* M. de Pontes, R.L. Espinoza, C.P. Nicolaides, J.H. Scholtz, and M.S. Scurrell (Editors). Elsevier B.V., Amsterdam, Netherlands. Page 207-212.

11
Synthesis Gas in the Refinery

11.1 Introduction

Refineries worldwide are facing growing difficulties in finding market outlets for the viscous high-sulfur resid because the crude oils themselves are higher in sulfur and, because of stringent environmental legislation, refiners are required to meet higher product quality standards. The trend of crude oil sulfur content in different various regions shows the progressive deterioration of crude oil qualities (Speight, 2011b, 2014a, 2017). To counteract this trend, refineries have the option to invest in more severe conversion processes, which, however, will leave the refiners with a reduced volume of much more contaminated bottoms, even more difficult to handle and dispose.

Gasification is a technology which, along with other processes (Table 11.1), makes it possible for the refineries to attain a zero residue target, contrary to all conversion technologies (such as thermal cracking, catalytic cracking, cooking, deasphalting, and hydroprocessing) which can reduce the volume of the resid. In addition, the flexibility of gasification permits to handle any kind of refinery residue, including crude oil coke, tank bottoms and refinery sludges and make available a range of value-added products, electricity, steam, hydrogen and various chemicals based on synthesis gas chemistry (such as ammonia, methanol, ammonia, acetic acid, and formaldehyde).

Thus, gasification is an appealing process for the utilization of relatively inexpensive feedstocks that might otherwise be declared as waste and sent to a landfill (where the production of methane – a so-called greenhouse gas – will be produced) or combusted which may not (depending upon the feedstock) be energy efficient. Overall, use of a gasification technology (Furimsky, 1999; Higman and Van der Burgt, 2008; Wolff and Vliegenthart, 2011; Khosravi and Khadse, 2013; Speight, 2013a, 2014b; Luque and Speight, 2015) with the necessary gas cleanup options can have a smaller environmental footprint and lesser effect on the environment than landfill operations or combustion of the waste. In fact, there are strong indications that gasification is a technically viable option for the waste conversion, including residual waste from separate collection of municipal solid waste. The process can meet existing emission limits and can have a significant effect on the reduction of landfill disposal using known gasification technologies (Arena, 2012; Speight, 2014b; Luque and Speight, 2015) or thermal plasma (Fabry *et al.*, 2013).

Indeed, the mounting interest in gasification technology reflects a convergence of two changes in the electricity generation marketplace: (i) the maturity of gasification technology, and (ii) the extremely low emissions from integrated gasification combined cycle (IGCC) plants, especially air emissions, and the potential for lower cost control of greenhouse gases than other coal-based systems (Speight, 2014b). Another advantage of gasification is the use of synthesis gas, which is potentially more efficient as compared to direct combustion of the

Table 11.1 Options for resid processing in a refinery.

Resid	Gasification		
	Solvent deasphalting	Deasphalted oil	Visbreaking
			Delayed coking
			Fluid catalytic cracking
		Deasphalter bottoms	Visbreaking
			Delayed coking
			Fuel oil blend stock
			Road asphalt
			Gasification

original fuel because it can be (i) combusted at higher temperatures, (ii) used in fuel cells, (iii) used to produce methanol and hydrogen, (iv) converted via the Fischer-Tropsch process into a range of synthesis liquid fuels suitable for use gasoline engines or diesel engines. It is the latter two options that have increased the interest in the production of synthesis gas in a refinery (Chadeesingh, 2011; Speight, 2019b).

11.2 Processes and Feedstocks

Gasification is an established, tried-and-true method that can be used to convert crude oil coke (petcoke) and other refinery waste streams and residuals (vacuum residual, visbreaker tar, and deasphalter pitch) into power, steam, and hydrogen for use in the production of cleaner transportation fuels (Gray and Tomlinson, 2000; Luque and Speight, 2015) (Figure 11.1) and the synthesis gas produced can give rise to a variety of chemical products

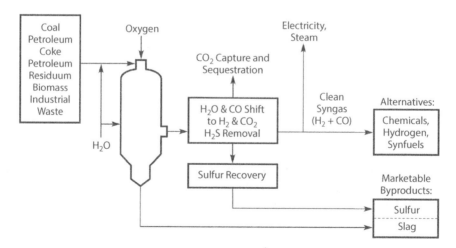

Figure 11.1 The gasification process can accommodate a variety of feedstocks.

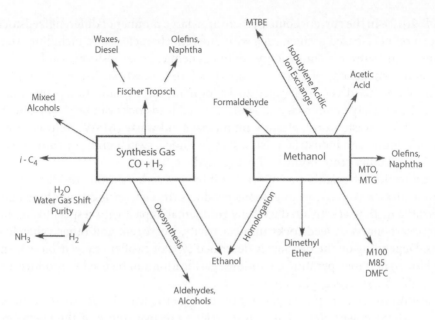

Figure 11.2 Potential products from synthesis gas.

(Figure 11.2). The main requirement for a gasification feedstock is that it contains both hydrogen and carbon. Thus, a number of factors have increased the interest in gasification applications in crude oil refinery operations: (i) coking capacity has increased with the shift to heavier, more sour crude oils being supplied to the refiners, (ii) hazardous waste disposal has become a major issue for refiners in many countries, (iii) there is strong emphasis on the reduction of emissions of criteria pollutants and greenhouse gases, (iv) requirements to produce ultra-low-sulfur fuels are increasing the hydrogen needs of the refineries, and (v) the requirements to produce low-sulfur fuels and other regulations could lead to refiners falling short of demand for light products such as gasoline and jet and diesel fuel. The typical gasification system incorporated into the refinery consists of several process plants including (i) feedstock preparation, (ii) the type of gasifier, (iii) gas cleaning, and (iv) a sulfur recovery unit as well as downstream process options depending on the desired products.

The gasification process can provide high-purity hydrogen for a variety of uses within the refinery. Hydrogen is used in the refinery to remove sulfur, nitrogen, and other impurities from intermediate to finished product streams (Chapter 10) and in hydrocracking operations for the conversion of high-boiling distillates and oils into low-boiling products such as naphtha, kerosene and diesel (Chapter 11). Furthermore, electric power and high-pressure steam can be generated by the gasification of petcoke and residuals to drive mostly small and intermittent loads such as compressors, blowers, and pumps. Steam can also be used for process heating, steam tracing, partial pressure reduction in fractionation systems, and stripping low-boiling components to stabilize process streams. Also, the gasification system and refinery operations can share common process equipment. This usually includes an amine stripper or sulfur plant, waste water treatment, and cooling water systems (Mokhatab *et al.*, 2006; Speight, 2014a, 2019a).

The gasification of carbonaceous feedstocks has been used for many years to convert organic solids and liquids into useful gaseous, liquid and cleaner solid fuels (Speight, 2011a;

Brar *et al.*, 2012). In the current context, there are a large number of different feedstock types for use in a refinery-based gasifier, each with different characteristics, including size, shape, bulk density, moisture content, energy content, chemical composition, ash fusion characteristics, and homogeneity of all these properties (Speight, 2013a, 2014a, 2014b). Coal and crude oil coke are used as primary feedstocks for many large gasification plants worldwide. Additionally, a variety of biomass and waste-derived feedstocks can be gasified, with wood pellets and chips, waste wood, plastics, municipal solid waste (MSW), refuse-derived fuel (RDF), agricultural and industrial wastes, sewage sludge, switch grass, discarded seed corn, corn stover and other crop residues all being used.

The gasification of coal, biomass, crude oil or any carbonaceous residues is generally aimed to feedstock conversion to gaseous products. In fact, gasification offers one of the most versatile methods (with a reduced environmental impact with respect to combustion) to convert carbonaceous feedstocks into electricity, hydrogen, and other valuable energy products. Depending on the previously described type of gasifier (e.g., air-blown, enriched oxygen-blown) and the operating conditions, gasification can be used to produce a fuel gas that is suitable for several applications.

Gasification for electric power generation enables the use of a common technology in modern gas-fired power plants (*combined cycle*) to recover more of the energy released by burning the fuel. The use of these two types of turbines in the combined cycle system involves (i) a combustion turbine and (ii) a steam turbine. The increased efficiency of the combined cycle for electrical power generation results in a 50% v/v decrease in carbon dioxide emissions compared to conventional coal plants. Gasification units could be modified to further reduce their climate change impact because a large part of the carbon dioxide generated can be separated from the other product gas *before* combustion (for example, carbon dioxide can be separated/sequestered from gaseous byproducts by using adsorbents (e.g., MOFs) to prevent its release to the atmosphere). Gasification has also been considered for many years as an alternative to combustion of solid or liquid fuels. Gaseous mixtures are simpler to clean as compared to solid or high-viscosity liquid fuels. Cleaned gases can be used in internal combustion-based power plants that would suffer from severe fouling or corrosion if solid or low-quality liquid fuels were burned inside them.

In fact, the hot synthesis gas produced by gasification of carbonaceous feedstocks can then be processed to remove sulfur compounds, mercury, and particulate matter prior to its use as fuel in a combustion turbine generator to produce electricity. The heat in the exhaust gases from the combustion turbine is recovered to generate additional steam. This steam, along with the steam produced by the gasification process, drives a steam turbine generator to produce additional electricity. In the past decade, the primary application of gasification to power production has become more common due to the demand for high efficiency and low environmental impact.

As anticipated, the quality of the gas generated in a system is influenced by feedstock characteristics, gasifier configuration as well as the amount of air, oxygen or steam introduced into the system. The output and quality of the gas produced is determined by the equilibrium established when the heat of oxidation (combustion) balances the heat of vaporization and volatilization plus the sensible heat (temperature rise) of the exhaust gases. The quality of the outlet gas (BTU/ft.^{3}) is determined by the amount of volatile gases (such as hydrogen, carbon monoxide, water, carbon dioxide, and methane) in the gas stream. With some feedstocks, the higher the amounts of volatile produced in the early stages of the process, the higher the heat content of the product gas. In some cases, the highest gas quality may be produced at lower temperatures. However, char oxidation reaction is suppressed when the temperature is too low, and the overall heat content of the product gas is diminished.

Gasification agents are normally air, oxygen-enriched air or oxygen. Steam is sometimes added for temperature control, heating value enhancement or to allow the use of external heat (*allothermal gasification*). The major chemical reactions break and oxidize hydrocarbon derivatives to give a product gas containing carbon monoxide, carbon dioxide, hydrogen and water. Other important components include hydrogen sulfide, various compounds of sulfur and carbon, ammonia, low-boiling hydrocarbon derivatives and high-boiling hydrocarbon derivatives (including tar products).

Depending on the employed gasifier technology and operating conditions, significant quantities of water, carbon dioxide and methane can be present in the product gas, as well as a number of minor and trace components. Under reducing conditions in the gasifier, most of the feedstock sulfur converts to hydrogen sulfide (H_2S), but 3-10% converts to carbonyl sulfide (COS). Organically bound nitrogen in the coal feedstock is generally converted to gaseous nitrogen (N_2), but some ammonia (NH_3) and a small amount of hydrogen cyanide (HCN) are also formed. Any chlorine in the coal is converted to hydrogen chloride (HCl), with some chlorine present in the particulate matter (fly ash). Trace elements, such as mercury and arsenic, are released during gasification and partition among the different phases (e.g., fly ash, bottom ash, slag, and product gas).

11.2.1 Gasification of Residua

The gasification process can be used to convert heavy feedstocks such as vacuum residua and deasphalter bottoms with coal into synthesis gas which is primarily hydrogen and carbon monoxide (Wallace *et al.*, 1998). The heat generated by the gasification reaction is recovered as the product gas is cooled. For example, when the quench version of Texaco gasification is employed, the steam generated is of medium and low pressure. Note that the low-level heat used for deasphalting integration is the last stage of synthesis gas cooling. In non-integrated cases, much of this heat is uneconomical to recover and is lost to air fans and to cooling water exchangers.

In addition, integration of solvent deasphalting/gasification facility is an alternative for upgrading heavy oils economically (Wallace *et al.*, 1998). The products of deasphalting and gasification can also be beneficially integrated. The deasphalted oil (DAO) requires hydro treating and cat cracking to become diesel. Hydrogen is required for this treating, which is a primary product of gasification. The hydrogen can be generated from the asphaltene to eliminate the need for any externally supplied hydrogen. Thus, an integrated solvent deasphalting/gasification unit can increase the throughput or the crude flexibility of the refinery without creating a new, highly undesirable heavy oil stream producing such streams as: (i) deasphalted oil to a hydrotreating unit, (ii) hydrogen to hydrotreater, (iii) synthesis gas, (iv) fuel gas, (v) carbon dioxide tertiary recovery operations, and (vi) sulfur to sales.

Typically, the addition of a solvent deasphalting unit to process vacuum tower bottoms increase the production of diesel oil in the refinery. The deasphalted oil is converted to diesel using hydro treating and catalytic cracking (Chapter 9). Unfortunately, the deasphalter bottoms often need to be blended with product diesel oil to produce a viable outlet for these bottoms. A gasification process is capable of converting these deasphalter bottoms to synthesis gas which can then be converted to hydrogen for use in hydrotreating and hydrocracking processes. The synthesis gas may also be used in cogeneration facilities to provide low-cost power and steam to the refinery. If the refinery is part of a petrochemical complex, the synthesis gas can be used as chemical feedstock. The heat generated by the gasification reaction is recovered as the product gas is cooled.

11.2.2 Gasification of Residua with Coal

The gasification process can be used to convert heavy feedstocks such as vacuum residua and deasphalter bottoms with coal into synthesis gas (syngas) which is primarily hydrogen and carbon monoxide (Wallace *et al.*, 1998). The heat generated by the gasification reaction is recovered as the product gas is cooled. For example, when the quench version of Texaco gasification is employed, the steam generated is of medium and low pressure. Note that the low-level heat used for deasphalting integration is the last stage of syngas cooling. In non-integrated cases, much of this heat is uneconomical to recover and is lost to air fans and to cooling water exchangers.

In addition, integration of solvent deasphalting/gasification facility is an alternative for upgrading heavy oils economically (Wallace *et al.*, 1998). An integrated solvent deasphalting/gasification unit can increase the throughput or the crude flexibility of the refinery without creating a new, highly undesirable heavy oil stream.

11.2.3 Gasification of Residua with Biomass

Gasification is an established technology (Hotchkiss, 2003; Speight, 2013a). Comparatively, biomass gasification has been the focus of research in recent years to estimate efficiency and performance of the gasification process using various types of biomass such as sugarcane residue (Gabra *et al.*, 2001), rice hulls (Boateng *et al.*, 1992), pine sawdust (Lv *et al.*, 2004), almond shells (Rapagnà and Latif, 1997; Rapagnà *et al.*, 2000), wheat straw (Ergudenler and Ghali, 1993), food waste (Ko *et al.*, 2001), and wood biomass (Pakdel and Roy, 1991; Bhattacharaya *et al.*, 1999; Chen *et al.*, 1992; Hanaoka *et al.*, 2005). Recently, co-gasification of various biomass and coal mixtures has attracted a great deal of interest from the scientific community. Feedstock combinations including Japanese cedar wood and coal (Kamabe *et al.*, 2007), coal and saw dust (Vélez *et al.*, 2009), coal and pine chips (Pan *et al.*, 2000), coal and silver birch wood (Collot *et al.*, 1999), and coal and birch wood (Brage *et al.*, 2000) have been reported in gasification practices. Co-gasification of coal and biomass has some synergy – the process not only produces a low carbon footprint on the environment, but also improves the H_2/CO ratio in the produced gas which is required for liquid fuel synthesis (Sjöström *et al.*, 1999; Kumabe *et al.*, 2007). In addition, the inorganic matter present in biomass catalyzes the gasification of coal. However, co-gasification processes require custom fittings and optimized processes for the coal and region-specific wood residues.

While co-gasification of coal and biomass is advantageous from a chemical viewpoint, some practical problems are present on upstream, gasification, and downstream processes. On the upstream side, the particle size of the coal and biomass is required to be uniform for optimum gasification. In addition, moisture content and pretreatment (torrefaction) are very important during upstream processing.

While upstream processing is influential from a material handling point of view, the choice of gasifier operation parameters (temperature, gasifying agent, and catalysts) dictate the product gas composition and quality. Biomass decomposition occurs at a lower temperature than coal and therefore different reactors compatible to the feedstock mixture are required (Brar *et al.*, 2012). Furthermore, feedstock and gasifier type along with operating parameters not only decide product gas composition but also dictate the amount of impurities to be handled downstream. Downstream processes need to be modified if coal is co-gasified with biomass. Heavy metal and impurities such as sulfur and mercury present in coal can make synthesis gas difficult to use and unhealthy for the environment. Alkali metals present in biomass can also cause corrosion problems high temperatures in downstream pipes. An alternative option to downstream gas cleaning would be to process coal to remove mercury and sulfur prior to feeding into the gasifier.

However, first and foremost, coal and biomass require drying and size reduction before they can be fed into a gasifier. Size reduction is needed to obtain appropriate particle sizes; however, drying is required to achieve moisture content suitable for gasification operations. In addition, biomass densification may be conducted to prepare pellets and improve density and material flow in the feeder areas. It is recommended that biomass moisture content should be less than 15% w/w prior to gasification. High moisture content reduces the temperature achieved in the gasification zone, thus resulting in incomplete gasification. Forest residues or wood has a fiber saturation point at 30 to 31% moisture content (dry basis) (Brar *et al.*, 2012). Compressive and shear strength of the wood increases with decreased moisture content below the fiber saturation point. In such a situation, water is removed from the cell wall leading to shrinkage. The long-chain molecules constituents of the cell wall move closer to each other and bind more tightly. A high level of moisture, usually injected in form of steam in the gasification zone, favors formation of a water-gas shift reaction that increases hydrogen concentration in the resulting gas.

The torrefaction process is a thermal treatment of biomass in the absence of oxygen, usually at 250 to 300°C to drive off moisture, decompose hemicellulose completely, and partially decompose cellulose (Speight, 2011a). Torrefied biomass has reactive and unstable cellulose molecules with broken hydrogen bonds and not only retains 79 to 95% of feedstock energy but also produces a more reactive feedstock with lower atomic hydrogen-carbon and oxygen-carbon ratios to those of the original biomass. Torrefaction results in higher yields of hydrogen and carbon monoxide in the gasification process.

Finally, the presence of mineral matter in the coal-biomass feedstock is not appropriate for fluidized-bed gasification. Low melting point of ash present in woody biomass leads to agglomeration which causes defluidization of the ash and sintering, deposition as well as corrosion of the gasifier construction metal bed (Vélez *et al.*, 2009). Biomass containing alkali oxides and salts are likely to produce clinkering/slagging problems from ash formation (McKendry, 2002). Thus, it is imperative to be aware of the melting of biomass ash, its chemistry within the gasification bed (no bed, silica/sand, or calcium bed), and the fate of alkali metals when using fluidized-bed gasifiers. Most small to medium-sized biomass/waste gasifiers are air blown, operate at atmospheric pressure and at temperatures in the range 800 to 100°C (1470 to 2190°F). They face very different challenges to large gasification plants – the use of small-scale air separation plant should oxygen gasification be preferred. Pressurized operation, which eases gas cleaning, may not be practical.

Biomass fuel producers, coal producers and, to a lesser extent, waste companies are enthusiastic in relation to supplying co-gasification power plants and realize the benefits of co-gasification with alternate fuels (Lee, 2007; Lee *et al.*, 2007; Speight, 2008, 2011a; Lee and Shah, 2013; Speight, 2013a). The benefits of a co-gasification technology involving coal and biomass include the use of a reliable coal supply with gate-fee waste and biomass which allows the economies of scale from a larger plant to be supplied just with waste and biomass. In addition, the technology offers a future option of hydrogen production and fuel development in refineries. In fact, oil refineries and petrochemical plants are opportunities for gasifiers when the hydrogen is particularly valuable (Speight, 2011b, 2014a).

11.2.4 Gasification of Residua with Waste

Waste may be municipal solid waste (MSW) which had minimal presorting, or refuse-derived fuel (RDF) with significant pretreatment, usually mechanical screening and shredding. Other more specific waste sources (excluding hazardous waste) and possibly including crude oil coke, may provide niche opportunities for co-utilization.

The traditional waste-to-energy plant, based on mass-burn combustion on an inclined grate, has a low public acceptability despite the very low emissions achieved over the last decade with modern flue gas clean-up equipment. This has led to difficulty in obtaining planning permissions to construct needed new waste-to-energy plants. After much debate, various governments have allowed options for advanced waste conversion technologies (gasification, pyrolysis and anaerobic digestion), but will only give credit to the proportion of electricity generated from non-fossil waste.

Co-utilization of waste and biomass with coal may provide economies of scale that help achieve the above-identified policy objectives at an affordable cost. In some countries, governments propose co-gasification processes as being *well suited for community-sized developments* suggesting that waste should be dealt with in smaller plants serving towns and cities, rather than moved to large, central plants (satisfying the so-called *proximity principle*).

In fact, neither biomass nor wastes are currently produced, or naturally gathered at sites in sufficient quantities to fuel a modern large and efficient power plant. Disruption, transport issues, fuel use, and public opinion all act against gathering hundreds of megawatts (MWe) at a single location. Biomass or waste-fired power plants are therefore inherently limited in size and hence in efficiency (labor costs per unit electricity produced) and in other economies of scale. The production rates of municipal refuse follow reasonably predictable patterns over time periods of a few years. Recent experience with the very limited current *biomass for energy* harvesting has shown unpredictable variations in harvesting capability with long periods of zero production over large areas during wet weather.

The situation is very different for coal. This is generally mined or imported and thus large quantities are available from a single source or a number of closely located sources, and supply has been reliable and predictable. However, the economics of new coal-fired power plants of any technology or size have not encouraged any new coal-fired power plant in the gas generation market.

The potential unreliability of biomass, longer-term changes in refuse and the size limitation of a power plant using only waste and/or biomass can be overcome combining biomass, refuse and coal. It also allows benefit from a premium electricity price for electricity from biomass and the gate fee associated with waste. If the power plant is gasification-based, rather than direct combustion, further benefits may be available. These include a premium price for the electricity from waste, the range of technologies available for the gas to electricity part of the process, gas cleaning prior to the main combustion stage instead of after combustion and public image, which is currently generally better for gasification as compared to combustion. These considerations lead to current studies of co-gasification of wastes/biomass with coal (Speight, 2008).

For large-scale power generation (>50 MWe), the gasification field is dominated by plant based on the pressurized, oxygen-blown, entrained-flow or fixed-bed gasification of fossil fuels. Entrained gasifier operational experience to date has largely been with well-controlled fuel feedstocks with short-term trial work at low co-gasification ratios and with easily handled fuels.

Use of waste materials as co-gasification feedstocks may attract significant disposal credits. Cleaner biomass materials are renewable fuels and may attract premium prices for the electricity generated. Availability of sufficient fuel locally for an economic plant size is often a major issue, as is the reliability of the fuel supply. Use of more-predictably available coal alongside these fuels overcomes some of these difficulties and risks. Coal could be regarded as the "flywheel" which keeps the plant running when the fuels producing the better revenue streams are not available in sufficient quantities.

Coal characteristics are very different from younger hydrocarbon fuels such as biomass and waste. Hydrogen-to-carbon ratios are higher for younger fuels, as is the oxygen content. This means that reactivity is very different under gasification conditions. Gas cleaning issues can also be very different, sulfur being a major concern for coal gasification and chlorine compounds and tars more important for waste and biomass gasification. There are no current proposals for adjacent gasifiers and gas cleaning systems, one handling biomass or waste and one coal, alongside each other and feeding the same power production equipment. However, there are some advantages to such a design as compared with mixing fuels in the same gasifier and gas cleaning systems.

Electricity production or combined electricity and heat production remain the most likely area for the application of gasification or co-gasification. The lowest investment cost per unit of electricity generated is the use of the gas in an existing large power station. This has been done in several large utility boilers, often with the gas fired alongside the main fuel. This option allows a comparatively small thermal output of gas to be used with the same efficiency as the main fuel in the boiler as a large, efficient steam turbine can be used. It is anticipated that addition of gas from a biomass or wood gasifier into the natural gas feed to a gas turbine to be technically possible but there will be concerns as to the balance of commercial risks to a large power plant and the benefits of using the gas from the gasifier.

The use of fuel cells with gasifiers is frequently discussed but the current cost of fuel cells is such that their use for mainstream electricity generation is uneconomic. Furthermore, the disposal of municipal and industrial waste has become an important problem because the traditional means of disposal, landfill, are much less environmentally acceptable than previously. Much stricter regulation of these disposal methods will make the economics of waste processing for resource recovery much more favorable.

One method of processing waste streams is to convert the energy value of the combustible waste into a fuel. One type of fuel attainable from waste is a low heating value gas, usually 100-150 Btu/scf, which can be used to generate process steam or to generate electricity. Co-processing such waste with coal is also an option (Speight, 2008).

In summary, coal may be co-gasified with waste or biomass for environmental, technical or commercial reasons. It allows larger, more efficient plants than those sized for grown biomass or arising waste within a reasonable transport distance; specific operating costs are likely to be lower and fuel supply security is assured.

Co-gasification technology varies, being usually site specific and high feedstock dependent. At the largest scale, the plant may include the well-proven fixed-bed and entrained-flow gasification processes. At smaller scales, emphasis is placed on technologies which appear closest to commercial operation. Pyrolysis and other advanced thermal conversion processes are included where power generation is practical using the on-site feedstock produced. However, the needs to be addressed are (i) core fuel handling and gasification/pyrolysis technologies, (ii) fuel gas clean-up, and (iii) conversion of fuel gas to electric power (Ricketts *et al.*, 2002).

11.3 Synthetic Fuel Production

The varying prices of crude oil, the politics of crude oil and other variable economic factors has led to a strong interest in the production of liquid fuels from coal, natural gas, and biomass (Speight, 2008, 2011a, 2011c; Hu *et al.*, 2012). The technology to produce fuels from such sources is varied but a tried-and-true technology involves the so-called *indirect* process in which the feedstock is first converted to gases (particularly synthesis gas) from which liquid products are generated by the Fischer-Tropsch process (Kreutz *et al.*, 2008).

Current conditions almost reprise the era of the 1970s when energy security concerns generated by oil embargoes stimulated federal spending in synthetic fuels. Despite considerable investment, federal support in many countries was withdrawn after supply concerns eased in the 1980s. The currently favored approach to producing synthetic fuels – the Fischer-Tropsch process – uses synthesis gas (mixtures of carbon monoxide and hydrogen from the gasification of carbonaceous materials – fossil fuels or organically derived feedstocks) number (Gary *et al.* 2007; Speight, 2008, 2013a, 2013b, 2014a; Hsu and Robinson, 2017; Speight, 2017).

Many countries have attempted to capitalize on the gasification-with-Fischer-Tropsch method but the up-and-down prices of crude oil – especially when crude oil prices are lower – tend to discourage such efforts on the basis of poor economic return. Nevertheless, several private ventures in the United States and throughout the world are now studying the feasibility of constructing Fischer-Tropsch synthetic fuel plants based on crude oil residua, coal, natural gas, and biomass. It is required that governments make the decision to support such efforts rather than react at a time when the occasion has passed and fuel shortages are endemic. Perhaps this is too much to ask – that a government will use foresight instead of hindsight, which is always 20-20.

The Fischer-Tropsch process is well suited to producing naphtha – the precursor to gasoline) as well as middle-distillate range fuels like diesel fuel and jet fuel. The diesel produced is superior to conventionally refined diesel in terms of higher cetane-number and low sulfur content. Overall, middle distillate fuels represent roughly a quarter of many refinery operations, which are typically driven by the demand for gasoline. In order for a synthetic fuels industry (whether coal, natural gas, or biomass based) to begin rivaling or even supplanting conventional crude oil refining, a major shift in political outlook would have to occur.

In addition, recent energy legislation promotes research on capturing and storing greenhouse gas emissions and improving vehicle fuel efficiency, among other goals. Fisher-Tropsch fuels present the paradox of high carbon emissions associated with production versus lower carbon emissions associated with their use. Hence, as crude oil production decreases and its price increases, the Fischer-Tropsch technology which enables the production of synthetic hydrocarbon derivatives from carbonaceous feedstocks, is becoming an increasingly attractive technology in the energy mix. In fact, coupled with this is the fact that Fischer-Tropsch products are ultraclean fuels in that they contain no aromatics, no sulfur compounds, and no nitrogen compounds.

In essence, compared to crude oil-derived gasoline and diesel fuel, the analogous product produced by the Fischer-Tropsch process will burn to produce considerably less polynuclear aromatic hydrocarbon derivatives (PNAs), and no sulfur oxides (SOx) and no nitrogen oxides (NOx). With global pressures to reduce greenhouse gas emissions intensifying, legislative frameworks in Europe and the United States have already been put in place to force producers of liquid transportation fuels to comply with stricter emission standards. The impact of such legislation is that dilution of crude oil derived fuels with the cleaner Fischer-Tropsch derived fuels is becoming an increasingly important way to achieve environmental compliance. It is thus not surprising that Fischer-Tropsch technology now occupies a visible place in the energy mix required for sustainable global development.

The intent of this section is to provide the reader with a broad perspective of thermal decompositon pyrolysis technology as it relates to converting a variety of feedstocks to distillate products for the reader to compare with gasification technologies and Fischer-Tropsch technologies. Thus:

> Feedstock preparation → Gasification
> Gasification → Gas cleaning
> Gas cleaning → Fischer-Tropsch process
> Fischer-Tropsch process → Synthetic crude oil
> Synthetic crude oil → Hydrocarbon fuels

In addition, recent developments in thermal technology for synthetic fuel production are also presented and, for comparison, the chapter also provides presentation of the means by which non-Fischer-Tropsch synthetic crude oil is converted to specification-grade fuels.

11.3.1 Fischer-Tropsch Synthesis

The Fischer-Tropsch process (Chapter 12) is well known and has been commercially demonstrated in many countries (Chadeesingh, 2011). The process uses synthesis gas which, in addition to carbon monoxide and hydrogen may also contain water, carbon dioxide, nitrogen (when air is used as the gasification oxidant), and methane, and has been used in the Sasol process for the commercial production of liquid fuels.

As an abundant resource in many non-oil-producing countries, coal has long been exploited as a solid fossil fuel. As oil and natural gas supplanted coal throughout the last two centuries, technologies developed to convert coal into other fuels. Proponents of expanding the use of the Fischer-Tropsch process argue that the United States and many other countries could alleviate their dependence on imported crude oil and strained refinery capacity by converting non-crude oil feedstocks to transportation fuels.

Fischer-Tropsch synthesis, particularly the coal-based process, poses several challenges: (i) the process is criticized as inefficient and costly, (ii) carbon dioxide – a greenhouse gas associated with global climate change – is a byproduct of the process, (iii) the use of coal and natural gas as Fischer-Tropsch feedstocks would compete with electric power generation, and (iv) the fuels produced, primarily diesel fuel and jet fuel, would not substitute widely for the preferred transportation fuel – gasoline, and (v) the use of biomass as feedstock would compete with cellulosic ethanol production, as it is now envisioned. Each of these items is part reality and part mythology that can be overcome by judicious planning to silence (if that is possible) the nay-sayers.

11.3.2 Fischer-Tropsch Liquids

In principle, synthesis gas (primarily consisting of carbon monoxide and hydrogen) can be produced from any carbonaceous feedstock, including natural gas, naphtha, residual oil, crude oil coke, coal, and biomass leading to a host of reactions and products (Wender, 1996). Current economic considerations dictate that the current production of liquid fuels from synthesis gas translates into the use of coal or natural gas as the hydrocarbon source with the economics of the use of other feedstocks continuing to improve. Nevertheless, the synthesis gas production operation in a gas-to-liquids plant amounts to greater than half of the capital cost of the plant (Spath and Dayton, 2003). The choice of technology for synthesis gas production also depends on the scale of the synthesis operation. Improving the economics of a feedstock-to-liquids plant is through: (i) decreasing capital costs associated with synthesis gas production and (ii) improving the thermal efficiency with better heat integration and utilization. Improved thermal efficiency can be obtained by combining the gas-to-liquids plant with a power generation plant to take advantage of the availability of low-pressure steam.

Two main characteristics of Fischer-Tropsch synthesis are the unavoidable production of a wide range of hydrocarbon products (olefins, paraffins, and oxygenated products) and the liberation of a large amount of heat from the highly exothermic synthesis reactions. Product distributions are influenced by (i) temperature, (ii) feed gas composition (H_2/CO), (iii) pressure, (iv) catalyst type, and (v) catalyst composition. Fischer-Tropsch products are produced in four main steps: synthesis gas generation, gas purification, Fischer-Tropsch synthesis, and product upgrading. Depending on the types and quantities of Fischer-Tropsch products desired, either low (200 to 240°C, 390 to 465°F) or high temperature (300–350°C, 570 to 660°F) synthesis is used with either an iron (Fe) or cobalt catalyst (Co).

The process for producing synthesis gas can be described as comprising three components: (i) synthesis gas generation, (ii) waste heat recovery, and (iii) gas processing. Within each of the above three listed systems are several options. For example, synthesis gas can be generated to yield a range of compositions ranging from high-purity hydrogen to high-purity carbon monoxide. Two major routes can be utilized for high-purity gas production: (i) pressure-swing adsorption and (ii) utilization of a cold box, where separation is achieved by distillation at low temperatures. In fact, both processes can also be used in combination as well. Unfortunately, both processes require high capital expenditure.

However, to address these concerns, research and development is ongoing and successes can be measured by the demonstration and commercialization of technologies such as permeable membrane for the generation of high-purity hydrogen, which in itself can be used to adjust the H_2/CO ratio of the synthesis gas produced.

Essentially, the Fischer-Tropsch synthesis is the formation of straight-chain hydrocarbon derivatives and relies on the potential for carbon monoxide to exchange oxygen with hydrogen in the presence of a catalyst (Chadeesingh, 2011). The carbonaceous feedstock is gasified in the presence of a calculated amount of oxygen (or air) to produce carbon monoxide and hydrogen, while at the same time steam reacts with the carbonaceous feedstock to produce water gas, coal is burned to produce the carbon monoxide and steam reacting with hot coal disassociates to produce hydrogen – other gases are also produced as byproducts:

$$C_{coal} + O2 \rightarrow CO + H_2 + H_2O + CO_2 + CH_4$$

$$C_{coal} + H_2O \rightarrow CO + H_2$$

$$CO + H_2O \rightarrow CO_2 + H_2$$

Then:

$$2H_2 + CO \rightarrow H(CH_2)_nH + H_2O$$

$$CO + H_2O \rightarrow CO_2 + H_2$$

$$2CO + H_2 \rightarrow H(CH_2)_nH + CO_2$$

Unfortunately these simple equations must suffice but they are not a true representation of the complexity of the gasification process followed by the Fischer-Tropsch synthesis. A key issue is the composition of the hydrocarbon product $[H(CH_2)_nH]$ which will vary depending upon: (i) the configuration of the reactor, (ii) the process parameter, and (iii) the catalyst.

Catalysts used for the Fischer-Tropsch reaction are generally based on iron and cobalt (Khodakov *et al.*, 2007). Ruthenium is an active catalyst for Fischer-Tropsch but is not economically feasible due to its high cost and insufficient reserves worldwide. Iron has been the traditional catalyst of choice for Fischer-Tropsch reaction. It is reactive and the most economical catalyst for synthesis of clean fuel from the synthesis gas mixture. Compared to cobalt, iron tends to produce more olefins and also catalyzes the water-gas shift reaction. An iron-based catalyst is usually employed in high-temperature operations (300 to 350°C, 570 to 660°F) (Steynberg *et al.*, 1999).

Cobalt has higher activity for Fischer-Tropsch reaction but is more expensive compared to iron. The low-temperature (200 to 240°C; 390 to 465°F) Fischer-Tropsch process usually employs cobalt-based catalysts due to their stability and high hydrocarbon productivity. Catalyst supports that have been utilized include silica (SiO_2), alumina (Al_2O_3), titania (TiO_2), zirconia (ZrO_2), magnesia (MgO), carbon, and molecular sieves. The cost of catalyst support, metal and catalyst preparation contributes to the cost of Fischer-Tropsch catalyst, which represents a significant part of the cost for the Fischer-Tropsch technology. Various types of reactors have been installed in the FT industry such as fixed-bed reactor, multi-tubular reactor, adiabatic fixed-bed reactor, slurry reactor, fluidized-bed reactor and circulating fluid-bed reactor systems (Bao *et al.*, 2004; Steynberg *et al.*, 1999; Chadeesingh, 2011). Since the Fischer-Tropsch reaction is highly exothermic, temperature control and heat removal constitute the two most important design factors for the Fischer-Tropsch reactors (Hu *et al.*, 2012).

11.3.3 Upgrading Fischer-Tropsch Liquids

Typically, the Fischer-Tropsch process produces four streams (i) low molecular weight gases in the raw Fischer-Tropsch product (unconverted synthesis gas and C1-C4 gases), which are separated from the liquid fraction in a hydrocarbon recovery step, (ii) naphtha – light and heavy, (iii) middle distillate, and (iv) wax – soft and hard, which are co-produced as synthetic crude oil (Table 11.2) (Chadeesingh, 2011). Fractions 2-4 form the basis of the synthetic crude oil which is distilled to produce separate streams and each fraction is then processed through a series of refining steps suitable to the boiling range of the fraction (Speight, 2014a).

Product upgrading processes for the synthetic fuel directly originate from the refining industry and are highly optimized using appropriate catalysts (De Klerk and Furimsky, 2011; De Klerk, 2011). Thus, the naphtha stream is first hydrotreated, resulting in the production of hydrogen-saturated liquids (primarily paraffins), a portion of which are converted by isomerization from normal paraffins to iso-paraffins to boost the octane value. Another fraction of the hydrotreated naphtha is catalytically reformed to provide some aromatic content to (and further boost the octane value of) the final gasoline blending stock. The middle distillate stream is also hydrotreated, resulting directly in a finished diesel blending stock. The wax fraction is hydrocracked into a finished distillate stream and naphtha streams that augment the hydrotreated naphtha streams sent for isomerization and for catalytic cracking. In some scenarios, any unconverted wax is recycled to extinction within the hydroprocessing section (Collins *et al.*, 2006).

Table 11.2 Range of products from the Fischer-Tropsch process.

Product	Carbon number
SNG (Synthetic Natural Gas)	C1 – C2
LPG (liquefied petroleum gas)	C3 – C4
Light naphtha	C5 – C7
Heavy naphtha	C8 – C10
Middle distillate	C11 – C20
Soft Wax	C21 – C30
Hard Wax	C31 – C60

Generally, the Fischer-Tropsch synthesis is well suited to produce synthetic naphtha and diesel fuel since Fischer-Tropsch products are free from sulfur, nitrogen and metals (such as nickel and vanadium) and the levels of naphthenes and aromatics are very low. In fact, the Fischer-Tropsch liquids (*synthetic crude oil*) can be refined into end products in current refineries or integrated refining units. The synthetic crude oil is sulfur-free, nitrogen-free, and contains little or no aromatic constituents. Possible products from the synthetic crude oil include liquefied petroleum gas (LPG), gasoline (from the naphtha product), diesel fuel (from the kerosene product), and jet fuel (from the naphtha and kerosene product). These products are fully compatible with the comparable crude oil-based products and fit into the current distribution network. Furthermore, Fischer-Tropsch products are very well suited for use as vehicle fuels from an environmental point of view and future market demands will determine the product emphasis.

Finally, the product distribution of hydrocarbon derivatives formed during the Fischer–Tropsch process follows an Anderson–Schulz–Flory distribution, which can be expressed as:

$$W_n/n = (1 - \alpha)^2 \alpha^{n-1}$$

W_n is the weight fraction of hydrocarbon derivatives containing *n* carbon atoms, α is the chain growth probability or the probability that a molecule will continue reacting to form a longer chain, which is dependent on the catalyst type and the process parameters.

In addition, a value of α close to unity increases production of long-chain hydrocarbon derivatives – typically waxes, which are solid at room temperature. Therefore, for production of liquid transportation fuels it will be necessary to thermally decompose (crack) these waxes. There are suggestions that the use of zeolite catalysts (or other catalysts with fixed sized pores) that can restrict the formation of hydrocarbon derivatives longer than some characteristic size (usually n<10).

11.3.3.1 Gasoline Production

From synthesis gas, the Fischer-Tropsch process produces a wide range of hydrocarbon products:

$$(2n+1)H_2 + nCO \rightarrow C_nH_{(2n+2)} + nH_2O$$

The alkanes, or saturated hydrocarbon derivatives ($C_nH_{(2n+2)}$), tend to be *normal*, or straight-chain isomers. The mean value of n is determined by catalyst, process conditions, and residence time, which are usually selected to maximize formation of alkanes in the range C_5-C_{21}.- the lower-boiling fraction (C_5-C_{12}), is separated as naphtha, which may be further refined into gasoline (which typically contains aromatic and branched hydrocarbon fractions).

High-temperature circulating fluidized-bed reactors (Synthol reactors) have been developed for gasoline and light olefin production and these reactors operate at 350°C (660°F) and up to 400 psi. The combined gas feed (fresh and recycled) enters at the bottom of the reactor and entrains catalyst that is flowing down the standpipe and through the slide valve. The high gas velocity carries the entrained catalyst into the reaction zone where heat is removed through heat exchangers. Product gases and catalyst are then transported into a large diameter catalyst hopper where the catalyst settles out and the product gases exit through a cyclone. These Synthol reactors have been successfully used for many years; however, they have a number of limitations – they are physically very complex reactors that involve circulation of large amounts of catalyst that leads to considerable erosion in particular regions of the reactor.

The higher-boiling fraction (C_8-C_{21}), being straight-chain hydrocarbon derivatives, is suitable for direct blending into the diesel fuel pool. Higher molecular weight alkanes (waxes) may also be formed but are usually undesirable. One inescapable aspect of Fischer-Tropsch chemistry is that more water will be produced than hydrocarbon derivatives (by mass). This produced water must be considered an undesirable sink for expensive and valuable hydrogen and also an unwanted waste stream.

The naphtha fraction is not generally marketable and must be shipped to a refinery for further processing into gasoline blending stock – it is advisable to have commercial Fischer-Tropsch plant associated with a refinery complexes. In fact, for instance, the composition of the –naphtha-gasoline fraction [$H(CH_2)_nH$, where n = approximately 5 to 12] is an issue – the Fischer-Tropsch synthesis produces primarily straight-chain paraffins, thus any gasoline produced has a relatively low octane rating (<85).

The naphtha fraction contains components that are equivalent to the crude oil counterparts produced in a typical refinery. Alkylate, produced from reacting C_3, C_4, and C_5 olefins with isobutane, is the highest octane component in the gasoline. Isomerate is produced from isomerizing normal pentane and hexane – it has a moderate octane rating but is relatively volatile. The reformate, on the other hand, has a high octane rating but contains undesirable aromatic components. All of the gasoline blending components have zero sulfur and olefins, which is of considerable benefit when manufacturing specification-grade and environmentally mandated fuels.

Modern automobile gasoline as sold to the consumer ranges in octane from 87 to 93 which is achieved by blending various crude oil streams distillates, reforming gasoline-range hydrocarbon derivatives, and ethanol or other additives increase the octane-number (Gary *et al.*, 2007; Speight, 2007, 2014a; Hsu and Robinson, 2017; Speight, 2017). Branched paraffin series like iso-octane cannot be directly produced in Fischer-Tropsch synthesis. Consequently, when Fischer-Tropsch synthesis has been used to produce gasoline, it has been blended with conventionally refined crude oil to achieve the desired octane-number. On the other hand, the methanol-to-gasoline (MTG) process developed by Mobil Oil Corporation involves the conversion of methanol to hydrocarbon derivatives over zeolite catalysts offers a better quality naphtha-gasoline (Hindman, 2013).

Methanol synthesis also enjoys a long history, actually preceding the Fischer-Tropsch process. In 1923, BASF first synthesized methanol on an industrial scale, also from coal-produced synthesis gas:

$$H_2 + CO - CH_3OH$$

The methanol-to-gasoline process occurs in two steps. In the first step, crude methanol (containing 17% v/v water) is super-heated to 300°C (570°F) and partially dehydrated over an alumina catalyst at 400 psi to yield an equilibrium mixture of methanol, dimethyl ether, and water (75% of the methanol is converted). In the second step, this effluent is then mixed with heated recycled synthesis gas and introduced into a reactor containing ZSM-5 zeolite catalyst at 350 to 365°C (660 to 690°F) and 280 to 340 psi to produce hydrocarbon derivatives (44%) and water (56%) (Spath and Dayton, 2003). The overall process usually contains multiple gasoline conversion reactors in parallel because the zeolites have to be regenerated frequently to burn off the coke formed during the reaction. The reactors are then cycled so that individual reactors can be regenerated without stopping the process (Kam *et al.*, 1984). The process reactions may be summarized simply as:

$$2CH_3OH \rightarrow CH_3OCH_3 + H_2O$$

$$CH_3OCH_3 \rightarrow C_2\text{-}C_5 \text{ olefins}$$

$$C_2\text{-}C_5 \text{ olefins} \rightarrow \text{paraffins, cycloparaffins, aromatics}$$

The selectivity of the process for gasoline range hydrocarbon derivatives is greater than 85% with the remainder of the product being primarily low-boiling hydrocarbon derivatives (such as LPG constituents) (Wender, 1996). Approximately 40% of the gasoline produced from the process is aromatic hydrocarbon derivatives with the following distribution: 4% v/v benzene, 26% v/v toluene, 2% v/v ethylbenzene, 43% v/v mixed xylenes, 14% v/v trimethyl-substituted benzenes, plus 12% v/v other aromatics (Wender, 1996). The shape selectivity of the zeolite catalyst results in a relatively high durene (1,2,4,5-tetramethyl benzene) concentration, 3 to 5% v/v of the gasoline produced (MacDougall, 1991).

11.3.3.2 Diesel Production

The Fischer-Tropsch synthesis is well suited to producing middle-distillate range fuels such as diesel fuel and jet fuel. The diesel produced is superior to conventionally refined diesel in terms of higher cetane-number and low sulfur content. Thus, the Fischer-Tropsch process is more amenable to the production of diesel fuel [$H(CH_2)_nH$, where n = approximately 7 to 24] and the various types of jet fuel [$H(CH_2)_nH$, where n = approximately 5 to 18]. Diesel produced from conventional upgrading of Fischer-Tropsch synthetic fuel consists of hydrotreated straight-run distillate blended with distillate from wax hydrocracking. Like the naphtha/gasoline, Fischer-Tropsch diesel has rather unique properties relative to crude oil-derived diesels – it is sulfur free, almost completely paraffinic, and typically has an acceptable-to-high cetane rating.

The standard for diesel fuel rates the ease of which auto-ignition occurs during compression in the engine cylinder, thus eliminating the need for a spark plug. The number 100 was assigned to cetane (n-hexadecane, $C_{16}H_{34}$) to represent a straight-chain hydrocarbon in the paraffin series – the hydrocarbon type and molecular weight that the Fischer-Tropsch synthesis is best suited to produce. Diesel fuel cetane-numbers range from 40 to 45, and as high as 55 in Europe, where high-speed diesel engines are prevalent in light-duty passenger vehicles.

Recent efforts to improve the Fischer-Tropsch process tend to focus on increasing selectivity for the diesel fraction and minimizing the naphtha fraction. With certain modifications and modest post-processing, currently the Fischer-Tropsch process can typically claim selectivity for the diesel fraction with the distribution of the hydrocarbon fraction as: diesel (kerosene) 75% v/v, naphtha (gasoline) 20% v/v, and LPG 5% v/v (Lewis, 2013).

11.4 Sabatier-Senderens Process

The synthesis of hydrocarbon derivatives from hydrogenation of carbon monoxide was discovered in 1902 by Sabatier and Senderens (sometimes written as Sabatier and Sanderens) who produced methane by passing carbon monoxide and hydrogen over nickel-, iron-, and cobalt-containing catalysts. At about the same time, the first commercial hydrogen from synthesis gas produced from steam methane reforming was commercialized. The production of liquid hydrocarbon derivatives and oxygenates from synthesis gas conversion over iron catalysts was discovered in 1923 by Fischer and Tropsch. Variations on this synthesis pathway were soon to follow for the selective production of methanol, and mixed alcohols. Another outgrowth of Fischer-Tropsch Synthesis (FTS) was the hydroformylation of olefins discovered in 1938.

The Sabatier reaction (Sabatier process) involves the reaction of hydrogen with carbon dioxide at elevated temperatures (optimal 300 to 400°C, 570 to 750°F) and pressures in the presence of a nickel-based catalysts to produce methane and water:

$$CO_2 + 4H_2 \rightarrow CH_4 + 2H_2O$$

Ruthenium on alumina (Al_2O_3) has been shown to be a more (aluminum oxide) makes a more efficient catalyst. The reaction is exothermic and some initial energy/heat has to be added to start the reaction.

Interest in the Sabatier reaction has increased recently because of growing concerns about global climate change and the reaction/process represents a means to reduce emissions of carbon dioxide. Considerable efforts are currently underway to develop practical and affordable ways to capture carbon dioxide from major point sources, such as gasification plants and dispose of this carbon dioxide by means of geologic sequestration.

Common applications of the Sabatier reaction include scrubbing traces of carbon dioxide from hydrogen-containing gas. Thus there is the potential of using the process to reduce the emissions of carbon dioxide from source such as power plants and gasification plants. The increased urgency in addressing greenhouse gas emissions warrant further investigation of the application of carbon dioxide recycling from power plant emissions and gasification plant emissions by the Sabatier reaction.

When used in the gasification industry, this reaction will take place in a specifically designed reactor in the presence of an efficient catalyst. The flue gas containing the carbon dioxide will have to be cooled by a heat exchanger to reach the optimum reaction temperature. The water formed during the combustion of methane and the Sabatier reaction will be removed from the stream coming from the methanation reactor. This water will be used to cool the flue gas and the methanation reactor. After recovering this heat the water will be sent to the water splitter where the generated hydrogen will be mixed with the flue gas from the reactor before it enters the methanation reactor. The methane generated will be mixed with any required make-up natural gas needed to operate the process at the desired capacity.

It is not necessary to separate isolate and compress the carbon dioxide – the reaction between the carbon dioxide and hydrogen will take place in the gaseous phase and the amount of methane produced will depend on the amount of hydrogen produced by the splitting of water (Fujita *et al.*, 1993; Zhilyaeva *et al.*, 2002; Takenaka *et al.*, 2004; Görke *et al.*, 2005; Brooks *et al.*, 2007; Du *et al.*, 2007. A high conversion (98% v/v) of carbon dioxide to methane has been achieved at a space velocity of more than 15,000 h^{-1} and at a temperature of 350°C (660°F).

11.4.1 Methanol Production

The first idea of using synthesis gas for producing methanol was found by Paul Sabatier in 1905. Eight years later, the first synthesis patent was given to the Badische Anilin und Soda Fabrik (BASF) (Cheng and Kung, 1994). The synthesis process developed by BASF operates at a temperature between 300°C and 400°C and a pressure between 100 and 250 bar over sulfur-resistant zinc oxide-chromia (ZnO-Cr2O$_3$) catalyst. Ten years later, the first commercial methanol synthesis plant was built. For many years this was the only, but not energy-efficient technique, to produce methanol. In this exothermic process, synthesis gas is converted into methanol:

$$CO + 2H_2 \rightarrow CH_3OH$$

In 1927, methanol was produced for the first time by using carbon dioxide instead of carbon monoxide, and hydrogen, both obtained as fermentation gases:

$$CO_2 + 3H_2 \rightarrow CH_3OH + H_2O$$

Also in 1927, DuPont improved the BASF process with a more efficient zinc/copper catalyst. Both processes, with coal as a feedstock, continued to produce methanol up to 1940, when natural gas became abundant From this time on, only the reforming of natural gas was used to produce methanol because natural gas as a feedstock was economically more beneficial (Lee, 1990). The first real breakthrough for energy-efficient production of methanol was in 1966 by Imperial Chemical Industries (ICI, now Synetix), which developed a CU/ZnO/Al$_2$O$_3$ catalyst (Weissermel, 2003). This process operates at relatively low pressures (700 to 1500 psi) and lower temperatures (250 to 300°C, 480 to 570°F).

In this process, the first time only 10 to 15% v/v of the new inlet gases will be converted into methanol and water, the rest remains unreacted. To achieve high conversion rates, and therefore a higher energy efficiency, ICI developed a process in which the unreacted gases were recycled and put back into the catalyst of the reactor. Another improvement was that the inlet gases and the recycled gases were preheated by a heat exchanger before they were inserted into the reactor vessel. The exothermic heat that was generated by the conversion process was recovered in the reactor vessel and used to preheat the feed water from the boiler. This new process of methanol synthesis was the end of the inefficient methanol production techniques developed by BASF and DuPont (Lee, 1990). A few years later, the Lurgi low-pressure process was developed which uses overall the same type of catalyst. The difference with the ICI process is that the temperature of the inlet gasses are regulated by boiling water in the reactor instead of preheating the synthesis gas outside the reactor vessel.

In 2006, 60% of the commercial methanol was produced by the process of ICI and 27% by the Lurgi process (Ohla et al., 2006). The rest was generally produced by the Kellogg process or in laboratories. According to the patents of the Icelandic company CRI, they are using the Lurgi methanol processes with hydrogen and carbon dioxide as feedstock. Hydrogen is produced by the electrolysis of water and carbon dioxide is recovered from a geothermal power plant in the Svartsengi power station, which is located in the Svartsengi geothermal field, about four km north of Grindavik, approximately 12 miles to the southeast of Keflavik International Airport and 26 miles from Reykjavík (Iceland). These two streams are compressed to approximately 750 psi and a temperature on the order of 225°C (435°F). After the reactor vessel, a mixture of unreacted hydrogen, carbon dioxide, methanol and water (byproduct), flows through a heat exchanger to preheat the inlet gasses. After that, this mixture flows to a preheater for the distillation system and then methanol is condensed in a condenser.

11.4.2 Dimethyl Ether Production

The synthesis of dimethyl ether (DME) from synthesis gas process can be carried out in the liquid phase at moderate temperature and pressure, 250°C (480°F) and 1000 psi. This single-stage process involves dual catalysts slurried in a liquid oil medium and the bi-functional catalyst consists of a mixture of methanol synthesis catalyst ($Cu/ZnO/Al_2O_3$) and methanol dehydration catalyst ($\gamma\text{-}Al_2O_3$). The process chemistry is represented by chemical equations that might belie the true more complex character of the process:

$$CO_2 + 3H_23 \rightarrow CH_3OH + H_2O$$

$$CO + H_2O \rightarrow CO_2 + H_2$$

$$2CH_3OH \rightarrow CH_3OCH_3 + H_2O$$

The single-stage, liquid phase process reduces the chemical equilibrium limitation that could be encountered in methanol synthesis from synthesis gas, especially in the areas of catalyst activity, per-pass conversion and reactor productivity. The process also offers considerable advantages over the conventional vapor phase synthesis of methanol in the areas of heat transfer, exothermic character, and selectivity toward methanol. However, this process suffers from the drawback that the methanol synthesis reaction is a thermodynamically governed equilibrium reaction and the concentration of methanol in the liquid phase in the vicinity of the catalytic sites is quite high due to its low solubility. Thus, the productivity of the liquid phase methanol synthesis as well as the conversion of synthesis gas could be limited by the chemical equilibrium barrier caused by high local methanol concentration in the liquid phase. One of the routes to alleviate this limitation is the in-situ dehydration of methanol into dimethyl ether which significantly improves the methanol reactor productivity. Two functionally different yet compatible catalysts are used in this dual catalytic mode of operation.

This single-step, liquid phase synthesis of dimethyl ether from synthesis gas is extremely significant from both scientific and commercial perspectives. Several key advantages of this process over methanol synthesis include higher methanol reactor productivity, higher synthesis gas conversion, and lesser dual catalyst deactivation and crystal growth. Furthermore, a process that can convert dimethyl ether to gasoline-range hydrocarbon derivatives or to lower olefins over zeolite catalysts has been developed (Lee *et al.*, 1995, 2007). This process when coupled with a single-stage process for the synthesis of dimethyl ether offers a ready route to gasoline-range hydrocarbon derivatives:

$$CH_3OCH_3 \rightarrow C_2\text{-}C_4 \text{ olefins} \rightarrow \text{Aromatics} + \text{Paraffins}$$

Selectivity towards light olefins can be enhanced by using low acidity catalysts (high SiO_2/Al_2O_3 ratio) and optimum operating conditions such as temperature, partial pressure and space velocity of dimethyl ether. Zeolite catalysts (such as ZSM-5) have pores and channels of molecular dimensions that impose spatial constraints on reactants/products of the reaction. Shape selectivity is an important property in terms of product distribution as well as the catalyst activity – zeolites exhibit product shape selectivity, which involves the limitation of diffusion of some of the hydrocarbon products out of the pores thereby enabling a tailored product spectrum. Another important aspect of the process is the transition state shape selectivity that offers constraints toward the formation of transition states based on molecular size and orientation – the formation of high molecular weight and sterically bulky molecular products (especially coke precursors that deactivate the catalyst) is hindered.

11.5 The Future

As energy demand continues to raise, so does concern over the future availability of conventional fuels. There is a growing need to find alternative fuel options, such as synthetic fuels. While conventional transport fuels are products of crude oil refining, synthetic fuels can be produced from various fossil fuels and biomass. In fact synthetic fuels derived from various sources are already available and the supplies are due to increase over the next few years.

Furthermore, many countries could eliminate the need for crude oil by using a combination of coal, natural gas, oil shale, non-food crops to make synthetic fuel, as well as waste carbonaceous materials. Synthetic fuels would be an easy fit for the transportation system because they could be used directly in automobile engines and are almost identical to fuels refined from crude oil. That sets them apart from currently available biofuels, such as ethanol, which have to be mixed with gas or require special engines.

A realistic approach would call for a gradual implementation of synthetic fuel technology, and it would take 30 to 40 years for the United States to fully adopt synthetic fuel production in a way that it could supplement crude oil supplies (Speight, 2008, 2011a, 2011c). The economics of synthetic fuel production is often quoted favorably and unfavorably but more realistically and even including the capital costs, synthetic fuels can still approach profitability depending on the feedstock and the processes required. It would take decisions by typically indecisive governments to support country-wide synthetic fuels industries when those same politicians might have to inform their constituents that gasoline/diesel process will increase. It is the perennial question: *what is a country willing to pay for energy independence?*

Over the years, the original Fischer-Tropsch method has been tweaked and improved to increase efficiency and acceptability (Table 11.3). The hydrocarbon product mixture leaving the Fischer-Tropsch reactor is frequently referred to as synthetic crude oil. This already illustrates that the standard product upgrading techniques that are used in refineries are also suitable for the upgrading of the Fischer-Tropsch wax (Marano, 2007). In fact, the refinery of the future could well be a gasification refinery would have, as the center piece, gasification technology as is the case of the Sasol refinery in South Africa (Couvaras, 1997). The refinery would produce synthesis gas (from the carbonaceous feedstock) from which liquid fuels would be manufactured using the Fischer-Tropsch synthesis technology.

Thus, advantages of producing fuels by means of gasification followed by the Fischer-Tropsch process include: (i) Fischer-Tropsch-based fuels are compatible with current diesel- and gasoline-powered vehicles and fuel distribution infrastructure – these fuels do not require new or modified pipelines, storage tanks, or retail station pumps, (ii) there is reduced reliance on imported crude oil and increase energy security, (iii) little or no particulate emissions exist because Fischer-Tropsch fuels have no sulfur and aromatics content, and there are fewer hydrocarbon and carbon monoxide emissions (Table 11.3) (Speight, 2008, 2013a; Chadeesingh, 2011).

In fact, in many ways, synthetic fuels from Fischer-Tropsch liquids are cleaner than fuels produced thermally from fossil fuel and biomass. The heavy metal and sulfur contaminants of fossil fuels can be captured in the synthetic plants before the fuel is shipped out. Fischer-Tropsch fuels also can be used in gasoline and diesel engines with no (or little) need for modifications. Fischer-Tropsch fuels do not have to compete with conventional crude oil-based fuels but can act as a valuable, less environmentally objectionable blend stock that would also allow carbon reduction with the fleet of cars currently on the road.

However, it must never be forgotten that the production of synthetic fuel from the Fischer-Tropsch process alone has a head-start insofar as the process commences with a *clean* (non-contaminated) feedstock – the gases have to be free of contaminants or the catalysts will be contaminated and rendered inefficient – to produce the clean (sulfur-free, nitrogen-free, metals-free) synthetic fuel.

The future depends very much on the effect of gasification processes on the surrounding environment. It is these environmental effects and issues that will direct the success of gasification. In fact, there is the distinct possibility that within the foreseeable future the gasification process will increase in popularity in crude oil refineries – some refineries may even be known as gasification refineries (Speight, 2011b). A gasification refinery would have, as the center piece, gasification technology as is the case of the Sasol refinery in South Africa (Couvaras, 1997). The refinery would produce synthesis gas (from the carbonaceous feedstock) from which liquid fuels would be manufactured using the Fischer-Tropsch synthesis technology.

In fact, gasification to produce synthesis gas can proceed from any carbonaceous material, including biomass. Inorganic components of the feedstock, such as metals and minerals, are trapped in an inert and environmentally safe form as char, which may have use as a fertilizer. Biomass gasification is therefore one of the most technically and economically convincing energy possibilities for a potentially carbon-neutral economy.

The manufacture of gas mixtures of carbon monoxide and hydrogen has been an important part of chemical technology for approximately a century. Originally, such mixtures were obtained by the reaction of steam with incandescent coke and were known as *water gas*. Eventually, steam

Table 11.3 Benefits of Fischer-Tropsch synthetic fuels.

Composition:
Sulfur-free
Low aromatics content
Odorless
Colorless
Local emissions:
Allow significant reduction of regulated and non-regulated vehicle pollutant emissions (NOx, SOx, PM, VOC, CO, CO_2)
CO_2 separation during synthesis gas production makes capture feasible
Diversification of energy supply:
Contribute to crude oil substitution
Diversification and security of energy supply
Distribution infrastructure:
Can be used in existing fuel infrastructure:
Compatibility with existing engines:
Can be used in existing automobile and diesel engines
Produces ultra-low sulfur, high cetane diesel
Produces low-octane gasoline that can be improved
Potential for future engines:
Enable the development of new generation of internal combustion engine technologies
Lead to improved engine efficiency
Further reduction of vehicle pollutant emissions
Impact on bio-sphere:
Readily biodegradable
Non-toxic
Not harmful to aquatic organisms

reforming processes, in which steam is reacted with natural gas (methane) or crude oil naphtha over a nickel catalyst, found wide application for the production of synthesis gas.

A modified version of steam reforming known as autothermal reforming, which is a combination of partial oxidation near the reactor inlet with conventional steam reforming further along the reactor, improves the overall reactor efficiency and increases the flexibility of the process. Partial oxidation processes using oxygen instead of steam also found wide application for synthesis gas manufacture, with the special feature that they could utilize low-value feedstocks such as heavy crude oil residues. In recent years, catalytic partial oxidation employing very short reaction times (milliseconds) at high temperatures (850 to 1000°C, 1560 to 1830°F) is providing still another approach to synthesis gas manufacture (Hickman and Schmidt, 1993).

Partial oxidation is the most commonly used process for the gasification of heavy oils and other refinery residues, although virtually all hydrocarbon mixtures (including gaseous hydrocarbon derivatives) regardless of their origin, are suitable feedstocks. In fact, gasification is replacing direct combustion in ever more countries due to environmental regulations, since ash removal and flue gas clean-up are more difficult and expensive than synthesis gas cleaning at elevated pressures.

In a gasifier, the carbonaceous material undergoes several different processes: (i) pyrolysis of carbonaceous fuels, (ii) combustion, and (iii) gasification of the remaining char. The process is very dependent on the properties of the carbonaceous material and determines the structure and composition of the char, which will then undergo gasification reactions.

The main advantages derived from application of gasification in a refinery are (i) the capability of processing low-quality, highly viscous and heavy feedstocks, and in addition (mostly in quench gasifiers) emulsions such as tank sludge, coke slurries and other liquid wastes, (ii) the capability of processing high-sulfur crude oils, because of the almost complete removal of sulfur compounds in the treating unit downstream of the gasification unit, (iii) the possibility of producing hydrogen for the various conversion and upgrading processes of the refinery, with increased production of gas oil, which is sought as feedstock for catalytic cracking units, (iv) the many outlets for synthesis gas, such as hydrogen for the refinery or for export, electricity via integrated gasification combined cycle (IGCC) operations and the production of chemicals such as ammonia, methanol, acetic acid, and oxo-alcohols (Liebner, 2000).

Thus, as crude oil supplies decrease, the desirability of producing gas from other carbonaceous feedstocks will increase, especially in those areas where natural gas is in short supply. It is also anticipated that costs of natural gas will increase, allowing coal gasification to compete as an economically viable process. Research in progress on a laboratory and pilot-plant scale should lead to the invention of new process technology by the end of the century, thus accelerating the industrial use of coal gasification.

The conversion of the gaseous products of gasification processes to synthesis gas, a mixture of hydrogen (H_2) and carbon monoxide (CO), in a ratio appropriate to the application, needs additional steps, after purification. The product gases – carbon monoxide, carbon dioxide, hydrogen, methane, and nitrogen – can be used as fuels or as raw materials for chemical or fertilizer manufacture.

Finally, gasification by means other than the conventional methods has also received some attention and has provided rationale for future processes (Rabovitser *et al.*, 2010). In the process, a carbonaceous material and at least one oxygen carrier are introduced into a non-thermal plasma reactor at a temperature in the range of approximately 300°C to approximately 700°C (570 to 1290°F) and a pressure in a range from atmospheric pressure to approximate 1030 psi and a non-thermal plasma discharge is generated within the non-thermal plasma reactor. The carbonaceous feedstock and the oxygen carrier are exposed to the non-thermal plasma discharge, resulting in the formation of a product gas which comprises substantial amounts of hydrocarbon derivatives, such as methane, hydrogen and/or carbon monoxide.

Finally, gasification and conversion of carbonaceous solid fuels to synthesis gas for application of power, liquid fuels and chemicals is practiced worldwide. Crude oil coke, coal, biomass, and refinery waste are major feedstocks for gasification. The concept of blending of coal, biomass, or refinery waste with crude oil coke is advantageous in order to obtain the highest value of products as compared to gasification of crude oil coke alone. Furthermore, based on gasifier type, co-gasification of carbonaceous feedstocks can be an advantageous and efficient process. In addition, the variety of upgrading and delivery options that are available for application to synthesis gas enable the establishment of an integrated energy supply system whereby synthesis gases can be upgraded, integrated, and delivered to a distributed network of energy conversion facilities, including power, combined heat and power, and combined cooling, heating and power (sometimes referred to as *tri-generation*) as well as used as fuels for transportation applications.

References

Arena, U. 2012. Process and technological aspects of municipal solid waste gasification. A review. *Waste Management*, 32: 625-639.

Bhattacharya, S., Md. Mizanur Rahman Siddique, A.H., and Pham, H-L. 1999. A Study in Wood Gasification on Low Tar Production. *Energy*, 24: 285-296.

Boateng, A.A., Walawender, W.P., Fan, L.T., and Chee, C.S. 1992. Fluidized-Bed Steam Gasification of Rice Hull. *Bioresource Technology*, 40(3): 235-239.

Brage, C., Yu, Q., Chen, G., and Sjöström, K. 2000. Tar Evolution Profiles Obtained from Gasification of Biomass and Coal. *Biomass and Bioenergy*, 18(1): 87-91.

Brar, J.S., Singh, K., Wang, J., and Kumar, S. 2012. Cogasification of Coal and Biomass: A Review. *International Journal of Forestry Research*, 2012: 1-10.

Brooks, K.P., Hu, J., Zhu, H., and Kee, R.J. 2007. Methanation of Carbon Dioxide by Hydrogen Reduction Using the Sabatier Process in Micro-channel Reactors. *Chemical Engineering Science*, 62(4): 1161-1170.

Chadeesingh, R. 2011. The Fischer-Tropsch Process. In: *The Biofuels Handbook*. J.G. Speight (Editor). The Royal Society of Chemistry, London, United Kingdom. Part 3, Chapter 5, Page 476-517.

Chen, G., Sjöström, K. and Bjornbom, E. 1992. Pyrolysis/Gasification of Wood in a Pressurized Fluidized Bed Reactor. *Ind. Eng. Chem. Research*, 31(12): 2764-2768.

Cheng, W.H., and Kung, H.H. (Editors). 1994. *Methanol Production and Use*. Marcel Dekker Inc., New York.

Collins, J.P., Joep, J.H.M., Freide, F., and Nay, B. 2006. A History of Fischer-Tropsch Wax Upgrading at BP – From Catalyst Screening Studies to Full Scale Demonstration in Alaska. *Journal of Natural gas Chemistry*, 15: 1-10.

Collot, A.G., Zhuo, Y., Dugwell, D.R., and Kandiyoti, R. 1999. Co-Pyrolysis and Cogasification of Coal and Biomass in Bench-Scale Fixed-Bed and Fluidized Bed Reactors. *Fuel*, 78: 667-679.

Couvaras, G. 1997. Sasol's Slurry Phase Distillate Process and Future Applications. *Proceedings. Monetizing Stranded Gas Reserves Conference, Houston. December 1997*.

De Klerk, A., and Furimsky, E. 2010. *Catalysis in the Refining of Fischer–Tropsch Syncrude*. RSC Catalysis Series, No. 4, Royal Society of Chemistry, London, United Kingdom.

De Klerk, A. 2011. *Fischer-Tropsch Refining*, Wiley-VCH, Weinheim, Germany.

Du, G., Lim, S., Yang, Y., Wang, C., Pfefferle, L., and Haller, G.L. 2007. Methanation of Carbon Dioxide on Ni-incorporated MCM-41 Catalysts: The Influence of Catalyst Pretreatment and Study of Steady-State Reaction. *Journal of Catalysis*, 249(2): 370-379.

Ergudenler, A., and Ghaly, A.E. 1993. Agglomeration of Alumina Sand in a Fluidized Bed Straw Gasifier at Elevated Temperatures. *Bioresource Technology*, 43(3): 259-268.

Fabry, F., Rehmet, C., Rohani, V-J., and Fulcheri, L. 2013. Waste Gasification by Thermal Plasma: A Review. *Waste and Biomass Valorization*, 4(3): 421-439.

Fujita, S., Nakamura, M., Doi, T., and Takezawa, N. 1993. Mechanisms Of Methanation Of Carbon Dioxide And Carbon Monoxide Over Nickel/Alumina Catalysts. *Appl. Catal. A-Gen.*, 104: 87-100.

Furimsky, E. 1999. Gasification in Petroleum Refinery of 21st Century. *Oil & Gas Science and Technology – Rev. IFP*, 54(5): 597-618.

Gabra, M., Pettersson, E., Backman, R., and Kjellström, B. 2001. Evaluation of Cyclone Gasifier Performance for Gasification of Sugar Cane Residue – Part 1: Gasification of Bagasse. *Biomass and Bioenergy*, 21(5): 351-369.

Gary, J.H., Handwerk, G.E., and Kaiser, M.J. 2007. *Petroleum Refining: Technology and Economics* 5th Edition. CRC Press, Taylor & Francis Group, Boca Raton, Florida.

Görke, O., Pfeifer, P., and Schubert, K., 2005. Highly Selective Methanation by the Use of a Microchannel Reactor. *Catalysis Today*, 110(1-2): 132-139.

Gray, D., and Tomlinson, G. 2000. Opportunities For Petroleum Coke Gasification Under Tighter Sulfur Limits For Transportation Fuels. *Proceedings. 2000 Gasification Technologies Conference, San Francisco, California. October 8-11.*

Hanaoka, T., Inoue, S., Uno, S., Ogi, T., and Minowa, T. 2005. Effect of Woody Biomass Components on Air-Steam Gasification. *Biomass and Bioenergy*, 28(1): 69-76.

Hickman, D.A., and Schmidt, L.D. 1993. Synthesis Gas Formation by Direct Catalytic Oxidation of Methane. *Science*. 259: 343-346.

Higman, C., and Van der Burgt, M. 2008. *Gasification* 2nd Edition. Gulf Professional Publishing, Elsevier, Amsterdam, Netherlands.

Hindman, M.L. 2013. Methanol to Gasoline Technology. *Proceedings. 23rd International Offshore and Polar Engineering Conference. Anchorage, Alaska. June 30-July 5.* Page 38.

Hotchkiss, R. 2003. Coal Gasification Technologies. *Proceedings. Institute of Mechanical Engineers Part A*, 217(1): 27-33.

Hsu, C.S., and Robinson, P.R. 2006. *Practical Advances in Petroleum Processing*, Volumes 1 and 2. Springer, New York.

Hu, J., Yu, F., and Lu, Y. 2012. Application of Fischer–Tropsch Synthesis in Biomass to Liquid Conversion. *Catalysts*, 2: 303-326.

Kam, A. Y., Schreiner, M., and Yurchak, S. 1984. Mobil Methanol-to-Gasoline (MTG) Process. In: *Handbook of Synfuels Technology*, R.A. Meyers (Editor). McGraw-Hill Book Company, New York. Chapter 2-3.

Khodakov, A.Y., Chu, W., and Fongarland, P. 2007. Advances in the Development of Novel Cobalt Fischer-Tropsch Catalysts for Synthesis of Long-Chain Hydrocarbon and Clean Fuels. *Chemical Reviews*, 107(7): 1692-1744.

Khosravi, M., and Khadse, A., 2013. Gasification of Petcoke and Coal/Biomass Blend: A Review. *International Journal of Emerging Technology and Advanced Engineering*, 3(12): 167-173.

Ko, M.K., Lee, W.Y., Kim, S.B., Lee, K.W., and Chun, H.S. 2001. Gasification of Food Waste with Steam in Fluidized Bed. *Korean Journal of Chemical Engineering*, 18(6): 961-964.

Kreutz, T.G. Larson, E.D., Liu, G., and Williams, R.H. 2008. Fischer-Tropsch Fuels from Coal and Biomass. *Proceedings. 25th Annual International Pittsburgh Coal Conference. Pittsburgh, Pennsylvania. September 29-October 2.*

Kumabe, K., Hanaoka, T., Fujimoto, S., Minowa, T., and Sakanishi, K. 2007. Cogasification of Woody Biomass and Coal with Air and Steam. *Fuel*, 86: 684-689.

Lahaye, J., and Ehrburger, P. (Editors). 1991. *Fundamental Issues in Control of Carbon Gasification Reactivity*. Kluwer Academic Publishers, Dordrecht, Netherlands.

Lee, S. 1990. Methanol Synthesis Technology. CRC Press, Taylor & Francis Group, Boca Raton, Florida.

Lee, S., Gogate, M.R., Fullerton, K.L., and Kulik, C.J. 1995. Catalytic Process for Production of Gasoline From Synthesis Gas. United States Patent 5,459,166. October 17.

Lee, S. 2007. Gasification of Coal. In: *Handbook of Alternative Fuel Technologies*. S. Lee, J.G. Speight, and S. Loyalka (Editors). CRC Press, Taylor & Francis Group, Boca Raton, Florida. 2007.

Lee, S., Speight, J.G., and Loyalka, S. 2007. *Handbook of Alternative Fuel Technologies*. CRC-Taylor & Francis Group, Boca Raton, Florida.

Lee, S., and Shah, Y.T. 2013. *Biofuels and Bioenergy*. CRC Press, Taylor & Francis Group, Boca Raton, Florida.

Lewis, P.E. 2013. Gas to Liquids: Beyond Fischer Tropsch. Paper No. SPE 165757. *Proceedings. SPE Asia Pacific Oil & Gas Conference and Exhibition held in Jakarta, Indonesia. October 22-14.*

Liebner, W. 2000. Gasification by Non-Catalytic Partial Oxidation of Refinery Residues. In: *Modern Petroleum Technology*. A.G. Lucas (Editor). John Wiley & Sons Inc., Hoboken, New Jersey.

Luque, R., and Speight, J.G. (Editors). 2015. *Gasification for Synthetic Fuel Production: Fundamentals, Processes, and Applications*. Woodhead Publishing, Elsevier, Cambridge, United Kingdom.

Lv, P.M., Xiong, Z.H., Chang, J., Wu, C.Z., Chen, Y., and Zhu, J.X. 2004. An Experimental Study on Biomass Air-Steam Gasification in a Fluidized Bed. *Bioresource Technology*, 95(1): 95-101.

MacDougall, L.V. 1991. Methanol to Fuels Routes – The Achievements and Remaining Problems. *Catalysis Today*, 8: 337-369.

Marano, J.J. 2007. Options for Upgrading and Refining Fischer-Tropsch Liquids. *Proceedings. 2nd International Freiberg Conference on IGCC and XtL Technologies, Freiberg, Germany. May 8-12.* http://www.iec.tu-freiberg.de/conference/conf07/pdf/8.2.pdf

McKendry, P. 2002. Energy Production from Biomass Part 3: Gasification Technologies. *Bioresource Technology*, 83(1): 55-63.

Mokhatab, S., Poe, W.A., and Speight, J.G. 2006. *Handbook of Natural Gas Transmission and Processing*. Elsevier, Amsterdam, Netherlands.

Pakdel, H., and Roy, C. 1991. Hydrocarbon Content of Liquid Products and Tar from Pyrolysis and Gasification of Wood. *Energy & Fuels*, 5: 427-436.

Pan, Y.G., Velo, E., Roca, X., Manyà, J.J., and Puigjaner, L. 2000. Fluidized-Bed Cogasification of Residual Biomass/Poor Coal Blends for Fuel Gas Production. *Fuel*, 79: 1317-1326.

Rabovitser, I.K., Nester, S., and Bryan, B. 2010. Plasma Assisted Conversion of Carbonaceous Materials into A Gas. United States Patent 7,736,400. June 25.

Rapagnà, N.J., and Latif, A. 1997. Steam Gasification of Almond Shells in a Fluidized Bed Reactor: The Influence of Temperature and Particle Size on Product Yield and Distribution. *Biomass and Bioenergy*, 12(4): 281-288.

Rapagnà, N.J., and, A. Kiennemann, A., and Foscolo, P.U. 2000. Steam-Gasification of Biomass in a Fluidized-Bed of Olivine Particles. *Biomass and Bioenergy*, 19(3): 187-197.

Ricketts, B., Hotchkiss, R., Livingston, W., and Hall, M. 2002. Technology Status Review of Waste/Biomass Co-Gasification with Coal. *Proceedings. Inst. Chem. Eng. Fifth European Gasification Conference. Noordwijk, Netherlands. April 8-10.*

Shen, C-H., Chen, W-H., Hsu, H-W., Sheu, J-Y., and Hsieh, T-H. 2012. Co-Gasification Performance of Coal and Petroleum Coke Blends in A Pilot-Scale Pressurized Entrained-Flow Gasifier. *Int. J. Energy Res.*, 36: 499-508.

Sjöström, K., Chen, G., Yu, Q., Brage, C., and Rosén, C. 1999. Promoted Reactivity of Char in Cogasification of Biomass and Coal: Synergies in the Thermochemical Process. *Fuel*, 78: 1189-1194.

Spath P.L., and Dayton. D.C. 2003. Preliminary Screening – Technical and Economic Assessment of Synthesis Gas to Fuels and Chemicals with Emphasis on the Potential for Biomass-Derived Syngas, National Renewable Energy Laboratory (NREL), Golden, Colorado.

Speight, J.G. 2008. *Synthetic Fuels Handbook: Properties, Processes, and Performance*. McGraw-Hill, New York.

Speight, J.G. 2011a. *An Introduction to Petroleum Technology, Economics, and Politics*. Scrivener Publishing, Salem, Massachusetts, 2011.

Speight, J.G. 2011b. *The Refinery of the Future*. Gulf Professional Publishing, Elsevier, Oxford, United Kingdom.

Speight, J.G. (Editor). 2011c. *The Biofuels Handbook*. Royal Society of Chemistry, London, United Kingdom.

Speight, J.G. 2013a. *The Chemistry and Technology of Coal* 3rd Edition. CRC Press, Taylor & Francis Group, Boca Raton, Florida.

Speight, J.G. 2013b. *Coal-Fired Power Generation Handbook*. Scrivener Publishing, Salem, Massachusetts.

Speight, J.G. 2014a. *The Chemistry and Technology of Petroleum* 5th Edition. CRC Press, Taylor & Francis Group, Boca Raton, Florida.

Speight, J.G. 2014b. *Gasification of Unconventional Feedstocks*. Gulf Professional Publishing, Elsevier, Oxford, United Kingdom.

Speight, J.G. 2017. *Handbook of Petroleum Refining*. CRC Press, Taylor & Francis Group, Boca Raton, Florida.

Speight, J.G. 2019a. *Natural Gas: A Basic Handbook* 2nd Edition. Gulf Publishing Company, Elsevier, Cambridge, Massachusetts.

Speight, J.G. 2019b. *Handbook of Petrochemical Processes*. CRC Press, Taylor & Francis Group, Boca Raton, Florida.

Steynberg, A.P., Espinoza, R.L., Jager, B., and Vosloo, A.C. 1999. High Temperature Fischer-Tropsch Synthesis in Commercial Practice. *Applied Catalysis A.*, 186(1-2): 41-54.

Takenaka, N., Shimizu, T., and Otsuka, K. 2004. Complete Removal of Carbon Monoxide in Hydrogen-Rich Gas Stream through Methanation over Supported Metal Catalysts. *Int. J. Hydrogen Energy*, 29: 1065-1073.

Wallace, P.S., Anderson, M.K., Rodarte, A.I., and Preston, W.E. 1998. Heavy Oil Upgrading by the Separation and Gasification of Asphaltenes. *Proceedings. Presented at the Gasification Technologies Conference. San Francisco, California. October.*

Weissermel, K. 2003. *Industrial Organic Chemistry* 4th Edition. Wiley-VCH, Weinheim, Germany.

Wender, I. 1996. Reactions of Synthesis Gas. *Fuel Processing Technology*, 48(3): 189-297.

Wolff, J., and Vliegenthart, E. 2011. Gasification of Heavy Ends. *Petroleum Technology Quarterly*, Q2: 1-5.

Zhilyaeva, N.A., Volnina, E.A., Kukuna, M.A., and Frolov, V.M. 2002." Carbon Dioxide Hydrogenation Catalysts (a Review). *Petrol. Chem.*, 42: 367-386.

12
Hydrogen Production

12.1 Introduction

Hydrogen is an important refinery commodity and, as a result, throughout the previous chapters (especially Chapter 10 and Chapter 11) there have been several references and/or acknowledgments of a very important property of crude oil and crude oil products. And that is the hydrogen content or the use of hydrogen during refining in hydrotreating processes, such as desulfurization (Chapter 10) and in hydroconversion processes, such as hydrocracking (Table 12.1) (Parkash, 2003; Gary *et al.*, 2007; Speight, 2011a, 2014; Hsu and Robinson, 2017; Speight, 2017). Although the hydrogen recycle gas in various streams may contain up to 40% v/v of other gases (usually hydrocarbon derivatives), hydrotreater catalyst life is a strong function of hydrogen partial pressure. Optimum hydrogen purity at the reactor inlet extends catalyst life by maintaining desulphurization kinetics at lower operating temperatures and reducing carbon laydown. Typical purity increases resulting from hydrogen purification equipment and/or increased hydrogen sulfide removal as well as tuning hydrogen circulation and purge rates, may extend catalyst life up to approximately 25%.

In fact, the typical refinery runs at a hydrogen deficit and a critical issue facing refiners is the influx of heavier feedstocks into refineries and the need to process the refinery feedstock into refined transportation fuels under an environment of increasingly more stringent clean fuel regulations, decreasing heavy fuel oil demand and increasingly heavy, sour crude supply. Thus, hydrogen is a major requirement not only for hydrogenating processes (such as hydrotreatment and hydrocracking) but is also necessary to complete the hydrogen balance in synthesis gas and, thus, the production of synthesis gas plays an integral role in hydrogen availability in the refinery (Parkash, 2003; Gary *et al.*, 2007; Speight, 2011a, 2014; Hsu and Robinson, 2017; Speight, 2016, 2017).

The hydrogen network optimization is at the forefront of world refineries' options to address clean fuel trends, to meet growing transportation fuel demands and to continue to make a profit from their crudes. A key element of a hydrogen network analysis in a refinery involves the capture of hydrogen in its fuel streams and extending its flexibility and processing options. Thus, innovative hydrogen network optimization will be a critical factor influencing the operating flexibility and profitability of the future refinery in a shifting world of crude feedstock supplies and ultra-low-sulfur (ULS) gasoline and diesel fuel.

As hydrogen use has become more widespread in refineries, hydrogen production has moved from the status of a high-tech specialty operation to an integral feature of most refineries (Raissi, 2001; Vauk *et al.*, 2008; Liu *et al.*, 2010; Speight, 2016). This has been made necessary by the increase in hydrotreating and hydrocracking, including the treatment of progressively heavier feedstocks. In fact, the use of hydrogen in thermal processes is perhaps the single most significant advance in refining technology during the 20th century

Table 12.1 Summary of typical hydrogen application and production process in a refinery.

Hydrogen Applications
Naphtha hydrotreater:
• Uses hydrogen to desulfurize naphtha from atmospheric distillation; must hydrotreat the naphtha before sending to a catalytic reformer unit.
Distillate hydrotreater:
• Desulfurizes distillates after atmospheric or vacuum distillation; in some units aromatics are hydrogenated to cycloparaffins or alkanes.
Hydrodesulfurization:
Sulfur compounds are hydrogenated to hydrogen sulphide H_2S as feed for Claus plants.
Hydroisomerization:
• Normal (straight-chain) paraffins are converted into iso-paraffins to improve the product properties (e.g., octane number).
Hydrocracker:
• Uses hydrogen to upgrade heavier fractions into lower boiling, more valuable products.
Hydrogen Production
Catalytic reformer:
• Used to convert the naphtha-boiling range molecules into higher octane reformate; hydrogen is a byproduct.
Steam-methane reformer:
• Produces hydrogen for the hydrotreaters or hydrocracker.
Steam reforming of higher molecular weight hydrocarbon derivatives:
• Produces hydrogen from low-boiling hydrocarbon derivatives other than methane.
Recovery from refinery off-gases:
• Process gas often contains hydrogen in the range up to 50% v/v.
Gasification of petroleum residua:
• Recovery from synthesis gas produced in gasification units.
Gasification of petroleum coke:
• Recovery from synthesis gas produced in gasification units.
Partial oxidation processes:
• Analogous to gasification process; produce synthesis gas from which hydrogen can be isolated.

(Scherzer and Gruia, 1996; Dolbear, 1998). The continued increase in hydrogen demand over the last several decades is a result of the conversion of crude oil to match changes in product slate and the supply of heavy, high-sulfur oil, and in order to make lower-boiling, cleaner, and more salable products. There are also many reasons other than product quality for using hydrogen in processes adding to the need to add hydrogen at relevant stages of the refining process and, most important according to the availability of hydrogen (Bezler, 2003; Miller and Penner, 2003; Ranke and Schödel, 2003).

With the increasing need for *clean* fuels, the production of hydrogen for refining purposes requires a major effort by refiners. In fact, the trend to increase the number of hydrogenation (*hydrocracking* and/or *hydrotreating*) processes in refineries coupled with the need to process the heavier oils, which require substantial quantities of hydrogen for upgrading because of the increased use of hydrogen in hydrocracking processes, has resulted in vastly increased demands for this gas. The hydrogen demands can be estimated to a very rough approximation using API gravity and the extent of the reaction, particularly the hydrodesulfurization reaction (Speight, 2000; Parkash, 2003; Gary *et al.*, 2007; Speight, 2011a, 2014; Hsu and Robinson, 2017; Speight, 2017). But accurate estimation requires equivalent process parameters and a thorough understanding of the nature of each process. Thus, as hydrogen production grows, a better understanding of the capabilities and requirements of a hydrogen plant becomes ever more important to overall refinery operations as a means of making the best use of hydrogen supplies in the refinery.

The chemical nature of the crude oil used as the refinery feedstock has always played the major role in determining the hydrogen requirements of that refinery. For example, the lower density, more paraffinic crude oils will require somewhat less hydrogen for upgrading to, say, a gasoline product than a heavier more asphaltic crude oil (Speight, 2000). It follows that the hydrodesulfurization of heavy oils and residua (which, by definition, is a hydrogen-dependent process) needs substantial amounts of hydrogen as part of the processing requirements.

In general, considerable variation exists from one refinery to another in the balance between hydrogen produced and hydrogen consumed in the refining operations. However, what is more pertinent to the present text is the excessive amounts of hydrogen that are required for hydroprocessing operations, whether these be hydrocracking or the somewhat milder hydrotreating processes. For effective hydroprocessing, a substantial hydrogen partial pressure must be maintained in the reactor and, in order to meet this requirement, an excess of hydrogen above that actually consumed by the process must be fed to the reactor. Part of the hydrogen requirement is met by recycling a stream of hydrogen-rich gas. However, the need still remains to generate hydrogen as makeup material to accommodate the process consumption of 500 to 3000 scf/bbl depending upon whether the heavy feedstock is being subjected to a predominantly hydrotreating process (hydrodesulfurization process) or to a predominantly hydrocracking process.

In some refineries, the hydrogen needs can be satisfied by hydrogen recovery from catalytic reformer product gases, but other external sources are required. However, for the most part, many refineries now require on-site hydrogen production facilities to supply the gas for their own processes. Most of this non-reformer hydrogen is manufactured either by steam-methane reforming or by oxidation processes.

An early use of hydrogen in refineries was in naphtha hydrotreating, as feed pretreatment for catalytic reforming (which in turn was producing hydrogen as a byproduct).

As environmental regulations tightened, the technology matured and heavier streams were hydrotreated. Thus in the early refineries, the hydrogen for hydroprocesses was provided as a result of catalytic reforming processes in which dehydrogenation is a major chemical reaction and, as a consequence, hydrogen gas is produced (Chapter 5, Chapter 13). The light ends from the catalytic reformer contain a high ratio of hydrogen to methane so the stream is freed from ethane and/or propane to get a high concentration of hydrogen in the stream.

The hydrogen is recycled though the reactors where the reforming takes place to provide the atmosphere necessary for the chemical reactions and also prevents the carbon from being deposited on the catalyst, thus extending its operating life. An excess of hydrogen above whatever is consumed in the process is produced, and, as a result, catalytic reforming processes are unique in that they are the only crude oil refinery processes to produce hydrogen as a byproduct. However, as refineries and refinery feedstocks evolved during the last four decades, the demand for hydrogen has increased and reforming processes are no longer capable of providing the quantities of hydrogen necessary for feedstock hydrogenation. Within the refinery, other processes are used as sources of hydrogen. Thus, the recovery of hydrogen from the byproducts of the coking units, visbreaker units and catalytic cracking units is also practiced in some refineries.

In coking units and visbreaker units, heavy feedstocks are converted to crude oil coke, oil, low-boiling hydrocarbon derivatives (benzene, naphtha, liquefied petroleum gas, LPG) and gas (Chapter 8). Depending on the process, hydrogen is present in a wide range of concentrations. Since coking processes need gas for heating purposes, adsorption processes are best suited to recover the hydrogen because they feature a very clean hydrogen product and an off-gas suitable as fuel.

Catalytic cracking is the most important process step for the production of low-boiling products from gas oil and increasingly from vacuum gas oil and heavy feedstocks (Chapter 9). In catalytic cracking the molecular mass of the main fraction of the feed is lowered, while another part is converted to coke that is deposited on the hot catalyst. The catalyst is regenerated in one or two stages by burning the coke off with air that also provides the energy for the endothermic cracking process. In the process, paraffins and naphthenes are cracked to olefins and to alkanes with shorter chain length, mono-aromatic compounds are dealkylated without ring cleavage, di-aromatics and poly-aromatics are dealkylated and converted to coke. Hydrogen is formed in the last type of reaction, whereas the first two reactions produce low-boiling hydrocarbon derivatives and therefore require hydrogen. Thus, a catalytic cracker can be operated in such a manner that enough hydrogen for subsequent processes is formed.

In reforming processes, naphtha fractions are reformed to improve the quality of gasoline (Speight, 2000; Parkash, 2003; Gary et al., 2007; Speight, 2011a, 2014; Hsu and Robinson, 2017; Speight, 2017). The most important reactions occurring during this process are the dehydrogenation of naphthenes to aromatics. This reaction is endothermic and is favored by low pressures and the reaction temperature lies in the range of 300 to 450°C (570 to 840°F). The reaction is performed on platinum catalysts, with other metals, such as rhenium, as promoters.

Hydrogen is generated in a refinery by the catalytic reforming process, but there may not always be the need to have a catalytic reformer as part of the refinery sequence. Nevertheless, assuming that a catalytic reformer is part of the refinery sequence, the hydrogen production from the reformer usually falls well below the amount required for hydroprocessing purposes. For example, in a 100,000-bbl/day hydrocracking refinery, assuming intensive reforming of hydrocracked gasoline, the hydrogen requirements of the refinery may still fall some 500 to 900 scf/bbl of crude charge below that necessary for the hydrocracking sequences. Consequently, an *external* source of hydrogen is necessary to meet the daily hydrogen requirements of any process where the heavier feedstocks are involved.

The trend to increase the number of hydrogenation (*hydrocracking* and/or *hydrotreating*) processes in refineries (Dolbear, 1998; Parkash, 2003; Gary *et al.*, 2007; Speight, 2011a, 2014; Hsu and Robinson, 2017; Speight, 2017) coupled with the need to process the heavier oils, which require substantial quantities of hydrogen for upgrading, has resulted in vastly increased demands for this gas.

Hydrogen has historically been produced during catalytic reforming processes as a byproduct of the production of the aromatic compounds used in gasoline and in solvents. As reforming processes changed from fixed-bed to cyclic to continuous regeneration, process pressures have dropped and hydrogen production per barrel of reformate has tended to increase. However, hydrogen production as a byproduct is not always adequate to the needs of the refinery and other processes are necessary. Thus, hydrogen production by steam reforming or by partial oxidation of residua has also been used, particularly where heavy oil is available. Steam reforming is the dominant method for hydrogen production and is usually combined with pressure-swing adsorption (PSA) to purify the hydrogen to greater than 99% by volume (Bandermann and Harder 1982).

The gasification of residua and coke to produce hydrogen and/or power may become an attractive option for refiners (Dickenson *et al.*, 1997; Gross and Wolff, 2000). The premise that the gasification section of a refinery will be the *garbage can* for deasphalter residues, high-sulfur coke, as well as other refinery wastes is worthy of consideration.

Several other processes are available for the production of the additional hydrogen that is necessary for the various heavy feedstock hydroprocessing sequences and it is the purpose of this chapter to present a general description of these processes. In general, most of the external hydrogen is manufactured by steam-methane reforming or by oxidation processes. Other processes such as ammonia dissociation, steam-methanol interaction, or electrolysis are also available for hydrogen production, but economic factors and feedstock availability assist in the choice between processing alternatives.

The processes described in this chapter are those gasification processes that are often referred to as the *garbage disposal units* of the refinery. Hydrogen is produced for use in other parts of the refinery as well as for energy and it is often produced from process byproducts that may not be of any use elsewhere. Such byproducts might be the highly aromatic, heteroatom, and metal containing reject from a deasphalting unit or from a mild hydrocracking process. However attractive this may seem, there will be the need to incorporate a gas cleaning operation to remove any environmentally objectionable components from the hydrogen gas.

12.2 Processes

The production of hydrogen as an integral part of the production of carbon monoxide, since the two gases make up the mixture known as synthesis gas. Hydrogen is indeed an important commodity in the refining industry because of its use in hydrotreating processes, such as desulfurization, and in hydroconversion processes, such as hydrocracking (Parkash, 2003; Gary et al., 2007; Speight, 2011a, 2014; Hsu and Robinson, 2017; Speight, 2017). Part of the hydrogen is produced during reforming processes but that source, once sufficient, is now insufficient for the hydrogen needs of a modern refinery (Speight, 2000; Parkash, 2003; Ancheyta and Speight, 2007; Gary et al., 2007; Speight, 2011a, 2014; Hsu and Robinson, 2017; Speight, 2017). In addition, optimum hydrogen purity at the reactor inlet extends catalyst life by maintaining desulphurization kinetics at lower operating temperatures and reducing carbon laydown. Typical purity increases resulting from hydrogen purification equipment and/or increased hydrogen sulfide removal as well as tuning hydrogen circulation and purge rates, may extend catalyst life up to approximately 25%. Indeed, since hydrogen use has become more widespread in refineries, hydrogen production has moved from the status of a high-tech specialty operation to an integral feature of most refineries (Raissi, 2001; Vauk et al., 2008).

While the gasification of residua and coke as well as other carbonaceous feedstocks to produce hydrogen and/or power may increase in use in refineries over the next two decades (Speight, 2011a, 2011b), several other processes are available for the production of the additional hydrogen that is necessary for the various heavy feedstock hydroprocessing sequences (Speight, 2014) and it is the purpose of this section to present a general description of these processes.

In addtion to the reforming processes (Chapter 8) by which hydrogen can be produced, there is a collection of other processes that are used on-site in the refinery to produce hydrogen. The processes described in this section are those gasification processes that are often referred to as the *garbage disposal units* of the refinery and that have not been described above.

12.2.1 Feedstocks

The most common, and perhaps the best, feedstocks for steam reforming are low-boiling saturated hydrocarbon derivatives that have a low sulfur content, including natural gas, refinery gas, liquefied petroleum gas, and low-boiling naphtha (Parkash, 2003; Gary et al., 2007; Speight, 2011a, 2014; Hsu and Robinson, 2017; Speight, 2017).

Natural gas is the most common feedstock for hydrogen production since it meets all the requirements for reformer feedstock. Natural gas typically contains more than 90% methane and ethane with only a few percent of propane and higher boiling hydrocarbon derivatives (Mokhatab et al., 2006; Speight, 2014). Natural gas may (or most likely will) contain traces of carbon dioxide with some nitrogen and other impurities. Purification of natural gas, before reforming, is usually relatively straightforward. Traces of sulfur must be removed to avoid poisoning the reformer catalyst; zinc oxide treatment in combination with hydrogenation is usually adequate.

Refinery gas, containing a substantial amount of hydrogen, can be an attractive feedstock for steam reforming since it is produced as a byproduct. Processing of refinery gas will depend on its composition, particularly the levels of olefins and of propane and heavier hydrocarbon derivatives. Olefins, that can cause problems by forming coke in the reformer, are converted to saturated compounds in the hydrogenation unit. Higher-boiling hydrocarbon derivatives in refinery gas can also form coke, either on the primary reformer catalyst or in the preheater. If there is more than a few percent of C_3 and higher compounds, a promoted reformer catalyst should be considered, in order to avoid carbon deposits.

Refinery gas from different sources varies in suitability as hydrogen plant feed. Catalytic reformer off-gas, for example, is saturated, very low in sulfur, and often has high hydrogen content. The process gases from a coking unit or from a fluid catalytic cracking unit are much less desirable because of the content of unsaturated constituents. In addition to olefins, these gases contain substantial amounts of sulfur that must be removed before the gas is used as feedstock. These gases are also generally unsuitable for direct hydrogen recovery, since the hydrogen content is usually too low. Hydrotreater off-gas lies in the middle of the range. It is saturated, so it is readily used as hydrogen plant feed. Content of hydrogen and heavier hydrocarbon derivatives depends to a large extent on the upstream pressure. Sulfur removal will generally be required.

12.2.2 Commercial Processes

Hydrogen is generated in a refinery by the catalytic reforming process, but there may not always be the need to have a catalytic reformer as part of the refinery sequence. Nevertheless, assuming that a catalytic reformer is part of the refinery sequence, the hydrogen production from the reformer usually falls well below the amount required for hydroprocessing purposes. Consequently, an *external* source of hydrogen is necessary to meet the daily hydrogen requirements of any process where the heavier feedstocks are involved which is accompanied by various energy changes and economic changes to the refinery balance sheet (Cruz and de Oliveira, 2008).

Thus, hydrogen production as a byproduct is not always adequate to the needs of the refinery and other processes are necessary (Lipman, 2011; Speight, 2011, 2014). Thus, hydrogen production by steam reforming or by partial oxidation of residua has also been used, particularly where heavy crude oil is available. Steam reforming is the dominant method for hydrogen production and is usually combined with pressure-swing adsorption (PSA) to purify the hydrogen to greater than 99% v/v. However, the process parameters need to be carefully defined in order to optimize capital cost. An unnecessarily stringent specification in the hydrogen purity may cause undesired and unnecessary capital cost – an example is the residual concentration of nitrogen that should not be less than 100 ppm.

The most common, and perhaps the best, feedstocks for steam reforming are low-boiling saturated hydrocarbon derivatives that have a low sulfur content; including natural gas, refinery gas, liquefied petroleum gas (LPG), and low-boiling naphtha. *Natural gas* is the most common feedstock for hydrogen production since it meets all the requirements for steam-methane reformer feedstock.

However, one of the key variables in operating the steam-methane reforming unit is maintaining the proper steam to carbon ratio, which can be difficult when natural gas containing various hydrocarbon derivatives (Mokhatab *et al.*, 2006; Speight, 2014) is used as the feedstock and if the feedstock is typically a mixture of refinery fuel gas and natural gas, the composition is not constant. If the steam to carbon ratio is too low, carbon will deposit on the catalyst, lowering the activity of the catalyst. In some cases, the catalyst can be completely destroyed, and the unit will need to be shut down to change the catalyst, thereby causing a disruption in the hydrogen supply. On the other hand, if the steam to carbon ratio is run too high, this wastes energy, increases steam consumption, and could also be affecting throughput.

The gasification of petroleum residua, petroleum coke, and other feedstocks such as biomass (Speight, 2008, 2011a, 2011b, 2014) to produce hydrogen and/or power may become an attractive option for refiners. The premise that the gasification section of a refinery will be the *garbage can* for deasphalter residues, high-sulfur coke, as well as other refinery wastes is worthy of consideration. Other processes such as ammonia dissociation, steam-methanol interaction, or electrolysis are also available for hydrogen production, but economic factors and feedstock availability assist in the choice between processing alternatives.

Hydrogen is produced for use in other parts of the refinery as well as for energy and it is often produced from process byproducts that may not be of any use elsewhere. Such byproducts might be the highly aromatic, heteroatom, and metal-containing reject from a deasphalting unit or from a mild hydrocracking process. However attractive this may seem, there will be the need to incorporate a gas cleaning operation to remove any environmentally objectionable components from the hydrogen gas.

When the hydrogen content of the refinery gas is greater than 50% v/v, the gas should first be considered for hydrogen recovery, using a membrane (Brüschke, 1995, 2003) or pressure-swing adsorption unit. The tail gas or reject gas that will still contain a substantial amount of hydrogen can then be used as steam reformer feedstock. Generally, the feedstock purification process uses three different refinery gas streams to produce hydrogen. First, high-pressure hydrocracker purge gas is purified in a membrane unit that produces hydrogen at medium pressure and is combined with medium pressure off-gas that is first purified in a pressure-swing adsorption unit. Finally, low-pressure off-gas is compressed, mixed with reject gases from the membrane and pressure-swing adsorption units, and used as steam reformer feed.

Various processes are available to purify the hydrogen stream but since the product streams are available as a wide variety of composition, flows, and pressures, the best method of purification will vary. And there are several factors that must also be taken into consideration in the selection of a purification method. These are: (1) hydrogen recovery, (2) product purity, (3) pressure profile, (4) reliability, and (5) cost, an equally important parameter is not considered here since the emphasis is on the technical aspects of the purification process.

Thus, hydrogen can be produced by separation and purification of the constituents of the gaseous mixture from the aforementioned gasification processes (Chapter 5). Other process options are presented below.

12.2.2.1 Hydrocarbon Gasification

The gasification of hydrocarbon derivatives to produce hydrogen is a continuous, non-catalytic process that involves partial oxidation of the hydrocarbon and one of several processes that are used for gasification of carbonaceous fuels to gaseous products (Breault, 2010).

As an example, the partial oxidation of methane to produce carbon monoxide and hydrogen has more potential than the steam reforming (SR) process. In the partial oxidation process, supported metal catalysts, including a noble metal catalyst or a transition metal catalysts are often used. In the process, air or oxygen (with steam or carbon dioxide) is used as the oxidant at 1095 to 1,480°C (2000 to 2700°F). Any carbon produced (2 to 3% w/w of the feedstock) during the process is removed as a slurry in a carbon separator and pelletized for use either as a fuel or as raw material for carbon-based products.

By way of explanation, the noble metals are metals that are resistant to oxidation and corrosion in moist air (unlike most base metals). The short list of chemically noble metals comprises ruthenium (Ru), rhodium (Rh), palladium (Pd), silver (Ag), osmium (Os), iridium (Ir), platinum (Pt), and gold (Au). More inclusive lists include one or more of the following mercury (Hg), rhenium (Re), and copper (Cu) as noble metals. On the other hand, a transition metal, any of various chemical elements that have valence electrons (i.e., the electrons that can participate in the formation of chemical bonds) in two shells instead of only one. While the term *transition* has no particular chemical significance, it is a convenient name by which to distinguish the similarity of the atomic structures and resulting properties of the elements so designated. They occupy the middle portions of the long periods of the Periodic Table of the elements between the groups on the left-hand side and the groups on the right (Table 12.2). For further clarification, the first main transition series begins with either scandium (Sc) or titanium (Ti) and ends with zinc (Zn). The second series includes the elements yttrium (Y) to cadmium (Cd). The third series extends from lanthanum (La) to mercury (Hg).

12.2.2.2 Hypro Process

Methane, due to its abundance and high H/C ratio (highest among all hydrocarbon derivatives) is an obvious source for hydrogen. Steam reforming of methane represents the current trend for hydrogen production. Other popular methods of hydrogen production include autothermal reforming and partial oxidation. However, all these processes involve the formation of large amounts of unwanted (if hydrogen is the desired product) COx as byproduct.

Table 12.2 The periodic table of the elements.

Hydrogen production routes, which do not require complex COx removal procedures, are therefore desired for production of high-purity hydrogen. Thus, there is much interest in the catalytic decomposition of natural gas (whose major constituent is methane) for production of hydrogen. Since only hydrogen and carbon are formed in the decomposition process, separation of products is not an issue. The other main advantage is the simplicity of the methane decomposition process as compared to conventional methods. For example, the high- and low-temperature water-gas shift reactions and carbon dioxide removal step (involved in the conventional methods) are completely eliminated. Catalyst regeneration is extremely important for the practical application of the clean hydrogen production process.

The Hypro process is a continuous catalytic method for hydrogen manufacture from natural gas or from refinery effluent gases, especially the decomposition of methane to hydrogen and carbon (Choudhary *et al.*, 2003; Choudhary and Goodman, 2006):

$$CH_4 \rightarrow C + 2H_2$$

Hydrogen is recovered by phase separation to yield hydrogen of approximately 93% purity; the principal contaminant is methane.

12.2.2.3 Hydrogen from Pyrolysis Processes

There has been recent interest in the use of pyrolysis processes to produce hydrogen. Specifically the interest has focused on the pyrolysis of methane (natural gas) and hydrogen sulfide.

Natural gas is readily available and offers relatively rich stream of methane with lower amounts of ethane, propane and butane also present. The thermocatalytic decompositon of natural gas hydrocarbon derivatives (c.f., Hypro process) offers an alternate method for the production of hydrogen (Uemura *et al.*, 1999; Weimer *et al.*, 2000):

$$C_nH_m \rightarrow nC + (m/2)H_2$$

The production of hydrogen by direct decomposition of hydrogen sulfide has also been proposed (Clark and Wassink, 1990; Zaman and Chakma, 1995; Donini, 1996; Luinstra 1996). Hydrogen sulfide decomposition is a highly endothermic process and equilibrium yields are poor (Clark *et al.*, 1995). At temperatures less than 1500°C (2730°F), the thermodynamic equilibrium is unfavorable toward hydrogen formation. However, in the presence of catalysts such as platinum-cobalt (at 1000°C; 1830°F), disulfides of molybdenum (Mo) or tungsten (W) at 800°C (1470°F) (Kotera *et al.*, 1976), or other transition metal sulfides supported on alumina (at 500 to 800°C; 930 to 1470°F), decomposition of hydrogen sulfide proceeds. In the temperature range of approximately 800 to 1500°C (1470 to 2730°F), thermolysis of hydrogen sulfide can be treated simply:

$$H_2S \rightarrow H_2 + 1/xS_x \qquad \Delta H_{298K} = +34,300 \text{ Btu/lb}$$

In this equation, $x = 2$.

Outside of this temperature range, multiple equilibria may be present depending on temperature, pressure, and relative abundance of hydrogen and sulfur (Clark and Wassink, 1990). In addition, the steam-iron process is an established process, which was used for the production of hydrogen from cokes at the beginning of the 20th century. However, the process could not compete with the later on developed steam reforming of methane and the process fell into disuse. There is renewed interest in the development of the steam-iron process that is mainly focused on the use of renewable energy sources, like biomass. In this thesis, the production of hydrogen by the steam-iron process from pyrolysis oil is studied. Pyrolysis oil, obtained from the pyrolysis of biomass, is used to facilitate transportation and to simplify gasification and combustion processes, before being processed to hydrogen. The benefit of the steam-iron process compared to other thermo-chemical routes of biomass is that hydrogen can be produced in a two-step redox cycle, without the need of any purification steps (such as pressure-swing adsorption) (Bleeker et al., 2007; Bleeker, 2009).

12.2.2.4 Hydrogen from Refinery Gas

Refinery gas is the non-condensable gas obtained during distillation of crude oil or treatment of oil products (such as cracking processes) in refineries. The gas consists mainly of hydrogen, methane, ethane, and olefins derivatives as well as gases which are returned from the petrochemical industry.

Recovering of hydrogen from refinery gas can help refineries to satisfy high hydrogen demand. Cryogenic separation is typically viewed as being the most thermodynamically efficient separation technology. The basic configuration for hydrogen recovery from refinery gases involves a two-stage partial condensation process, with post purification via pressure-swing adsorption (Dragomir et al., 2010). The major steps in this process involve first compressing and pretreating the crude refinery gas stream before chilling to an intermediate temperature (-60 to -120 °F). This partially condensed stream is then separated in a flash-drum after which the liquid stream is expanded through a Joule-Thompson valve to generate refrigeration and then is fed to the wash column. Optionally, the wash column can be replaced by a simple flash drum.

A crude liquefied petroleum gas stream is collected at the bottom of the column, and a methane rich vapor is obtained at the top. The methane rich vapor is sent to compression and then to fuel. The vapor from the flash drum is further cooled in a second heat exchanger before being fed to a second flash drum where it produces a hydrogen-rich stream and a methane rich liquid. The liquid is expanded in a Joule-Thomson valve to generate refrigeration, and then is sent for further cooling. The hydrogen rich gas is then sent to the pressure-swing adsorption unit for further purification – the tail gas from this unit is compressed and returned to fuel together with the methane-rich gas.

12.2.2.5 Other Options

Other processes, such as the steam-methanol interaction (Palo et al., 2007) or the dissociation of ammonia (Nagaoka et al., 2017), may also be used as sources of hydrogen.

$$CH_3OH + H_2O \rightarrow CO_2 + 2H_2$$

The endothermic ammonia dissociation reaction is simply the reverse of the synthesis reaction. Thus:

$$NH_3 (g) \rightarrow 1/2 N_2 (g) + 3/2 H_2 (g)$$

or

$$2NH_3 \rightarrow N_2 + 3H_2$$

The temperature required for efficient cracking depends on the catalyst. There are a wide variety of materials that have been found to be effective, but some (e.g., supported Ni catalysts) require temperatures above 1000°C (1830°F) while other catalysts have a high conversion efficiency at temperatures on the order of 650 to 700°C (1200 to 1290°F) or even at lower temperatures (Nagaoka et al., 2017).

The electrolysis of water produces high-purity hydrogen. The overall reaction is represented simply as:

$$2H_2O(l) \rightarrow 2H_2(g) + O_2(g)$$

If the above-described processes occur in pure water, H^+ cations will be consumed/reduced at the cathode and OH^- anions are consumed/oxidized at the anode. The negative hydroxide ions that approach the anode mostly combine with the positive hydronium ions (H_3O^+) to form water. The positive hydronium ions that approach the cathode mostly combine with negative hydroxide ions to form water.

12.2.3 Process Chemistry

Before the feedstock is introduced to a process, there is the need for application of a strict feedstock purification protocol. Prolonging catalyst life in hydrogen production processes is attributable to effective feedstock purification, particularly sulfur removal. A typical natural gas or other low-boiling hydrocarbon feedstock contains traces of hydrogen sulfide and organic sulfur.

In order to remove sulfur compounds, it is necessary to hydrogenate the feedstock to convert the organic sulfur to hydrogen that is then reacted with zinc oxide (ZnO) at approximately 370°C (700°F) that results in the optimal use of the zinc oxide as well as ensuring complete hydrogenation. Thus, assuming assiduous feedstock purification and removal of all of the objectionable contaminants, the chemistry of hydrogen production can be defined.

In *steam reforming*, low-boiling hydrocarbon derivatives such as methane are reacted with steam to form hydrogen:

$$CH_4 + H_2O \rightarrow 3H_2 + CO \qquad \Delta H_{298K} = +97,400 \text{ Btu/lb}$$

H is the heat of reaction. A more general form of the equation that shows the chemical balance for higher-boiling hydrocarbon derivatives is:

$$C_nH_m + nH_2O \rightarrow (n + m/2)H_2 + nCO$$

The reaction is typically carried out at approximately 815°C (1500°F) over a nickel catalyst packed into the tubes of a reforming furnace. The high temperature also causes the hydrocarbon feedstock to undergo a series of cracking reactions, plus the reaction of carbon with steam:

$$CH_4 \rightarrow 2H_2 + C$$

$$C + H_2O \rightarrow CO + H_2$$

Carbon is produced on the catalyst at the same time that hydrocarbon is reformed to hydrogen and carbon monoxide. With natural gas or similar feedstock, reforming predominates and the carbon can be removed by reaction with steam as fast as it is formed. When higher-boiling feedstocks are used, the carbon is not removed fast enough and builds up thereby requiring catalyst regeneration or replacement. Carbon buildup on the catalyst (when high-boiling feedstocks are employed) can be avoided by addition of alkali compounds, such as potash, to the catalyst thereby encourage or promoting the carbon-steam reaction.

However, even with an alkali-promoted catalyst, feedstock cracking limits the process to hydrocarbon derivatives with a boiling point less than of 180°C (350°F). Natural gas, propane, butane, and low-boiling naphtha are most suitable. Pre-reforming, a process that uses an adiabatic catalyst bed operating at a lower temperature, can be used as a pretreatment to allow heavier feedstocks to be used with lower potential for carbon deposition (coke formation) on the catalyst.

After reforming, the carbon monoxide in the gas is reacted with steam to form additional hydrogen (the *water-gas shift* reaction):

$$CO + H_2O \rightarrow CO_2 + H_2 \qquad \Delta H_{298K} = -16,500 \text{ Btu/lb}$$

This leaves a mixture consisting primarily of hydrogen and carbon monoxide that is removed by conversion to methane:

$$CO + 3H_2O \rightarrow CH_4 + H_2O$$

$$CO_2 + 4H_2 \rightarrow CH_4 + 2H_2O$$

The critical variables for steam reforming processes are (i) temperature, (ii) pressure, and (iii) the steam/hydrocarbon ratio. Steam reforming is an equilibrium reaction, and conversion of the hydrocarbon feedstock is favored by high temperature, which in turn requires higher fuel use. Because of the volume increase in the reaction, conversion is also favored by low pressure, which conflicts with the need to supply the hydrogen at high pressure. In practice, materials of construction limit temperature and pressure.

On the other hand, and in contrast to reforming, shift conversion is favored by low temperature. The gas from the reformer is reacted over iron oxide catalyst at 315 to 370°C (600 to 700°F) with the lower limit being dictated activity of the catalyst at low temperature.

Hydrogen can also be produced by *partial oxidation* (POX) of hydrocarbon derivatives in which the hydrocarbon is oxidized in a limited or controlled supply of oxygen:

$$2CH_4 + O_2 \rightarrow CO + 4H_2 \qquad \Delta H_{298K} = -10,195 \text{ Btu/lb}$$

The shift reaction also occurs and a mixture of carbon monoxide and carbon dioxide is produced in addition to hydrogen. The catalyst tube materials do not limit the reaction temperatures in partial oxidation processes and higher temperatures may be used that enhance the conversion of methane to hydrogen. Indeed, much of the design and operation of hydrogen plants involves protecting the reforming catalyst and the catalyst tubes because of the extreme temperatures and the sensitivity of the catalyst. In fact, minor variations in feedstock composition or operating conditions can have significant effects on the life of the catalyst or the reformer itself. This is particularly true of changes in molecular weight of the feed gas, or poor distribution of heat to the catalyst tubes. Since the high temperature takes the place of a catalyst, partial oxidation is not limited to the lower-boiling feedstocks that are required for steam reforming. Partial oxidation processes were first considered for hydrogen production because of expected shortages of lower-boiling feedstocks and the need to have available a disposal method for higher-boiling, high-sulfur streams such as asphalt or crude oil coke.

Catalytic partial oxidation, also known as auto-thermal reforming, reacts oxygen with a low-boiling feedstock and by passing the resulting hot mixture over a reforming catalyst. The use of a catalyst allows the use of lower temperatures than in non-catalytic partial oxidation and which causes a reduction in oxygen demand.

The feedstock requirements for catalytic partial oxidation processes are similar to the feedstock requirements for steam reforming and low-boiling hydrocarbon derivatives from refinery gas to naphtha are preferred. The oxygen substitutes for much of the steam in preventing coking and a lower steam/carbon ratio is required. In addition, because a large excess of steam is not required, catalytic partial oxidation produces more carbon monoxide and less hydrogen than steam reforming. Thus, the process is more suited to situations where carbon monoxide is the more desirable product such as, for example, as synthesis gas for chemical feedstocks.

12.3 Hydrogen Purification

When the hydrogen content of the refinery gas is greater than 50% by volume, the gas should first be considered for hydrogen recovery, using a membrane (Brüschke, 1995, 2003) or pressure-swing adsorption unit (Mokhatab *et al.*, 2006; Speight, 2014). The tail gas or reject gas that will still contain a substantial amount of hydrogen can then be used as steam reformer feedstock. Generally, the feedstock purification process uses three different refinery gas streams to produce hydrogen. First, high-pressure hydrocracker purge gas is purified in a membrane unit that produces hydrogen at medium pressure and is combined with medium pressure off-gas that is first purified in a pressure-swing adsorption unit. Finally, low-pressure off-gas is compressed, mixed with reject gases from the membrane and pressure-swing adsorption units, and used as steam reformer feed.

Various processes are available to purify the hydrogen stream but since the product streams are available as a wide variety of composition, flows, and pressures, the best method of purification will vary. And there are several factors that must also be taken into consideration in the selection of a purification method. These are: (i) hydrogen recovery, (ii) product purity, (iii) pressure profile, (iv) reliability, and (v) cost, an equally important parameter is not considered here since the emphasis is on the technical aspects of the purification process.

12.3.1 Wet Scrubbing

A wet scrubber is one of several devices that is used to remove pollutants from a gas stream. In a wet scrubber, the polluted gas stream is brought into contact with the scrubbing liquid, by spraying it with the liquid, by forcing it through a pool of liquid, or by some other contact method, so as to remove the pollutants.

Wet scrubbing systems, particularly amine or potassium carbonate systems, are used for removal of acid gases such as hydrogen sulfide or carbon dioxide (Mokhatab *et al.*, 2006; Speight, 2014, 2019). Most systems depend on chemical reaction and can be designed for a wide range of pressures and capacities. They were once widely used to remove carbon dioxide in steam reforming plants, but have generally been replaced by pressure-swing adsorption units except where carbon monoxide is to be recovered. Wet scrubbing is still used to remove hydrogen sulfide and carbon dioxide in partial oxidation plants.

Wet scrubbing systems remove only acid gases or heavy hydrocarbon derivatives but they do not remove methane or other hydrocarbon gases, hence have little influence on product purity. Therefore, wet scrubbing systems are most often used as a pretreatment step, or where a hydrogen-rich stream is to be desulfurized for use as fuel gas.

12.3.2 Pressure-Swing Adsorption

The pressure-swing adsorption (PSA) technology is a technology that is used to separate some gas species from a mixture of gases under pressure according to the molecular characteristics of the constituents and the relative affinity for an adsorbent. The technology operates at near-ambient temperatures and differs significantly from cryogenic distillation techniques of gas separation. A specific adsorbent (such as a zeolite, activate carbon, or a molecular sieve) is used as a trap, preferentially adsorbing the target gas species at high pressure. The process then swings to low pressure to desorb the adsorbed material.

Put simply, the technology operates by alternating the flow of the gas stream through two columns which are packed with an adsorbent material (in the form of beads) which acts as a molecular sieve. While the hydrogen is being passed through one column, a small bleed taken from the dry gas is passed down the other column. No further adsorption capacity is available; at this point the adsorbent material is forced to regenerate. This action completely regenerates the adsorbent material in the column so that no replacement of the material is required. The vessel is ready for another production cycle after a small amount of the product hydrogen flushes away the waste.

Thus, in the process, the pressure-swing adsorption units use beds of solid adsorbent to separate impurities from hydrogen streams leading to high-purity high-pressure hydrogen and a low-pressure tail gas stream containing the impurities and some of the hydrogen. The beds are then regenerated by depressurizing and purging. Part of the hydrogen (up to 20%) may be lost in the tail gas. Pressure-swing adsorption is generally the purification method of choice for steam reforming units because of its production of high-purity hydrogen and is also used for purification of refinery off-gases, where it competes with membrane systems.

Many hydrogen plants that formerly used a *wet scrubbing* process for hydrogen purification are now using the *pressure-swing adsorption* (PSA) for purification (Mokhatab et al., 2006; Speight, 2014, 2019). The pressure-swing adsorption process is a cyclic process that uses beds of solid adsorbent to remove impurities from the gas and generally produces higher purity hydrogen (99.9% v/v percent purity compared to less than 97% v/v purity). The purified hydrogen passes through the adsorbent beds with only a tiny fraction absorbed and the beds are regenerated by depressurization followed by purging at low pressure.

When the beds are depressurized, a waste gas (or *tail gas*) stream is produced and consists of the impurities from the feed (carbon monoxide, carbon dioxide, methane, and nitrogen) plus some hydrogen. This stream is burned in the reformer as fuel and reformer operating conditions in a pressure-swing adsorption plant are set so that the tail gas provides no more than approximately 85% of the reformer fuel. This gives good burner control because the tail gas is more difficult to burn than regular fuel gas and the high content of carbon monoxide can interfere with the stability of the flame. As the reformer operating temperature is increased, the reforming equilibrium shifts, resulting in more hydrogen and less methane in the reformer outlet and hence less methane in the tail gas.

For hydrogen recovery/purification applications where the gas stream is at low to intermediate pressure (<1000 psi) and where downstream process requirements require minimum pressure reduction and high-purity hydrogen product, pressure-swing adsorption (PSA) technology is often the process of choice.

12.3.3 Membrane Systems

Membrane separation is a technology which selectively separates (fractionates) materials by way of pore size and/or minute gaps in the molecular arrangement of a continuous structure. The membrane acts as a selective barrier allowing relatively free passage of one component while retaining another. In membrane contactors the membrane function is to provide an interface between two phases but not to control the rate of passage of permeants across the membrane. Membrane separations are classified by pore size and by the separation driving force and these classifications are (i) microfiltration (MF), (ii) ultrafiltration (UF), (iii) ion-exchange (IE), and (v) reverse osmosis (RO).

Thus, membrane systems separate gases by taking advantage of the difference in rates of diffusion through membranes (Brüschke, 1995, 2003). Gases that diffuse faster (including hydrogen) become the permeate stream and are available at low pressure whereas the slower-diffusing gases become the non-permeate and leave the unit at a pressure close to the pressure of the feedstock at entry point. Membrane systems contain no moving parts or switch valves and have potentially very high reliability. The major threat is from components in the gas (such as aromatics) that attack the membranes, or from liquids, which plug them.

Membranes are fabricated in relatively small modules; for larger capacity more modules are added. Cost is therefore virtually linear with capacity, making them more competitive at lower capacities. The design of membrane systems involves a tradeoff between pressure drop (and diffusion rate) and surface area, as well as between product purity and recovery. As the surface area is increased, the recovery of fast components increases; however, more of the slow components are recovered, which lowers the purity.

As an example, hydrogen purifiers that use palladium membranes operate via pressure driven diffusion across palladium membranes. Only hydrogen can diffuse through the palladium diffuser. The palladium diffuser can take a number of forms, including an array of tubes, a coiled tube or membrane foil. The membrane actually is made up of a palladium and silver alloy material possessing the unique property of only allowing monatomic hydrogen to pass through its crystal lattice when it is heated above nominally 300°C (570°F). The hydrogen gas molecule coming into contact with the palladium membrane surface dissociates into monatomic hydrogen and passes through the membrane. On the other surface of the palladium membrane, the monatomic hydrogen is recombined into diatomic hydrogen.

12.3.4 Cryogenic Separation

Cryogenic separation units operate by cooling the gas and condensing some, or all, of the constituents for the gas stream. Depending on the product purity required, separation may involve flashing or distillation. Cryogenic units offer the advantage of being able to separate a variety of products from a single feed stream. One specific example is the separation of low-boiling olefins from a hydrogen stream. Hydrogen recovery is in the range of 95%, with purity above 98% obtainable.

Cryogenic separation is suitable currently for separating carbon dioxide from a gas stream with high concentration of carbon dioxide (usually more than 50% v/v). This technique may not be used for the removal of carbon dioxide from streams such as exhaust gas from coal-based or natural gas-fired plants because the amount of carbon dioxide in these streams is in relatively small quantities, and the required energy to bring the stream to the subzero temperatures of cryogenic application would be uneconomical for the entire process.

The gas mixture is compressed and cooled in several stages to bring about phase changes of carbon dioxide in exhaust gases and ultimately other constituents of the gas mixture. Subject to the operating parameters, the carbon dioxide can come out as a liquid or solid together with other components from which it can be recovered. The major benefit of the cryogenic capture of carbon dioxide is the fact that there is no chemical sorbent required and the process can be carried out at atmospheric pressure. It also reduces the cost of compression because the carbon dioxide produced in this way can come out as a liquid which is the phase needed to transport the carbon dioxide by way of a pipeline.

Caution is advised insofar as the shipper should make sure that water should be absent from the gas stream to the cooling units. If water is present, ice can form thereby causing plugging and bring about an increase in pressure drop. As a result, a number of steps will be needed to remove all traces of water from the gas stream before cryogenic operations.

A combination of processes (a hybrid process) comprising a membrane unit with a cryogenic process has been found of value for producing hydrogen from a rather dilute source such as the off-gas from a fluid catalyst cracking unity. The success of the hybrid stems from the use of each component part of the process in its region of greatest efficiency. The judicious combination of processes leads to an overall process that is more efficient than either of the component processes (Agrawal *et al.*, 1988).

Another option that can be combined with one if the above processes is the desiccant option. If water remains in the hydrogen stream, it can be removed by use of silica (SiO_2) desiccant columns. In the process, the hydrogen flows through a stainless steel desiccant cartridge for moisture removal. The desiccant column is most commonly made up of silica gel beads which act as a drying agent in hydrogen to produce high-purity hydrogen, meeting industry purity requirements.

12.4 Hydrogen Management

Typically, considerable variation exists from one refinery to another in the balance between hydrogen produced and hydrogen consumed in the refining operations. However, what is more pertinent to the present chapter is the amounts of hydrogen that are required for hydroprocessing operations, whether these be hydrocracking or the somewhat milder hydrotreating processes. For effective hydroprocessing, a substantial hydrogen partial pressure must be maintained in the reactor and, in order to meet this requirement, an excess of hydrogen above that actually consumed by the process must be fed to the reactor. Part of the hydrogen requirement is met by recycling a stream of hydrogen-rich gas. However, the need still remains to generate hydrogen as makeup material to accommodate the process consumption of 500 to 3000 scf/bbl depending upon whether the heavy feedstock is being subjected to a predominantly hydrotreating (hydrodesulfurization) or to a predominantly hydrocracking process.

Many existing refinery hydrogen plants use a conventional process, which produces a medium-purity (94% to 97%) hydrogen product by removing the carbon dioxide in an absorption system and methanation of any remaining carbon oxides. Since the 1980s, most hydrogen plants are built with pressure-swing adsorption (PSA) technology to recover and purify the hydrogen to purities above 99.9%. Since many refinery hydrogen plants utilize refinery off-gas feeds containing hydrogen, the actual maximum hydrogen capacity that can be synthesized via steam reforming is not certain since the hydrogen content of the off-gas can change due to operational changes in the hydrotreaters.

Hydrogen management has become a priority in current refinery operations and when planning to produce lower sulfur gasoline and diesel fuels as well as a component for the increasingly important synthesis gas (Zagoria *et al.*, 2003; Luckwal and Mandal, 2009; Speight, 2014, 2016). Along with increased hydrogen consumption for deeper hydrotreating, additional hydrogen is needed for processing heavier and higher sulfur crude slates. In many refineries, hydroprocessing capacity and the associated hydrogen network is limiting refinery throughput and operating margins. Furthermore, higher hydrogen purities within the refinery network are becoming more important to boost hydrotreater capacity, achieve product value improvements and lengthen catalyst life cycles.

Improved hydrogen utilization and expanded or new sources for refinery hydrogen and hydrogen purity optimization are now required to meet the needs of the future transportation fuel market and the drive towards higher refinery profitability. Many refineries developing hydrogen management programs fit into the two general categories of either a catalytic reformer supplied network or an on-purpose hydrogen supply.

Some refineries depend solely on catalytic reformer(s) as their source of hydrogen for hydrotreating. Often, they are semi-regenerative reformers where off-gas hydrogen quantity, purity, and availability change with feed naphtha quality, as octane requirements change seasonally, and when the reformer catalyst progresses from start-of-run to end-of-run conditions and then goes offline for regeneration. Typically, during some portions of the year, refinery margins are reduced as a result of hydrogen shortages.

Multiple hydrotreating units compete for hydrogen – either by selectively reducing throughput, managing intermediate tankage logistics, or running the catalytic reformer sub-optimally just to satisfy downstream hydrogen requirements. Part of the operating year still runs in hydrogen surplus, and the network may be operated with relatively low hydrogen utilization (consumption/production) at 70 to 80%. Catalytic reformer off-gas hydrogen supply may swing from 75 to 85% hydrogen purity. Hydrogen purity upgrade can be achieved through some hydrotreaters by absorbing heavy hydrocarbon derivatives. But without supplemental hydrogen purification, critical control of hydrogen partial pressure in hydroprocessing reactors is difficult, which can affect catalyst life, charge rates, and/or gasoline yields.

More complex refineries, especially those with a high demand for hydrogen, also have on-purpose hydrogen production, typically with a steam methane reformer steam methane reformer that utilizes refinery off gas and supplemental natural gas as feedstock. The steam methane reformer plant provides the swing hydrogen requirements at higher purities (92% to more than 99% hydrogen) and serves a hydrogen network configured with several purity and pressure levels. Multiple purities and existing purification units allow for more optimized hydroprocessing operation by controlling hydrogen partial pressure for maximum benefit as well as a pure stream of hydrogen that can be used to maintain the necessary balance for use in synthesis gas.

References

Agrawal, R., Auvil, S.R., DiMartino, P.D., Choe, J.S., and Hopkins, J.A. 1988. Membrane/Cryogenic Hybrid Processes for Hydrogen Purification. *Gas Separation & Purification*, 2(1): 9-15.

Ancheyta, J., and Speight, J.G. 2007. *Hydroprocessing of Heavy Oils and Residua*. CRC Press, Taylor & Francis Group, Boca Raton, Florida.

Bandermann, F., Harder, K.B. 1982. Production of Hydrogen via Thermal Decomposition of Hydrogen Sulfide and. Separation of Hydrogen and Hydrogen Sulfide by Pressure Swing Adsorption. *Int. J. Hydrogen Energy*. 7(6): 471-475.

Bezler, J. 2003. Optimized Hydrogen Production – A Key Process Becoming Increasingly Important in Refineries. *Proceedings. DGMK Conference on Innovation in the Manufacture and Use of Hydrogen. Dresden, Germany. October 15-17*. Page 65.

Bleeker, M.F., Kersten, S.R.S., and Veringa, H.J. 2007. Pure Hydrogen from Pyrolysis Oil Using the Steam-Iron Process. *Catalysis Today*, 127(1-4): 278-290.

Bleeker, M.F. 2009. *Pure Hydrogen from Pyrolysis Oil by the Steam-Iron Process*. Ipskamp Drukkers B.V., Enschede, Netherlands.

Breault, R.W. 2010. Gasification Processes Old and New: A Basic Review of the Major Technologies. *Energies*, 3(2), 216-240.

Brüschke, H. 1995. Industrial Application of Membrane Separation Processes. *Pure Appl. Chem.*, 345 67(6): 993-1002.

Brüschke, H. 2003. Separation of Hydrogen from Dilute Streams (e.g. Using Membranes). *Proceedings. DGMK Conference on Innovation in the Manufacture and Use of Hydrogen. Dresden, Germany. October 15-17*. Page 47.

Choudhary, T.V., Sivadinarayana, C., and Goodman, D.W. 2003. Production of CO_x-free Hydrogen for Fuel Cells via Step-wise Hydrocarbon Reforming and Catalytic Dehydrogenation of Ammonia. *Chem. Eng. J.*, 2003, 93: 69-80.

Choudhary, T.V., and Goodman, D.W. 2006. Methane Decomposition: Production of Hydrogen and Carbon Filaments. *Catalysis*, 19: 164-183.

Clark, P.D., Wassink B. 1990. A Review of Methods for the Conversion of Hydrogen Sulfide to Sulfur and Hydrogen. *Alberta Sulfur Res. Quart. Bull.*, 26(2/3/4): 1.

Clark, P.D., N.I. Dowling, J.B. Hyne, D.L. Moon. 1995. Production of Hydrogen and Sulfur from Hydrogen Sulfide in Refineries and Gas Processing Plants. *Quarterly Bulletin* (Alberta Sulfur Research Ltd.), 32(1): 11-28.

Cruz, F.E., and De Oliveira, S. 2008. Petroleum Refinery Hydrogen Production Unit: Exergy and Production Cost Evaluation. *Int. J. of Thermodynamics*, 11(4): 187-193.

Dalrymple, D.A., Trofe, T.W., and Leppin, D. Gas Industry Assesses New Ways to Remove Small Amounts of Hydrogen Sulfide. *Oil & Gas Journal*, May 23.

Dickenson, R.L., Biasca, F.E., Schulman, B.L., and Johnson, H.E. 1997. Refiner Options for Converting and Utilizing Heavy Fuel Oil. *Hydrocarbon Processing*. 76(2): 57.

Dolbear, G.E. 1998. *Hydrocracking: Reactions, Catalysts, and Processes. In Petroleum Chemistry and Refining*. J.G. Speight (Editor). Taylor & Francis, Washington, DC. Chapter 7.

Donini, J. C. 1996. Separation and Processing of Hydrogen Sulfide in the Fossil Fuel Industry. *Minimum Effluent Mills Symposium*. 357-363.

Dragomir, R., Drnevich, R.F., Morrow, J., Papavassiliou, V., Panuccio, G., and Watwe, R. 2010. Technologies for Enhancing Refinery Gas Value. *Proceedings. AIChE 2010 SPRING Meeting. San Antonio, Texas. November 7-12.*

Gary, J.G., Handwerk, G.E., and Kaiser, M.J. 2007. *Petroleum Refining: Technology and Economics*, 5th Edition. CRC Press, Taylor & Francis Group, Boca Raton, Florida.

Gross, M., and Wolff, J. 2000. Gasification of Residue as a Source of Hydrogen for the Refining Industry in India. *Proceedings. Gasification Technologies Conference. San Francisco, California. October 8-11.*

Hsu, C.S., and Robinson, P.R. (Editors). 2017. *Handbook of Petroleum Technology*. Springer International Publishing AG, Cham, Switzerland.

Kotera, Y., Todo, N., and Fukuda, K. 1976. Process for Production of Hydrogen and Sulfur from Hydrogen Sulfide as Raw Material. U.S. Patent No. 3,962,409. June 8.

Lipman, T. 2011. *An Overview of Hydrogen Production and Storage Systems with Renewable Hydrogen Case Studies*. A Clean Energy States Alliance Report. Conducted under US DOE Grant DE-FC3608GO18111, Office of Energy Efficiency and Renewable Energy Fuel Cell Technologies Program, United States Department of Energy, Washington, DC.

Liu, K., Song, C., and Subramani, V. (Editors) 2010. *Hydrogen and Syngas Production and Purification Technologies*. John Wiley & Sons Inc. Hoboken, New Jersey.

Luckwal, K., and Mandal, K.K. 2009. Improve Hydrogen Management of Your Refinery. *Hydrocarbon Processing*, 88(2): 55-61.

Luinstra, E. 1996. Hydrogen from Hydrogen Sulfide – A Review of the Leading Processes. *Proceedings. 7th Sulfur Recovery Conference. Gas Research Institute, Chicago.* Page 149-165.

Miller G.Q., and Penner, D.W. 2003. Meeting Future Needs for Hydrogen – Possibilities and Challenges. *Proceedings. DGMK Conference on Innovation in the Manufacture and Use of Hydrogen. Dresden, Germany. October 15-17.* Page 7.

Mokhatab, S., Poe, W.A., and Speight, J.G. 2006. *Handbook of Natural Gas Transmission and Processing*. Elsevier, Amsterdam, Netherlands.

Nagaoka K., Eboshi T., Takeishi Y., Tasaki R., Honda, K., Imamura K., and Sato K. 2017. Carbon-free H_2 Production from Ammonia Triggered at Room Temperature With an Acidic RuO_2/γ-Al_2O_3 Catalyst. *Science Advances*, **3(4):** e1602747.

Palo, D.R., Dagle, R.A., and Holladay, J.D. 2007. Methanol Steam Reforming for Hydrogen Production. *Chemical Reviews*, 107: 3992-4021.

Parkash, S. 2003. *Refining Processes Handbook*. Gulf Professional Publishing, Elsevier, Amsterdam, Netherlands.

Raissi, A.T. 2001. Technoeconomic Analysis of Area II Hydrogen Production. Part 1. *Proceedings. US DOE Hydrogen Program Review Meeting, Baltimore, Maryland.*

Ranke, H., and Schödel, N 2003. Hydrogen Production Technology – Status and New Developments. *Proceedings. DGMK Conference on Innovation in the Manufacture and Use of Hydrogen. Dresden, Germany. October 15-17.* Page 19.

Scherzer, J., and Gruia, A.J. *Hydrocracking Science and Technology.* Marcel Dekker Inc., New York.

Speight, J.G. 2000. *The Desulfurization of Heavy Oils and Residua.* 2nd Edition. Marcel Dekker Inc., New York.

Speight, J.G. (Editor). 2011a. *The Biofuels Handbook.* Royal Society of Chemistry, London, United Kingdom.

Speight, J.G. 2011b. *The Refinery of the Future.* Gulf Professional Publishing, Elsevier, Oxford, United Kingdom.

Speight, J.G. 2014. *The Chemistry and Technology of Petroleum* 5th Edition. CRC Press, Taylor & Francis Group, Boca Raton, Florida.

Speight, J.G. 2016. Hydrogen in Refineries. In: *Hydrogen Science and Engineering: Materials, Processes, Systems, and Technology.* D. Stolten and B. Emonts (Editors). Wiley-VCH Verlag GmbH & Co., Weinheim, Germany. Chapter 1. Page 3-18.

Speight, J.G. 2017. *Handbook of Petroleum Refining.* CRC Press, Taylor and Francis Group, Boca Raton, Florida.

Speight, J.G. 2019. *Natural Gas: A Basic Handbook* 2nd Edition. Gulf Publishing Company, Elsevier, Cambridge, Massachusetts.

Uemura, Y., Ohe, H., Ohzuno, Y., and Hatate, Y. 1999. Carbon and Hydrogen from Hydrocarbon Pyrolysis. *Proceedings. Int. Conf. Solid Waste Technology Management.* 15: 5E/25-5E/30.

Vauk, D., Di Zanno, P., Neri, B., Allevi, C., Visconti, A., and Rosanio, L. 2008. What Are Possible Hydrogen Sources for Refinery Expansion? *Hydrocarbon Processing,* 87(2): 69-76.

Weimer, A.W., Dahl, J., Tamburini, J., Lewandowski, A., Pitts, R., Bingham, C., and Glatzmaier, G.C. 2000. Thermal Dissociation of Methane Using a Solar Coupled Aerosol Flow Reactor. *Proceedings. Hydrogen Program Review, NREL/CP-570-28890.*

Zagoria, A., Huychke, R., and Boulter, P. H. 2003. Refinery Hydrogen Management – The Big Picture. *Proceedings. DGMK Conference on Innovation in the Manufacture and Use of Hydrogen. Dresden, Germany. October 15-17.* Page 95.

Zaman, J., and Chakma. 1995. Production of Hydrogen and Sulfur from Hydrogen Sulfide. *Fuel Processing Technology.* 41: 159-198.

13
Chemicals from Synthesis Gas

13.1 Introduction

The constant demand for hydrocarbon products such as liquid fuels is one of the major driving forces behind the crude oil industry. However, the other driving force is the production of a major group of hydrocarbon products – petrochemicals – that are the basis of the chemicals industry. In fact, there is a myriad of products that have evolved through the short life of the crude oil industry, either as bulk fractions or as single hydrocarbon products (Table 13.1, Table 13.2). And the complexities of product composition have matched the evolution of the products. In fact, it is the complexity of product composition that has served the industry well and, at the same time, had an adverse effect on product use.

A *petrochemical* is a chemical product developed from crude oil that has become an essential part of the modern chemical industry (Table 13.3) (Speight, 1987; Parkash, 2003; Gary *et al.*, 2007; Speight, 2014; Hsu and Robinson, 2017; Speight, 2017, 2019). The *chemical industry* is, in fact, the *chemical process industry* by which a variety of chemicals are manufactured. The chemical process industry is, in fact, sub-divided into other categories: (i) the chemicals and allied products industries in which chemicals are manufactured from a variety of feedstocks and may then be put to further use, (ii) the rubber and miscellaneous products industries which focus on the manufacture of rubber and plastic materials, and (iii) crude oil refining and related industries which, on the basis of prior chapters in this text, is now self-explanatory. Thus, the petrochemical industry falls under the sub-category of *crude oil and related industries*.

In the context of this book, the definition of petrochemicals excludes fuel products, lubricants, asphalt, and crude oil coke but does include chemicals produced from other feedstocks such as coal, oil shale, and biomass which could well be the sources of chemicals in the future. Thus, petrochemicals are, in the strictest sense, different from crude oil products insofar as the petrochemicals are the basic building blocks of the chemical industry. Petrochemicals are found in products as diverse as plastics, polymers, synthetic rubber, synthetic fibers, detergents, industrial chemicals, and fertilizers (Table 13.3). Petrochemicals are used for production of several feedstocks and monomers and monomer precursors. The monomers after polymerization process creates several polymers which ultimately are used to produce gels, lubricants, elastomers, plastics and fibers.

By way of definition and clarification as it applies to the petrochemical and chemical industry, primary raw materials are naturally occurring substances that have not been subjected to chemical changes after being recovered. Currently, through a variety of intermediates natural gas and crude oil are the main sources of the raw materials because they are the least expensive, most readily available, and can be processed most easily into the primary petrochemicals. An aromatic petrochemical is also an organic chemical compound but one that contains, or is derived from, the basic benzene ring system.

Table 13.1 The various distillation fractions of crude oil.

Product	Lower carbon number*	Upper carbon number*	Lower b.p. °C*	Upper b.p. °C*	Lower b.p. °F*	Upper b.p. °F*
Liquefied petroleum gas	C3	C4	-42	-1	-44	31
Naphtha	C5	C17	36	302	97	575
Kerosene	C8	C18	126	258	302	575
Light gas oil	C12	>C20	216	421	>345	>650
Heavy gas oil	>C20		>345		>650	
Residuum	>C20		>345		>660	

*The carbon number and boiling point are difficult to assess accurately because of variations in production parameters from refinery-to-refinery and are inserted for illustrative purposes only

Table 13.2 Properties of hydrocarbon products from crude oil (excluding liquid fuels).

Product	Molecular weight	Specific gravity	Boiling point, °F	Ignition temperature, °F	Flash, point, °F	Flammability limits in air, % v/v
Benzene	78.1	0.879	176.2	1040	12	1.35-6.65
n-Butane	58.1	0.601	31.1	761	-76	1.86-8.41
iso-Butane	58.1		10.9	864	-117	1.80-8.44
n-Butene	56.1	0.595	21.2	829	Gas	1.98-9.65
iso-Butene	56.1		19.6	869	Gas	1.8-9.0
Ethane	30.1	0.572	-127.5	959	Gas	3.0-12.5
Ethylene	28.0		-154.7	914	Gas	2.8-28.6
n-Hexane	86.2	0.659	155.7	437	-7	1.25-7.0
n-Heptane	100.2	0.668	419.0	419	25	1.00-6.00
Kerosene	154.0	0.800	304-574	410	100-162	0.7-5.0
Methane	16.0	0.553	-258.7	900-1170	Gas	5.0-15.0
Naphthalene	128.2		424.4	959	174	0.90-5.90
Neohexane	86.2	0.649	121.5	797	-54	1.19-7.58
Neopentane	72.1		49.1	841	Gas	1.38-7.11
n-Octane	114.2	0.707	258.3	428	56	0.95-3.2
iso-Octane	114.2	0.702	243.9	837	10	0.79-5.94

(*Continued*)

Table 13.2 Properties of hydrocarbon products from crude oil (excluding liquid fuels). (*Continued*)

Product	Molecular weight	Specific gravity	Boiling point, °F	Ignition temperature, °F	Flash, point, °F	Flammability limits in air, % v/v
n-Pentane	72.1	0.626	97.0	500	-40	1.40-7.80
iso-Pentane	72.1	0.621	82.2	788	-60	1.31-9.16
n-Pentene	70.1	0.641	86.0	569	–	1.65-7.70
Propane	44.1		-43.8	842	Gas	2.1-10.1
Propylene	42.1		-53.9	856	Gas	2.00-11.1
Toluene	92.1	0.867	321.1	992	40	1.27-6.75
Xylene	106.2	0.861	281.1	867	63	1.00-6.00

Table 13.3 Examples of products from the petrochemical industry.

Group	Areas of use
Plastics and polymers	Agricultural water management
	Packaging
	Automobiles
	Telecommunications
	Health and hygiene
	Transportation
Synthetic rubber	Transportation industry
	Electronics
	Adhesives
	Sealants
	Coatings
Synthetic fibers	Textile
	Transportation
	Industrial fabrics
Detergents	Health and hygiene
Industrial chemicals	Pharmaceuticals
	Pesticides
	Explosives
	Surface coating
	Dyes
	Lubricating oil additives
	Adhesives
	Oil field chemicals
	Antioxidants
	Printing ink
	Paints
	Corrosion inhibitors
	Solvents

(*Continued*)

Table 13.3 Examples of products from the petrochemical industry. (*Continued*)

Group	Areas of use
	Perfumes
	Food additives
Fertilizers	Agriculture

Primary petrochemicals include: (i) olefin derivatives such as ethylene, propylene, and butadiene, (ii) aromatic derivatives such as benzene, toluene, and the isomers of xylene, and (iii) methanol. However, although crude oil contains different types of hydrocarbon derivatives, not all hydrocarbon derivatives are used in producing petrochemicals. Petrochemical analysis has made it possible to identify some major hydrocarbon derivatives used in producing petrochemicals (Speight, 2015).

From the multitude of hydrocarbon derivatives, those hydrocarbon derivatives serving as major raw materials used by petrochemical industries in the production of petrochemicals are: (i) the raw materials obtained from natural gas processing such as methane, ethane, propane and butane, (ii) the raw materials obtained from crude oil refineries such as naphtha and gas oil, and (iii) the raw materials such as benzene, toluene and the xylene isomers (BTX) obtained when extracted from reformate (the product of reforming processes through catalysts called catalytic reformers in crude oil refineries (Parkash, 2003; Gary *et al.*, 2007; Speight, 2008, 2014; Hsu and Robinson, 2017; Speight, 2017, 2019). Thus, petrochemicals are chemicals derived from natural gas and crude oil and, for convenience of identification, petrochemicals can be divided into two groups: (i) primary petrochemicals and (ii) intermediates and derivatives (Figure 13.1) (Speight, 2014, 2019).

Primary petrochemicals include olefins (ethylene, propylene and butadiene) aromatics (benzene, toluene, and xylenes), and methanol. Petrochemical intermediates are generally

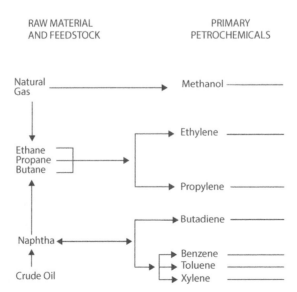

Figure 13.1 Raw materials and primary petrochemicals.

produced by chemical conversion of primary petrochemicals to form more complicated derivative products. Petrochemicals can be made in a variety of ways: (i) directly from primary petrochemicals and (ii) through intermediate products which still contain only carbon and hydrogen; and, through intermediates which incorporate chlorine, nitrogen or oxygen in the finished derivative. In some cases, they are finished products; in others, more steps are needed to arrive at the desired composition.

Moreover, petrochemical feedstocks can be classified into several general groups: olefins, aromatics, and methanol; a fourth group includes inorganic compounds and synthesis gas (mixtures of carbon monoxide and hydrogen). In many instances, a specific chemical included among the petrochemicals may also be obtained from other sources, such as coal, coke, or vegetable products. For example, materials such as benzene and naphthalene can be made from either crude oil or coal, while ethyl alcohol may be of petrochemical or vegetable origin.

Thus, primary petrochemicals are not end products but are the chemical building blocks for a wide range of chemical and manufactured materials. For example, *petrochemical intermediates* are generally produced by chemical conversion of primary petrochemicals to form more complicated derivative products (Parkash, 2003; Gary *et al.*, 2007; Speight, 2008, 2014; Hsu and Robinson, 2017; Speight, 2017, 2019). Petrochemical derivative products can be made in a variety of ways: (i) directly from primary petrochemicals, (ii) through intermediate products which still contain only carbon and hydrogen, (iii) and, through intermediates which incorporate chlorine, nitrogen or oxygen in the finished derivative. In some cases, they are finished products; in others, more steps are needed to arrive at the desired composition. Some typical petrochemical intermediates are: (i) vinyl acetate, $CH_3CO_2CH=CH_2$, for paint, paper and textile coatings, (ii) vinyl chloride, $CH_2=CHCl$, for polyvinyl chloride, PVC, (iii) ethylene glycol, $HOCH_2CH_2OH$, for polyester textile fibers, and (iv) styrene, $C_6H_5CH=CH_2$, which is important in rubber and plastic manufacturing. Of all the processes used, one of the most important is polymerization. It is used in the production of plastics, fibers and synthetic rubber, the main finished petrochemical derivatives.

Following from this, secondary raw materials, or intermediate chemicals, are obtained from a primary raw material through a variety of different processing schemes. The intermediate chemicals may be low-boiling hydrocarbon compounds such as methane, ethane, propane, and butane or higher-boiling hydrocarbon mixtures such as naphtha, kerosene, or gas oil. In the latter cases (naphtha, kerosene, and gas oil), these fractions are used (in addition to the production of fuels) as feedstocks for cracking processes to produce a variety of petrochemical products (e.g., ethylene, propylene, benzene, toluene, and the xylene isomers) which are identified by the relative placement of the two methyl groups on the aromatic ring:

1-2-dimethylbenzene
(*ortho*-xylene)

1-3-dimethylbenzene
(*meta*-xylene)

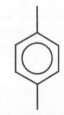
1-4-dimethylbenzene
(*para*-xylene)

Also, by way of definition, petrochemistry is a branch of chemistry in which the transformation of natural gas and crude oil into useful products or feedstock for other processes is studied. A petrochemical plant is a plant that uses chemicals from crude oil as a raw material (the feedstock); such plants are usually located adjacent to (or within the precinct of) a crude oil refinery in order to minimize the need for transportation of the feedstocks produced by the refinery (Figure 13.2). On the other hand, specialty chemical plants and fine chemical plants are usually much smaller than petrochemical plants and are not as sensitive to location.

Furthermore, a paraffinic petrochemical is an organic chemical compound but one that does not contain any ring systems such as a cycloalkane (naphthene) ring or an aromatic ring. A naphthenic petrochemical is an organic chemical compound that contains one or more cycloalkane ring systems. An aromatic petrochemical is also an organic chemical compound but one that contains, or is derived from, the basic benzene ring system.

Crude oil products (in contrast to *petrochemicals*) are those hydrocarbon fractions that are derived from crude oil and have commercial value as a bulk product (Table 13.1, Table 13.2) (Parkash, 2003; Gary *et al.*, 2007; Speight, 2014; Hsu and Robinson, 2017; Speight, 2017, 2019). These products are generally not accounted for in petrochemical production or use statistics. Thus, in the context of this definition of petrochemicals, this book focuses on chemicals that are produced from crude oil as distinct from *crude oil products* which are organic compounds (typically hydrocarbon compounds) that are burned as a fuel. In the strictest sense of the definition, a petrochemical is any chemical that is manufactured from natural gas and crude oil as distinct from fuels and other products, derived from natural gas and crude oil by a variety

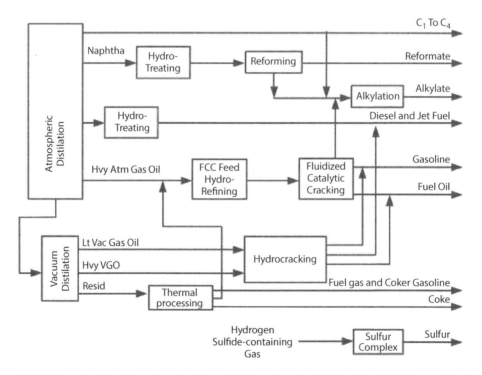

Figure 13.2 Schematic of a refinery showing the production of products during the distillation and during thermal processing (e.g., visbreaking, coking, and catalytic cracking).

of processes and used for a variety of commercial purposes (Chenier, 2002; Meyers 2005; Naderpour, 2008; Speight, 2014). Petrochemical products include such items as plastics, soaps and detergents, solvents, drugs, fertilizers, pesticides, explosives, synthetic fibers and rubbers, paints, epoxy resins, and flooring and insulating materials.

Moreover, the classification of materials as petrochemicals is used to indicate the source of the chemical compounds, but it should be remembered that many common petrochemicals can be made from other sources, and the terminology is therefore a matter of source identification. However, in the setting of modern industry, the term petrochemicals is often used in an expanded form to include chemicals produced from other fossil fuels such as coal or natural gas, oil shale, and renewable sources such as corn or sugar cane as well as other forms of biomass (Chapter 1, Chapter 6). It is in the expanded form of the definition that the term petrochemicals is used in this book.

In fact, in the early days of the chemical industry, coal was the major source of chemicals (it was not then called the *petrochemical industry*) and it was only after the discovery of crude oil and the recognition that crude oil could produce a variety of products other than fuels that the petrochemical industry came into being (Spitz, 1988; Speight, 2013, 2014, 2017, 2019). For several decades both coal and crude oil served as the primary raw materials for the manufacture of chemicals. Then during World War II, crude oil began to outpace coal as a source of chemicals – the exception being the manufacture of synthetic fuels from coal because of the lack of access to crude oil by the wartime German industry.

The *chemical industry*, of which a petrochemical plant is a part, is in fact the *chemical process industry* by which a variety of chemicals are manufactured. Thus, the petrochemical industry falls under the sub-category of *crude oil and related industries*.

To complete this series of definitions and to reduce the potential for any confusion that might occur later in this text, specialty chemicals (also called *specialties* or *effect chemicals*) are particular chemical products which provide a wide variety of effects on which many other industry sectors rely. Specialty chemicals are materials used on the basis of their performance or function. Consequently, in addition to *effect chemicals* they are sometimes referred to as *performance chemicals* or *formulation chemicals*. The physical and chemical characteristics of the single molecules or the formulated mixtures of molecules and the composition of the mixtures influences the performance end product.

On the other hand, the term *fine chemicals* is used in distinction to *heavy chemicals*, which are produced and handled in large lots and are often in a crude state. Since their inception in the late 1970s, fine chemicals have become an important part of the chemical industry. Fine chemicals are typically single, but often complex pure chemical substances, produced in limited quantities in multipurpose plants by multistep batch chemical or biotechnological processes and are described by specifications to which the chemical producers must strictly adhere. Fine chemicals are used as starting materials for specialty chemicals, particularly pharmaceutical chemicals, biopharmaceutical chemicals, and agricultural chemicals.

To return to the subject of petrochemicals, a crude oil refinery converts raw crude oil into useful products such as liquefied petroleum gas (LPG), naphtha (from which gasoline is manufactured), kerosene from which diesel fuel is manufactured, and a variety of gas oil fractions – of particular interest is the production of naphtha that serves a feedstock for several process that produce petrochemical feedstocks (Table 13.4). However, each refinery has its own specific arrangement and combination of refining processes largely determined by the market demand (Parkash, 2003; Gary *et al.*, 2007; Speight, 2014; Hsu and

Table 13.4 Naphtha production.

Primary process	Primary product	Secondary process	Secondary product
Atmospheric distillation	naphtha		light naphtha
			heavy naphtha
	gas oil	catalytic cracking	naphtha
	gas oil	hydrocracking	naphtha
Vacuum distillation	gas oil	catalytic cracking	naphtha
		hydrocracking	naphtha
	residuum	coking	naphtha
		hydrocracking	naphtha

Robinson, 2017; Speight, 2017, 2019). The most common petrochemical precursors are various hydrocarbon derivatives, olefin derivatives, aromatic derivatives (including benzene, toluene and xylene isomers), and synthesis gas (also called syngas), which is a mixture of carbon monoxide and hydrogen.

A typical crude oil refinery produces a variety of hydrocarbon derivatives, olefin derivatives, aromatic derivatives by processes such as coking and of fluid catalytic cracking of various feedstocks. Chemical plants produce olefin derivatives by steam cracking natural gas liquids, such as ethane (CH_3CH_3) and propane (CH_3CH_2CH3) to produce ethylene ($CH_2=CH_2$) and propylene ($CH_3CH=CH_2$), respectively. A steam cracking unit is, in theory, one of the simplest operations is a refinery – essentially a hot reactor into which steam and the feedstock are introduced, but in reality the steam cracking is one of the most technically complex and energy-intensive plants in the refining industry and in the petrochemical industry (Parkash, 2003; Gary et al., 2007; Speight, 2014; Hsu and Robinson, 2017; Speight, 2017, 2019). The equipment typically operates over the range 175 to 1125°C (345 to 2055°F) and from a near vacuum, i.e., <14.7 psi to high pressure (1500 psi). Whilst the fundamentals of the process have not changed in recent decades, improvements continue to be made to the energy efficiency of the furnace, ensuring that the cost of production is continually reduced.

In more general terms, steam cracking units use a variety of feedstocks, for example (i) ethane, propane and butane from natural gas, (ii) naphtha, a mixture of C_5 to C_8 or C_5 to C_{10} hydrocarbon derivatives, from the distillation of crude oil, (iii) gas oil, and (iv) resids – also called residue or residua – from the primary distillation of crude oil. In the process, cracking, a gaseous or liquid hydrocarbon feedstock is diluted with steam and then briefly heated in a furnace, obviously without the presence of oxygen (Parkash, 2003; Gary et al., 2007; Speight, 2014; Hsu and Robinson, 2017; Speight, 2017, 2019). Typically, the reaction temperature is high (up to 1125°C, 2055°F) but the reaction is only allowed to take place very briefly (short residence time). The residence time is even reduced to milliseconds (resulting in gas velocities reaching speeds beyond the speed of sound) in order to improve the yield of desired products. After the cracking temperature has been reached, the gas is quickly quenched to stop the reaction in a transfer line exchanger. The product-type and product yield produced in the cracking unit depend on (i) the composition of the feed, (ii) the hydrocarbon to steam ratio, (iii) the cracking temperature, and (iv) the residence time of the feedstock in the hot zone.

The advantages of steam cracking are that the process reduces the need for repeated product distillation and that a wider range of products are produced. However, the disadvantage is that the process may not produce the product that is needed in high enough yield. In fact, aromatic derivatives, such as benzene (C_6H_6), toluene ($C_6H_5CH_3$) and the xylenes (ortho-, meta, and para-isomers, $H_3CC_6H_4CH_3$) are produced by reforming naphtha, which is a low-boiling liquid product obtained by distillation from crude oil (Table 13.1, Table 13.4) (Parkash, 2003; Gary et al., 2007; Speight, 2014; Hsu and Robinson, 2017; Speight, 2017, 2019). With higher molecular weight (higher-boiling) feedstocks, such as gas oil, it is important to ensure that the feedstock does not crack to form carbon, which is normally formed at this temperature. This is avoided by passing the gaseous feedstock very quickly and at very low pressure through the pipes which run through the furnace.

On the basis of chemical structure, petrochemicals are categorized into a variety of petrochemical products, which are named according to the chemical character of the constituents:

- Paraffin derivatives: such as methane (CH_4), ethane (C_2H_6), propane (C_3H_8), the butane isomers (C_4H_{10}), and higher molecular weight hydrocarbon derivatives up to and including low-boiling mixtures such as naphtha.
- Olefin derivatives: such as ethylene ($CH_2=CH_2$) and propylene ($CH_3CH=CH_2$) which are important sources of industrial chemicals and plastics; the diolefin derivative butadiene ($CH_2=CHCH=CH_2$) is used in making synthetic rubber.
- Aromatic derivatives: such as benzene (C6H6), toluene (C6H5CH3), and the xylene xylenes isomers ($CH_3C_6H_4CH_3$) which are identified by the relative placement of the two methyl groups on the aromatic ring (1,2-$CH_3C_6H_4CH_3$, 1,3-$CH_3C_6H_4CH_3$, and 1,4-$CH_3C_6H_4CH_3$) and which have a variety of uses – benzene is a raw material for dyes and synthetic detergents, and benzene and toluene for isocyanates while the xylene isomers are used in the manufacture of plastics and synthetic fibers.
- Synthesis gas: a mixture of carbon monoxide (CO) and hydrogen (H_2) that is sent to a Fischer-Tropsch reactor to produce naphtha-range and kerosene-range hydrocarbon derivative as well as methanol (CH_3OH) and dimethyl ether (CH_3OCH_3).

Ethylene and propylene, the major part of olefins are the basic source in preparation of several industrial chemicals and plastic products whereas butadiene is used to prepare synthetic rubber. Benzene, toluene and the xylene isomers are major components of aromatic chemicals. These aromatic petrochemicals are used in manufacturing of secondary products like synthetic detergents, polyurethanes, plastic and synthetic fibers. Synthesis gas comprises carbon monoxide and hydrogen which basically is used to produce ammonia and methanol which are further used to produce other chemical and synthetic substances.

Low-boiling olefins (light olefins) – ethylene and propylene – are the most important intermediates in the production of plastics and other chemical products. The current end use of ethylene worldwide is for (i) the manufacture of polyethylene, which is used plastics, (ii) the manufacture of ethylene oxide/glycol, which is used in fibers and plastic, (iii) the manufacture of ethylene dichloride, which is used in polyvinyl chloride polymers, and (iv) the manufacture of ethylbenzene, which is in styrene polymers. The current end use of propylene worldwide is in (i) the manufacture of polypropylene, which is used in plastics, (ii) the manufacture of acrylonitrile, (iii) the manufacture of cumene, which is used in phenolic resin, (iv) the manufacture of propylene oxide, and (v) the manufacture of 8% oxo-alcohol derivatives.

Also, non-olefin petrochemicals are typically aromatic derivatives (benzene, toluene and the xylene isomers or simply BTX). Most of the benzene is used to make (i) styrene, which is used in the manufacture of polymers and plastics, (ii) phenol, which is used in the manufacture of resins and adhesives by way of cumene, and (iii) cyclohexane, which is used in the manufacture of nylon. Benzene is also used in the manufacture of rubber, lubricants, dyes, detergents, drugs, explosives, and pesticides. Toluene is used as a solvent (for making paint, rubber and adhesives), a gasoline additive or for making toluene isocyanate (toluene isocyanate is used for making polyurethane foam), phenol and trinitrotoluene (generally known as TNT). Xylene is used as a solvent and as an additive for making fuels, rubber, leather and terephthalic acid [,14-HO$_2$CC$_6$H$_4$CO$_2$H, also written as 1,4-CO2HC$_6$H$_4$CO$_2$H or 1,4-C$_6$H$_4$(CO$_2$H)$_2$]) which is used in the manufacture of polymers.

Terephthalic acid

The primary focus of this book is on chemical products derived from natural gas and crude oil but chemicals from sources such as coal and biomass are also included as alternate feedstocks (Table 13.5). However, while petrochemicals in the strictest sense are chemical products derived from crude oil, many of the same chemical compounds are also obtained from other fossil fuels such as coal and natural gas or from renewable sources such as corn, sugar cane, and other types of biomass (Matar and Hatch, 2001; Meyers, 2005; Speight, 2008, 2013, 2014; Clark and Deswarte, 2015). But first there is the need to understand the origins of the industry and, above all, the continuing need for the petrochemical industry.

13.2 Historical Aspects and Overview

When coal came to prominence as a fuel during the Industrial Revolution, there was a parallel development relating to the use of coal for the production of chemicals. Byproduct liquids and gases from coal carbonization processes became the basic raw materials for the organic chemical industry and the production of metallurgical coke from coal was essential to the development of steel manufacture (Speight, 2013). Coal tar constituents were

Table 13.5 Alternative feedstocks for the production of petrochemicals.

Chemicals	Natural gas-crude oil feedstock	Alternate feedstock
Methane	Natural gas	Coal, as byproduct of separation of coke gases
	Refinery gas	Coal hydrogenation
Ammonia	Methane	From coal via water gas
Methyl alcohol	Methane	From coal via water-gas reaction
Ethylene	Pyrolysis of low-boiling hydrocarbon derivatives	Dehydration of ethyl alcohol
Acetylene	Ethylene	Calcium carbide
Ethylene glycol	Ethylene	From coal via carbon-monoxide and formaldehyde
Acetaldehyde	Paraffin gas oxidation	Fermentation of ethyl alcohol
	Oxidation of ethylene	Acetylene
Acetone	Propylene	Destructive distillation of wood
		Pyrolysis of acetic acid
		Acetylene-steam reaction
Glycerol	Propylene	Byproduct of soap manufacture
Butadiene	1- and 2-Butenes	Ethyl alcohol; acetaldehyde via 1:3-butanediol
	Butane	Acetylene and formaldehyde from coal
Aromatic hydrocarbons	Aromatic-rich fractions by catalytic reforming	Byproducts of coal-tar distillation
	Naphthene-rich fractions by catalytic reforming	

used for the industrial syntheses of dyes, perfumes, explosives, flavorings, and medicines. Processes were also developed for the conversion of coal to fuel gas and to liquid fuels.

By the time the 1930s had dawned, the direct and indirect liquefaction technologies became available for the substantial conversion of coals to liquid fuels and chemicals. Subsequently, the advent of readily available natural gas and crude oil, and the decline of the steel industry, reduced dependence on coal as a resource for the production of chemicals and materials. For the last several decades as the 20th century came to a close and the 21st century dawned, the availability of coal tar chemicals has depended on the production of metallurgical coke which is, in turn, tied to the fortunes and future of the steel industry. Although the crude oil era was ushered in by the discovery of crude oil at Titusville,

Pennsylvania, in 1859, the production of chemicals from natural gas and crude oil has been a recognized industry only since the early 20th century. Nevertheless, the petrochemical industry has made quantum leaps in the production of a wide variety of chemicals (Chenier, 2002), which being based on starting feedstocks from crude oil are termed *petrochemicals*.

Following from this, the production of chemicals from natural gas and crude oil has been a recognized industry since the early decades of the 20th century. Nevertheless, the lead up to and onset of World War II led to the development and expansion of the petrochemical industry which, since that time, has made quantum leaps in terms of the production of a wide variety of chemicals (Chenier, 2002; Meyers, 2005; Naderpour, 2008; Speight, 2014; EPCA, 2016; Speight, 2017, 2019). At this time, coal alone could no longer satisfy the demand for basic chemicals that had increased by the demands of World War II. The production of chemicals from coal tar or some agricultural products was not sufficient and led to major development of chemicals production from crude oil. During the 1950s and 1960s, the increased demand for liquid fuels increased phenomenally and, paralleling the demand for fuels, the onset of the *age of plastics* (which also included demand for rubber, fibers, surfactants, pesticides, fertilizers, pharmaceuticals, dyes, solvents, lubricating oils, and food additives) caused an increase in the demand for chemicals from natural gas and crude oil.

This trend has continued until the present decade and demand for the manufacture of chemicals will continue for the foreseeable future.

13.3 The Petrochemical Industry

The petrochemical industry, as the name implies, is based upon the production of chemicals from crude oil. However, there is more to the industry than just crude oil products. The petrochemical industry also deals with chemicals manufactured from the byproducts of crude oil refining, such as natural gas, natural gas liquids, and (in the context of this book) other feedstocks such as coal, oil shale and biomass. The structure of the industry is extremely complex, involving thousands of chemicals and processes and there are many interrelationships within the industry with products of one process being the feedstocks of many others. For most chemicals, the production route from feedstock to final products is not unique, but includes many possible alternatives. As complicated as it may seem, however, this structure is comprehensible, at least in general form.

At the beginning of the production chain are the raw feedstocks: crude oil, natural gas, and alternate carbonaceous feedstocks tar. From these are produced a relatively small number of important building blocks which include primarily, but not exclusively, the lower-boiling olefins and aromatic derivatives, such as ethylene, propylene, butylene isomers, butadiene, benzene, toluene, and the xylene isomers. These building blocks are then converted into a complex array of thousands of intermediate chemicals. Some of these intermediates have commercial value in and of themselves, and others are purely intermediate compounds in the production chains. The final products of the petrochemical industry are generally not consumed directly by the public, but are used by other industries to manufacture consumer goods.

Thus, on a scientific basis, as might be expected, the petrochemical industry is concerned with the production and trade of petrochemicals that have a wide influence on lifestyles though the production of commodity chemicals and specialty chemicals that have a marked influence on lifestyles:

Crude oil/natural gas → bulk chemicals (commodity chemicals) → specialty chemicals

The basis of the petrochemical industry and, therefore, petrochemicals production, consists of two steps: (i) feedstock production from primary energy sources to feedstocks and (ii) and petrochemicals production from feedstocks.

Crude oil/natural gas → feedstock production → petrochemical products

This simplified equation encompasses the multitude of production routes available for most chemicals. In the actual industry, many chemicals are products of more than one method, depending upon local conditions, corporate polices, and desired byproducts. There are also additional methods available, which have either become obsolete and are no longer used, or which have never been used commercially but could become important as technology, supplies and other factors change. Such versatility, adaptability, and dynamic nature are three of the important features of the modern petrochemical industry.

Thus, the petrochemical industry began as suitable byproducts became available through improvements in the refining processes. As the decades of the 1920s and 1930s closed, the industry had developed in parallel with the crude oil industry and has continued to expand rapidly since the 1940s as the crude oil refining industry was able to provide relatively cheap and plentiful raw materials (Speight, 2002; Hsu and Robinson, 2006; Gary et al., 2007; Lee et al., 2007; Speight, 2011). The supply-demand scenario as well as the introduction of many innovations have resulted in basic chemicals and plastics becoming the key building blocks for manufacture of a wide variety of durable and nondurable consumer goods. Chemicals and plastic materials provide the fundamental building blocks that enable the manufacture of the vast majority of consumer goods. Moreover, the demand for chemicals and plastics is driven by global economic conditions, which are directly linked to demand for consumer goods.

At the start of the production chain is the selection and preparation of the feedstock from which the petrochemicals will be produced. Typically, the feedstock is a primary energy source (such as crude oil, natural gas, coal and biomass) extracted and then converted into feedstocks (such as naphtha, gas oil, and/or methanol). In the production of petrochemicals, the feedstocks are converted into basic petrochemicals, such as ethylene ($CH_2=CH_2$) and aromatic derivatives, which are then separated from each other. Thus, petrochemicals or products derived from these feedstocks, along with other raw materials, are converted to a wide range of products (Table 13.3).

Therefore, the history of the industry has always been strongly influenced by the supply of primary energy sources and feedstocks. Thus, the petrochemical industry directly interfaces with the crude oil industry and the natural gas industry, which provides the feedstocks (Speight, 2014, 2017, 2019), and especially the downstream sector, as well as the potential for the introduction and use of non-conventional feedstocks (Chapter 2). A major part of the petrochemical industry is made up of the polymer (plastics) industry (Speight, 2014, 2019).

The petrochemical industry is currently the biggest of the industrial chemicals sectors and petrochemicals represent the majority of all chemicals shipped between the continents of the world (EPCA, 2016). The history of petrochemicals began in the 19th century, and the industry has gone through many changes. However, from the beginning there have been underlying trends which shaped the evolution of the industry to modern times. From the start, it was an industry that was destined to become a global sector because of the contribution the product made to raising the standards of living of much of the population of the world. These same influences have also shaped the rate and nature of the expansion and the structure of the industry as it exists in the 21st century.

In the petrochemical industry, the organic chemicals produced in the largest volumes are methanol (methyl alcohol, CH_3OH), ethylene ($CH_2=CH_2$), propylene ($CH_3CH=CH_2$), butadiene ($CH_2=CHCH=CH_2$), benzene (C_6H_6), toluene ($C_6H_5CH_3$), and the xylene isomers ($H_3CC_6H_4CH_3$). Ethylene, propylene, and butadiene, along with butylene isomers, are collectively called olefins, which belong to a class of unsaturated aliphatic hydrocarbon derivatives having the general formula C_nH_{2n}. Olefin derivatives contain one or more double bonds (>C=C<), which make them chemically reactive and, hence, starting materials for many products. Benzene, toluene, and xylenes, commonly referred to as aromatics, are unsaturated cyclic hydrocarbon derivatives containing one or more rings.

As stated above, some of the chemicals and compounds produced in a refinery are destined for further processing and as raw material feedstocks for the fast-growing petrochemical industry. Such non-fuel uses of crude oil products are sometimes referred to as its non-energy uses. Crude oil products and natural gas provide two of the basic starting points for this industry; methane, naphtha, including benzene, toluene, and the xylene isomers and refinery gases which contain olefin derivatives such as ethylene, propylene, and, potentially, all of the butylene isomers (Figure 13.3) (Parkash, 2003; Gary *et al.*, 2007; Speight, 2014; Hsu and Robinson, 2017; Speight, 2017, 2019).

Petrochemical intermediates are generally produced by chemical conversion of primary petrochemicals to form more complicated derivative products. Petrochemical derivative

IUPAC name	Common name	Structure	Skeletal formula
But-1-ene	1-butylene	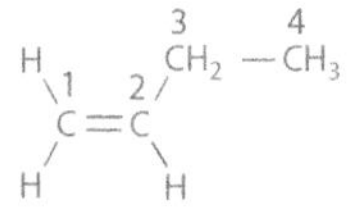	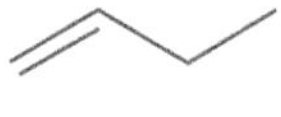
cis-But-2-ene	cis-2-butylene	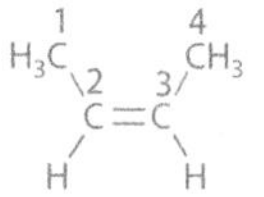	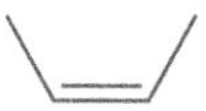
cis-But-2-ene	trans-3-butylene	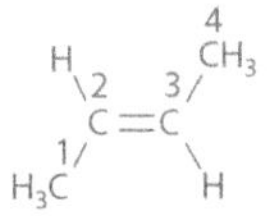	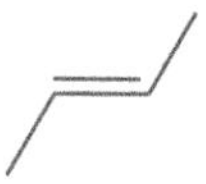
2-methylprop-1-ene	Isobutylene	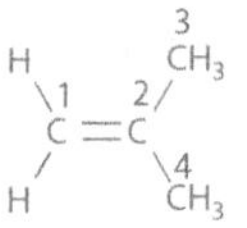	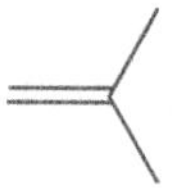

Figure 13.3 Representation of the various isomers of butylene (C_4H_8).

products can be made in a variety of ways: directly from primary petrochemicals; through intermediate products which still contain only carbon and hydrogen; and, through intermediates which incorporate chlorine, nitrogen or oxygen in the finished derivative. In some cases, they are finished products; in others, more steps are needed to arrive at the desired composition.

Of all the processes used, one of the most important is polymerization (Speight, 2014, 2019). It is used in the production of products such as plastics, fibers, and synthetic rubber, the main finished petrochemical derivatives. Some typical petrochemical intermediates are: (i) vinyl acetate, $CH_2=CHOCOCH_3$, that is used in the manufacture of paint, paper and textile coatings, (ii) vinyl chloride, $CH_2=CHCl$, that is used in the manufacture of polyvinyl chloride polymer that is generally known as PVC, resin manufacture, ethylene glycol, CH_2OHCH_2OH, for polyester textile fibers, styrene which is important in rubber, and the manufacture of plastics. The end products number is in the thousands, some going on as inputs into the chemical industry for further processing. The more common products made from petrochemicals include adhesives, plastics, soaps, detergents, solvents, paints, drugs, fertilizer, pesticides, insecticides, explosives, synthetic fibers, synthetic rubber, and flooring and insulating materials.

Petrochemical products include such items as plastics, soaps and detergents, solvents, drugs, fertilizers, pesticides, explosives, synthetic fibers and rubbers, paints, epoxy resins, and flooring and insulating materials. Petrochemicals are found in products as diverse as aspirin, luggage, boats, automobiles, aircraft, polyester clothes, and recording discs and tapes.

The petrochemical industry has grown with the crude oil industry (Goldstein, 1949; Steiner, 1961; Hahn, 1970) and is considered by some to be a mature industry. However, as is the case with the latest trends in changing crude oil types, it must also evolve to meet changing technological needs (Speight, 2011). The manufacture of chemicals or chemical intermediates from a variety of raw materials is well established (Wittcoff and Reuben, 1996). And the use of natural gas and crude oil is an excellent example of the conversion of such raw materials to more valuable products. The individual chemicals made from natural gas and crude oil are numerous and include industrial chemicals, household chemicals, fertilizers, and paints, as well as intermediates for the manufacture of products, such as synthetic rubber and plastics.

Petrochemicals are generally considered chemical compounds derived from crude oil either by direct manufacture or indirect manufacture as byproducts from the variety of processes that are used during the refining of crude oil. Gasoline, kerosene, fuel oil, lubricating oil, wax, asphalt, and the like are excluded from the definition of petrochemicals, since they are not, in the true sense, chemical compounds but are in fact intimate mixtures of hydrocarbon derivatives.

The classification of materials as petrochemicals is used to indicate the source of the chemical compounds, but it should be remembered that many common petrochemicals can be made from other sources, and the terminology is therefore a matter of source identification.

The starting materials for the petrochemical industry are obtained from crude oil in one of two general ways. They may be present in the raw crude oil and, as such, are isolated by physical methods, such as distillation or solvent extraction (Parkash, 2003; Gary et al., 2007;

Speight, 2014; Hsu and Robinson, 2017; Speight, 2017, 2019). On the other hand, they may be present, if at all, in trace amounts and are synthesized during the refining operations. In fact, unsaturated (olefin) hydrocarbon derivatives, which are not usually present in crude oil, are nearly always manufactured as intermediates during the various refining sequences (Parkash, 2003; Gary et al., 2007; Speight, 2014; Hsu and Robinson, 2017; Speight, 2017, 2019).

The manufacture of chemicals from crude oil is based on the ready response of the various compound types to basic chemical reactions, such as oxidation, halogenation, nitration, dehydrogenation addition, polymerization, and alkylation. The low-molecular-weight paraffins and olefins, as found in natural gas and refinery gases, and the simple aromatic hydrocarbon derivatives have so far been of the most interest because it is individual species that can be readily be isolated and dealt with. A wide range of compounds is possible, many are being manufactured and we are now progressing to the stage in which a sizable group of products is being prepared from the heavier fractions of crude oil. For example, the various reactions of asphaltene constituents (Speight, 1994, 2014) indicate that these materials may be regarded as containing chemical functions and are therefore different and are able to participate in numerous chemical or physical conversions to, perhaps, more useful materials. The overall effect of these modifications is the production of materials that either afford good-grade aromatic cokes comparatively easily or the formation of products bearing functional groups that may be employed as a non-fuel material.

Petrochemicals are made, or recovered from, the entire range of crude oil fractions, but the bulk of petrochemical products are formed from the lighter (C_1 to C_4) hydrocarbon gases as raw materials. These materials generally occur in natural gas, but they are also recovered from the gas streams produced during refinery, especially cracking, operations. Refinery gases are also particularly valuable because they contain substantial amounts of olefins that, because of the double bonds, are much more reactive then the saturated (paraffin) hydrocarbon derivatives. Also important as raw materials are the aromatic hydrocarbon derivatives (benzene, toluene, and xylene), that are obtained in rare cases from crude oil and, more likely, from the various product streams. By means of the catalytic reforming process, non-aromatic hydrocarbon derivatives can be converted to aromatics by dehydrogenation and cyclization (Parkash, 2003; Gary et al., 2007; Speight, 2014; Hsu and Robinson, 2017; Speight, 2017, 2019).

A highly significant proportion of these basic petrochemicals is converted into plastics, synthetic rubbers, and synthetic fibers. Together these materials are known as polymers, because their molecules are high-molecular-weight compounds made up of repeated structural units that have combined chemically. The major products are polyethylene, polyvinyl chloride, and polystyrene, all derived from ethylene, and polypropylene, derived from monomer propylene. Major raw materials for synthetic rubbers include butadiene, ethylene, benzene, and propylene. Among synthetic fibers the polyesters, which are a combination of ethylene glycol and terephthalic acid (made from xylene), are the most widely used. They account for about one-half of all synthetic fibers. The second major synthetic fiber is nylon, its most important raw material being benzene. Acrylic fibers, in which the major raw material is the propylene derivative acrylonitrile, make up most of the remainder of the synthetic fibers.

13.4 Petrochemicals

For the purposes of this text, there are four general types of petrochemicals: (i) aliphatic compounds, (ii) aromatic compounds, (iii) inorganic compounds, and (iv) synthesis gas (carbon monoxide and hydrogen). Synthesis gas is used to make ammonia (NH_3) and methanol (methyl alcohol, CH_3OH) as well as a variety of other chemicals (Parkash, 2003; Gary *et al.*, 2007; Speight, 2014; Hsu and Robinson, 2017; Speight, 2017, 2019). Ammonia is used primarily to form ammonium nitrate (NH_4NO_3), a source of fertilizer. Much of the methanol produced is used in making formaldehyde (HCH=O). The rest is used to make polyester fibers, plastics, and silicone rubber.

An aliphatic petrochemical compound is an organic compound that has an open chain of carbon atoms, be it normal (straight), e.g., *n*-pentane ($CH_3CH_2CH_2CH_2CH_3$) or branched, e.g., *iso*-pentane [2-methylbutane, $CH_3CH_2CH(CH_3)CH_3$]. The unsaturated compounds, olefins, include important starting materials such as ethylene ($CH_2=CH_2$), propylene ($CH_3.CH=CH_2$), butene-1 ($CH_3CH_2CH_2=CH_2$), *iso*-butene also known as 2-methylpropene [$CH_3(CH_3)C=CH_2$] and butadiene ($CH_2=CHCH=CH_2$).

As already defined, a *petrochemical* is any chemical (as distinct from fuels and crude oil products) manufactured from crude oil (and natural gas as well as other carbonaceous sources) and used for a variety of commercial purposes (Chenier, 2002). The definition, however, has been broadened to include the whole range of aliphatic, aromatic, and naphthenic organic chemicals, as well as carbon black and such inorganic materials as sulfur and ammonia. Gasoline, kerosene, fuel oil, lubricating oil, wax, asphalt, and the like are excluded from the definition of petrochemicals, since they are not, in the true sense, chemical compounds but are in fact intimate mixtures of hydrocarbon derivatives. The classification of materials as petrochemicals is used to indicate the source of the chemical compounds, but it should be remembered that many common petrochemicals can be made from other sources, and the terminology is therefore a matter of source identification.

Natural gas and crude oil are made up of (predominantly) hydrocarbon constituents, which are comprised of one or more carbon atoms, to which hydrogen atoms are attached – in some cases, crude oil contains a considerable proportion of non-hydrocarbon constituents such as organic compounds containing one or more heteroatoms (such as nitrogen, oxygen, sulfur, and metals). Currently, through a variety of intermediates, natural gas and crude oil are the main sources of the raw materials because they are the least expensive, most readily available, and can be processed most easily into the primary petrochemicals. An aromatic petrochemical is also an organic chemical compound but one that contains, or is derived from, the basic benzene ring system. Furthermore, petrochemicals are often made in clusters of plants in the same area. These plants are often operated by separate companies, and this concept is known as *integrated manufacturing*. Groups of related materials are often used in adjacent manufacturing plants, to use common infrastructure and minimize transport.

13.4.1 Primary Petrochemicals

The primary petrochemicals are not the raw materials for the petrochemical industry. Primary raw materials are naturally occurring substances that have not been subjected to chemical changes after being recovered. Natural gas and crude oil are the basic raw materials for the manufacture of petrochemicals. Secondary raw materials, or intermediates, are obtained from natural gas and crude oils through different processing schemes. The intermediate chemicals may be low-boiling hydrocarbon compounds such as methane and ethane, or heavier hydrocarbon mixtures such as naphtha or gas oil. Both naphtha and gas oil are crude oil fractions with different boiling ranges. Coal, oil shale, and biomass are complex carbonaceous raw materials and possible future energy and chemical sources. However, they must undergo lengthy and extensive processing before they yield fuels and chemicals similar to those produced from crude oils (substitute natural gas (SNG) and synthetic crudes from coal, oil shale, and bio-oil). The term *primary petrochemicals* is more specific and includes olefins (ethylene, propylene and butadiene) aromatics (benzene, toluene, and the isomers of xylene), and methanol from which petrochemical products are manufactured.

The two most common petrochemical classes are olefin derivatives (including ethylene, $CH_2=CH_2$, and propylene, $CH_3CH=CH_2$) and aromatic derivatives such as benzene (C_6H_6), toluene ($C_6H_5CH_3$) and the xylene isomers ($H_3CC_6H_4CH_3$). Olefin derivaitves and aromatic derivatives are typically produced in a crude oil refinery by fluid catalytic cracking of the various crude oil distillate fractions (Speight, 1987; Parkash, 2003; Gary *et al.*, 2007; Speight, 2014; Hsu and Robinson, 2017; Speight, 2017, 2019). Olefins are also produced by steam cracking of methane (CH_4), ethane (CH_3CH_3), and propane ($CH_3CH_2CH_3$) and aromatic derivatives are produced by steam reforming of naphtha (Speight, 1987; Parkash, 2003; Gary *et al.*, 2007; Speight, 2014; Hsu and Robinson, 2017; Speight, 2017, 2019). Olefin derivatives and aromatic derivatives are the intermediate chemicals that lead to a substantial number (some observes would say *an innumerable number*) of products such as solvents, detergents, plastics, fibers, and elastomers.

In many instances, a specific chemical included among the petrochemicals may also be obtained from other sources, such as coal, coke, or vegetable products. For example, materials such as benzene and naphthalene can be made from either crude oil or coal, while ethyl alcohol may be of petrochemical or vegetable origin (Matar and Hatch, 2001; Meyers, 2005; Speight, 2008, 2013, 2014).

13.4.2 Products and End Use

Petrochemical products include such items as plastics, soaps and detergents, solvents, drugs, fertilizers, pesticides, explosives, synthetic fibers and rubbers, paints, epoxy resins, and flooring and insulating materials (Table 13.3). Petrochemicals use is also found in products as diverse as aspirin, luggage, boats, automobiles, aircraft, polyester clothes, and recording discs and tapes.

Although the petrochemical industry was showing steady growth (some observers would say "rapid growth"), the onset of World War II, which increased the demand for synthetic materials to replace costly and sometimes less efficient products, was a catalyst for the development of petrochemicals. Before the 1940s it was an experimental sector, starting with basic materials: (i) synthetic rubber in the 1900s, (ii) Bakelite, the first petrochemical-derived plastic in 1907, (iii) the first petrochemical solvents in the 1920s, and (iv) polystyrene in the 1930s. After this, the industry moved into an variety of areas – from household goods (kitchen appliances, textile, furniture) to medicine (heart pacemakers, transfusion bags), from leisure (such as running shoes, computers), to highly specialized fields like archaeology or crime detection.

Thus, the petrochemical industry has grown with the crude oil industry (Goldstein, 1949; Steiner, 1961; Hahn, 1970; Chenier, 1992) and is considered by some to be a mature industry. However, as is the case with the latest trends in changing crude oil types, the refining industry must also evolve to meet changing technological needs (Speight, 2011, 2014, 2017). The manufacture of chemicals or chemical intermediates from a variety of raw materials is well established (Wittcoff and Reuben, 1996; Speight, 2014) – the use of natural gas and crude oil is an excellent example of the conversion of such raw materials to more valuable products. The individual chemicals made from natural gas and crude oil are numerous and include industrial chemicals, household chemicals, fertilizers, and paints, as well as intermediates for the manufacture of products, such as synthetic rubber and plastics.

The crude oil and petrochemical industries have revolutionized modern life and by providing the major basic needs of rapidly growing, expanding and highly technical civilization. They provide source products such as fertilizers, synthetic fibers, synthetic rubbers, polymers, intermediates, explosives, agrochemicals, dyes, and paints. The petrochemical industry fulfills the requirements of a large number of industries, including uses in fields such as automobile manufacture, telecommunication, pesticides, fertilizers, textiles, dyes, pharmaceuticals, and explosives (Table 13.3).

13.4.3 Production of Petrochemicals

For approximately 100 years, chemicals obtained as byproducts in the primary processing of coal to metallurgical coke have been the main source of a multitude of chemicals used as intermediates in the synthesis of dyes, drugs, antiseptics, and solvents. Historically, producing chemicals from coal through gasification has been used since the 1950s and, as such, dominated a large share of the chemicals industry.

Because the slate of chemical products that can be made via coal gasification can in general also use feedstocks derived from natural gas and crude oil, the chemical industry tends to use whatever feedstocks are most cost-effective. Therefore, interest in using coal tends to increase for higher oil and natural gas prices and during periods of high global economic growth that may strain oil and gas production. Also, production of chemicals from coal is of much higher interest in countries like South Africa, China, India and the United States where there are abundant coal resources. However, in recent decades, largely due to the supply of relatively cheap natural gas and crude oil, the use of coal as a source of chemicals has been superseded by the production of the chemicals from crude oil-related sources. The use of coal has also decreased because of environmental concerns without any acknowledgement that with the installation of modern process controls, coal can be a clean fuel (Speight, 2013).

Katz, 1959; Kohl and Riesenfeld, 1985; Maddox *et al.*, 1985; Newman. 1985; Kohl and Nielsen, 1997; Kidnay and Parrish, 2006. Crude oil refining (Parkash, 2003; Gary *et al.*, 2007; Speight, 2014; Hsu and Robinson, 2017; Speight, 2017, 2019) begins with the distillation, or fractionation of crude oils into separate fractions hydrocarbon groups. The resultant products are directly related to the characteristics of the natural gas and crude oil being processed. Most of these products of distillation are further converted into more useable products by changing their physical and molecular structures through cracking, reforming and other conversion processes. These products are subsequently subjected to various treatment and separation processes, such as extraction, hydrotreating and sweetening, in order to produce finished products. Whereas the simplest refineries are usually limited to atmospheric and vacuum distillation, integrated refineries incorporate fractionation, conversion, treatment and blending with lubricant, heavy fuels and asphalt manufacturing; they may also include petrochemical processing. It is during the refining process that other products are also produced. These products include the gaseous constituents dissolved in the crude oil that are released during the distillation processes as well as the gases produced during the various refining processes, and both of these gaseous streams provide feedstocks for the petrochemical industry.

The gas (often referred to as *refinery gas* or *process gas*) varies in composition and volume, depending on the origin of the crude oil and on any additions (i.e., other crude oils blended into the refinery feedstock) to the crude oil made at the loading point. It is not uncommon to reinject light hydrocarbon derivatives such as propane and butane into the crude oil before dispatch by tanker or pipeline. This results in a higher vapor pressure of the crude, but it allows one to increase the quantity of light products obtained at the refinery. Since light ends in most crude oil markets command a premium, while in the oil field itself propane and butane may have to be reinjected or flared, the practice of *spiking* crude oil with liquefied petroleum gas is becoming fairly common. These gases are recovered by distillation (Figure 13.2). In addition to distillation, gases are also produced in the various *thermal cracking processes* (Figure 13.2) (Parkash, 2003; Gary et al., 2007; Speight, 2014; Hsu and Robinson, 2017; Speight, 2017, 2019).

Thus, in processes such as coking or visbreaking a variety of gases is produced. Another group of refining operations that contributes to gas production is that of the *catalytic cracking processes*. Both catalytic and thermal cracking processes result in the formation of unsaturated hydrocarbon derivatives, particularly ethylene ($CH_2=CH_2$), but also propylene (propene, $CH_3CH=CH_2$), *iso*-butylene [*iso*-butene, $(CH_3)_2C=CH_2$] and the *n*-butenes ($CH_3CH_2CH=CH_2$, and $CH_3CH=CHCH_3$) in addition to hydrogen (H_2), methane (CH_4) and smaller quantities of ethane (CH_3CH_3), propane ($CH_3CH_2CH_3$), and butanes [$CH_3CH_2CH_2CH_3$, $(CH_3)_3CH$]. Diolefins such as butadiene ($CH_2=CHCH=CH_2$) and are also present. A further source of refinery gas is *hydrocracking*, a catalytic high-pressure pyrolysis process in the presence of fresh and recycled hydrogen. The feedstock is again heavy gas oil or residual fuel oil, and the process is mainly directed at the production of additional middle distillates and gasoline. Since hydrogen is to be recycled, the gases produced in this process again have to be separated into lighter and heavier streams; any surplus recycle gas and the liquefied petroleum gas from the hydrocracking process are both saturated (Parkash, 2003; Gary et al., 2007; Speight, 2014; Hsu and Robinson, 2017; Speight, 2017, 2019).

In a series of *reforming processes* (Parkash, 2003; Gary et al., 2007; Speight, 2014; Hsu and Robinson, 2017; Speight, 2017, 2019), commercialized under names such as *Platforming*, paraffin and naphthene (cyclic non-aromatic) hydrocarbon derivatives are converted in the presence of hydrogen and a catalyst are converted into aromatics, or isomerized to more highly branched hydrocarbon derivatives. Catalytic reforming processes thus not only result in the formation of a liquid product of higher octane number, but also produce substantial quantities of gases. The latter are rich in hydrogen, but also contain hydrocarbon derivatives from methane to butane isomers, with a preponderance of propane ($CH_3CH_2CH_3$), n-butane ($CH_3CH_2CH_2CH_3$) and iso-butane [$(CH_3)_3CH$].

As might be expected, the composition of the process gas varies in accordance with reforming severity and reformer feedstock. All catalytic reforming processes require substantial recycling of a hydrogen stream. Therefore, it is normal to separate reformer gas into a propane ($CH_3CH_2CH_3$) and/or a butane stream [$CH_3CH_2CH_2CH_3$ plus $(CH_3)_3CH$], which becomes part of the refinery liquefied petroleum gas production, and a lighter gas fraction, part of which is recycled. In view of the excess of hydrogen in the gas, all products of catalytic reforming are saturated, and there are usually no olefin gases present in either gas stream.

In many refineries, naphtha in addition to other finery gases, is also used as the source of petrochemical feedstocks. In the process, naphtha crackers convert naphtha feedstock (produced by various process) (Table 13.4) into ethylene, propylene, benzene, toluene, and xylenes as well as other byproducts in a two-step process of cracking and separating. In some cases, a combination of naphtha, gas oil, and liquefied petroleum gas may be used. The feedstock, typically naphtha, is introduced into the pyrolysis section of the naphtha where it is cracked in the presence of steam. The naphtha is converted into lower-boiling fractions, primarily ethylene and propylene. The hot gas effluent from the furnace is then quenched to inhibit further cracking and to condense higher molecular weight products. The higher molecular weight products are subsequently processed into fuel oil, light cycle oil and pyrolysis gas byproducts. The pyrolysis gas stream can then be fed to the aromatics plants for benzene and toluene production.

In addition to recovery of gases in the distillation section of a refinery distillation, gases are also produced in the various thermal processes, thermal cracking processes and catalytic cracking processes (Figure 13.2), are also available and in processes such as visbreaking and coking and visbreaking (Speight, 1987; Parkash, 2003; Gary et al., 2007; Speight, 2014; Hsu and Robinson, 2017; Speight, 2017, 2019). Thermal cracking processes were first developed for crude oil refining, starting in 1913 and continuing the next two decades, were focused primarily on increasing the quantity and quality of gasoline components (Parkash, 2003; Gary et al., 2007; Speight, 2014; Hsu and Robinson, 2017; Speight, 2017, 2019). As a byproduct of this process, gases were produced that included a significant proportion of lower-molecular-weight olefins, particularly ethylene ($CH_2=CH_2$), propylene ($CH_3CH=CH_2$), and butylenes (butenes, $CH_3CH=CH.CH_3$ and $CH_3CH_2CH=CH_2$). Catalytic cracking, introduced in 1937, is also a valuable source of propylene and butylene, but it does not account for a very significant yield of ethylene, the most important of the petrochemical building blocks (Parkash, 2003; Gary et al., 2007; Speight, 2014; Hsu and Robinson, 2017; Speight, 2017, 2019). Ethylene is polymerized to produce polyethylene or, in combination with propylene, to produce copolymers that are used extensively in food-packaging wraps, plastic household goods, or building materials. Prior to the use of natural gas and crude oil as sources of chemicals, coal was the main source of chemicals (Speight, 2013).

Once produced and separated from other product streams, the cooled gases are then compressed, treated to remove acid gases, dried over a desiccant and fractionated into separate components at low temperature through a series of refrigeration processes (Parkash,

2003; Gary et al., 2007; Speight, 2014; Hsu and Robinson, 2017; Speight, 2017, 2019). Hydrogen and methane are removed by way of a compression/expansion process after which the methane is distributed to other processes as deemed appropriate for fuel gas. Hydrogen is collected and further purified in a pressure swing unit for use in the hydrogenation (hydrotreating and hydrocracking) processes (Parkash, 2003; Gary et al., 2007; Speight, 2014; Hsu and Robinson, 2017; Speight, 2017, 2019). Polymer grade ethylene and propylene are separated in the cold section after which the ethane and propane streams are recycled back to the furnace for further cracking while the mixed butane (C_4) stream is hydrogenated prior to recycling back to the furnace for further cracking.

The composition of the process gas varies in accordance with reforming severity and reformer feedstock. All catalytic reforming processes require substantial recycling of a hydrogen stream. Therefore, it is normal to separate reformer gas into a propane ($CH_3CH_2CH_3$) and/or a butane stream [$CH_3CH_2CH_2CH_3$ plus $(CH_3)_3CH$], which becomes part of the refinery liquefied petroleum gas production, and a lighter gas fraction, part of which is recycled. In view of the excess of hydrogen in the gas, all products of catalytic reforming are saturated, and there are usually no olefin gases present in either gas stream.

In many refineries, naphtha in addition to other finery gases, is also used as the source of petrochemical feedstocks. In the process, naphtha crackers convert naphtha as well as gas oil feedstocks (produced by various process) (Table 13.4) into ethylene, propylene, benzene, toluene, and xylenes as well as other byproducts in a two-step process of cracking and separating. In some cases, a combination of naphtha, gas oil, and liquefied petroleum gas may be used. The feedstock, typically naphtha, is introduced into the pyrolysis section of the naphtha where it is cracked in the presence of steam. The naphtha is converted into lower-boiling fractions, primarily ethylene and propylene. The hot gas effluent from the furnace is then quenched to inhibit further cracking and to condense higher molecular weight products. The higher molecular weight products are subsequently processed into fuel oil, light cycle oil and pyrolysis gas byproducts. The pyrolysis gas stream can then be fed to the aromatics plants for benzene and toluene production.

The refinery gas (or the process gas) stream and he products of naphtha cracking are the source of a variety of *petrochemicals*. For example, thermal cracking processes (Parkash, 2003; Gary et al., 2007; Hsu and Robinson, 2017; Speight, 2017, 2019) developed for crude oil refining, starting in 1913 and continuing the next two decades, were focused primarily on increasing the quantity and quality of gasoline components. As a byproduct of this process, gases were produced that included a significant proportion of lower-molecular-weight olefins, particularly ethylene ($CH_2=CH_2$), propylene ($CH_3CH=CH_2$), and butylenes (butenes, $CH_3CH=CHCH_3$ and $CH_3CH_2CH=CH_2$). Catalytic cracking (Parkash, 2003; Gary et al., 2007; Hsu and Robinson, 2017; Speight, 2017, 2019), introduced in 1937, is also a valuable source of propylene and butylene, but it does not account for a very significant yield of ethylene, the most important of the petrochemical building blocks. Ethylene is polymerized to produce polyethylene or, in combination with propylene, to produce copolymers that are used extensively in food-packaging wraps, plastic household goods, or building materials. Prior to the use of natural gas and crude oil as sources of chemicals, coal was the main source of chemicals (Speight, 2013).

The petrochemical industry has grown with the crude oil industry (Goldstein, 1949; Steiner, 1961; Hahn, 1970) and is considered by some to be a mature industry. However, as is the case with the latest trends in changing crude oil types, it must also evolve to meet changing technological needs. The manufacture of chemicals or chemical intermediates from a variety of raw materials is well established (Wittcoff and Reuben, 1996). And the use of natural gas and crude oil is an excellent example of the conversion of such raw materials to more valuable products. The individual chemicals made from natural gas and crude oil are numerous and include industrial chemicals, household chemicals, fertilizers, and paints, as well as intermediates for the manufacture of products, such as synthetic rubber and plastics.

The starting materials for the petrochemical industry are obtained from crude natural gas and crude oil in one of two general ways. They may be present in the raw crude oil or natural gas and, as such, are isolated by physical methods, such as distillation or solvent extraction. On the other hand, they may be present, if at all, in trace amounts and are synthesized during the refining operations. In fact, unsaturated (olefin) hydrocarbon derivatives, which are not typically present in crude oil, are manufactured as intermediates during the various refining sequences.

The manufacture of chemicals from crude oil is based on the ready response of the various compound types to basic chemical reactions, such as oxidation, halogenation, nitration, dehydrogenation addition, polymerization, and alkylation. The low-molecular-weight paraffins and olefins, as found in natural gas and refinery gases, and the simple aromatic hydrocarbon derivatives have so far been of the most interest because it is individual species that can be readily be isolated and dealt with. A wide range of compounds is possible, many are being manufactured and we are now progressing to the stage in which a sizable group of products is being prepared from the higher molecular weight fractions of crude oil.

The various reactions of asphaltene constituents indicate that these materials may be regarded as containing chemical functions and are therefore different and are able to participate in numerous chemical or physical conversions to, perhaps, more useful materials. The overall effect of these modifications is the production of materials that either affords good-grade aromatic cokes comparatively easily or the formation of products bearing functional groups that may be employed as a non-fuel material.

For example, the sulfonated and sulfomethylated materials and their derivatives have satisfactorily undergone tests as drilling mud thinners, and the results are comparable to those obtained with commercial mud thinners (Moschopedis and Speight, 1971, 1974, 1976a, 1978). In addition, these compounds may also find use as emulsifiers for the *in situ* recovery of heavy oils. These are also indications that these materials and other similar derivatives of the asphaltene constituents, especially those containing such functions as carboxylic or hydroxyl, readily exchange cations and could well compete with synthetic zeolites. Other uses of the hydroxyl derivatives and/or the chloro-asphaltenes include high-temperature packing or heat transfer media (Moschopedis and Speight, 1976b).

Reactions incorporating nitrogen and phosphorus into the asphaltene constituents are particularly significant at a time when the effects on the environment of many materials containing these elements are receiving considerable attention. In this case, there is the potential slow-release soil conditioners that only release the nitrogen or phosphorus after considerable weathering or bacteriological action. One may proceed a step further and suggest that the carbonaceous residue remaining after release of the hetero-elements may be a benefit to humus-depleted soils, such as the gray-wooded soils. It is also feasible that coating a conventional quick-release inorganic fertilizer with a water-soluble or water-dispersible derivative will provide a slower release fertilizer and an organic humus-like residue. In fact, variations on this theme are multiple (Moschopedis and Speight, 1974, 1976a).

Nevertheless, the main objective in producing chemicals from crude oil is the formation of a variety of well-defined chemical compounds that are the basis of the petrochemical industry. It must be remembered, however, that ease of separation of a particular compound from crude oil does not guarantee its use as a petrochemical building block. Other parameters, particularly the economics of the reaction sequences, including the costs of the reactant equipment, must be taken into consideration.

Petrochemicals are made, or recovered from, the entire range of crude oil fractions, but the bulk of petrochemical products are formed from the lighter (C_1 to C_4) hydrocarbon gases as raw materials. These materials generally occur in natural gas, but they are also recovered from the gas streams produced during refinery, especially cracking, operations. Refinery gases are also particularly valuable because they contain substantial amounts of olefins that, because of the double bonds, are much more reactive then the saturated (paraffin) hydrocarbon derivatives. Also important as raw materials are the aromatic hydrocarbon derivatives (benzene, toluene, and xylene), that are obtained in rare cases from crude oil and, more likely, from the various product streams. By means of the catalytic reforming process, non-aromatic hydrocarbon derivatives can be converted to aromatics by dehydrogenation and cyclization.

A highly significant proportion of these basic petrochemicals are converted into plastics, synthetic rubbers, and synthetic fibers. Together these materials are known as polymers, because their molecules are high-molecular-weight compounds made up of repeated structural units that have combined chemically. The major products are polyethylene, polyvinyl chloride, and polystyrene, all derived from ethylene, and polypropylene, derived from monomer propylene. Major raw materials for synthetic rubbers include butadiene, ethylene, benzene, and propylene. Among synthetic fibers the polyesters, which are a combination of ethylene glycol and terephthalic acid (made from xylene), are the most widely used. They account for about one-half of all synthetic fibers. The second major synthetic fiber is nylon, its most important raw material being benzene. Acrylic fibers, in which the major raw material is the propylene derivative acrylonitrile, make up most of the remainder of the synthetic fibers.

The main objective in producing chemicals from crude oil is the formation of a variety of well-defined chemical compounds that are the basis of the petrochemical industry. It must be remembered, however, that ease of separation of a particular compound from crude oil does not guarantee its use as a petrochemical building block. Other parameters, particularly the economics of the reaction sequences, including the costs of the reactant equipment, must be taken into consideration.

Petrochemical production relies on multi-phase processing of oil and associated crude oil gas. Key raw materials in the petrochemical industry include products of crude oil refining (primarily gases and naphtha). Petrochemical goods include ethylene, propylene and benzene; source monomers for synthetic rubbers, and inputs for technical carbon.

The petrochemical industry has grown with the crude oil industry and is considered by some to be a mature industry. However, as is the case with the latest trends in changing crude oil types, it must also evolve to meet changing technological needs. The manufacture of chemicals or chemical intermediates from a variety of raw materials is well established. And the use of natural gas and crude oil is an excellent example of the conversion of such raw materials to more valuable products. The individual chemicals made from natural gas and crude oil are numerous and include industrial chemicals, household chemicals, fertilizers, and paints, as well as intermediates for the manufacture of products, such as synthetic rubber and plastics.

The main objective in producing chemicals from crude oil is the formation of a variety of well-defined chemical compounds that are the basis of the petrochemical industry. It must be remembered, however, that ease of separation of a particular compound from crude oil does not guarantee its use as a petrochemical building block. Other parameters, particularly the economics of the reaction sequences, including the costs of the reactant equipment, must be taken into consideration.

13.4.4 Gaseous Fuels and Chemicals

Synthesis gas is a very important chemical intermediate for many relevant processes including the production of methanol and Fischer-Tropsch synthesis of synthetic fuels (Chapter 10) and, thus, is an essential industrial feedstock in several catalytic processes.

13.4.4.1 Ammonia

In the process, natural gas is first natural gas (methane) or liquefied petroleum gases (such as propane) or crude oil naphtha into hydrogen by steam reforming (Chapter 8). The hydrogen is then combined with nitrogen to produce ammonia by the Haber-Bosch process:

Steam reforming process:

$$CH_4 + H_2O \rightarrow CO + 3H_2$$

$$C_3H_8 + 3H_2O \rightarrow 3CO + 7H_2$$

$$C_nH_m + nH_2O \rightarrow nCO + (0.5m + n)H_2$$

The Haber-Bosch process, also called the Haber process, which was developed it in the first decade of the 20th century, is a nitrogen fixation process and is the main industrial procedure for the production of ammonia. The process converts atmospheric nitrogen (N_2) to ammonia (NH_3) by a reaction with hydrogen (H_2) using a metal catalyst under high temperatures and pressures:

$$N_2 + 3H_2 \rightarrow 2NH_3$$

Starting with the feedstock (natural gas), the first process leading to hydrogen production is the removal of sulfur compounds from the feedstock because sulfur – sulfur compounds deactivate the catalysts used in subsequent steps. Sulfur removal requires catalytic hydrogenation to convert sulfur compounds in the feedstock to hydrogen sulfide:

$$H_2 + RSH \rightarrow RH + H_2S(gas)$$

The gaseous hydrogen sulfide is then adsorbed (Chapter 9) and removed by passing it through beds of zinc oxide where it is converted to solid zinc sulfide:

$$H_2S + ZnO \rightarrow ZnS + H_2O$$

Catalytic steam reforming of the sulfur-free feedstock is then used to form hydrogen plus carbon monoxide (synthesis gas):

$$CH_4 + H_2O \rightarrow CO + 3H_2$$

The next step is the shift conversion to convert the carbon monoxide to carbon dioxide and more hydrogen:

$$CO + H_2O \rightarrow CO_2 + H_2$$

The carbon dioxide is then removed either by absorption in aqueous ethanolamine solution or by adsorption in a pressure-swing adsorption unit using a solid adsorbent (Chapter 9). The final step in producing the hydrogen is to use catalytic methanation to remove any small residual amounts of carbon monoxide or carbon dioxide from the hydrogen:

$$CO + 3H_2 \rightarrow CH_4 + H_2O$$

$$CO_2 + 4H_2 \rightarrow CH_4 + 2H_2O$$

To produce the desired end-product ammonia, the hydrogen is then catalytically reacted with nitrogen (derived from process air) to form anhydrous liquid ammonia – the Haber-Bosch process (also known as the ammonia synthesis loop):

$$3H_2 + N_2 \rightarrow 2NH_3$$

Due to the nature of the (typically multi-promoted magnetite) catalyst used in the ammonia synthesis reaction, only very low levels of oxygen-containing compounds (especially carbon monoxide, carbon dioxide, and water) can be tolerated in the synthesis gas. Relatively pure nitrogen can be obtained by an air separation step, but additional oxygen removal may be required.

The steam reforming, shift conversion, carbon dioxide removal and methanation steps each operate at absolute pressures of approximately 375 to 525 psi, and the ammonia synthesis loop operates at absolute pressures ranging from 900 to 2,700 psi depending upon the process design.

13.4.4.2 Hydrogen

Hydrogen can be produced from the gasification product gas through the steam-methane reforming and water gas shift reaction.

Steam-methane reforming reaction:

$$CH_4 + H_2O \rightarrow CO + 3H_2$$

Water gas shift reaction:

$$CO + H_2O \rightleftharpoons CO_2 + H_2$$

Using a dual fluidized-bed (DFB) gasification system with carbon dioxide adsorption along with suitable catalysts, it is possible to achieve a hydrogen yield up to 70 vol% direct in the gasifier (Soukup *et al.*, 2009) Another sources of hydrogen is glycerol, a byproduct derived from the production of biodiesel. One approach to alleviate this problem is to transform glycerol into valuable chemicals such as hydrogen and syngas (Lin, 2013).

13.4.4.3 Synthetic Natural Gas

Natural gas that is produced from coal or biomass is known as *synthetic natural gas* or *substitute natural gas* (SNG). The typical catalyst for methanation is nickel and the main reaction is represented by the equation:

$$CO + 3H_2 \rightarrow CH_4 + H_2O$$

Nickel-based catalysts are also active in water-gas shift and hydration of higher hydrocarbon derivatives, such as olefin derivatives.

Typically a hydrogen-to-carbon monoxide ratio of 3 is necessary, which is obtained by a water-gas shift reactor before methanation. In some types of reactors, such as fluidized-bed reactors, the water-gas shift can be carried out also in parallel with the methanation, so no external adjustment of the hydrogen-to-carbon monoxide ratio is necessary. Methanation can be done at atmospheric pressure, although according to the thermodynamics higher pressure is preferred.

Since nickel-based catalysts are sensitive to poisoning by sulfur-containing compounds, gas treatment (Chapter 9) is quite important before the methanation, and sulfur compounds have to be removed to below 0.1 ppm.

13.4.5 Liquid Fuels and Chemicals

Liquid transportation hydrocarbon fuels and various other chemical products can be produced from syngas via the well-known and established catalytic Fischer-Tropsch process (Chapter 10). Depending on the source of the synthesis gas, the technology is often referred to as coal-to-liquids (CTL) technology and/or gas-to-liquids (GTL) technology.

13.4.5.1 Fischer-Tropsch Liquids

Fischer-Tropsch liquids are well known for their being virtually sulfur, nitrogen, and heteroatom free, and they are said to carry a product premium in excess of conventional crude oil derived liquid products such as diesels and gas oils.

Conventional refinery processes can be used for upgrading Fischer-Tropsch liquids and Fischer-Tropsch waxes. The low-temperature Fischer-Tropsch (LTFT) process is employed in the production of waxes, which are converted into naptha or kerosene (diesel oil) after a hydroprocessing step. The high-temperature Fischer–Tropsch (HTFT) process is employed in the production of alpha-olefins (i.e., $RCH=CH_2$) (Chapter 10). Fuels produced with Fischer-Tropsch synthesis are of a high quality due to a very low content of aromatic constituents and zero content of sulfur constituents. The product stream consists of various fuel types: liquefied petroleum gas (LPG), gasoline, diesel fuel, and jet fuel.

13.4.5.2 Methanol

Methanol (CH_3OH, MeOH, methyl alcohol, wood alcohol, or wood spirits) can be produced from fossil or renewable resources and can be used either directly as a transportation fuel or can be converted further to hydrocarbons.

Methanol is produced from synthesis gas by the hydrogenation of carbon oxides over a suitable (copper oxide, zinc oxide, or chromium oxide-based) catalyst:

$$CO + 2H_2 \leftrightarrow CH_3OH$$

$$CO_2 + 3H_2 \leftrightarrow CH_3OH + H_2O$$

The first reaction is the primary methanol synthesis reaction and a small amount of carbon dioxide (2-10% v/v) in the feedstock acts as a promoter of this primary reaction and helps maintain catalyst activity. The reactions are exothermic and give a net decrease in molar volume. Therefore, the equilibrium is favored by high pressure and low temperature. During production, heat is released and has to be removed to keep optimum catalyst life and reaction rate; 0.3% of the produced methanol reacts further to form side products such as dimethyl ether, formaldehyde, or higher molecular weight alcohols.

The catalyst deactivates primarily because of loss of active copper due to physical blockage of the active sites by large byproduct molecules; poisoning by halogens or sulfur in the synthesis gas, which irreversibly form inactive copper salts; and sintering of the copper crystallites into larger crystals, which then have a lower surface to volume ratio. Conventionally, methanol is produced in two-phase systems in which the reactants and products forming the gas phase and the catalyst being the solid phase.

13.4.5.3 Dimethyl Ether

Dimethyl ether (DME, CH_3OCH_3) is generally produced by dehydration of methanol. The methanol production and dehydration processes are combined in one reactor, such that dimethyl ether is produced directly from synthesis gas slightly more efficient than methanol. In addition, the direct synthesis of dimethyl ether allows a hydrogen-carbon monoxide ratio of approximately 1, which is an advantage for oxygen blown gasifiers.

Like methanol, dimethyl ether has promising features as fuel candidate with both the Otto and the diesel engine. With adaptations to the engine and the fuel system, dimethyl ether can be used in diesel engines, leading to higher fuel efficiency and lower emissions. In Otto engines, dimethyl ether can be used with liquefied petroleum gas (LPG). Also, since dimethyl ether is as easily reformed as methanol, it has a big potential as fuel for fuel cell vehicles. Dimethyl ether has similar physical properties as liquefied petroleum gas (LPG) and can be handled as a liquid, using the same infrastructure as liquefied petroleum gas.

13.4.5.4 Methanol-to-Gasoline and Olefins

In the 1970s, at the time of the oil embargoes, Mobil developed and commercialized a methanol-to-gasoline (MtG) process. The methanol-to-olefins (MtO) synthesis is a commercially attractive process because of the high demand of ethylene ($CH_2=CH_2$) and propylene ($CH_3CH=CH_2$).

Currently, these compounds are produced mainly through noncatalytic steam cracking of fossil fuels (naphtha). The methanol-to-olefins process uses zeolite-based catalysts that efficiently convert methanol into ethylene and propylene.

13.4.5.5 Other Processes

The *methanol-to-diesel process* (MtD process) first converts methanol into propylene, which is followed by olefin oligomerization (conversion to distillates), then product separation-plus-hydrogenation. The process would yield mostly kerosene and diesel, along with a small yield of naphtha (or gasoline) and low-boiling hydrocarbon derivatives. The near-zero sulfur/polyaromatics diesel fuel resulting from this process would differ from the more conventional Fischer-Tropsch diesel product only in cetane number (>52 via methanol-to-synfuel comparted to >70 cetane for Fischer-Tropsch diesel). The incidental gasoline stream not only would be near zero sulfur but also have commercially acceptable octane ratings (92 RON, 80 MON) and maximally 11% v/v aromatic derivatives.

The catalytic conversion of synthesis gas to *mixed alcohols* is of great importance because mixed alcohols are valuable additives to gasoline to increase the octane number and reduce the environmental pollution. Furthermore, a benefit of the mixed alcohol synthesis is the high resistance of the catalysts against sulfur poisoning and the fact that the gas cleaning facilities can be simpler as in other syntheses. Mixed alcohols can also be converted to high-quality fuels via dehydration and oligomerization. Typically, the alkali-doped oxides (zinc and chromium oxides) and alkali-doped sulfides (molybdenum sulfides) are used as catalysts for mixed alcohols synthesis.

Depending on process conditions and catalysts, the main primary products are generally methanol and water. The reaction mechanism of the mixed alcohol synthesis is represented simply as:

$$nCO + 2nH_2 \rightarrow C_nH_{2n+1}OH + (n-1)H_2O$$

Owing to reaction stoichiometry the proposed carbon monoxide ratio is 2, but the optimal ratio is in practice closer to 1 because of water-gas shift reaction that occurs in parallel with the alcohol formation.

13.5 The Future

The petrochemical industry is concerned with the production and trade of petrochemicals and has a direct relationship with the crude oil industry, especially the downstream sector of the industry. The petrochemical industries are specialized in the production of petrochemicals that have various industrial applications. The petrochemical industry can be considered to be a sub-sector of the crude oil industry since without the crude oil industry the petrochemical industry cannot exist. Thus, crude oil is the major prerequisite raw material for the production of petrochemicals either in qualities or quantities. In addition, the petrochemical industry is subject to the geopolitics of the crude oil industry, with each industry being reliant upon the other for sustained survival.

In the 1970s, as a result of various oil embargos, coal liquefaction processes seemed on the point of commercialization and would have provided new sources of coal liquids for chemical use, as well as fulfilling the principal intended function of producing alternate fuels. Because of the varying price of crude oil, this prospect is unlikely to come to fruition in the immediate future, due to the question of economic viability rather than technical feasibility. The combination of these and other factors has contributed to sharpening the focus on the use of coal for the production of heat and power, and lessening or eclipsing its possible use as a starting point for other processes.

The growth and development of petrochemical industries depends on a number of factors and also varies from one country to another either based on technical know-how, marketability and applicability of these petrochemicals for manufacture of petrochemical products through petrochemical processes which are made feasible by knowledge and application of petrochemistry. Moreover, petrochemistry is a branch of chemistry (chemistry being a branch of natural science concerned with the study of the composition and constitution of substances and the changes such substances undergo because of changes in the molecules that make up such substances) that deals with crude oil, natural gas and their derivatives.

However, not all of the petrochemical or commodity chemical materials produced by the chemical industry are made in one single location; groups of related materials are often made in adjacent manufacturing plants to induce industrial symbiosis as well as material and utility efficiency and other economies of scale (integrated manufacturing). Specialty and fine chemical companies are sometimes found in similar manufacturing locations as petrochemicals but, in most cases, they do not need the same level of large scale infrastructure (e.g., pipelines, storage, ports and power, etc.) and therefore can be found in multi-sector business parks. This will continue as long as the refining industry continues to exist in its present form (Favennec, 2001; Speight, 2011).

The petrochemical industry continues to be impacted by the globalization and integration of the world economy. For example, world-scale petrochemical plants built during the past several years are substantially larger than those built over two decades ago. As a result, smaller, older, and less efficient units are being shut down, expanded, or, in some cases, retrofitted to produce different chemical products. In addition, crude oil prices had been on the rise during the past decade and petrochemical markets are impacted during sharp price fluctuations, creating a cloud of uncertainty in upstream and downstream investments. Also, increasing concerns over fossil fuel supply and consumption, with respect to their impact on health and the environment, have led to the passage of legislation globally that will affect chemical and energy production and processing for the foreseeable future.

The recent shift from local markets to a large global market led to an increase in the competitive pressures on petrochemical industries. Further, because of fluctuations in products' price and high price of feedstocks, economical attractiveness of petrochemical plants can be considered as a main challenge. The ever-increasing cost of energy and more stringent environmental regulations impacted the operational costs. When cheap feedstocks are not available, the best method of profitability is to apply integration and optimization in petrochemical complexes with adjacent refineries. This is valid for installed plants and plants under construction. Petrochemical-refinery integration is an important factor in reducing costs and increasing efficiencies because integration guarantees the supply of feedstock for petrochemical industries. Also, integrated schemes take the advantage of the economy of scale as well and an integrated complex can produce more diverse products. Petrochemical-refinery integration avoids selling crude oil, optimizes products, economizes costs and increases benefits.

On an innovation and technological basis (Hassani *et al.*, 2017), manufacturing processes introduced in recent years have resulted in raw material replacement, shifts in the ratio of coproduct(s) produced, and cost. This has led to a supply/demand imbalance, particularly for smaller downstream petrochemical derivatives. In addition, growing environmental concerns and higher crude oil prices have expedited the development and commercialization of renewably derived chemical products and technologies previously considered economically impractical. Among the various technological advances, the combination of vertical hydraulic fracturing ("fracking") and horizontal drilling in multistage hydraulic fracturing resulted in a considerable rise in natural gas production in the United States. This new potential has caused many countries to reexamine their natural gas reserves and pursue development of their own gas plays.

Currently, natural gas and crude oil are the main sources of the raw materials for the production of petrochemicals because they (natural gas and crude oil) are the least expensive, most readily available, and can be processed most easily into the primary petrochemicals. However, as the current century progresses and, given the changes to crude oil supply that might be anticipated during the next five decades (Speight, 2011), there is a continuing need to assess the potential of other sources of petrochemicals.

For example, coal could well see a revitalization of use, understanding that there is the need to adhere to the various environmental regulations that apply to the use of any fossil fuel. Coal carbonization was the earliest and most important method to produce chemicals. For many years, chemicals that have been used for the manufacture of such diverse materials as nylon, styrene, fertilizers, activated carbon, drugs, and medicine, as well as many others have been made from coal. These products will expand in the future as natural gas and crude oil resources become strained to supply petrochemical feedstocks and coal becomes a predominant chemical feedstock once more. The ways in which coal may be converted to chemicals include carbonization, hydrogenation, oxidation, solvent extraction, hydrolysis, halogenation, and gasification (followed by conversion of the synthesis gas to chemical products) (Speight, 2013, 2014). In some cases such processing does not produce chemicals in the sense that the products are relatively pure and can be marketed as even industrial grade chemicals. Thus, although many traditional markets for coal tar chemicals have been taken over by the petrochemical industry, the position can change suddenly as oil prices fluctuate upwards. Therefore, the concept of using coal as a major source of chemicals can be very real indeed.

Compared to crude oil crude, shale oils obtained by retorting of the world's oil shales in their multitude and dissimilarity, are characterized by wide boiling range and by large concentrations of heteroelements and also by high content of oxygen-, nitrogen-, or sulfur-containing compounds. The chemical potential of oil shales as retort fuel to produce shale oil and from that liquid fuel and specialty chemicals has been used so far to a relatively small extent. While the majority of countries are discovering the real practical value of shale oil, in Estonia retorting of its national resource kukersite oil obtained for production of a variety of products has been in use for 75 years already. Using stepwise cracking motor fuels have been produced and even exported before World War II. At the same time, shale oils possess molecular structures of interest to the specialty chemicals industry and also a number of non-fuel specialty products have been marketed based on functional group, broad range concentrate, or even pure compound values.

Based on large quantity of oxygen-containing compounds in heavy fraction, asphalt blending material, road asphalt and road oils, construction mastics, anticorrosion oils, and rubber softeners. are produced, Benzene and toluene for production of benzoic acid as well as solvent mixtures on pyrolysis of lighter fractions of shale oil are produced. Middle shale oil fractions having antiseptic properties are used to produce effective oil for the impregnation of wood as a major shale oil derived specialty product. Water soluble phenols are selectively extracted from shale oil, fractionated and crystallized for production of pure 5-methylresorcinol and other alkyl resorcinol derivatives and high-value intermediates to produce tanning agents, epoxy resins and adhesives, diphenyl ketone and phenol-formaldehyde adhesive resins, rubber modifiers, chemicals, and pesticides. Some conventional products such as coke and distillate boiler fuels are produced from shale oil as byproducts. New market opportunities for shale oil and its fractions may be found improving the oil conversion and separation techniques. In the petrochemical industry, the organic chemicals produced in the largest volumes are methanol, ethylene, propylene, butadiene, benzene, toluene, and xylene isomers.

Basic chemicals and plastics are the key building blocks for manufacture of a wide variety of durable and nondurable consumer goods. The demand for chemicals and plastics is driven by global economic conditions, which are directly linked to demand for consumer goods. The petrochemical industry continues to be impacted by the globalization and integration of the world economy. In the future, manufacturing processes introduced in recent years will continue to result in the adaptation of the industry to new feedstocks which will chase shifts in the ratio of products produced. This, in turn, will lead to the potential for a supply/demand imbalance, particularly for smaller downstream petrochemical derivatives. In addition, growing environmental concerns and the variability of crude oil prices (usually upward) will expedite the development and commercialization of chemical products from source other than natural gas and crude oil. As a result, feedstocks and technologies previously considered economically impractical will rise to meet the increasing demand.

There is however, the ever-present political uncertainty that arises from the occurrence of natural gas and crude oil resources in countries other than user countries (i.e., provider countries). This has serious global implications for the supply and demand of petrochemicals and raw materials. In addition, the overall expansion of the population and an increase in individual purchasing power has resulted in an increase in demand for finished goods and greater consumption of energy in China, India, and Latin America.

However, the continued development of shale gas (tight gas) resources as well as crude oil from tight formation as well as the various technological advances to recover these resources (such as the combination of vertical hydraulic fracturing and horizontal drilling) will lead to a considerable rise in natural gas production and crude oil production. This new potential will cause many countries to reexamine their natural gas reserves and crude oil reserves to pursue development of their own nationally occurring gas plays and crude oil plays.

The production of chemicals from biomass is becoming an attractive area of investment for industries in the framework of a more sustainable economy. From a technical point of view, a large fraction of industrial chemicals and materials from fossil resources can be replaced by their bio-based counterparts. Nevertheless, fossil-based chemistry is still dominant because of optimized production processes and lower costs. The best approach to maximize the valorization of biomass is the processing of biological feedstocks in integrated biorefineries where both bio-based chemicals and energy carriers can be produced, similar to a traditional crude oil refinery. The challenge is to prove, together with the technical and economic feasibility, an environmental feasibility, in terms of lower impact over the entire production chain.

Biomass is essentially a rich mixture of chemicals and materials and, as such, has a tremendous potential as feedstock for making a wide range of chemicals and materials with applications in industries from pharmaceuticals to furniture. Various types of available biomass feedstocks, including waste, and the different pre-treatment and processing technologies being developed to turn these feedstocks into platform chemicals, polymers, materials and energy.

There are several viable biological and chemical transformation pathways from sugars to building blocks. A large number of sugar to building block transformations can be done by aerobic fermentation employing fungi, yeast or bacteria. Chemical and enzymatic transformations are also important process options. It should be noted however, that pathways with more challenges and barriers are less likely be considered as viable industrial processes. In addition to gasification followed by Fischer-Tropsch chemistry of the gaseous product (synthesis gas), chemical reduction, oxidation, dehydration, bond cleavage, and direct polymerization predominated. Enzymatic biotransformations comprise the largest group of biological conversions and some biological conversions can be accomplished without the need for an intermediate building block. 1,3-Propanediol ($HOCH_2CH_2CH_2OH$) is an example where a set of successive biological processes convert sugar directly to an end product. Each pathway has its own set of advantages and disadvantages. Biological conversions, of course, can be tailored to result in a specific molecular structure but the operating conditions must be relatively mild. Chemical transformations can operate at high throughput but, unfortunately, less conversion specificity is achieved.

Bio-based feedstocks may present a sustainable alternative to petrochemical sources to satisfy the ever-increasing demand for chemicals. However, the conversion processes needed for these future bio-refineries will likely differ from those currently used in the petrochemical industry. Biotechnology and chemo-catalysis offer routes for converting biomass into a variety of chemicals that can serve as starting-point chemicals. While a host of technologies can be leveraged for biomass upgrading, the outcome can be significant because there is the potential to upgrade the bio-derived feedstocks while minimizing the loss of carbon and the generation of byproducts.

In fact, biomass offers a source of carbon from the biosphere as an alternative to fossilized carbon laid down tens of millions of years ago. Anything that grows and is available in non-fossilized form can be classified as biomass, including arable crops, trees, bushes, animal byproducts, human and animal waste, waste food and any other waste stream that rots quickly and which can be replenished on a rolling timeframe of years or decades. One of the attractions of biomass is its versatility: under the right circumstances, it can be used to provide a sustainable supply of electricity, heat, transport fuels or chemical feedstocks in addition to its many other uses. One of the drawbacks of biomass, especially in the face of so many potential end uses, is its limited availability, even though the precise limitation is the subject of debate. Compared with the level of attention given to biomass as a source of electricity or heat, relatively little attention has been paid to biomass as a chemical feedstock. However, in a world in which conventional feedstocks are becoming constrained and countries are endeavoring to meet targets for reducing carbon dioxide emissions, there is a question as to whether biomass is too good to burn.

Developments in homogeneous and heterogeneous catalysis have led the way to effective approaches to utilizing renewable sources; however, further advances are needed to realize technologies that are competitive with established petrochemical processes. Catalysis will play a key role, with new reactions, processes, and concepts that leverage both traditional and emerging chemo- and bio-catalytic technologies.

Thus, new knowledge and better technologies are needed in dealing with chemical transformations that involve milder oxidation conditions, selective reduction, and dehydration, better control of bond cleavage, and improvements to direct polymerization of multifunctional monomers. For biological transformations, better understanding of metabolic pathways and cell biology, lower downstream recovery costs, increased utility of mixed sugar streams, and improved molecular thermal stability are necessary. While it is possible to prepare a very large number of molecular structures from the top building blocks, there is a scarcity of information about the behavior of the molecular products and industrial processing properties. A comprehensive database on biomolecular performance characteristics would prove extremely useful to both the public sector and private sector. Nevertheless, here is a significant market opportunity for the development of bio-based products from the four-carbon building blocks. In order to be competitive with petrochemicals derived products, there is a significant technical challenge and should be undertaken with a long-term perspective.

The petrochemical industry, which is based on natural gas and crude oil, competes with the energy-providing industry for the same fossil raw material. Dwindling oil and gas reserves, concern regarding the greenhouse effect (carbon dioxide emissions) and worldwide rising energy demand raise the question of the future availability of fossil raw materials. Biotechnological, chemical and engineering solutions are needed for utilization of this second-generation bio-renewable-based supply chain. One approach consists of the concept of a bio-refinery. Also gasification followed by Fischer-Tropsch chemistry is a promising pathway. Short and medium term, a feedstock mix with natural gas and crude oil dominating can most likely be expected. In the long term, due to the final limited availability of oil and gas, biomass will prevail. For this change to occur, great research and developments efforts must be carried out to have the necessary technology available when needed.

In summary, the petrochemical industry gives a series of added-value products to the natural gas and crude oil industry but, like any other business, suffers from issues relating to maturity. The reasons relating to the maturity of the industry are (i) expired patents, (ii) varying demand, (iii) matching demand with capacity, and intense competition. Actions to combat the aches and pains of maturity are to restructure capacity achieving mega sizes, downstream, and restructuring business practices. Strategies followed by some companies to combat maturity include exit, focus on core business and exploit a competitive advantage.

Nevertheless, the petrochemical industry is and will remain a necessary industry for the support of modern and emerging lifestyles (Table 13.6). In order to maintain an established petrochemical industry, strategic planning is the dominating practice to maintain the industry (replace imports, export, new products, alternate feedstocks such as the return to the chemicals-from-coal concept and the acceptance of feedstocks, such as oil shale and biomass) including developing criteria for selecting products/projects. After the oil crises of the 1970s (even though it is now four decades since these crises), it is necessary to cope with the new environment of product demand through the response to new growth markets and security of feedstock supply. Mergers, alliances and acquisitions could well be the dominating practice to combat industry maturity and increased market demand as one of the major activities. Other strategies are the focus on core business (the production of chemicals) and, last but not least, the emergence (or, in the case of coal, the reemergence) of alternate feedstocks to ensure industry survival.

In summary, synthesis gas offers many routes to industrial chemicals. They can be classified in a direct and indirect path. The direct path involves methanation, Fischer-Tropsch chemistry and synthesis of oxygenates. The direct conversion deals with the straight hydrogenation of carbon monoxide to paraffins, olefins and oxygen containing products. Best known in the direct hydrogenation of carbon monoxide is the Fischer-Tropsch synthesis yielding mixtures of mainly linear alkanes and/or alkenes.

Table 13.6 Illustration of the production of petrochemical starting materials from natural gas and crude oil.

Feedstock	Process	Product
Petroleum	Distillation	Light ends
		methane
		ethane
		propane
		butane
	Catalytic cracking	ethylene
		propylene
		butylenes
		higher olefins
	Catalytic reforming	benzene
		toluene
		xylenes
	coking	ethylene
		propylene
		butylenes
		higher olefins
Natural gas	refining	methane
		ethane
		propane
		butane

Mechanistically it can be described as a reductive oligomerization of carbon monoxide following a geometric progression (Schulz-Flory distribution). With a-values close to one broad product distributions are obtained, whereas small a-values predominantly yield methane. The indirect path embraces carbonylation, methanol and methyl formate chemistry (Keim, 1986).

The great desire for the synthesis of ethylene glycol based on carbon monoxide chemistry is also underlined by the research efforts being reported by various companies. Besides the direct hydrogenation also routes via methanol or formaldehyde are possible. Also, the reductive carbonylation of formaldehyde yields glycol dialdehyde, which can be hydrogenated to ethylene glycol (Keim, 1986).

References

Chenier, P.J. 2002. *Survey of Industrial Chemicals* 3rd Edition. Springer Publishing Company, New York.

Clark, J.H., and Deswarte, F. (Editors). 2015. *Introduction to Chemicals from Biomass* 2nd Edition. John Wiley & Sons Inc., Hoboken, New Jersey.

EPCA. 2016. *50 Years of Chemistry for You*. European Petrochemical Association. Brussels, Belgium. https://epca.eu/

Favennec, J-P. (Editor) (2001). *Petroleum Refining: Refinery Operation and Management*. Editions Technip, Paris, France.

Gary, J.H., Handwerk, G.E., and Kaiser, M.J. 2007. *Petroleum Refining: Technology and Economics*, Fifth Edition. CRC Press, Taylor & Francis Group, Boca Raton, Florida.

Goldstein, R.F. 1949. *The Petrochemical Industry*. E. & F. N. Spon, London.

Hahn, A.V. 1970. *The Petrochemical Industry: Market and Economics*. McGraw-Hill, New York.

Hassani, H., Silva, E.S., and Al Kaabi, A.M. 2017. The Role of Innovation and Technology in Sustaining the Petroleum and Petrochemical Industry. *Technological Forecasting and Social Change*, 119 (June): 1-17.

Hsu, C.S., and Robinson, P.R. (Editors). 2017. *Handbook of Petroleum Technology*. Springer International Publishing AG, Cham, Switzerland.

Katz, D.K. 1959. *Handbook of Natural Gas Engineering*. McGraw-Hill Book Company, New York.

Keim, W. 1986. C1 Chemistry: Potential and Developments. *Pure & Appl. Chem.*, 58(6); 825-832.

Kidnay, A.J., and Parrish, W.R. 2006. *Fundamentals of Natural Gas Processing*. CRC Press, Taylor & Francis Group, Boca Raton, Florida.

Kohl, A.L., and Nielsen, R.B., 1997. *Gas Purification*. Gulf Publishing Company, Houston, Texas.

Kohl, A.L., and Riesenfeld, F.C. 1985. *Gas Purification*. 4th Edition, Gulf Publishing Company, Houston, Texas.

Lee, S., Speight, J.G., and Loyalka, S. 2007. *Handbook of Alternative Fuel Technologies*. CRC Press, Taylor & Francis Group, Boca Raton, Florida.

Lin, Y-C. 2013. Catalytic Valorization of Glycerol to Hydrogen and Syngas. *International Jounral of Hydrogen Energy*, 38(6): 2678-2700.

Maddox, R.N., Bhairi, A., Mains, G.J., and Shariat, A. 1985. In *Acid and Sour Gas Treating Processes*. S.A. Newman (Editor). Gulf Publishing Company, Houston, Texas. Chapter 8.

Matar, S., and Hatch, L.F. 2001. *Chemistry of Petrochemical Processes* 2nd Edition. Butterworth-Heinemann, Woburn, Massachusetts.

Meyers, R.A. 2005. *Handbook of Petrochemicals Production Processes*. McGraw-Hill, New York.

Mokhatab, S., Poe, W.A., and Speight, J.G. 2006. *Handbook of Natural Gas Transmission and Processing*. Elsevier, Amsterdam, Netherlands.

Moschopedis, S.E., and Speight, J.G. 1971. Water-Soluble Derivatives of Athabasca Asphaltenes. *Fuel*, 50: 34.

Moschopedis, S.E., and Speight, J.G. 1974. The Chemical Modification of Bitumen and Its Non-Fuel Uses. *Preprints. Div. Fuel Chem., Am. Chem. Soc.*: 1974, 19(2): 291.

Moschopedis, S.E., and Speight, J.G. 1976a. The Chemical Modification of Bitumen Heavy Ends and Their Non-Fuel Uses. In: *Shale Oil, Tar Sands and Related Fuels Sources*. Adv. in Chem. Series No. 151, Am. Chem. Soc. T.F. Yen (Editor). Page 144.

Moschopedis, S.E., and Speight, J.G. 1976b. The Chlorinolysis of Petroleum Asphaltenes. *Chemika Chronika*, 5: 275.

Moschopedis, S.E., and Speight, J.G. 1978. Sulfoxidation of Athabasca Bitumen. *Fuel*, 857: 647.

Naderpour, N. 2008. *Petrochemical Production Processes*. SBS Publishers, Delhi, India.

Newman, S.A. 1985. *Acid and Sour Gas Treating Processes*. Gulf Publishing, Houston, Texas.

Parkash, S. 2003. *Refining Processes Handbook*. Gulf Professional Publishing, Elsevier, Amsterdam, Netherlands.

Soukup, G., Pfeifer, C., Kreuzeder, A., and Hofbauer, H. 2009. In Situ Carbon Dioxide Capture in a Dual Fluidized Bed Biomass Steam Gasifier – Bed Material and Fuel Variation. *Chem. Eng. Technol.*, 32: 348-354.

Spitz, P.H. 1988. *Petrochemicals: The Rise Of An Industry*. Wiley InterScience, Hoboken, New Jersey.

Speight, J.G. 1987. Petrochemicals. *Encyclopedia of Science and Technology*, Sixth Edition, McGraw-Hill, New York. Volume 13, page 251.

Speight, J.G. 1994. Chemical and Physical Studies of Petroleum Asphaltene constituents. In: *Asphaltene constituents and Asphalts. I. Developments in Petroleum Science,* 40. T.F. Yen and G.V. Chilingarian (Editors). Elsevier, Amsterdam, Netherlands. Chapter 2.

Speight, J.G. 2002. *Chemical Process and Design Handbook*. McGraw-Hill Publishers, New York.

Speight, J.G. 2008. *Handbook of Synthetic Fuels Handbook: Properties, Processes, and Performance*. McGraw-Hill, New York.

Speight. J.G. 2011. *The Refinery of the Future*. Gulf Professional Publishing, Elsevier, Oxford, United Kingdom.

Speight, J.G. 2013. *The Chemistry and Technology of Coal*. 3rd Edition. CRC Press, Taylor & Francis Group, Boca Raton, Florida.

Speight, J.G. 2014. *The Chemistry and Technology of Petroleum* 5th Edition. CRC Press, Taylor & Francis Group, Boca Raton, Florida.

Speight, J.G. 2015. *Handbook of Petroleum Product Analysis* 2nd Edition. John Wiley & Sons Inc., Hoboken, New Jersey.

Speight, J.G. 2017. *Handbook of Petroleum Refining*. CRC Press, Taylor & Francis Group, Boca Raton, Florida.

Speight, J.G. 2019. *Handbook of Petrochemical Processes*. CRC Press, Taylor & Francis Group, Boca Raton, Florida.

Steiner, H. 1961. *Introduction to Petroleum Chemicals*. Pergamon Press, New York.

Wittcoff, H.A., and Reuben, B.G. 1996. *Industrial Organic Chemicals*. John Wiley & Sons Inc., New York.

14
Technology Integration

14.1 Introduction

The projections for the continued use of fossil fuels indicate that there will be at least another five decades of fossil fuel use (especially in the cases of coal and crude oil) before biomass and other forms of alternate energy take hold and are able to provide sufficient energy not only to subsidize the production of energy from fossil fuels but to replace the fossil fuels (Speight, 2011a, 2011b, 2013a, 2013b). Furthermore, estimations that the era of fossil fuels (natural gas, crude oil, coal, and biomass, and carbonaceous wastes) will be almost over when the cumulative production of the fossil resources reaches 85% of their initial total reserves (Hubbert, 1973) may or may not have some merit. In fact, the relative scarcity (compared to a few decades ago) of crude oil was real but it seems that the remaining reserves make it likely that there will be an adequate supply of energy for several decades (Martin, 1985; MacDonald, 1990; Banks 1992; Krey *et al.*, 2009; Speight, 2011c, 2013a, 2013b, 2014). The environmental issues are very real and require serious and continuous attention.

Technologies which ameliorate the effects of fossil fuels combustion on acid rain deposition, urban air pollution, and global warming must be pursued vigorously (Vallero, 2008). There is a challenge that must not be ignored and the effects of acid rain in soil and water leave no doubt about the need to control its causes (Mohnen, 1988). Indeed, recognition of the need to address these issues is the driving force behind recent energy strategies as well as a variety of research and development programs (Stigliani and Shaw, 1990; United States Department of Energy, 1990; United States General Accounting Office, 1990).

While regulations on the greenhouse gas (GHG) carbon dioxide (CO_2) would be an immediate hurdle to deployment of coal plants, gasification plants are in the best position compared to other coal-based alternatives to capture carbon dioxide. However, with the continued uncertainty of carbon dioxide regulation, there is industry reluctance to make large investments in projects with high emissions of carbon dioxide since a cost-effective solution for reducing such emissions is not yet available. Nevertheless, the reduction in greenhouse gas emissions can be an enhancing factor for gasification in the long run because the carbon dioxide from a gasification plant is more amenable to capture.

As new technology is developed, emissions may be reduced by repowering in which aging equipment is replaced by more advanced and efficient substitutes. Such repowering might, for example, involve an exchange in which an aging unit is exchanged for a newer combustion chamber, such as the atmospheric fluidized-bed combustor (AFBC) or the pressurized fluidized-bed combustor (PFBC).

Indeed, recognition of the production of these atmospheric pollutants in considerable quantities every year has led to the institution of national emission standards for many

440   Synthesis Gas

pollutants. Using sulfur dioxide as the example, the various standards are not only very specific but will become more stringent with the passage of time. Atmospheric pollution is being taken very seriously and there is also the threat, or promise, of heavy fines and/or jail terms for any pollution-minded miscreants who seek to flaunt the laws (Vallero, 2008). Nevertheless, a trend to the increased use of fossil fuels will require more stringent approaches to environmental protection issues than we have ever known at any time in the past. The need to protect the environment is strong.

14.2 Applications and Products

Hydrogen and carbon monoxide, the major components of synthesis gas, are the basic building blocks of a number of other products, such as fuels, chemicals and fertilizers. In addition, a gasification plant can be designed to produce more than one product at a time (*co-production* or *polygeneration*), such as electricity, and chemicals (e.g., methanol or ammonia) (Table 14.1).

Table 14.1 Examples of feedstocks for, and products from, the gasification process.

Feedstocks*	Products*
Biomass	Acetic acid
Black liquor	Acetic anhydride
Coal	Ammonia
Crude oil coke	Ammonium nitrate
Municipal solid waste	Carbon Dioxide
Natural gas	Carbon Monoxide
Refinery gas	Dimethyl ether (DME)
Resids	Electric Power
Sewage sludge	Ethanol
Waste oil	Hot Water
	Hydrogen
	Industrial chemicals
	Methanol
	Steam
	Sulfur
	Sulfuric Acid
	Urea

*Listed alphabetically rather than by preference.

14.2.1 Chemicals and Fertilizers

The process of producing energy using the gasification method has been in use for more than 100 years. During that time coal was used to power these plants and, initially developed to produce town gas, for lighting and heating in the 1800s, this was replaced by electricity and natural gas as well as use in blast furnaces. But the bigger role for gasification (of coal) was in the production of synthetic chemicals where it has been in use since the 1920s. The concept is again being considered as a means of producing much-needed chemicals with the added benefit that not only coal but also low-value carbonaceous and/or hydrocarbonaceous feedstocks are gasified in a large chemical reactor. The resulting synthesis gas is cleansed and then converted into high-value products such as synthetic fuels, chemicals, and fertilizers.

Typically, the chemical industry uses gasification to produce methanol as well as a variety of other chemicals – such as ammonia and urea – which form the foundation of nitrogen-based fertilizers and to produce a variety of plastics. The majority of the operating gasification plants worldwide are designed to produce chemicals and fertilizers.

14.2.2 Substitute Natural Gas

Gasification can also be used to create substitute natural gas (SNG) from coal by using a *methanation* reaction in which the coal-based synthesis gas – mostly carbon monoxide and hydrogen – can be converted to methane.

In the early 19th century, coal gasification was used for the production of so-called town gas for street lamps. Later, gasification was used during World War II to produce liquid fuels from coal, and the process underwent a revival during the oil crises of the 1970s and 1980s. Prominent developments of gasification units from this period include the British Gas–Lurgi (BGL) fixed-bed gasifier, the Kellogg–Rust–Westinghouse (KRW) fluidized-bed gasifier and the Reinbraun AG hydrogasifier. However, dramatic changes in world energy markets rendered coal gasification economically unfeasible and research activities rapidly decreased alongside declining interest from industry.

Substitute natural gas (*synthetic natural gas*) (SNG) is an artificially produced version of natural gas which can be produced from coal, biomass, crude oil coke, or solid waste. The carbon-containing mass can be gasified and the resulting synthesis gas converted to methane, the major component of natural gas. There are several advantages associated with producing substitute natural gas. In times when natural gas is in short supply, substitute natural gas from coal could be a major driver for energy security by diversifying energy options and reducing imports of natural gas, thus helping to stabilize fuel prices.

Biomass and other low-cost feedstocks (such as municipal waste) can also be used along with coal to produce substitute natural gas. The use of biomass would reduce greenhouse gas emissions, as biomass is a carbon-neutral fuel. In addition, the development of substitute natural gas technology would also boost the other gasification-based technologies such as hydrogen generation and integrated gasification combined cycle (IGCC).

Whereas high-temperature gasification processes yield bio-synthesis gas with high concentrations of carbon monoxide and little methane, interest in synthetic natural gas (SNG) production is concentrated on gasification processes that yield product gases with high methane contents. SNG is a gas with similar properties as natural gas but produced by methanation of H_2 and CO in gasification product gas. Methanation is the catalytic reaction of carbon monoxide and/or carbon dioxide with hydrogen, forming methane and water:

$$CO + 3H_2 \rightarrow CH_4 + H_2O$$

Consecutive and side reactions (shift conversion, Boudouard equilibrium, hydrogenation of carbon) make the calculation of equilibrium conditions very complex. The methanation reactions of both carbon monoxide and carbon dioxide are highly exothermic. Such high heat releases strongly affect the process design of the methanation plant since it is necessary to prevent excessively high temperatures in order to avoid catalyst deactivation and carbon deposition. The highly exothermic reaction generally creates a problem for the design of methane synthesis plants: either the temperature increase must be limited by recycling of reacted gas or steam dilution, or special techniques such as isothermal reactors or fluidized beds, each with indirect cooling by evaporating water, must be used.

The methanation process is very well known and technically important as a catalytic purification step for the removal of trace carbon monoxide (typically below a few vol%) from process gases, especially for hydrogen production. Methanation of gases with a high CO content, like gasification product gas is not well established and there are no commercial catalysts available (i.e., the Dakota coal-to-SNG plant uses a non-commercial catalyst).

For the removal of carbon monoxide from synthesis gases, catalysts with usually <15 wt% nickel are used predominantly. For SNG production, catalysts with high nickel content are preferred, similar to those used in reforming naphtha to a methane-rich gas. Catalysts based on ruthenium have been tried repeatedly for methanation but have not found the broad application of nickel-based catalysts. Catalytic activity is affected seriously even by very low concentrations of catalyst poisons in gases to be reformed. Such catalyst poisons are sulphur, arsenic, copper, vanadium, lead, and chlorine or halogens in general. Precaution must be taken with nickel-containing catalysts to prevent formation of highly poisonous nickel carbonyl [$Ni(CO)_4$]. In practical operation of methanation plants, temperatures below 200°C at the nickel catalyst must be avoided.

Identical to conventional natural gas (methane, CH_4), the resulting substitute natural gas can be transported in existing natural gas pipeline networks and used to generate electricity, produce chemicals/fertilizers, or heat homes and businesses. For many countries that lack natural gas resources, substitute natural gas enhances domestic fuel security by displacing imported natural gas that is likely to be supplied in the form of *liquefied natural gas* (LNG).

14.2.3 Hydrogen for Crude Oil Refining

The use of hydrogen in thermal processes is perhaps the single most significant advance in refining technology during the 20th century and is now an inclusion in most refineries. In fact, a critical issue facing the refineries at present and in the future is the changing slate of crude oil feedstocks and the conversion of these feedstocks into refined transportation fuels under an environment of increasingly more stringent clean fuel regulations, decreasing heavy fuel oil demand, and increasing supply of heavy, high-sulfur crude oils. Hydrogen network optimization is at the forefront of world refineries options to address clean fuel trends, to meet growing transportation fuel demands and to continue to make a profit from their crudes (Long *et al.*, 2011). A key element of a hydrogen network analysis in a refinery involves the capture of hydrogen in its fuel streams and extending its flexibility and processing options. Thus, innovative hydrogen network optimization will be a critical factor influencing future refinery operating flexibility and profitability in a shifting world of crude feedstock supplies and ultra-low-sulfur (ULS) gasoline and diesel fuel.

Upgrading feedstocks such as heavy oils and residua evolved after the introduction of hydrodesulfurization processes (Speight, 2000; Rana *et al.*, 2007; Ancheyta and Speight, 2007). In the early days, the goal was desulfurization but, in later years, the processes were adapted to a 10 to 30% partial conversion operation, as intended to achieve desulfurization and obtain low-boiling fractions simultaneously, by increasing severity in operating conditions. However, as refineries have evolved and feedstocks have changed, refining heavy feedstocks has become a major issue in modern refinery practice and several process configurations have evolved to accommodate the heavy feedstocks (Khan and Patmore, 1997; Speight, 2011a, 2014, 2017).

Hydrogen, one of the two major components of synthesis gas, is used to produce high-quality gasoline, diesel fuel, and jet fuel, meeting the requirements for clean fuels in state and federal clean air regulations. Hydrogen is also used to upgrade heavy crude oil and tar sand bitumen. Refineries can gasify low-value residuals, such as crude oil coke, asphalts, tars, and some oily wastes from the refining process to generate both the required hydrogen and the power and steam needed to run the refinery.

Thus, the gasification of crude oil residua, crude oil coke, and other feedstocks such as biomass (Speight, 2008, 2011a, 2011b, 2014) to produce hydrogen and/or power may become an attractive option for refiners. The premise that the gasification section of a refinery will be the *garbage can* for deasphalter residues, high-sulfur coke, as well as other refinery wastes is worthy of consideration. Other processes such as ammonia dissociation, steam-methanol interaction, or electrolysis are also available for hydrogen production, but economic factors and feedstock availability assist in the choice between processing alternatives.

Gasification of biomass has been identified as a possible system for producing renewable hydrogen, which is beneficial to exploit biomass resources, to develop a highly efficient clean way for large-scale hydrogen production. Biomass gasification can be considered as a form of pyrolysis, which takes place in higher temperatures and produces a mixture of gases with a hydrogen content on the order of 6 to 6.5% v/v. The synthetic gas produced by the gasification of biomass is made up of hydrogen, carbon monoxide, methane, nitrogen, carbon dioxide, oxygen, and volatile tar. When gasifying biomass, tar that is formed together with the synthetic gas is difficult to remove with a physical dust removal method. The product distribution and gas composition depends on many factors including the gasification temperature and the reactor type. The most important gasifier types are fixed-bed (updraft or downdraft fixed beds), fluidized-bed, and entrained-flow gasifiers. All these gasifiers need to include significant gas conditioning along with the removal of tars and inorganic impurities and the subsequent conversion of carbon monoxide to hydrogen by the water gas shift reaction.

14.2.4 Transportation Fuels

Gasification is the foundation for converting coal and other solid feedstocks and natural gas into transportation fuels, such as gasoline, ultra-clean diesel fuel, jet fuel, naphtha, and synthetic oils. Two options are available for converting carbonaceous feedstocks to motor fuels via gasification.

In the *first option*, the synthesis gas undergoes an additional process, the Fischer-Tropsch (FT) reaction, to convert it to a liquid crude oil product. The Fischer-Tropsch process, with coal as a feedstock, was invented in the 1920s, used by Germany during World War II, and has been utilized in South Africa for decades. Currently, it is also used in Malaysia and the Middle East with natural gas as the feedstock. In the *second option*, the *methanol-to-gasoline* process (MTG process), the synthesis gas is first converted to methanol (a commercially used process) and the methanol is then converted to gasoline by reacting it over catalysts.

In the non-selective catalytic Fischer-Tropsch (FT) synthesis one mole of CO reacts with two moles of H_2 to form mainly paraffin straight-chain hydrocarbon derivatives (C_xH_{2x}) with minor amounts of branched and unsaturated hydrocarbon derivatives (i.e., 2-methyl paraffins and α-olefins), and primary alcohols. Undesirable side reactions include methanation, the Boudouard reaction, coke deposition, oxidation of the catalyst, or carbide formation. Typical operation conditions for FT synthesis are temperatures of 200-350°C and pressures between 25 and 60 bar [22]. In the exothermic FT reaction about 20% of the chemical energy is released as heat. The process is represented simply as:

$$nCO + (2n+1)H_2 \rightarrow H[-CH_2-]_nH + nH_2O$$

Fischer-Tropsch processes can be used to produce either a light synthetic crude oil (syncrude) and light olefins or heavy waxy hydrocarbon derivatives. The syncrude can be refined to a high-quality sulphur and aromatic liquid product and specialty waxes or, if hydrocracked and/or isomerized, to produce excellent diesel fuel, lube oils, and naphtha, which is an ideal feedstock for cracking to olefins. For direct production of gasoline and light olefins, the FT process is operated at high temperature (330–350°C), for production of waxes and/or diesel fuel, at low temperatures (220–250°C).

Several types of catalysts can be used for the Fischer-Tropsch synthesis – the most important are based on iron (Fe) or cobalt (Co). Cobalt catalysts have the advantage of a higher conversion rate and a longer life (over five years). The Co catalysts are in general more reactive for hydrogenation and produce therefore less unsaturated hydrocarbon derivatives (olefins) and alcohols compared to iron catalysts. Iron catalysts have a higher tolerance for sulphur, are cheaper, and produce more olefin products and alcohols. The lifetime of the Fe catalysts is short and in commercial installations generally limited to eight weeks.

Catalysts for Fischer-Tropsch synthesis can be damaged with impurities as NH_3, HCN, H_2S, and COS. These impurities poison the catalysts. HCl causes corrosion of catalysts. Alkaline metals are deposited on the catalyst. Tars are deposited, cause poisoning of catalyst and contaminate the products. And particles (dust, soot, ash) cause fouling of the reactor. The removal limit is based on an economic optimum determined by catalyst stand-time and investment in gas cleaning. But generally, all these impurities should be removed to a concentration below 1 ppm v/v.

The Fischer-Tropsch synthesis produces hydrocarbon derivatives of different chain lengths from a gaseous mixture of hydrogen and carbon monoxide. The higher molecular weight hydrocarbon derivatives can be hydrocracked to form mainly diesel of excellent quality. The fraction of short-chain hydrocarbon derivatives is used in a combined cycle plant with the remainder of the synthesis gas. The prognosis for the route to fuels and the use of gasification of biomass and conversion of the gaseous product(s) to Fischer-Tropsch-fuels in the transport sector is very promising. However, large-scale (pressurized) biomass gasification-based systems are necessary with particular attention given to the gas cleaning section.

14.2.5 Transportation Fuels from Tar Sand Bitumen

Tar sand deposits (oil sands deposits) can be found in many countries throughout the world, and may comprise more than 65% v/v of the total world oil reserve. The two largest deposits are in Canada and Venezuela – the tar sand deposits in Canada comprise three major deposits covering a region estimated to be more than 54,000 square miles (approximately 140,000 square kilometers). Estimates from the Alberta Energy and Utilities Board indicate that approximately 1.6 trillion barrels (1.6×10^{12} bbls) of crude oil equivalent are contained within the tar sand deposits of Canada. Of this amount, more than 170 billion barrels (170×10^9 bbls) are considered recoverable, but this amount is dependent on current oil prices.

Gasification, a commercially proven technology that can be used to convert crude oil coke into synthesis gas, is being recognized as a means to economically generate hydrogen, power, and steam for tar sand operators in northeastern Alberta, Canada. The tar sand deposits in Alberta (Canada) are estimated to contain as much recoverable bitumen as the crude oil available from the vast oil fields of Saudi Arabia. However, to convert the raw bitumen to saleable products requires extracting the bitumen from the sand and refining the separated bitumen to transportation fuels. The mining process requires massive amounts of steam to separate the bitumen from the sand and the refining process demands large quantities of hydrogen to upgrade the raw distillates to saleable products. Residual materials from the bitumen upgrading process include crude oil coke, deasphalted residua, vacuum residua, all of which contain unused energy that can be released and captured for use by gasification. Traditionally, tar sand operators have utilized natural gas to produce the steam and hydrogen needed for the mining, upgrading, and refining processes.

Traditionally, oil sands operators have utilized natural gas to produce the steam and hydrogen needed for the mining, upgrading, and refining processes. However, a number of operators will soon gasify coke to supply the necessary steam and hydrogen. Not only will gasification displace expensive natural gas as a feedstock, it will also enable the extraction of useable energy from what is otherwise a very low-value product (coke). In addition, black water from the mining and refining processes can be recycled to the gasifiers using a wet feed system, reducing fresh water usage and waste water management costs – traditional oil sand operations consume large volumes of water.

14.2.6 Power Generation

Coal-to-power through gasification technology allows the continued use of domestic supplies of coal without the high level of air emissions associated with conventional coal-burning technologies. One of the advantages of the coal gasification technology is that it offers the poly-generation: co-production of electric power, liquid fuels, chemicals, hydrogen and from the syngas generated from gasification.

Furthermore, an *Integrated Gasification Combined Cycle* power plant (IGCC power plant) combines the gasification process with a *combined cycle* power block (consisting of one or more gas turbines and a steam turbine). Clean synthesis gas is combusted in high-efficiency gas turbines to produce electricity. The excess heat from the gas turbines and from the gasification reaction is then captured, converted into steam, and sent to a steam turbine to produce additional electricity.

In the IGCC power plant – where power generation is the focus – the clean synthesis gas is combusted in high-efficiency gas turbines to generate electricity with very low emissions. The gas turbines used in these plants are slight modifications of proven, natural gas combined-cycle gas turbines that have been specially adapted for use with synthesis gas. For IGCC power plants that include carbon dioxide capture, these gas turbines are adapted to operate on synthesis gas with higher levels of hydrogen. Although state-of-the-art gas turbines are commercially ready for the high-hydrogen synthesis gas, there is a movement to develop the next generation of even more efficient gas turbines ready for carbon dioxide capture-based IGCC power plants.

The heat recovery steam generator (HRSG) captures heat in the hot exhaust from the gas turbines and uses it to generate additional steam that is used to make more power in the steam turbine portion of the combined-cycle unit. In most IGCC power plant designs, steam recovered from the gasification process is superheated in the heat recovery steam generator to increase overall efficiency output of the steam turbines. This IGCC combination, which includes a gasification plant, two types of turbine generators (gas and steam), and the heat recovery steam generator is clean and efficient.

Biomass fuel producers, coal producers and, to a lesser extent, waste companies are enthusiastic about supplying co-gasification power plant and realize the benefits of co-gasification with alternate fuels. The benefits of a co-gasification technology involving coal and biomass include use of a reliable coal supply with gate-fee waste and biomass which allows the economies of scale from a larger plant than could be supplied just with waste and biomass. In addition, the technology offers a future option for refineries to for hydrogen production and fuel development. In fact, oil refineries and petrochemical plants are opportunities for gasifiers when the hydrogen is particularly valuable (Speight, 2011a).

14.2.7 Waste-to-Energy Gasification

Municipalities are spending millions of dollars each year to dispose of solid waste that, in fact, contains valuable unused energy. In addition to the expense of collecting this waste, they must also contend with increasingly limited landfill space, the environmental impacts of landfilling, and stringent bans on the use of incinerators. As a result of these challenges, municipalities are increasingly looking to gasification as a solution to help transform this waste into energy.

The traditional waste-to-energy plant, based on mass-burn combustion on an inclined grate, has a low public acceptability despite the very low emissions achieved over the last decade with modern flue gas clean-up equipment. This has led to difficulty in obtaining planning permissions to construct the new waste-to-energy plants that are needed. After much debate, various governments have allowed options for advanced waste conversion technologies (gasification, pyrolysis and anaerobic digestion), but will only give credit to the proportion of electricity generated from non-fossil waste.

Gasification can convert municipal and other wastes (such as construction and demolition wastes) that cannot be recycled into electric power or other valuable products, such as chemicals, fertilizers, and substitute natural gas. Instead of paying to dispose of these wastes, municipalities are generating income from these waste products, since they are valuable feedstocks for a gasifier. Gasifying municipal and other waste streams reduces the need for landfill space, decreases the generation of methane (a potent greenhouse gas) as the landfill matures through bacterial action, and reduces the potential for groundwater contamination from landfill sites.

Co-utilization of waste and biomass with coal may provide economies of scale that help achieve the policy objectives identified above at an affordable cost. In some countries, governments propose cogasification processes as being *well suited for community-sized developments* suggesting that waste should be dealt with in smaller plants serving towns and cities, rather than moved to large, central plants (satisfying the so-called *proximity principle* – the tendency to band similar entities together to achieve a common goal.

The installation of gasifiers by municipalities to dispose of waste and create energy is almost a return to the days when gasification originally became commercial – perhaps déjà vu – and every small town has a gasification plant to produce gas (hence the name *town gas*) for heating and lighting purposes.

However, it is important to add that gasification does not compete with recycling. In fact, gasification complements existing recycling programs by the creation of an added-value product (energy). Many materials, including a wide range of plastics, cannot currently be recycled or recycled any further and are ideal candidates for feedstocks to the gasification process. As the amount of waste generated increases (in line with an increase in the population), recycling rates increase to the point of overburdening the system, gasification will alleviate any potential bottlenecks through the generation of energy.

14.2.8 Biomass Gasification

Biomass includes a wide range of materials, including energy crops such as switch grass and miscanthus, agricultural sources such as corn husks, wood pellets, lumbering and timbering wastes, yard wastes, construction and demolition waste, and bio-solids (treated sewage sludge). Gasification helps recover the energy locked in these materials. Gasification can convert biomass into electricity and products, such as ethanol, methanol, fuels, fertilizers, and chemicals. Thus, in addition to using the traditional feedstocks such as coal and crude oil coke, gasifiers can be designed to utilize biomass, such as yard and crop waste, bio-solids, energy crops, such as switch grass, and waste and residual pulp/paper plant materials as feedstock.

Biomass usually contains a high percentage of moisture (along with carbohydrates and sugars). The presence of high levels of moisture in the biomass reduces the temperature inside the gasifier, which then reduces the efficiency of the gasifier. Therefore, many biomass gasification technologies require that the biomass be dried to reduce the moisture content prior to feeding into the gasifier.

Like many solid feedstocks, biomass can come in a range of sizes. In many biomass gasification systems, the biomass must be processed to a uniform size or shape to feed into the gasifier at a consistent rate and to ensure that as much of the biomass is gasified as possible. However, beyond the issue of biomass availability, including the seasonal factors associated with many of the biomass feedstocks, another major concern is that more energy is expended in the collection and preparation stages than is generated through processing the biomass and although technical hurdles to biomass use remain. In general, there is the general perception in many countries that increased use of biomass feedstocks is driven more by environmental and regulatory factors, if anything, than by free-market forces. Without tax credits or similar incentives, biomass is unlikely to be used as a base-load feedstock and market entry is likely to occur as a blend with other feedstocks (Clayton *et al.*, 2002).

Most biomass gasification systems use air instead of oxygen for the gasification reactions (which is typically used in large-scale industrial and power gasification plants). Gasifiers that use oxygen require an air separation unit to provide the gaseous/liquid oxygen; this is usually not cost-effective at the smaller scales used in biomass gasification plants. Air-blown gasifiers use the oxygen in the air for the gasification reactions.

In general, biomass gasification plants are much smaller than the typical coal or crude oil coke gasification plants used in the power, chemical, fertilizer and refining industries. As such, they are less expensive to build and have a smaller *facility footprint*. While a large industrial gasification plant may take up 150 acres of land and process 2,500 to 15,000 tons per day of feedstock (such as coal or crude oil coke), the smaller biomass plants typically process 25 to 200 tons of feedstock per day and take up less than 10 acres.

Currently, most ethanol in the United States is produced from the fermentation of corn. Vast amounts of corn (and land, water and fertilizer) are needed to produce the ethanol. As more corn is being used, there is an increasing concern about less corn being available for food. Gasifying biomass, such as corn stalks, husks, and cobs, and other agricultural waste products to produce ethanol and synthetic fuels such as diesel and jet fuel can help break this energy-food competition.

Biomass, such as wood pellets, yard and crop wastes, and energy crops such as switch grass and waste from pulp and paper mills can be used to produce ethanol and synthetic diesel. The biomass is first gasified to produce synthesis gas (synthesis gas), and then converted via catalytic processes to these downstream products.

Each year, municipalities spend millions of dollars collecting and disposing of wastes, such as yard wastes (grass clippings and leaves) and construction and demolition debris. While some municipalities compost yard wastes, this takes a separate collection by a city – an expense many cities just can't afford. Yard waste and the construction and demolition debris can take up valuable landfill space – shortening the life of a landfill. With gasification, this material is no longer a waste, but a feedstock for a biomass gasifier. And, instead of paying to dispose of and manage a waste for years in a landfill, using it as a feedstock reduces disposal costs and landfill space, and converts those wastes to power and fuels.

Thus, the benefits of biomass gasification include (i) the conversion of a waste product into high-value products, (ii) the reduced need for landfill space for disposal of solid wastes, (iii) the decrease in the emission of methane from landfills, (iv) the reduced risk of the contamination of the groundwater by runoff from landfills, and (v) the production of ethanol from non-food sources. Thus, municipalities, as well as the paper and agricultural industries, would be well advised to use gasification to reduce the disposal costs associated with these wastes as well to produce electricity and other valuable products from these waste materials.

While still relatively new, biomass gasification shows a great deal of promise and a key advantage of biomass gasification is that the process can convert non-food biomass materials, such as corn stalks and wood wastes, to useful products such as alcohol derivatives that can be used as sources of useful chemicals (Chapter 13). Furthermore, biomass gasification does not diminish or remove the availability of food-based biomass, such as corn, from the economy, unlike typical fermentation processes for making alcohol derivatives.

14.3 Environmental Benefits

The careless combustion of fossil fuels can account for the large majority of the sulfur oxides and nitrogen oxides released to the atmosphere. Whichever technologies succeed in reducing the amounts of these gases in the atmosphere should also succeed in reducing the amounts of urban smog, those notorious brown and grey clouds that are easily recognizable at some considerable distances from urban areas, not only by their appearance but also by their odor.

$$SO_2 + H_2O \rightarrow H_2SO_4 \text{ (sulfurous acid)}$$

$$2SO_2 + O_2 \rightarrow 2SO_3$$

$$SO_3 + H_2O \rightarrow H_2SO_4 \text{ (sulfuric acid)}$$

$$2NO + H_2O \rightarrow HNO_2 + HNO_3 \text{ (nitrous acid + nitric acid)}$$

$$2NO + O_2 \rightarrow 2NO_2$$

$$NO_2 + H_2O \rightarrow HNO_3 \text{ (nitric acid)}$$

Current awareness of these issues by a variety of levels of government has resulted, in the United States, of the institution of the Clean Fossil fuels Program to facilitate the development of pollution abatement technologies. And it has led to successful partnerships between government and industry (United States Department of Energy, 1993). In addition, there is the potential that new laws, such as the passage in 1990 of the Clean Air Act Amendments in the United States (United States Congress, 1990; Stensvaag, 1991) will be a positive factor and supportive of the controlled clean use of fossil fuels. However, there will be a cost but industry is supportive of the measure and confident that the goals can be met.

Besides fuel and product flexibility, gasification-based systems offer significant environmental advantages over competing technologies, particularly coal-to-electricity combustion systems. Gasification plants can readily capture carbon dioxide, the leading greenhouse gas, much more easily and efficiently than coal-fired power plants. In many instances, this carbon dioxide can be sold, creating additional value from the gasification process.

Carbon dioxide captured during the gasification process can be used to help recover oil from otherwise depleted oil fields. The Dakota Gasification plant in Beulah, North Dakota, captures its carbon dioxide while making substitute natural gas and sells it for enhanced oil recovery. Since 2000, this plant has captured and sent the carbon dioxide via pipeline to EnCana's Weyburn oil fields in Saskatchewan, Canada, where it is used for enhanced oil recovery. More than five million tons of carbon dioxide has been sequestered.

14.3.1 Carbon Dioxide

In a gasification system, carbon dioxide can be captured using commercially available technologies (such as the *water gas shift reaction*) before it would otherwise be vented to the atmosphere. Converting the carbon monoxide to carbon dioxide and capturing it prior to combustion is more economical than removing carbon dioxide after combustion, effectively "de-carbonizing" or, at least, reducing the carbon in the synthesis gas.

Gasification plants manufacturing ammonia, hydrogen, fuels, or chemical products routinely capture carbon dioxide as part of the manufacturing process. According to the Environmental Protection Agency, the higher thermodynamic efficiency of the IGCC cycle minimizes carbon dioxide emissions relative to other technologies. IGCC plants offer a least-cost alternative for capturing carbon dioxide from a coal-based power plant. In addition, IGCC will experience a lower energy penalty than other technologies if carbon dioxide capture is required. While carbon dioxide capture and sequestration will increase the cost of all forms of power generation, an IGCC plant can capture and compress carbon dioxide at one-half the cost of a traditional pulverized coal plant. Other gasification-based options, including the production of motor fuels, chemicals, fertilizers or hydrogen, have even lower carbon dioxide capture and compression costs, which will provide a significant economic and environmental benefit in a carbon-constrained world.

14.3.2 Air Emissions

Gasification can achieve greater air emission reductions at lower cost than other coal-based power generation, such as supercritical pulverized coal. Coal-based IGCC offers the lowest emissions of sulfur dioxide nitrogen oxides and particulate matter (PM) of any coal-based power production technology. In fact, a coal IGCC plant is able to achieve low air-emissions rates that approach those of a natural gas combined cycle (NGCC) power plant. In addition, mercury emissions can be removed from an IGCC plant at one-tenth the cost of removal from a coal combustion plant. Technology exists today to remove more than 90% w/w of the volatile mercury from the synthesis gas in a coal-based gasification-based plant.

14.3.3 Solids Generation

During gasification, virtually all of the carbon in the feedstock is converted to synthesis gas. The mineral material in the feedstock separates from the gaseous products, and the ash and other inert materials melt and fall to the bottom of the gasifier as a non-leachable, glass-like solid or other marketable material. This material can be used for many construction and building applications. In addition, more than 99% w/w of the sulfur can be removed using commercially proven technologies and converted into marketable elemental sulfur or sulfuric acid.

14.3.4 Water Use

Gasification uses approximately 14 to 24% v/v less water to produce electric power from coal compared to other coal-based technologies, and water losses during operation are about 32 to 36% v/v less than other coal-based technologies. This is a major issue in many countries – including the United States – where water supplies have already reached critical levels in certain regions.

Finally for this section, one issue that cannot (and must not) be ignored relates to plant safety insofar as the gasification process produces a highly flammable gaseous mixture, including hydrogen, and the extremely toxic, carbon monoxide. In plant sections where pressure buildup exists, there is a risk of gas escape to atmosphere and precautions are necessary to prevent such escape of toxic or environmentally destructive gas. The areas outside the equipment must be adequately ventilated to prevent buildup of an explosive atmosphere, but also to ensure that there is no toxic atmosphere buildup to cause carbon monoxide poisoning – carbon monoxide detection equipment should be provided to detect possible leaks. Applying the relevant techniques to the gasification process is necessary to ensure safety and that the plant meets environmental standards.

14.4 A Process for Now and the Future

Gasification differs from more traditional energy-generating schemes in that it is not a combustion process, but rather it is a conversion process. Instead of the carbonaceous feedstock being wholly burned in air to create heat to raise steam, which is used to drive turbines, the feedstock to be gasified is combined with steam and limited oxygen in a heated, pressurized vessel. The atmosphere inside the vessel is deficient in oxygen leading to a complex series of reactions of the feedstocks to produce synthesis gas. Moreover, using current technologies, the synthesis gas can be cleaned beyond environmental regulatory requirements (current and proposed), as demonstrated by currently commercial chemical production plants that require ultra-clean synthesis gas to protect the integrity of expensive catalysts. The clean synthesis gas can be combusted in turbines or engines using higher temperature (more efficient) cycles than the conventional steam cycles associated with burning carbonaceous fuels, allowing possible efficiency improvements. Synthesis gas can also be used in fuel cells and fuel cell-based cycles with yet even higher efficiencies and exceptionally low emissions of pollutants.

Furthermore, one of the major challenges of the 21st century is finding a way to meet national and global energy needs while minimizing the impact on the environment. There is extensive debate surrounding this issue, but there are certain areas of focus: (i) production of cleaner energy, both from conventional fuel sources and alternative technologies, (ii) use of energy sources that are not only environmentally sound, but also economically viable, (iii) investment in a variety of technologies, and (iv) resources to produce clean energy to meet energy needs. Gasification technologies will help to answer these challenges.

14.4.1 The Process

Gasification, a time-tested, reliable and flexible technology, will be an increasingly important component of this new energy equation, even to the point of the evolution of the crude oil refinery as more gasification units are added to current refineries (Speight, 2011a). Any investment in gasification will yield valuable future returns in clean, abundant, and affordable energy from a variety of sources (Speight, 2008, 2011b).

Gasification is an environmentally sound way to transform any carbon-based material, such as coal, refinery byproducts, biomass, or even waste into energy by producing synthesis gas that can be converted into electricity and valuable products, such as transportation fuels, fertilizers, substitute natural gas, or chemicals (Chadeesingh, 2011; Speight, 2013a).

Gasification has been used on a commercial scale for approximately 100 years by the coal refining, chemical refining, and lighting industries. It is currently playing an important role in meeting energy needs in many countries and will continue to play an increasingly important role as one of the economically attractive manufacturing technologies that will allow production of clean, abundant energy. And, while gasification has typically been used in industrial applications, the technology is increasingly being adopted in smaller-scale applications to convert biomass and waste to energy and products.

Gasification is the cleanest, most flexible and reliable way of using fossil fuels and a variety of other carbonaceous (carbon-containing) or hydrocarbonaceous (carbon-containing and hydrogen-containing) feedstocks. It can convert low-value materials into high-value products, such as chemicals and fertilizers, substitute natural gas, transportation fuels, electric power, steam, and hydrogen. The process can be used to convert biomass, municipal solid waste and other materials (that are normally burned as fuel) into a clean gas. In addition, gasification provides the most cost-effective means of capturing carbon dioxide, a greenhouse gas, when generating power using fossil fuels as a feedstock. And most important for many countries dependent upon high-cost imported crude oil and natural gas from politically unstable regions of the world, gasification allows use of domestic resources to generate energy.

In fact, gasifiers can be designed to use a single material or a blend of feedstocks: (i) solids, such as coal, crude oil coke, biomass, wood waste, agricultural waste, household waste, and hazardous waste, (ii) liquids, such as crude oil resids, including used/recovered road asphalts, tar sand bitumen and liquid wastes from chemical plants and other processing plants, and (iii) natural gas or refinery/chemical off-gas.

The major sought-after products of gasification – synthesis gas and hydrogen – are dependent upon the specific gasification technology; smaller quantities of methane, carbon dioxide, hydrogen sulfide, and water vapor are also produced – typically, 70 to 85% of the carbon in the feedstock is converted into the synthesis gas. The ratio of carbon monoxide to hydrogen depends in part upon the hydrogen and carbon content of the feedstock and the type of gasifier used, but can also be adjusted downstream of the gasifier through use of catalysts. This inherent flexibility of the gasification process means that it can produce one or more products from the same process.

Another benefit of gasification is carbon dioxide removal in the synthesis gas clean-up stage using a number of proven commercial technologies (Mokhatab *et al.*, 2006; Speight, 2007). In fact, carbon dioxide is routinely removed in gasification-based ammonia, hydrogen, and chemical manufacturing plants. Gasification-based ammonia plants already capture/separate approximately 90% v/v of the carbon dioxide and gasification-based methanol plants capture approximately 70% v/v of the carbon dioxide. The gasification process offers the most cost-effective and efficient means of capturing carbon dioxide during the energy production process.

Other byproducts include slag – a glass-like product – composed primarily of sand, rock, and minerals originally contained in the gasifier feedstock. This slag is non-hazardous and can be used in roadbed construction, cement manufacturing or in roofing materials.

Also, in most gasification plants, more than 99% w/w of the feedstock sulfur is removed and recovered either as elemental sulfur or sulfuric acid.

In addition, plasma gasification is increasingly being used to convert all types of waste, including municipal solid waste and hazardous waste, into electricity and other valuable products. Plasma is an ionized gas that is formed when an electrical charge passes through a gas. Plasma torches generate extremely high temperatures which, when used in a gasification plant, which initiate and intensify the gasification reaction, increasing the rate of those reactions and making gasification more efficient. The plasma system allows different types of mixed feedstocks, such as municipal solid waste and hazardous waste, to avoid the expensive step of having to sort the feedstock by type before it is fed into the gasifier. These significant benefits make plasma gasification an attractive option for managing different types of wastes.

14.4.2 Refinery of the Future

With the entry into the 21st century, crude oil refining technology is experiencing great innovation driven by the increasing supply of heavy oils with decreasing quality and the fast increases in the demand for clean and ultra-clean vehicle fuels and petrochemical raw materials. As feedstocks to refineries change, there must be an accompanying change in refinery technology. This means a movement from conventional means of refining heavy feedstocks using (typically) coking technologies to more innovative processes (including hydrogen management) that will produce the ultimate amounts liquid fuels from the feedstock and maintain emissions within environmental compliance (Penning, 2001; Davis and Patel, 2004; Speight, 2008, 2011a).

During the next 20 to 30 years, the evolution future of crude oil refining and the current refinery configuration (Speight, 2014) will be primarily on process modification with some new innovations coming on-stream. The industry will move predictably on to (i) deep conversion of heavy feedstocks, (ii) higher hydrocracking and hydrotreating capacity, and (iii) more efficient processes currently under development.

High conversion refineries will move to gasification of feedstocks for the development of alternative fuels and to enhance equipment usage. A major trend in the refining industry market demand for refined products will be in synthesizing fuels from simple basic reactants (e.g., synthesis gas) when it becomes uneconomical to produce super clean transportation fuels through conventional refining processes. Fischer-Tropsch plants together with IGCC systems will be integrated with or even into refineries, which will offer the advantage of high-quality products (Stanislaus *et al.*, 2000).

A gasification refinery would have, as the center piece, gasification technology as is the case of the Sasol refinery in South Africa (Couvaras, 1997). The refinery would produce synthesis gas (from the carbonaceous feedstock) from which liquid fuels would be manufactured using the Fischer-Tropsch synthesis technology. In fact, gasification to produce synthesis gas can proceed from any carbonaceous material, including biomass. Inorganic components of the feedstock, such as metals and minerals, are trapped in an inert and environmentally safe form as char, which may have use as a fertilizer. Biomass gasification is therefore one of the most technically and economically convincing energy possibilities for a potentially carbon neutral economy.

A modified version of steam reforming known as autothermal reforming, which is a combination of partial oxidation near the reactor inlet with conventional steam reforming further along the reactor, improves the overall reactor efficiency and increases the flexibility of the process. Partial oxidation processes using oxygen instead of steam also found wide application for synthesis gas manufacture, with the special feature that they could utilize low-value feedstocks such as heavy crude oil residua. In recent years, catalytic partial oxidation employing very short reaction times (milliseconds) at high temperatures (850 to 1000°C, 1560 to 1830°F) is providing still another approach to synthesis gas manufacture.

As crude oil supplies decrease, the desirability of producing gas from other carbonaceous feedstocks will increase, especially in those areas where natural gas is in short supply. It is also anticipated that costs of natural gas will increase, allowing coal gasification to compete as an economically viable process. Research in progress on a laboratory and pilot-plant scale should lead to the invention of new process technology by the end of the century, thus accelerating the industrial use of coal gasification.

The conversion of the gaseous products of gasification processes to synthesis gas, a mixture of hydrogen (H_2) and carbon monoxide (CO), in a ratio appropriate to the application, needs additional steps, after purification. The product gases – carbon monoxide, carbon dioxide, hydrogen, methane, and nitrogen – can be used as fuels or as raw materials for chemical or fertilizer manufacture.

14.4.3 Economic Aspects

While a gasification plant is capital intensive (like any manufacturing unit), its operating costs can be lower than many other manufacturing processes or coal combustion plants. Because a gasification plant can use low-cost feedstocks, such as crude oil coke or high-sulfur coal, converting them into high-value products, it increases the use of available energy in the feedstocks while reducing disposal costs. Ongoing research, development, and demonstration investment efforts show potential to substantially decrease current gasification costs even further, driving the economic attractiveness of gasification.

In addition, gasification has a number of other significant economic benefits such as (i) the principal gasification byproducts – sulfur, sulfuric acid, and slag – are marketable, (ii) the gasification process can produce a number of high-value products at the same time thereby helping a facility to offset the capital cost and the operating costs by diversifying the risks, (iii) the gasification process offers wide feedstock flexibility – a gasification plant can be designed to vary the mix of the solid feedstocks or operate on natural gas or liquid feedstocks when desirable, (iv) the gasifiers require less emission control equipment because they generate fewer emissions, thereby further reducing the operating costs.

Investment in gasification injects capital into the economy (by building large-scale plants using domestic labor and suppliers), and creates domestic jobs (construction to build, well-paid jobs to run) that cannot be outsourced to overseas workers. While a gasification plant is capital intensive (like any manufacturing unit), its operating costs can be lower than many other manufacturing processes or coal combustion plants. Because a gasification plant can use low-cost feedstocks, such as crude oil coke or high-sulfur coal, converting them into high-value products, it increases the use of available energy in the feedstocks while reducing disposal costs. Ongoing research, development, and demonstration investment efforts show potential to substantially decrease current gasification costs even further, driving the economic attractiveness of gasification.

14.4.4 Market Outlook

The forecast for growth of gasification capacity focuses on two areas: large-scale industrial and power generation plants and the smaller-scale biomass and waste-to-energy area.

Worldwide industrial and power generation gasification capacity is projected to grow 70% by 2015, with 81% of the growth occurring in developing markets. The prime movers behind this expected growth are the chemical, fertilizer, and coal-to-liquids industries in China, tar sands in Canada, polygeneration (hydrogen and power or chemicals) in the United States, and refining in Europe. In fact, China has focused on gasification as part of the overall energy strategy. The industrial and power gasification industry in the United States faces a number of challenges, including rising construction costs and uncertainty about policy incentives and regulations. Despite these challenges, the industrial and power gasification capacity in the United States is expected to grow.

A number of factors will contribute to this expansion and include (i) the price of natural gas and conventional crude oil which will make low-cost and abundant domestic resources with stable prices increasingly attractive as feedstocks, and (ii) gasification processes will be able to comply with more stringent environmental regulations by having emission profiles that are substantially lower than the emission profiles of the more conventional technologies.

In fact, there is a growing consensus that carbon dioxide management will be required in power generation and energy production. Since the gasification process allows carbon dioxide to be captured in a cost-effective and efficient manner, it will be an increasingly attractive choice for the continued use of fossil fuels. The greatest area of growth in terms of the number of plants in the United States is likely to be in the biomass and waste-to-energy gasification areas. Because they are smaller in scale, these plants are easier to finance, easier to permit and take less time to construct. In addition, municipal and state restrictions on landfills and incineration and a growing recognition that these materials contain valuable sources of energy are driving the demand for these plants.

Furthermore, a number of factors will contribute to the growing interest in waste and biomass gasification and include (i) the restrictions on landfill space, (ii) the efforts to reduce costs associated with waste management, (iii) the growing recognition that waste and biomass contain unused energy that can be captured and converted into energy and valuable products, and (iv) the ability to use non-food biomass materials and convert them into valuable energy products.

14.5 Conclusions

The most obvious issue with fossil fuel use relates to the effects on the environment. As technology evolves, the means to reduce the damage done by fossil fuel use also evolves and the world is on the doorstep of adapting to alternative energy sources. In the meantime, gasification offers alternatives to meet the demand for fuels of the future and to reduce the potentially harmful emissions.

Recent policy to tackle climate change and resource conservation, such as the Kyoto Protocol, the deliberations at Copenhagen in 2009 and the Landfill Directive of the European Union, stimulated the development of renewable energy and landfill diversion technology, so providing gasification technology development a renewed impetus. However, even though they are the fastest-growing source of energy, renewable sources of energy will still represent only 15% of the world energy requirements in 2035 (up from the current estimation of 10%) and divesting from fossil fuels does not mean an end to environmental emissions. Crude oil, tar sand bitumen, coal, natural gas, and perhaps oil shale will still be dominant energy sources – and will grow at a relatively robust rate over, at least, the next two decades. These estimates are a reality check on the challenge ahead for clean technologies if they are to make an impact in reducing greenhouse gas emissions and satisfy future energy demands (EIA, 2013).

Gasification could now be proposed as a viable alternative solution for waste treatment with energy recovery. On the other hand, the gasification process still faces some technical and economic problems, mainly related to the highly heterogeneous nature of feeds like municipal solid wastes and the relatively limited number of plants (approximately 100) worldwide based on this technology that have continuous operating experience under commercial conditions. In the aggressive working environment of municipal solid waste management, with its uncompromising demand for reasonable cost, high reliability and operational flexibility, it could be premature to indicate the gasification as the thermal processing strategy of the future or even as a strong competitor for combustion systems, at least for any size of waste-to-energy plants.

The success of any advanced thermal technology is determined by its technical reliability, environmental sustainability and economic convenience. The first, and then mainly the on-line availability, is supported by years of successful continuous operations of about 100 gasification-based WtE plants, mainly in Japan but now also in South Korea and Europe. The environmental performance is one of the greatest strengths of gasification technology, which often is considered a sound response to the increasingly restrictive regulations applied around the world: independently verified emissions tests indicate that gasification is able to meet existing emissions limits and can have a great effect on the reduction of landfill disposal option.

Economic aspects are probably the crucial factor for a relevant market penetration, since gasification-based WtE tends to have ranges of operating and capital costs higher than those of conventional combustion-based waste-to-energy (in the order of about 10%), mainly as a consequence of the ash melting system or, in general, of the added complexity of the technology.

The greatest technical challenges to overcome for a wider market penetration of commercial advanced gasification technologies appears still to be that of an improved and cheaper synthesis gas cleaning, able to conveniently meet defined specifications and to obtain higher electric energy conversion efficiencies. It is essential that the performance and experience from several commercial waste gasifiers in operation will point to the gasification process as a strong competitor of conventional moving grate or fluidized-bed combustion systems.

References

Ancheyta, J., and Speight, J.G. 2007. *Hydroprocessing of Heavy Oils and Residua*. CRC Press, Taylor & Francis Group, Boca Raton, Florida.

Banks, F.E. 1992. Some Aspects of Natural Gas and Economic Development – A Short Note. *OPEC Bulletin*. 16(2): 235-240.

Chadeesingh, R. 2011. The Fischer-Tropsch Process. In *The Biofuels Handbook*. J.G. Speight (Editor). The Royal Society of Chemistry, London, United Kingdom. Part 3, Chapter 5, Page 476-517.

Clayton, S.J., Stiegel, G.J., and Wimer, J.G. 2002. *Gasification Technologies: Gasification Markets and Technologies – Present and Future, An Industry Perspective*. Report No. DOE/FE-0447. Office of Fossil Energy, United States Department of Energy, Washington, DC.

Couvaras, G. 1997. Sasol's Slurry Phase Distillate Process and Future Applications. *Proceedings. Monetizing Stranded Gas Reserves Conference, Houston, December 1997*.

Davis, R.A., and Patel, N.M. 2004. Refinery Hydrogen Management. *Petroleum Technology Quarterly*, Spring: 29-35.

EIA. 2013. *International Energy Outlook 2013: World Energy Demand and Economic Outlook*. International Energy Agency, Paris, France. http://www.eia.gov/forecasts/ieo/world.cfm, accessed September 13, 2013.

Hubbert, M.K. 1962. *Energy Resources*. Report to the Committee on Natural Resources, National Academy of Sciences, Washington, DC.

Khan, M.R., and Patmore, D.J. 1997. Heavy Oil Upgrading Processes. In *Petroleum Chemistry and Refining*. J.G. Speight (Editor). Taylor & Francis, Washington, DC. Chapter 6.

Krey, V., Canadell, J.G., Nakicenovic, N., Abe, Y., Andruleit, H., Archer, D., Grubler, A., Hamilton, N.T.M., Johnson, A., Kostov, V., Lamarque, J-F., Langhorne, N., Nisbet, E.G., O'Neill, B., Riahi, K., Riedel, M., Wang, W., and Yakushev, V. 2009. Gas Hydrates: Entrance to a Methane Age or Climate Threat? *Environ. Res. Lett.*, 4(3): 034007.

Long, R., Picioccio, K., and Zagoria, A. 2011. Optimizing Hydrogen Production and Use. *Petroleum Technology Quarterly*. Autumn. Page 1-12.

MacDonald, G.J. 1990. The Future of Methane as an Energy Resource. *Annual Reviews of Energy*. 15: 53-83.

Martin, A.J. 1985. The Prediction of Strategic Reserves. In *Prospects for the World Oil Industry*. T. Niblock and R. Lawless (Editors). Croom Helm Publishers, Beckenham, Kent. Chapter 1.

Mohnen, V.A. 1988. The Challenge of Acid Rain. *Scientific American*. 259(2) 30-38.

Mokhatab, S., Poe, W.A., and Speight, J.G. 2006. *Handbook of Natural Gas Transmission and Processing*. Elsevier, Amsterdam, Netherlands, 2006.

Penning, R.T. 2001. *Petroleum Refining: A Look at the Future*. Hydrocarbon Processing, 80(2): 45-46.

Rana, M.S., Sámano, V., Ancheyta, J., and Diaz, J.A.I. 2007. A Review of Recent Advances on Process Technologies for Upgrading of Heavy Oils and Residua. *Fuel*, 86: 1216-1231.

Speight, J.G. 2007. *Natural Gas: A Basic Handbook*. GPC Books, Gulf Publishing Company, Houston, Texas, 2007.

Speight, J.G. 2008. *Synthetic Fuels Handbook: Properties, Processes, and Performance*. McGraw-Hill, New York.

Speight, J.G. 2011a. *The Refinery of the Future*. Gulf Professional Publishing, Elsevier, Oxford, United Kingdom.

Speight, J.G. (Editor). 2011b. *The Biofuels Handbook*. Royal Society of Chemistry, London, United Kingdom.

Speight, J.G. 2011c. *An Introduction to Petroleum Technology, Economics, and Politics*. Scrivener Publishing, Salem, Massachusetts.

Speight, J.G. 2013a. *The Chemistry and Technology of Coal*. 3rd Edition. CRC Press, Taylor & Francis Group, Boca Raton, Florida.

Speight, J.G. 2013b. *Coal-Fired Power Generation Handbook*. Scrivener Publishing, Salem, Massachusetts.

Speight, J.G. 2014. *The Chemistry and Technology of Petroleum*, 5th Edition. CRC Press, Taylor & Francis Group, Boca Raton, Florida.

Speight, J.G. 2017. *Handbook of Petroleum Refining*. CRC Press, Taylor & Francis Group, Boca Raton, Florida.

Stanislaus, A., Qabazard, H., and Absi-Halabi, M.. 2000. Refinery of the Future. *Proceedings. 16th World Petroleum Congress, Calgary, Alberta, Canada. June 11-15*.

Stensvaag, J-M. 1991. *Clean Air Act Amendments: Law and Practice*. John Wiley and Sons Inc., New York.

Stigliani, W.M., and Shaw, R.W. 1990. Energy Use and Acid Deposition: The View from Europe. *Annual Reviews of Energy*. 15: 201-216.

United States Congress, 1990. Public Law 101-549. An Act to Amend the Clean Air Act to Provide for Attainment and Maintenance of Health protective National Ambient Air Quality Standards, and for Other Purposes. November 15.

United States Department of Energy, 1990. Gas Research Program Implementation Plan. DOE/FE-0187P. United States Department of Energy, Washington, D.C. April.

United States Department of Energy. 1993. Clean Fossil Fuels Technology Demonstration Program. DOE/FE-0272. United States Department of Energy, Washington, D.C. February.

United States General Accounting Office, 1990. *Energy Policy: Developing Strategies for Energy Policies in the 1990s*. Report to Congressional Committees. GAO/RCED-90-85. United States General Accounting Office, Washington, D.C. June.

Vallero, D. 2008. *Fundamentals of Air Pollution* 4th Edition. Elsevier, London, United Kingdom.

Conversion Factors

1. General

1 acre = 43,560 sq ft
1 acre foot = 7758.0 bbl
1 atmosphere = 760 mm Hg = 14.696 psia = 29.91 in. Hg
1 atmosphere = 1.0133 bars = 33.899 ft. H_2O
1 barrel (oil) = 42 gal = 5.6146 *cu* ft
1 barrel (water) = 350 lb at 60°F
1 barrel per day = 1.84 cu cm per second
1 Btu = 778.26 ft-lb
1 centipoise × 2.42 = lb mass/(ft) (hr), viscosity
1 centipoise × 0.000672 = lb mass/(ft) (sec), viscosity
1 cubic foot = 28,317 cu cm = 7.4805 gal
Density of water at 60°F = 0.999 gram/cu cm = 62.367 lb/cu ft = 8.337 lb/gal
1 gallon = 231 cu in. = 3,785.4 cu cm = 0.13368 cu ft
1 horsepower-hour - 0.7457 kwhr = 2544.5 Btu
1 horsepower = 550 ft-lb/sec = 745.7 watts
1 inch = 2.54 cm
1 meter = 100 cm = 1,000 mm = 10 microns = 10 angstroms (A)
1 ounce = 28.35 grams
1 pound = 453.59 grams = 7,000 grains
1 square mile = 640 acres

2. Concentration Conversions

1 part per million (1 ppm) = 1 microgram per liter (1 μg/L)
1 microgram per liter (1 μg/L) = 1 milligram per kilogram (1 mg/kg)
1 microgram per liter (μg/L) × 6.243×10^8 = 1 lb per cubic foot (1 lb/ft^3)
1 microgram per liter (1 μg/L) × 10^{-3} = 1 milligram per liter (1 mg/L)
1 milligram per liter (1 mg/L) × 6.243×10^5 = 1 pound per cubic foot (1 lb/ft^3)
I gram mole per cubic meter (1 g mol/m^3) × 6.243×10^5 = 1 pound per cubic foot (1 lb/ft^3)
10,000 ppm = 1% w/w
1 ppm hydrocarbon in soil × 0.002 = 1 lb of hydrocarbons per ton of contaminated soil

3. Weight Conversion

1 ounce (1 oz) = 28.3495 grams (18.2495 g)
1 pound (1 lb) = 0.454 kilogram
1 pound (1 lb) = 454 grams (454 g)
1 kilogram (1 kg) = 2.20462 pounds (2.20462 lb)
1 stone (English, 1st) = 14 pounds (14 lb)
1 ton (US; 1 short ton) = 2,000 lbs
1 ton (English; 1 long ton) = 2,240 lbs
1 metric ton = 2204.62262 pounds
1 tonne = 2204.62262 pounds

4. Temperature Conversions

°F = (°C × 1.8) + 32
°C = (°F − 32)/1.8
(°F − 32) × 0.555 = °C
Absolute zero = −273.15°C
Absolute zero = −459.67°F

5. Area

1 square centimeter (1 cm^2) = 0.1550 square inches
1 square meter 1 (m^2) = 1.1960 square yards
1 hectare = 2.4711 acres
1 square kilometer (1 km^2) = 0.3861 square miles
1 square inch (1 inch^2) = 6.4516 square centimeters
1 square foot (1 ft^2) = 0.0929 square meters
1 square yard (1 yd^2) = 0.8361 square meters
1 acre = 4046.9 square meters
1 square mile (1 mi^2) = 2.59 square kilometers

6. Other Approximations

14.7 pounds per square inch (14.7 psi) = 1 atmosphere (1 atm)
1 kiloPascal (kPa) × 9.8692 × 10^{-3} = 14.7 pounds per square inch (14.7 psi)
1 yd^3 = 27 ft^3
1 US gallon of water = 8.34 lbs
1 imperial gallon of water − 10 lbs
1 yd^3 = 0.765 m^3
1 acre-inch of liquid = 27,150 gallons = 3.630 ft^3
1 ft depth in 1 acre (in-situ) = 1,613 × (20 to 25 % excavation factor) = ~2,000 yd^3
1 yd^3 (clayey soils-excavated) = 1.1 to 1.2 tons (US)
1 yd^3 (sandy soils-excavated) = 1.2 to 1.3 tons (US)

7. SI Metric Conversion Factors

(E = exponent; i.e. E + 03 = 10^3 and E − 03 = 10^{-3}

acre-foot × 1.233482   E + 03 = meters cubed
barrels × 1.589873   E − 01 = meters cubed
centipoise × 1.000000   E − 03 = pascal seconds
Darcy × 9.869233   E − 01 = micro meters squared
feet × 3.048000   E − 01 = meters
pounds/acre-foot × 3.677332   E − 04 = kilograms/meters cubed
pounds/square inch × 6.894757   E + 00 = kilo pascals
dyne/cm × 1.000000   E + 00 = mN/m
parts per million × 1.000000   E + 00 = milligrams/kilograms

8. Converting Volumes of Gas

Multiply flow of	By	To obtain flow of
Natural Gas	0.625	Propane
	0.547	Butane
	0.775	Air
Propane	1.598	Natural gas
	0.874	Butane
	1.237	Air
Butane	1.826	Natural gas
	1.140	Propane
	1.414	Air
Air	1.290	Natural gas
	0.808	Propane
	0.707	Butane

9. Natural Gas Conversion Table

ft³	Ccf	Mcf	MMcf	therm	dekatherm	BTU	MMBTU	kJ	kWh
0.0009	0.00001	0.000001	0	0.00001	0.0001	0.9482	0.000001	1	0
0.001	0.00001	0.000001	0.000000001	0.00001	0.0001	1	0.000001	1.055	0
1	0.01	0.001	0.000001	0.01	0.001	1000	0.001	1054	0.293
100	1	0.1	0.0001	1	0.1	100000	0.1	105461.5	29
1000	10	1	0.001	10	1	1000000	1	1054615	293
1000000	10000	1000	1	10000	1000	1.00E+09	1000	1054615000	293071

* Assumed 1 cf = 1000 BTU, for approximate reference only.

Glossary

Abiotic: A process that is not associated with living organisms; synonymous with *abiological*.

Abiotic transformation: The process in which a substance in the environment is modified by non-biological mechanisms.

Absorption: The penetration of atoms, ions, or molecules into (within) the bulk mass of a substance.

Acetic acid (CH_3CO_2H): Trivial name for ethanoic acid, formed by the oxidation of ethanol with potassium permanganate.

Acetone (CH_3COCH_3: Trivial name for propanone, formed by the oxidation of 2-propanol with potassium permanganate.

Acid: Any of a class of substances whose aqueous solutions are characterized by a sour taste, the ability to turn blue litmus red, and the ability to react with bases and certain metals to form salts; a substance that yields hydrogen ions when dissolved in water and which can act as a proton (h^+) donor.

Acid deposition (acid rain): A mixture of wet and dry *deposition* (deposited material) from the atmosphere containing higher than typical amount of nitric and sulfuric acids.

Acid gas: Hydrogen sulfide (H_2S) or carbon dioxide (CO_2).

Acid hydrolysis: A chemical process in which acid is used to convert cellulose or starch to sugar.

Acid insoluble lignin: Lignin is mostly insoluble in mineral acids, and therefore can be analyzed gravimetrically after hydrolyzing the cellulose and hemicellulose fractions of the biomass with sulfuric acid; standard test method ASTM E1721 describes the standard method for determining acid insoluble lignin in biomass; see American Society for Testing and Materials.

Acid rain: A solution of acidic compounds formed when sulfur and nitrogen oxides react with water droplets and airborne particles.

Acid sludge: The residue left after treating crude oil with sulfuric acid for the removal of impurities; a black, viscous substance containing the spent acid and impurities.

Acid soluble lignin: A small fraction of the lignin in a biomass sample is solubilized during the hydrolysis process of the acid insoluble lignin method. This lignin fraction is referred to as acid soluble lignin and may be quantified by ultraviolet spectroscopy; see lignin and acid insoluble lignin.

Acid treating: A process in which unfinished crude oil products, such as gasoline, kerosene, and lubricating-oil stocks, are contacted with sulfuric acid to improve their color, odor, and other properties.

Acid/base reaction: A reaction in which an acidic hydrogen atom is transferred from one molecule to another.

Acidic: A solution with a high concentration of H^+ ions.

Acidity: The capacity of the water to neutralize OH⁻.

Additive: A material added to another (usually in small amounts) in order to enhance desirable properties or to suppress undesirable properties.

Adsorbent (sorbent): The solid phase or substrate onto which the sorbate adsorbs.

Adsorption: The retention of atoms, ions, or molecules on to the surface of another substance; the two-dimensional accumulation of an adsorbate at a solid surface. In the case of surface precipitation; also used when there is diffusion of the sorbate into the solid phase.

Aerobic: In the presence of, or requiring, oxygen; an environment or process that sustains biological life and growth, or occurs only when free (molecular) oxygen is present.

Aerobic bacteria: Any bacteria requiring free oxygen for growth and cell division.

Aerobic conditions: Conditions for growth or metabolism in which the organism is sufficiently supplied with oxygen.

Aerosol: A dispersion of a liquid or solid in a gas.

Agglomerate: If operating temperature of the reactor is higher than the initial deformation temperature of ash, the ash commences melting and forms agglomerates.

Agglomerating: Coal which, during volatile matter determinations, produces either an agglomerate button capable of supporting a 500-gram weight without pulverizing, or a button showing swelling or cell structure.

Agglomeration: Formation of larger coal or ash particles by smaller particles sticking together.

Agitator: A device such as a stirrer that provides complete mixing and uniform dispersion of all components in a mixture; are generally used continuously during the thermal processes and intermittently during fermentation.

Agricultural residue: Agricultural crop residues are the plant parts, primarily stalks and leaves, not removed from the fields with the primary food or fiber product; examples include corn stover (stalks, leaves, husks, and cobs); wheat straw; and rice straw.

Air gasification; A gasification process that uses a minimal quantity of air and steam to convert the char to gas in a single unit.

Air quality maintenance area: Specific populated area where air quality is a problem for one or more pollutants.

Alcohol: The family name of a group of organic chemical compounds composed of carbon, hydrogen, and oxygen. The molecules in the series vary in chain length and are composed of a hydrocarbon plus a hydroxyl group. Alcohol includes methanol and ethanol.

Aldehyde: An organic compound with a carbon bound to a -(C=O)-H group; a compound in which a carbonyl group is bonded to one hydrogen atom and to one alkyl group [RC(=O)H,].

Alicyclic hydrocarbon: A compound containing carbon and hydrogen only which has a cyclic structure (e.g., cyclohexane); also collectively called naphthenes.

Aliphatic hydrocarbon: A compound containing carbon and hydrogen only which has an open-chain structure (e.g., as ethane, butane, octane, butene) or a cyclic structure (e.g., cyclohexane); any non-aromatic hydrocarbon having an open-chain structure.

Alkali: A soluble mineral salt.

Alkali lignin: Lignin obtained by acidification of an alkaline extract of wood.

Alkali metal: A metal in Group IA on the periodic table; an active metal which may be used to react with an alcohol to produce the corresponding metal alkoxide and hydrogen gas.

Alkali wash: See caustic wash.

Alkalinity: The capacity of water to accept H⁺ ions (protons).

Alkanes: Hydrocarbons that contain only single carbon-hydrogen bonds. The chemical name indicates the number of carbon atoms and ends with the suffix "ane".

Alkenes: Hydrocarbons that contain carbon-carbon double bonds. The chemical name indicates the number of carbon atoms and ends with the suffix "ene".

Alkylate: The product of an alkylation (q.v.) process.

Alkylation: A process for manufacturing high-octane blending components used in unleaded petrol or gasoline.

Allothermal gasification: The heat required for gasification reactions is afforded by an external source.

Alternative fuel: As defined in the United States Energy Policy Act of 1992 (EPACT): methanol, denatured ethanol and other alcohols, separately or in blends of at least 10% by volume with gasoline or other fuels; compressed natural gas; liquefied natural gas, liquefied propane gas, hydrogen, coal-derived liquid fuels, fuels other than alcohols derived from biological materials, electricity, biodiesel, and any other fuel deemed to be substantially not crude oil and yielding potential energy security benefits and substantial environmental benefits.

Ambient air quality: The condition of the air in the surrounding environment.

American Society for Testing and Materials (ASTM): An international standards organization that develops and produces technical standards for materials, products, systems and services; also known as ASTM International.

Amine washing: A method of gas cleaning whereby acidic impurities such as hydrogen sulfide and carbon dioxide are removed from the gas stream by washing with an amine (usually an alkanolamine).

Ammoniated ash: Ash that contains ammonia and/or ammonium salts as a result of the addition of ammonia or ammonium salts to the flue gas at the power plant.

Anaerobic: Biological processes that occur in the absence of oxygen.

Anaerobic bacteria: Any bacteria that can grow and divide in the partial or complete absence of oxygen.

Anaerobic digestion (anaerobic decompositon): Decomposition of biological wastes by micro-organisms, usually under wet conditions, in the absence of air (oxygen), to produce a gas comprising mostly methane and carbon dioxide.

Anhydrous: Without water; transesterification of biodiesel must be an anhydrous process; water in the vegetable oil causes either no reaction or cloudy biodiesel, and water in lye or methanol renders it less useful or even useless, depending on how much water is present.

Anoxic: An environment without oxygen.

Anthracene oil: The heaviest distillable coal tar fraction, with distillation range 270 to 400°C (520 to 750°F), containing creosote oil, anthracene, phenanthrene, carbazole, and so on.

Anthracite (hard coal): A hard, black, shiny coal very high in fixed carbon and low in volatile matter, hydrogen, and oxygen; a rank class of non-agglomerating coals as defined by the American Society for Testing and Materials having more than 86% fixed carbon and less than 14% volatile matter on a dry, mineral-matter-free basis; this class of coal is divisible into the semi-anthracite, anthracite, and meta-anthracite groups on the basis of increasing fixed carbon and decreasing volatile matter. The heat content of anthracite ranges from 22 to 28 million Btu per short ton on a moist, mineral-matter-free basis.

The heat content of anthracite coal consumed in the United States averages 25 million Btu per short ton, on the as-received basis (i.e., containing both inherent moisture and mineral matter); since the 1980s, anthracite refuse or mine waste has been used for steam electric power generation. This fuel typically has a heat content up to 15 million Btu per short ton.

Anticline: An upward fold or arch of rock strata.

API gravity: A measure of the lightness or heaviness of crude oil that is related to density and specific gravity; °API = (141.5/sp. Gr @ 60°F) - 131.5,

Aquifer: A water-bearing formation through which water moves more readily than in adjacent formations with lower permeability.

Aromatic ring: An exceptionally stable planar ring of atoms with resonance structures that consist of alternating double and single bonds, such as benzene:

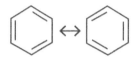

Aromatics: A range of hydrocarbons which have a distinctive sweet smell and include benzene and toluene, occur naturally in crude oil and are also extracted as a petrochemical feedstock, as well as for use as solvents.

Ash: The noncombustible residue remaining after complete coal combustion; the final form of the mineral matter present in coal.

Ash analysis: Percentages of inorganic oxides present in an ash sample. Ash analyses are used for evaluation of the corrosion, slagging, and fouling potential of coal ash. The ash constituents of interest are silica (SiO_2) alumina (Al_2O_3), titania (TiO_2), ferric oxide (Fe_2O_3), lime (CaO), magnesia (MgO), potassium oxide (k_mo), sodium oxide (na_no), and sulfur trioxide (so). An indication of ash behavior can be estimated from the relative percentages of each constituent.

Ash deformation temperature: The temperature at which ash begins to fuse and become soft.

Ash free: A theoretical analysis calculated from basic analytical data expressed as if the total ash had been removed.

Ash-fusion temperature: A temperature that characterizes the behavior of ash as it is heated. These temperatures are determined by heating cones of ground, pressed ash in both oxidizing and reducing atmospheres.

Asphaltene fraction: The brown to black powdery material produced by treatment of crude oil, heavy oil, bitumen, or residuum with a low-boiling liquid hydrocarbon, such as n-heptane.

As-received basis: The analysis of a sample as received at a laboratory.

As-received moisture: The moisture present in a coal sample when delivered.

Atmospheric pressure: Pressure of the air and atmosphere surrounding us which changes from day to day; equal to 14.7 psia.

Autothermal gasification: The heat required for gasification reactions is supplied by partial oxidation of the synthesis gas; air or a steam/oxygen mixture is used as oxidant agents.

Available production capacity: The biodiesel production capacity of refining facilities that are not specifically designed to produce biodiesel.

Average megawatt (Mwa or aMw): One megawatt of capacity produced continuously over a period of one year; 1 aMw = 1 Mw × 8760 hours/year = 8,760 Mwh = 8,760,000 kMh.

Bagasse: Sugar cane waste.

Baghouse: An air pollution control device that removes particulate matter from flue gas, usually achieving a removal rate above 99.9%.

Bark: The outer protective layer of a tree outside the cambium comprising the inner bark and the outer bark; the inner bark is a layer of living bark that separates the outer bark from the cambium and in a living tree is generally soft and moist; the outer bark is a layer of dead bark that forms the exterior surface of the tree stem; the outer bark is frequently dry and corky.

Barrel (bbl): The unit of measure used by the crude oil industry; equivalent to approximately 42 US gallons or approximately 34 (33.6) imperial gallons or 159 liters; 7.2 barrels are equivalent to one tonne of oil (metric).

Barrel of oil equivalent (boe, BoE): A unit of energy equal to the amount of energy contained in a barrel of crude oil; approximately 5.78 million Btu or 1,700 kwh; one barrel equals 5.6 cubic feet or 0.159 cubic meters; for crude oil, one barrel is about 0.136 metric tons, 0.134 long tons, and 0.150 short tons; a barrel is a liquid measure equal to 42 gallons or about 306 pounds.

Batch distillation: A process in which the liquid feed is placed in a single container and the entire volume is heated, in contrast to continuous distillation in which the liquid is fed continuously through the still.

Batch fermentation: Fermentation conducted from start to finish in a single vessel; see fermentation.

Batch process: Unit operation where one cycle of feedstock preparation, cooking, fermentation and distillation is completed before the next cycle is started.

Benzene: A toxic, six-carbon aromatic component of gasoline; a known carcinogen.

Billion: 1×10^9

Biochemical conversion: The use of fermentation or anaerobic digestion to produce fuels and chemicals from organic sources.

Biochemical conversion process: The use of living organisms or their products to convert organic material to fuels, chemicals or other products.

Biodegradable: Capable of decomposing rapidly under natural conditions.

Biodiesel: A fuel derived from biological sources that can be used in diesel engines instead of crude oil-derived diesel; through the process of transesterification, the triglycerides in the biologically derived oils are separated from the glycerin, creating a clean-burning, renewable fuel.

Bioenergy: Useful, renewable energy produced from organic matter – the conversion of the complex carbohydrates in organic matter to energy; organic matter may either be used directly as a fuel, processed into liquids and gases, or be a residual of processing and conversion.

Bioethanol: Ethanol produced from biomass feedstocks; includes ethanol produced from the fermentation of crops, such as corn, as well as cellulosic ethanol produced from woody plants or grasses.

Biofuels: A generic name for liquid or gaseous fuels that are not derived from crude oil-based fossils fuels or contain a proportion of non-fossil fuel; fuels produced from plants, crops such as sugar beet, rape seed oil or reprocessed vegetable oils or fuels made from gasified biomass; fuels made from renewable biological sources and include ethanol, methanol, and biodiesel; sources include, but are not limited to: corn, soybeans, flaxseed, rapeseed, sugarcane, palm oil, raw sewage, food scraps, animal parts, and rice.

Biogas: A combustible gas derived from decomposing biological waste under anaerobic conditions. Biogas normally consists of 50 to 60% methane. See also landfill gas.

Biomass: Any organic matter that is available on a renewable or recurring basis, including agricultural crops and trees, wood and wood residues, plants (including aquatic plants), grasses, animal manure, municipal residues, and other residue materials. Biomass is generally produced in a sustainable manner from water and carbon dioxide by photosynthesis. There are three main categories of biomass – primary, secondary, and tertiary.

Biomass fuel: Any liquid, solid or gaseous fuel produced by the conversion of biomass.

Biomass gasifier: Any one of a variety of gasifiers for the conversion of biomass to products; can use gas-solid contact method (updraft, downdraft, fluidized bed, or suspended flow); the feedstock can be in the form of residues, pellets, powders to give products such as low- or medium-energy gas, char, or pyrolysis oil through the use of various heating rates and residence times (slow and fast pyrolysis).

Biomass processing residues: Byproducts from processing all forms of biomass that have significant energy potential; the residues are typically collected at the point of processing; they can be convenient and relatively inexpensive sources of biomass for energy.

Biomass to liquid (BTL, BtL): The process of converting biomass to liquid fuels.

Biopower: The use of biomass feedstock to produce electric power or heat through direct combustion of the feedstock, through gasification and then combustion of the resultant gas, or through other thermal conversion processes. Power is generated with engines, turbines, fuel cells, or other equipment.

Biorefinery: A facility that processes and converts biomass into value-added products. These products can range from biomaterials to fuels such as ethanol or important feedstocks for the production of chemicals and other materials.

Bitumen: Also, on occasion, incorrectly referred to as native asphalt, and extra heavy oil; a naturally occurring material that has little or no mobility under reservoir conditions and which cannot be recovered through a well by conventional oil well production methods including currently used enhanced recovery techniques; current methods involve mining for bitumen recovery.

Bituminous (soft) coal: A relatively soft dark brown to black coal, lower in fixed carbon than anthracite but higher in volatile matter, hydrogen, and oxygen; a rank class of coals as defined by the American Society for Testing and Materials (ASTM) high in carbonaceous matter, having less than 86% fixed carbon, and more than 14% volatile matter on a dry, mineral-matter-free basis and more than 10,500 Btu on a moist, mineral-matter-free basis. This class may be either agglomerating or non-agglomerating and is divisible into the high-volatile c, b, a; medium; and low-volatile bituminous coal groups on the basis of increasing heat content and fixed carbon and decreasing volatile matter. Bituminous coal is the most abundant coal in active mining regions in the United States. Its moisture content usually is less than 20%. The heat content of bituminous coal ranges from 21 to 30 million Btu per ton on a moist, mineral-matter-free basis. The heat content of bituminous coal consumed in the United States averages 24 million Btu per ton, on the as- received basis (i.e., containing both inherent moisture and mineral matter).

Black acid(s): A mixture of the sulfonates found in acid sludge which are insoluble in naphtha, benzene, and carbon tetrachloride; very soluble in water but insoluble in 30% sulfuric acid; in the dry, oil-free state, the sodium soaps are black powders.

Black liquor: Solution of lignin-residue and the pulping chemicals used to extract lignin during the manufacture of paper.

Blue gas: A mixture consisting chiefly of carbon monoxide and hydrogen formed by action of steam on hot coal or coke.

Boiler: Any device used to burn biomass fuel to heat water for generating steam.

Boiler horsepower: A measure of the maximum rate of heat energy output of a steam generator; one boiler horsepower equals 33,480 Btu/hr output in steam.

Boiler slag: A molten ash collected at the base of slag tap and cyclone boilers that is quenched with water and shatters into black, angular particles having a smooth glassy appearance.

Bone dry: Having zero percent moisture content. Wood heated in an oven at a constant temperature of 100°C (212°F) or above until its weight stabilizes is considered bone dry or oven dry.

Bottom ash: Consists of agglomerated ash particles formed in combustors or gasification reactors that are too large to be carried in the flue gases and impinge on the reactor boiler walls or fall through open grates to an ash hopper at the bottom of the reactor; typically gray to black in color, is quite angular, and has a porous surface.

Bottoming cycle: A cogeneration system in which steam is used first for process heat and then for electric power production.

Briquetting: A process of applying pressure to feedstock fines, with or without a binder, to form a compact or agglomerate.

British thermal unit (Btu): The quantity of heat required to raise the temperature of 1 pound of water 1°F at, or near, its point of maximum density of 39.1°F (equivalent to 251.995 gram calories; 1,054.35 joules; 1.05435 kilojoules; 0.25199 kilocalorie).

Brown grease: Waste grease that is the least expensive of the various grades of waste grease.

BTEX: The collective name given to benzene, toluene, ethylbenzene and the xylene isomers (*p*-, *m*-, and *o*-xylene); a group of volatile organic compounds (VOCs) found in crude oil hydrocarbons, such as gasoline, and other common environmental contaminants.

BTX: The collective name given to benzene, toluene, and the xylene isomers (*p*-, *m*-, and *o*-xylene); a group of volatile organic compounds (VOCs) found in crude oil hydrocarbons, such as gasoline, and other common environmental contaminants.

Bunker: A storage tank.

Butanol: Though generally produced from fossil fuels, this four-carbon alcohol can also be produced through bacterial fermentation of alcohol.

Byproduct: A substance, other than the principal product, generated as a consequence of creating a biofuel.

C_1, C_2, C_3, C_4, C_5 fractions: A common way of representing fractions containing a preponderance of hydrocarbons having 1, 2, 3, 4, or 5 carbon atoms, respectively, and without reference to hydrocarbon type.

Calorie: The quantity of heat required to raise 1 gram of water from 15 to 16°C; a calorie is also termed gram calorie or small calorie (equivalent to 0.00396832 Btu; 4.184 Joules; 0.001 kilogram calorie).

Calorific value: The quantity of heat that can be liberated from one pound of coal or oil measured in Btu/lb.

Capital cost: The total investment needed to complete a project and bring it to a commercially operable status; the cost of construction of a new plant; the expenditures for the purchase or acquisition of existing facilities.

Carbohydrate: A chemical compound made up of carbon, hydrogen, and oxygen; includes sugars, cellulose, and starches.

Carbon chain: The atomic structure of hydrocarbons in which a series of carbon atoms, saturated by hydrogen atoms, form a chain; volatile oils have shorter chains while fats have longer chain lengths, and waxes have extremely long carbon chains.

Carbon dioxide (CO_2): A product of combustion that acts as a greenhouse gas in the Earth's atmosphere, trapping heat and contributing to climate change.

Carbon monoxide (CO): A lethal gas produced by incomplete combustion of carbon-containing fuels in internal combustion engines; it is colorless, odorless, and tasteless and poisons by displacing the oxygen in hemoglobin (the oxygen carrier in the blood).

Carbon sequestration: The absorption and storage of carbon dioxide from the atmosphere; naturally occurring in plants.

Carbon sink: A geographical area whose vegetation and/or soil soaks up significant carbon dioxide from the atmosphere; such areas, typically in tropical regions, are increasingly being sacrificed for energy crop production.

Carbonization: A process whereby a carbonaceous feedstock is converted to char or coke by devolatilization.

Carbureted blue gas: See water gas.

Catalyst: A substance that accelerates a chemical reaction without itself being affected. In refining, catalysts are used in the cracking process to produce blending components for fuels.

Catalytic reforming: Rearranging hydrocarbon molecules in a gasoline-boiling-range feedstock to produce other hydrocarbons having a higher antiknock quality; isomerization of paraffins, cyclization of paraffins to naphthenes, dehydrocyclization of paraffins to aromatics.

Caustic wash: The process of treating a product with a solution of caustic soda to remove minor impurities; often used in reference to the solution itself.

Cellulose: Fiber contained in leaves, stems, and stalks of plants and trees; most abundant organic compound on Earth; it is a polymer of glucose with a repeating unit of $C_6H_{10}O_5$ strung together by ß-glycoside linkages – the ß-linkages in cellulose form linear chains that are highly stable and resistant to chemical attack because of the high degree of hydrogen bonding that can occur between chains of cellulose; hydrogen bonding between cellulose chains makes the polymers more rigid, inhibiting the flexing of the molecules that must occur in the hydrolytic breaking of the glycoside linkages – hydrolysis can reduce cellulose to a cellobiose repeating unit, $C_{12}H_{22}O_{11}$, and ultimately to the six-carbon sugar glucose, $C_6H_{12}O_6$.

Cetane number: A measure of the ignition quality of diesel fuel; the higher the number the more easily the fuel is ignited under compression.

Cetane rating: Measure of the combustion quality of diesel fuel.

Chips: Small fragments of wood chopped or broken by mechanical equipment – total tree chips include wood, bark, and foliage while pulp chips or clean chips are free of bark and foliage.

Class C fly ash: Fly ash that meets criteria defined in ASTM C618 for use in concrete.

Class F fly ash: Fly ash that meets criteria defined in ASTM C618 for use in concrete.

Clastic dike: A vertical or near-vertical seam of sedimentary material that fills a crack in and cuts across sedimentary strata – the dikes are found in sedimentary basin deposits worldwide; dike thickness varies from millimeters to meters and the length is usually many times the width.

Clastic rocks: Rocks composed of fragments (clasts) of pre-existing rock; may include sedimentary rocks as well as to transported particles whether in suspension or in deposits of sediment.

Clay: A very fine-grained soil that is plastic when wet but hard when fired; typical clay minerals consist of silicate and aluminosilicate minerals that are the products of weathering reactions of other minerals; the term is also used to refer to any mineral of very small particle size.

Clay vein: A fissure that has been infilled as a result of gravity, downward-percolating ground waters, or compactional pressures which cause unconsolidated clays or thixotropic sand to flow into the fissure.

Clean Air Act (CAA): US national law establishing ambient air quality emission standards to be implemented by participating states; originally enacted in 1963, the Clean Air Act has been amended several times, most recently in 1990 and includes vehicle emission standards regulating the emission of criteria pollutants (lead, ozone, carbon monoxide, sulfur dioxide, nitrogen oxides and particulate matter); the 1990 amendments added reformulated gasoline (RFG) requirements and oxygenated gasoline provisions.

Clean coal technologies: A number of innovative, new technologies designed to use coal in a more efficient and cost-effective manner while enhancing environmental protection; technologies include fluidized-bed combustion, integrated gasification combined cycle, limestone injection multi-stage burner, enhanced flue gas desulfurization (or *scrubbing*), coal liquefaction and coal gasification.

Clean fuels: Fuels such as e-10 (unleaded) that burn cleaner and produce fewer harmful emissions compared to ordinary gasoline.

Closed-loop biomass: Crops grown, in a sustainable manner, for the purpose of optimizing their value for bioenergy and bio-product uses. This includes annual crops such as maize and wheat, and perennial crops such as trees, shrubs, and grasses such as switch grass.

Cloud point: The temperature at which the first wax crystals appear and a standardized ASTM test protocol is used to determine this temperature.

Coal: A readily combustible black or brownish-black organic rock whose composition, including inherent moisture, consists of more than 50% w/w and more than 70% v/v of carbonaceous material. It is formed from plant remains that have been compacted, hardened, chemically altered, and metamorphosed by heat and pressure over geologic time.

Coal ash: A collective term referring to any solid materials or residues (such as fly ash, bottom ash, or boiler slag) produced primarily from the combustion of coal. Current usage of the coal ash collective term is synonymous with the term coal combustion ash and coal combustion residue. Also, coal ash is a component of the term coal utilization byproduct (cub) covering only the materials or residues associated with the combustion of coal.

Coal bed methane (coalbed methane): Methane adsorbed to the surface of coal; often considered to be a part of the coal seam.

Coal gas: The mixture of volatile products (mainly hydrogen, methane, carbon monoxide, and nitrogen) remaining after removal of water and tar, obtained from carbonization of coal, having a heat content of 400-600 Btu/ft^3.

Coal gasification: Production of synthetic gas from coal.

Coal liquefaction: Conversion of coal to a liquid.

Coal rank: Indicates the degree of coalification that has occurred for a particular coal. Coal is formed by the decomposition of plant matter without free access to air and under the influence of moisture, pressure, and temperature. Over the course of the geologic process that forms coal—coalification—the chemical composition of the coal gradually changes to compounds of lower hydrogen content and higher carbon content in aromatic ring structures. As the degree of coalification increases, the percentage of volatile matter decreases and the calorific value increases. The common ranks of coal are anthracite, bituminous, subbituminous, and brown coal/lignite.

Coal tar: The condensable distillate containing light, middle, and heavy oils obtained by carbonization of coal. About 8 gal of tar is obtained from each ton of bituminous coal.

Coarse materials: Wood residues suitable for chipping, such as slabs, edgings, and trimmings.

Cogeneration: The sequential production of electricity and useful thermal energy from a common fuel source.

Cogenerator: A generating facility that produces electricity and another form of useful thermal energy (such as heat or steam) used for industrial, commercial, heating, and cooling purposes. To receive status as a qualifying facility (QF) under the Public Utility Regulatory Policies Act (PURPA), the facility must produce electric energy and *another form of useful thermal energy through the sequential use of energy*, and meet certain ownership, operating, and efficiency criteria established by the Federal Energy Regulatory Commission (FERC).

Coke: A gray, hard, porous, and coherent cellular-structured combustible solid, primarily composed of amorphous carbon; produced by destructive distillation or thermal decomposition of certain carbonaceous feedstocks that passes through a plastic state in the absence of air.

Coke-oven gas: A medium-Btu gas, typically 550 Btu/ft^3, produced as a byproduct in the manufacture of coke by heating coal at moderate temperatures.

Coking: A thermal method used in refineries for the conversion of bitumen and residua to volatile products and coke (see delayed coking and fluid coking).

Combustion (burning): The transformation of biomass fuel into heat, chemicals, and gases through chemical combination of hydrogen and carbon in the fuel with oxygen in the air.

Combustion gases: The gases released from a combustion process.

Conditioned ash: Ash that has been moistened with water during the load out process at the temporary storage silo at the power plant to allow for its handling, transport, and placement without causing fugitive dusting.

Conservation: Efficiency of energy use, production, transmission, or distribution that results in a decrease of energy consumption while providing the same level of service.

Continuous flow process: A general term for any number of biodiesel production processes that involve the continuous addition of ingredients to produce biodiesel on a continual, round-the-clock basis, as opposed to the batch process.

Conventional biofuels: Biofuels such as bioethanol and biodiesel, which are typically made from corn, sugarcane and beet, wheat or oilseed crops such as soy and rapeseed oil.

Conventional crude oil (conventional crude oil): Crude oil that is pumped from the ground and recovered using the energy inherent in the reservoir; also recoverable by application of secondary recovery techniques.

Conversion efficiency: A comparison of the useful energy output to the potential energy contained in the fuel; the efficiency calculation relates to the form of energy produced and allows a direct comparison of the efficiency of different conversion processes can be made only when the processes produce the same form of energy output.

Cord: A stack of wood comprising 128 cubic feet (3.62 m^3); standard dimensions are $4 \times 4 \times 8$ feet, including air space and bark. One cord contains approx. 1.2 US Tons (oven-dry) = 2400 pounds = 1089 kg.

Corn stover: Residue materials from harvesting corn consisting of the cob, leaves and stalk.

Cracking: A refining process that uses heat and/or a catalyst to break down high molecular weight chemical components into lower molecular weight products which can be used as blending components for fuels.

Cropland: Total cropland includes five components: cropland harvested, crop failure, cultivated summer fallow, cropland used only for pasture, and idle cropland.

Crude oil: Also called petroleum; a hydrocarbon-based substance comprising a complex blend of hydrocarbons derived from crude oil through the process of separation, conversion, upgrading, and finishing, including motor fuel, jet oil, lubricants, crude oil solvents, and used oil.

Cyclone: A cone-shaped air-cleaning apparatus which operates by centrifugal separation that is used in particle collecting and fine grinding operations.

Cyclone collectors: Equipment in which centrifugal force is used to separate particulates from a gas stream.

Dehydration reaction (condensation reaction): A chemical reaction in which two organic molecules become linked to each other via covalent bonds with the removal of a molecule of water; common in synthesis reactions of organic chemicals.

Dehydrohalogenation: Removal of hydrogen and halide ions from an alkane resulting in the formation of an alkene.

Delayed coking: A coking process in which the thermal reactions are allowed to proceed to completion to produce gaseous, liquid, and solid (coke) products.

Density: The mass (or weight) of a unit volume of any substance at a specified temperature (see also "specific gravity").

Descending-bed system: Gravity down-flow of packed solids contacted with upwardly flowing gases – sometimes referred to as *fixed-bed* or *moving-bed* system.

Desorption: The release of ions or molecules from a solid adsorbent into solution.

Desulfurization: The removal of sulfur or sulfur compounds from a feedstock.

Devolatilization: The removal of vaporizable material by the action of heat.

Dewatering: The removal of water from a feedstock by mechanical equipment such as a vibrating screen, filter, or centrifuge.

Diesel #1 and diesel #2: Diesel #1 is also called kerosene and is not generally used as a fuel oil in diesel vehicles – it has a lower viscosity (it is thinner) than diesel #2, which is the typical diesel vehicle fuel. Biodiesel replaces diesel #2 or a percentage.

Diesel engine: Named for the German engineer Rudolph Diesel; this internal-combustion, compression-ignition engine operates by heating fuels through compression which and causes the fuel to ignite. It can use either crude oil or bio-derived fuel.

Diesel fuel: A distillate of fuel oil that has been historically derived from crude oil for use in internal combustion engines; can also be derived from plant and animal sources.

Diesel, Rudolph: German inventor famed for fashioning the diesel engine, which made its debut at the 1900 world's fair; he initially intended the engine to run on vegetable-derived fuels, with the hope that farmers would be able to grow their own fuel sources.

Diffusion: Blending of a gas and air, resulting in a homogeneous mixture; blending of two or more gases.

Direct-injection engine: A diesel engine in which fuel is injected directly into the cylinder; most of the newer models are *turbo direct injection*.

Distillate: Any crude oil product produced by boiling crude oil and collecting the vapors produced as a condensate in a separate vessel, for example gasoline (light distillate), gas oil (middle distillate), or fuel oil (heavy distillate).

Distillate oil: Any distilled product of crude oil; a volatile crude oil product used for home heating and most machinery.

Distillation: The primary distillation process which uses high temperature to separate crude oil into vapor and fluids which can then be fed into a distillation or fractionating tower.

Distillation: The process to separate the components of a liquid mixture by boiling the liquid and then condensing the resulting vapor.

Downdraft gasifier: A gasifier in which the product gases pass through a combustion zone at the bottom of the gasifier.

Dry ash: The ash has not melted because the operating temperature of the reactor is lower than the initial deformation temperature of the ash.

Dry, ash-free (daf) basis: An analysis basis calculated as if moisture and ash were removed.

Dry fly ash: Fly ash that has been collected by particulate removal equipment such as electrostatic precipitators, bag houses, mechanical collectors, or fabric filters.

Drying: The removal of water by thermal drying, screening, or centrifuging.

Dry ton: 2,000 pounds of material dried to a constant weight.

Ebullating-bed reactor: A system similar to a fluidized bed but operated at higher gas velocities, such that a portion of the solids is carried out with the up-flowing gas.

E diesel: A blend of ethanol and diesel fuel plus other additives designed to reduce air pollution from heavy equipment, city buses and other vehicles that operate on diesel engines.

Effluent: The liquid or gas discharged from a process or chemical reactor, usually containing residues from that process.

Electrostatic precipitation: Separation of liquid or solid particles from a gas stream by the action of electrically charged wires and plates.

Electrostatic precipitator (ESP): Collection of fly ash requires the application of an electrostatic charge to the fly ash, which then is collected on grouped plates in a series of hoppers. Fly ash collected in different hoppers may have differing particle size and chemical composition, depending on the distance of the hopper from the combustor. The ESP ash may also be collected as a composite.

Elemental analysis: The determination of carbon, hydrogen, nitrogen, oxygen, sulfur, chlorine, and ash in a sample.

Emissions: Substances discharged into the air during combustion, e.g., all that stuff that comes out of your car.

Endothermic reaction: A process in which heat is absorbed.

Energy balance: The difference between the energy produced by a fuel and the energy required to obtain it through agricultural processes, drilling, refining, and transportation.

Energy-efficiency ratio: A number representing the energy stored in a fuel as compared to the energy required to produce, process, transport, and distribute that fuel.

Entrained flow system: Solids suspended in a moving gas stream and flowing with it.

Environment: The external conditions that affect organisms and influence their development and survival.

Environmental assessment (EA): A public document that analyzes a proposed federal action for the possibility of significant environmental impacts – if the environmental impacts will be significant, the federal agency must then prepare an environmental impact statement.

Environmental impact statement (EIS): A statement of the environmental effects of a proposed action and of alternative actions. Section 102 of the National Environmental Policy Act requires an EIS for all major federal actions.

Enzymatic hydrolysis: A process by which enzymes (biological catalysts) are used to break down starch or cellulose into sugar.

Enzyme: A protein or protein-based molecule that speeds up chemical reactions occurring in living things; enzymes act as catalysts for a single reaction, converting a specific set of reactants into specific products.

Esters: Any of a large group of organic compounds formed when an acid and alcohol is mixed; methyl acetate (CH_3COOCH_3) is the simplest ester; biodiesel contains methyl stearate.

ETBE: See ethyl tertiary butyl ether.

Ethanol (ethyl alcohol, alcohol, or grain-spirit): A clear, colorless, flammable oxygenated hydrocarbon; used as a vehicle fuel by itself (E100 is 100% ethanol by volume), blended with gasoline (E85 is 85% ethanol by volume), or as a gasoline octane enhancer and oxygenate (10% by volume).

Ethers: Liquid fuel made from a blending an alcohol with isobutylene.

Ethyl tertiary butyl ether (ethyl t-butyl ether): Ether created from ethanol that can increase octane and reduce the volatility of gasoline, decreasing evaporation and smog formation.

Evaporation: The conversion of a liquid to the vapor state by the addition of latent heat or vaporization.

Exothermic reaction: A process in which heat is evolved.

Extractives: Any number of different compounds in biomass that are not an integral part of the cellular structure – the compounds can be extracted from wood by means of polar and non-polar solvents including hot or cold water, ether, benzene, methanol, or other solvents that do not degrade the biomass structure and the types of extractives found in biomass samples are entirely dependent upon the sample itself.

Fast pyrolysis: Thermal conversion of biomass by rapid heating to between 450 to 600°C (840 to 1110°F) in the absence of oxygen.

Fatty acid: A carboxylic acid (an acid with a -COOH group) with long hydrocarbon side chains; feedstocks are first converted to fatty acids and then to biodiesel by transesterification.

Feedstock: Raw material used in an industrial process; biomass used in the creation of a biofuel (e.g., corn or sugarcane for ethanol, soybeans or rapeseed for biodiesel).

Fermentation: Conversion of carbon-containing compounds by micro-organisms for production of fuels and chemicals such as alcohols, acids or energy-rich gases.

Fiber products: Products derived from fibers of herbaceous and woody plant materials; examples include pulp, composition board products, and wood chips for export.

Fischer-Tropsch process: Process for producing liquid fuels, usually diesel fuel, from natural gas or synthetic gas from gasified coal or biomass.

Fixed-bed system: See descending-bed system.

Fixed carbon: The carbonaceous residue remaining after heating in a prescribed manner to decompose thermally unstable components and to distill volatiles; part of the proximate analysis group.

Flash point: The temperature at which the vapor over a liquid will ignite when exposed to an ignition source. A liquid is considered to be flammable if its *flash point* is less than 60°C (140°F). *Flash point* is an extremely important factor in relation to the safety of spill cleanup operations. Gasoline and other light fuels can ignite under most ambient conditions and therefore are a serious hazard when spilled. Many freshly spilled crude oils also have low *flash points* until the lighter components have evaporated or dispersed.

Flash pyrolysis (fast pyrolysis): A process in which biomass is converted at approximately 500°C (930°F) into a liquid oil/char slurry which can be atomized in an entrained-flow gasifier. Another attractive aspect is that the high-energy density slurry can be transported to a central processing facility. See slow pyrolysis.

Flue gas desulfurization (FGD): Is removal of the sulfur gases from the flue gases – typically using a high-calcium sorbent such as lime or limestone. The three primary types of FGD processes commonly used by utilities are wet scrubbers, dry scrubbers, and sorbent injection.

Flue gas recirculation: A procedure in which part of the flue gas is recirculated to the furnace; can be used to modify conditions in the combustion zone (lowering the temperature and reducing the oxygen concentration) to reduce NOx formation; another use for flue gas recirculation is to use the flue gas as a carrier to inject fuel into a reburn zone to increase penetration and mixing.

Fluid coking: A continuous fluidised solids process that cracks feed thermally over heated coke particles in a reactor vessel to gas, liquid products, and coke.

Fluidized-bed boiler: A large, refractory-lined vessel with an air distribution member or plate in the bottom, a hot gas outlet in or near the top, and some provisions for introducing fuel; the fluidized bed is formed by blowing air up through a layer of inert particles (such as sand or limestone) at a rate that causes the particles to go into suspension and continuous motion; the super-hot bed material increased combustion efficiency by its direct contact with the fuel.

Fluidization: See fluidized-bed system.

Fluidized-bed system: Solids suspended in space by an upwardly moving gas stream.

Fluid temperature (ash fluid temperature): The temperature at which ash becomes fluid and flows in streams.

Fly ash: Ash that exits a combustion chamber in the flue gas; non-burnable ash that are carried into the atmosphere by stack gases; ash that exits a combustion chamber in the flue gas and is captured by air pollution control equipment such as electrostatic precipitators, baghouses, and wet scrubbers.

Forest residues: Material not harvested or removed from logging sites in commercial hardwood and softwood stands as well as material resulting from forest management operations such as pre-commercial thinnings and removal of dead and dying trees.

Fossil fuel: Solid, liquid, or gaseous fuels formed in the ground after millions of years by chemical and physical changes in plant and animal residues under high temperature and pressure. Oil, natural gas, and coal are fossil fuels.

Fouling: The accumulation of small, sticky molten particles of ash on a reactor surface.

Free moisture (surface moisture): The part of the moisture that is removed by air-drying under standard conditions approximating atmospheric equilibrium.

Fuel cell: A device that converts the energy of a fuel directly to electricity and heat, without combustion.

Fuel oil: A high-boiling crude oil fraction, typically black in color and used to generate power or heat by burning in furnaces.

Fuel wood: Wood used for conversion to some form of energy, primarily for residential use.

Gas engine: A piston engine that uses natural gas rather than gasoline – fuel and air are mixed before they enter cylinders; ignition occurs with a spark.

Gaseous emissions: Substances discharged into the air during combustion, typically including carbon dioxide, carbon monoxide, water vapor, and hydrocarbons.

Gasification: A chemical or heat process used to convert carbonaceous material (such as coal, crude oil, and biomass) into gaseous components such as carbon monoxide and hydrogen.

Gasifier: A device for converting solid fuel into gaseous fuel; in biomass systems, the process is referred to as pyrolytic distillation.

Gasohol: A mixture of 10% anhydrous ethanol and 90% gasoline by volume; 7.5% anhydrous ethanol and 92.5% gasoline by volume; or 5.5% anhydrous ethanol and 94.5% gasoline by volume.

Gasoline: A volatile, flammable liquid obtained from crude oil that has a boiling range of approximately 30 to 220°C 86 to 428°F) and is used for fuel for spark-ignition internal combustion engines.

Gas to liquids (GTL): The process of refining natural gas and other hydrocarbons into longer-chain hydrocarbons, which can be used to convert gaseous waste products into fuels.

Gas purification: Gas treatment to remove contaminants such as fly ash, tars, oils, ammonia, and hydrogen sulfide.

Gas shift process: A process in which carbon monoxide and hydrogen react in the presence of a catalyst to form methane and water.

Gas turbine (combustion turbine): A turbine that converts the energy of hot compressed gases (produced by burning fuel in compressed air) into mechanical power – often fired by natural gas or fuel oil.

Global climate change: The gradual warming of the Earth caused by the greenhouse effect; believed to be the result of man-made emissions of greenhouse gases such as carbon dioxide, chlorofluorocarbons (cfc) and methane, although there is no agreement among the scientific community on this controversial issue.

Greenhouse effect: The effect of certain gases in the Earth's atmosphere in trapping heat from the sun.

Greenhouse gases: Gases that trap the heat of the sun in the Earth's atmosphere, producing the greenhouse effect. The two major greenhouse gases are water vapor and carbon dioxide. Other greenhouse gases include methane, ozone, chlorofluorocarbons, and nitrous oxide.

Gross heating value (GHV): The maximum potential energy in the fuel as received, considering moisture content (mc).

GTL (GtL, gas to liquid): A refinery process which converts natural gas into longer-chain hydrocarbons; gas can be converted to liquid fuels via a direct conversion or using a process such as the Fischer-Tropsch process.

Hard coal: Coal with heat content greater than 10,260 Btu/lb. On a moist ash-free basis. Includes anthracite, bituminous, and the higher-rank subbituminous coals.

Hardwood: One of the botanical groups of dicotyledonous trees that have broad leaves in contrast to the conifers or softwoods – the term has no reference to the actual hardness of the wood – the botanical name for hardwoods is angiosperms; short-rotation, fast-growing hardwood trees are being developed as future energy crops which are uniquely developed for harvest from 5 to 8 years after planting. Examples include: hybrid poplars (*populus sp.*), hybrid willows (*salix sp.*), silver maple (*acer saccharinum*), and black locust (*robinia pseudoacacia*).

Heat value: The amount of heat obtainable from a feedstock expressed in British thermal units per pound, joules per kilogram, kilojoules or kilocalories per kilogram, or calories per gram: to convert Btu/lb to kcal/kg, divide by 1.8. To convert kcal/kg to Btu/lb, multiply by 1.8.

Heating value: The maximum amount of energy that is available from burning a substance; see gross heating value.

Heavy oil (heavy crude oil): Crude oil that is more viscous that conventional crude oil, has a lower mobility in the reservoir but can be recovered through a well from the reservoir by the application of a secondary or enhanced recovery methods.

Heavy oil product: A high-boiling tar fraction with distillation range usually 250 to 300°C (480 to 570°F), containing polynuclear aromatic compounds and tar bases.

Hemicellulose: Hemicellulose consists of short, highly branched chains of sugars in contrast to cellulose, which is a polymer of only glucose, hemicellulose is a polymer of five different sugars; contains five-carbon sugars (usually d-xylose and l-arabinose) and six-carbon sugars (d-galactose, d-glucose, and d-mannose) and uronic acid which are highly substituted with acetic acid; the branched nature of hemicellulose renders it amorphous and relatively easy to hydrolyze to its constituent sugars compared to cellulose; when hydrolyzed, the hemicellulose from hardwoods releases products high in xylose (a five-carbon sugar); hemicellulose contained in softwoods, by contrast, yields more six-carbon sugars.

Heteroatom compounds: Chemical compounds that contain nitrogen and/or oxygen and/or sulfur and/or metals bound within their molecular structure(s).

Hexose: Any of various simple sugars that have six carbon atoms per molecule (e.g., Glucose, mannose, and galactose).

High temperature tar: The high-boiling distillate from the pyrolysis of a carbonaceous feedstock at a temperature of about 800C (1470F).

High-volatile bituminous coal: Three related rank groups of bituminous coal as defined by the American Society for Testing and Materials which collectively contain less than 69% fixed carbon on a dry, mineral-matter-free basis; more than 31% volatile matter on a dry, mineral-matter-free basis; and a heat value of more than 10,500 Btu per pound on a moist, mineral-matter-free basis.

Higher heating value (HHV): The potential combustion energy when water vapor from combustion is condensed to recover the latent heat of vaporization; lower heating value (LHV) is the potential combustion energy when water vapor from combustion is not condensed.

Hydrocarbon: A chemical compound that contains a carbon backbone with hydrogen atoms attached to that backbone; an organic compound consisting of carbon and hydrogen only.

Hydrocarbon compounds: Chemical compounds containing only carbon and hydrogen.

Hydrocarbonaceous material: A material such as tar sand bitumen that is composed of carbon and hydrogen with other elements (heteroelements) such as nitrogen, oxygen, sulfur, and metals chemically combined within the structures of the constituents; even though carbon and hydrogen may be the predominant elements, there may be very few true hydrocarbons (q.v.).

Hydrocyclone: Hydraulic device for separating suspended solid particles from liquids by centrifugal action. Cyclone action splits the inlet flow, a small part of which exits via the lower cone, the remainder overflowing the top of the cylindrical section. Particles are separated according to their densities, so that the denser particles exit via the cone underflow and less dense particles exit with the overflow.

Hydrodesulfurization: The removal of sulfur by hydrotreating.

Hydrogasification: Gasification of a carbonaceous feedstock in the presence of hydrogen to produce methane; can produce high yields of methane.

Hydrogenation: Chemical reaction of a substance with molecular hydrogen, usually in the presence of a catalyst; a common hydrogenation is the hardening of animal fats or vegetable oils to make them solid at room temperature and improve their stability; hydrogen is added (in the presence of a nickel catalyst) to carbon-carbon double bonds in the unsaturated fatty acid portion of the fat or oil molecule.

Hydroprocesses: Refinery processes designed to add hydrogen to various products of refining.

Hydrotreating: The removal of heteroatomic (nitrogen, oxygen, and sulfur) species by treatment of a feedstock or product at relatively low temperatures in the presence of hydrogen.

Integrated Gasification Combined Cycle (IGCC): A power plant that combines the gasification process with an efficient "combined cycle" power generator (consisting of one or more gas turbines and a steam turbine). Clean synthesis gas is combusted in the gas turbines to generate electricity. The excess heat from the gasification reactions is then converted into steam. This is combined with steam produced from the gas turbines, and sent to a steam turbine generator to produce additional electricity.

Incinerator: Any device used to burn solid or liquid residues or wastes as a method of disposal.

Inclined grate: A type of furnace or gasifier in which fuel enters at the top part of a grate in a continuous ribbon, passes over the upper drying section where moisture is removed, and descends into the lower burning section; ash is removed at the lower part of the grate.

Included minerals: Minerals that are part of a feedstock particle and matrix.

Indirect hydrogenation: A feedstock is first gasified to make a synthesis gas. The gas is then passed over a catalyst to produce methanol or paraffinic hydrocarbons.

Indirect liquefaction: The conversion of coal, biomass, or any carbonaceous feedstock to a liquid fuel through a synthesis gas intermediate step.

Joule (J): Metric unit of energy, equivalent to the work done by a force of 1 newton applied over distance of 1 meter (= 1 kg m^2/s^2). 1 J = 0.239 cal (1 cal = 4.187 J).

Kerosene: A middle distillate that in various forms is used as aviation turbine fuel or for burning in heating boilers or as a solvent, such as white spirit.

Kilowatt (kw): A measure of electrical power equal to 1000 w. 1 kw = 3412 Btu/hr = 1.341 horsepower.

Kilowatt hour (kwh): A measure of energy equivalent to the expenditure of 1 kw for 1 hour. For example, 1 kwh will light a 100-w light bulb for 10 hours. 1 kwh = 3412 Btu.

Knock: Engine sound that results from ignition of the compressed fuel-air mixture prior to the optimal moment.

KOH: See potassium hydroxide.

Landfill gas: A type of biogas that is generated by decomposition of organic material at landfill disposal sites. Landfill gas is approximately 50% methane. See also biogas.

Light oil: A low-boiling fraction with distillation range between 80 and 210°C (175 to 410°F).

Lignin: Structural constituent of wood and (to a lesser extent) other plant tissues, which encrust the walls and cements the cells together; energy-rich material contained in biomass that can be used for boiler fuel.

Lignite: A brownish-black woody-structured coal, lower in fixed carbon and higher in volatile matter and oxygen than either anthracite or bituminous coal – similar to the *brown coal* of Europe and Australia; a class of brownish-black, low-rank coal defined by the American Society for Testing and Materials as having less than 8,300 Btu on a moist, mineral-matter-free basis; in the United States, lignite is separated into two groups: lignite a (6,300 to 8,300 Btu) and lignite b (<6,300 Btu).

Lignocellulose: Plant material made up primarily of lignin, cellulose, and hemicellulose.

Lipid: Any of a group of organic compounds, including the fats, oils, waxes, sterols, and triglycerides, that are insoluble in water but soluble in nonpolar organic solvents, are oily to the touch, and together with carbohydrates and proteins constitute the principal structural material of living cells.

Liquefaction: The conversion of feedstock into nearly mineral-free hydrocarbon liquids or low-melting solids by a process of direct or indirect hydrogenation at elevated temperatures and pressures and separation of liquid products from residue by either filtration or distillation or both.

Liquefied petroleum gas (LPG): A mixture of propane and butane.

Long ton: A unit of weight in the US Customary system and in the United Kingdom equal to 2,240 pounds (1.0160469 metric tons; 1.1200 short tons; 1,016.0469 kilograms).

Low-Btu gas: A nitrogen-rich gas with a heat content of 100-200 Btu/ft^3 produced in gasification processes using air as the oxygen source. The air-blown form of producer gas.

Lower heating value (LHV, net heat of combustion): The heat produced by combustion of one unit of a substance, at atmospheric pressure under conditions such that all water in the products remains in the form of vapor; the net heat of combustion is calculated from the gross heat of combustion at 20°C (68°F) by subtracting 572 cal/g (1030 Btu/lb) of water derived from one unit mass of sample, including both the water originally present as moisture and that formed by combustion – this subtracted amount is not equal to the latent heat of vaporization of water because the calculation also reduces the data from the gross value at constant volume to the net value at constant pressure and the appropriate factor for this reduction is 572 cal/g.

LPG: See liquefied petroleum gas.

Lye: See sodium hydroxide.

Magmavication (vitrification): The result of the interaction between plasma and inorganic materials, in presence of a coke bed or coke-like products in the cupola or reactor, a vitrified material is produced that can be used in the manufacture of architectural tiles and construction materials.

Megawatt (mw): A measure of electrical power equal to one million watts (1,000 kw).

Membrane separation: A technology which selectively separates (fractionates) materials by way of pore size and/or minute gaps in the molecular arrangement of a continuous structure. The membrane acts as a selective barrier allowing relatively free passage of one component while retaining another. In membrane contactors the membrane function is to provide an interface between two phases but not to control the rate of passage of permeants across the membrane. Membrane separations are classified by pore size and by the separation driving force and these classifications are (i) microfiltration (MF), (ii) ultrafiltration (UF), (iii) ion-exchange (IE), and (v) reverse osmosis (RO).

Methanation: A process for catalytic conversion of 1 mole of carbon monoxide and 3 moles of hydrogen to 1 mole of methane and 1 mole of water.

Methane: A potentially explosive gas formed naturally from the decay of vegetative matter; the principal component of natural gas.

Methanol: A fuel typically derived from natural gas, but which can be produced from the fermentation of sugars in biomass.

Methoxide (sodium methoxide, sodium methylate, $CH_3O^-\ Na^+$): An organic salt, in pure form a white powder; in biodiesel production, *methoxide* is a product of mixing methanol and sodium hydroxide, yielding a solution of sodium methoxide in methanol, and a significant amount of heat; making sodium methoxide is the most dangerous step when making biodiesel.

Methyl alcohol: See methanol.

Methyl esters: See biodiesel.

Million: 1×10^6

Mineral-matter: The solid inorganic material in a feedstock.

Mineral-matter-free basis: A theoretical analysis calculated from basic analytical data expressed as if the total mineral-matter had been removed.

Moisture: A measure of the amount of water and other components that are volatilized at 105°C (221°F) present in the biomass sample.

Moisture content: The weight of the water contained in wood, usually expressed as a percentage of weight, either oven-dry or as received.

Moisture content, dry basis: Moisture content expressed as a percentage of the weight of oven-wood, that is, [(weight of wet sample − weight of dry sample) / weight of dry sample] × 100.

Moisture content, wet basis: Moisture content expressed as a percentage of the weight of wood as-received, that is, [(weight of wet sample−weight of dry sample) / weight of wet sample] × 100.

Moisture-free basis: Biomass composition and chemical analysis data is typically reported on a moisture free or dry weight basis – moisture (and some volatile matter) is removed prior to analytical testing by heating the sample at 105°C (221°F) to constant weight; by definition, samples dried in this manner are considered moisture free.

Molten bath gasifier: A reaction system in which a carbonaceous feedstock and air or oxygen with steam are contacted underneath a pool of liquid iron, ash, or salt.

Monosaccharide: A simple sugar such as a five-carbon sugar (xylose, arabinose) or six-carbon sugar (glucose, fructose); sucrose, on the other hand is a disaccharide, composed of a combination of two simple sugar units, glucose and fructose.

Moving-bed system: See descending-bed system.

Municipal solid waste (MSW): Residential and commercial materials that are used and then discarded; includes paper, yard waste, food waste, and containers (such as plastic bottles and cans), tires and electronics; may include recyclable materials, depending on the amount of recycling provided by the local government.

Municipal waste: Residential, commercial, and institutional post-consumer waste that contain a significant proportion of plant-derived organic material that constitutes a renewable energy resource; waste paper, cardboard, construction and demolition wood waste, and yard wastes are examples of biomass resources in municipal wastes.

Natural gas: A naturally occurring gas with a heat content over 1000 Btu/ft^3, consisting mainly of methane but also containing smaller amounts of the C_2 to C_4 hydrocarbons as well as nitrogen, carbon dioxide, and hydrogen sulfide.

Nitrogen oxides (NOx): Products of combustion that contribute to the formation of smog and ozone.

Octane number: Measure of a fuel's resistance to self-ignition; the octane number of a fuel is indicated on the pump – the higher the number, the slower the fuel burns; bioethanol typically adds two to three octane numbers when blended with ordinary crude oil – making it a cost-effective octane-enhancer; see knock.

Oil from tar sand: Synthetic crude oil.

Oxygen gasification: A relatively simple process that produces a medium-energy gas composed primarily of carbon monoxide and hydrogen. While quite satisfactory for burning, it can also be used for chemical synthesis to make methanol, ammonia, hydrogen, methane, or naphtha.

Oxygenate: A substance which, when added to gasoline, increases the amount of oxygen in that gasoline blend; includes fuel ethanol, methanol, and methyl tertiary butyl ether (MTBE).

Oxygenated fuels: Ethanol is an oxygenate, meaning that it adds oxygen to the fuel mixture – more oxygen helps the fuel burn more completely thereby reducing the amount of harmful emissions from the tailpipe; a fuel such as ethanol-blended gasoline that contains a high oxygen content is called *oxygenated*.

Partial oxidation: A chemical reaction that occurs in a limited oxygen environment such as in a pressurized vessel with heat, feedstock (such as coal) and limited oxygen creating a synthesis gas consisting primarily of hydrogen and carbon monoxide.

Particulate: A small, discrete mass of solid or liquid matter that remains individually dispersed in gas or liquid emissions.

Particulate emissions: Particles of a solid or liquid suspended in a gas, or the fine particles of carbonaceous soot and other organic molecules discharged into the air during combustion.

Petrodiesel: Crude oil-based diesel fuel, usually referred to simply as diesel.

Petroleum: See crude oil.

pH: A measure of acidity and alkalinity of a solution on a scale with 7 representing neutrality; lower numbers indicate increasing acidity, and higher numbers increasing alkalinity; each unit of change represents a tenfold change in acidity or alkalinity.

Pipeline gas: A methane-rich gas with a heat content of 950 to 1050 Btu/ft^3 compressed to 1000 psi.

Pitch: The non-volatile portion of thermally produced tar.

Plasma: Often called the fourth state of matter (the other three are solid, liquid, and gas). Plasma is created when an electrical charge passes through a gas. The resultant flash of lightning is an example of plasma found in nature.

Plasma gasification: The use of plasma, generally in the form of a plasma "torch" or "arc" to provide the heat energy needed to initiate a gasification reaction. Plasma torches and arcs can reach temperatures of 5,000 to 10,000°F; the most typical use of plasma gasification is to convert various materials, such as municipal solid waste and hazardous waste, into a clean synthesis gas used to produce electricity and other products.

Potassium hydroxide (KOH): Used as a catalyst in the transesterification reaction to produce biodiesel.

Pour point: The lowest temperature at which oil will pour or flow when it is chilled without disturbance under definite conditions.

Pressure-swing adsorption (PSA): A technology that is used to separate some gas species from a mixture of gases under pressure according to the molecular characteristics of the constituents and the relative affinity for an adsorbent. The technology operates at near-ambient temperatures and differs significantly from cryogenic distillation techniques of gas separation. A specific adsorbent (such as a zeolite, activate carbon, or a molecular sieve) is used as a trap, preferentially adsorbing the target gas species at high pressure after which the process then swings to low pressure to desorb the adsorbed material.

Pretreatment: Preprocess treating of a feedstock to eliminate undesirable reaction tendencies.

Process heat: Heat used in an industrial process rather than for space-heating or other housekeeping purposes.

Producer gas: Mainly carbon monoxide with smaller amounts of hydrogen, methane, and variable nitrogen, obtained from partial combustion of a carbonaceous feedstock in air or oxygen, having a heat content of 110-160 Btu/ft^3 (air combustion) or 400-500 Btu/ft^3 (oxygen combustion).

Proximate analysis: The determination, by prescribed methods, of moisture, volatile matter, fixed carbon (by difference), and ash; the term proximate analysis does not include determinations of chemical elements or determinations other than those named and the group of analyses is defined in ASTM D3172.

Pyrolysis: The thermal decomposition of a feedstock at high temperatures (greater than 200°C, 400°F) in the absence of air; in the case of a carbonaceous feedstock, the end product of pyrolysis is a mixture of solids (char), liquids (oxygenated oils), and gases (methane, carbon monoxide, and carbon dioxide) with proportions determined by operating temperature, pressure, oxygen content, and other conditions.

Pyrolysis oil: A bio-oil produced by fast pyrolysis of biomass; typically a dark brown, mobile liquid containing much of the energy content of the original biomass, with a heating value about half that of conventional fuel oil; conversion of raw biomass to pyrolysis oil represents a considerable increase in energy density and it can thus represent a more efficient form in which to transport it.

Quad: One quadrillion Btu (10^{15} Btu) = 1.055 exajoules (eJ), or approximately 172 million barrels of oil equivalent.

Rank: A complex property of coals that is descriptive of their degree of coalification (i.e., the stage of metamorphosis of the original vegetal material in the increasing sequence peat, lignite, subbituminous, bituminous, and anthracite).

Ranks of coal: The classification of coal by degree of hardness, moisture and heat content. *Anthracite* is hard coal, almost pure carbon, used mainly for heating homes. *Bituminous coal* is soft coal. It is the most common coal found in the United States and is used to generate electricity and to make coke for the steel industry. *Subbituminous coal* is a coal with a heating value between bituminous and lignite. It has low fixed carbon and high percentages of volatile matter and moisture. *Lignite* is the softest coal and has the highest moisture content. It is used for generating electricity and for conversion into synthetic gas. In terms of Btu or heat content, anthracite has the highest value, followed by bituminous, subbituminous and lignite.

Rapeseed oil: Food grade oil produced from rape seed is called canola oil; see colza.

Refinery gas: The non-condensable gas obtained during distillation of crude oil or treatment of oil products (such as cracking processes) in refineries; the gas consists mainly of hydrogen, methane, ethane, and olefins derivatives as well as gases which are returned from the petrochemical industry.

Refractory lining: A lining which isolates the reactor and helps to keep the temperature; it can operate at 1600°C (2910°F); the main failure problems of such linings are due to chemical corrosion caused by the silica compounds contained in ash – in addition to sodium compounds, other alkali compounds diffuse into the refractory lining in addition to the chemical attack; physical erosion can be caused by the molten slag flowing down the wall.

Refuse-derived fuel (RDF): Fuel prepared from municipal solid waste; non-combustible materials such as rocks, glass, and metals are removed, and the remaining combustible portion of the solid waste is chopped or shredded.

Residuum (pl. Residua, also known as resid or resids): The non-volatile portion of crude oil that remains as residue after refinery distillation; hence, atmospheric residuum, vacuum residuum.

Saponification: The reaction of an ester with a metallic base and water (i.e., the making of soap); occurs when too much lye is used in biodiesel production.

Saturated steam: Steam at boiling temperature for a given pressure.
Scrubbers: Any of several forms of chemical/physical devices that remove sulfur compounds formed during combustion of a carbonaceous feedstock; technically known as flue gas desulfurization systems, combine the sulfur in gaseous emissions with another chemical medium to form inert sludge, which must then be removed for disposal; see flue gas desulfurization.
Shale: A rock formed by consolidation of clay, mud, or silt, having a laminated structure and composed of minerals essentially unaltered since deposition.
Short ton: A unit of weight equal to 2,000 pounds.
Slag: The nonmetallic product resulting from the interaction of flux and impurities in the smelting and refining of metals; a glass-like byproduct of the gasification process. Slag is carbon or inert material such as ash that was not converted to synthesis gas in the gasifier; inert and non-hazardous and can be used in roadbed construction, roofing materials and other applications.
Slag pile: A significant amount of dirt and rock excavated from the earth below that is dumped into a pile.
Slagging: The accumulation of ash on the wall tubes of a boiler furnace or gasifier, forming a solid layer of ash residue and interfering with heat transfer.
Sodium hydroxide (lye, caustic soda, NaOH): Strongly alkaline and extremely corrosive; mixing with fluids usually causes heat, and can create enough heat to ignite flammables (such as methanol); one of the main reactants for biodiesel production.
Specific energy: The energy per unit of throughput required to reduce feedstock material to a desired product size.
Specific gravity: The ratio of weight per unit volume of a substance to the weight of the same unit volume of water.
Stack gas: The product gas evolved during complete combustion of a fuel.
Steam turbine: A device for converting energy of high-pressure steam (produced in a boiler) into mechanical power which can then be used to generate electricity.
Subbituminous coal: A glossy-black-weathering and non-agglomerating coal that is lower in fixed carbon than bituminous coal, with more volatile matter and oxygen; a rank class of non-agglomerating coals having a heat value content of more than 8,300 Btu and less than 11,500 Btu on a moist, mineral-matter-free basis – this class of coal is divisible on the basis of increasing heat value into the subbituminous c, b, and a coal groups. The heat content of subbituminous coal ranges from 17 to 24 million Btu per ton on a moist, mineral-matter-free basis. The heat content of subbituminous coal consumed in the United States averages 17 to 18 million Btu per ton, on the as-received basis (i.e., containing both inherent moisture and mineral matter).
Substitute natural gas: See synthetic natural gas.
Sulfur (total sulfur): Sulfur found in a feedstock as mineral sulfur, mineral sulfates, and organic compounds. It is undesirable because the sulfur oxides formed when it burns contribute to air pollution, and sulfur compounds contribute to combustion-system corrosion and deposits.
Superheated steam: Steam which is at a higher temperature than boiling temperature for a given pressure.
Sustainable energy: The provision of energy that meets the needs of the present without compromising the ability of future generations to meet their needs; sources include

renewable energy sources and, in the near term because of the wealth of reserves, coal and oil shale.

Suspension: A dispersion of a solid in a gas, liquid, or solid.

Sweetened gas: A gas stream from which acid (sour) gases such as hydrogen sulfide and carbon dioxide have been removed.

Switch grass: Prairie grass native to the United States and known for its hardiness and rapid growth, often cited as a potentially abundant feedstock for ethanol.

Syncrude: Synthetic crude oil produced by pyrolysis or hydrogenation of tar sand bitumen, coal, or oil shale.

Syngas: See synthesis gas.

Synthesis gas: A mixture of carbon monoxide (CO) and hydrogen (H_2) which is the product of high-temperature gasification of organic carbonaceous materials, such as coal, crude oil resid, biomass, and waste; after clean-up to remove any impurities such as tars, synthesis gas can be used to synthesize organic molecules such as synthetic natural gas (SNG, methane (CH_4) or liquid biofuels such as gasoline and diesel fuel via the Fischer-Tropsch process.

Synthetic crude oil (syncrude): A hydrocarbon product produced by the conversion of coal, oil shale, or tar sand bitumen that resembles conventional crude oil; can be refined in a crude oil refinery.

Synthetic natural gas (substitute natural gas): Pipeline-quality gas that is interchangeable with natural gas (mainly methane).

Tail gas: Residual gas leaving a process; gas produced in a refinery and not usually required for further processing.

Tar sand (bituminous sand): A formation in which the bituminous material (bitumen) is found as a filling in veins and fissures in fractured rocks or impregnating relatively shallow sand, sandstone, and limestone strata; a sandstone reservoir that is impregnated with a heavy, extremely viscous, black hydrocarbonaceous, crude oil-like material that cannot be retrieved through a well by conventional or enhanced oil recovery techniques; (FE 76-4): the several rock types that contain an extremely viscous hydrocarbon which is not recoverable in its natural state by conventional oil well production methods including currently used enhanced recovery techniques.

Thermal conversion: A process that uses heat and pressure to break apart the molecular structure of organic solids.

Thermochemical conversion: Use of heat to chemically change substances from one state to another, e.g., to make useful energy products.

Ton: A short or net ton is equal to 2,000 pounds; a long or British ton is 2,240 pounds; a metric ton is approximately 2,205 pounds.

Tonne (imperial ton, long ton, shipping ton): 2,240 pounds; equivalent to 1,000 kilograms or in crude oil terms about 7.5 barrels of oil.

Torrefaction: A thermal method for the conversion of biomass operating in the low temperature range of 200 to 300°C (390 to 570°F); the process is carried out under atmospheric conditions in the absence of oxygen and is used as a pretreatment step for biomass conversion techniques such as gasification and cofiring.

Town gas: A gaseous mixture of coal gas and carbureted water gas manufactured from coal with a heat content of 600 Btu/ft^3.

Trace element: Any element present in minute quantities, such as lead and mercury.

Traveling grate: A type of furnace or gasifier in which assembled links of grates are joined together in a perpetual belt arrangement; fuel is fed in at one end and ash is discharged at the other.

Trillion: 1×10^{12}

Tumbling-bed gasifier: An apparatus in which a carbonaceous feedstock is lifted vertically in a revolving cylinder and dropped through an axially flowing stream of oxygen and steam.

Turbine: A machine for converting the heat energy in steam or high-temperature gas into mechanical energy. In a turbine, a high-velocity flow of steam or gas passes through successive rows of radial blades fastened to a central shaft.

Two-stage gasification: Partial gasification or pyrolysis in a first step followed by complete gasification of the resultant char in a second step.

Ultimate analysis: The determination of the elemental composition of the organic portion of carbonaceous materials, as well as the total ash and moisture; a description of a fuel's elemental composition as a percentage of the dry fuel weight as determined by prescribed methods.

Ultra-low sulfur diesel (ULSD): Ultra-low sulfur diesel describes a new EPA standard for the sulfur content in diesel fuel sold in the united states beginning in 2006 – the allowable sulfur content (15 ppm) is much lower than the previous US standard (500 ppm), which not only reduces emissions of sulfur compounds (blamed for acid rain), but also allows advanced emission control systems to be fitted that would otherwise be poisoned by these compounds.

Vacuum distillation: A secondary distillation process which uses a partial vacuum to lower the boiling point of residues from primary distillation and extract further blending components.

Viscosity: A measure of the ability of a liquid to flow or a measure of its resistance to flow; the force required to move a plane surface of area 1 square meter over another parallel plane surface 1 meter away at a rate of 1 meter per second when both surfaces are immersed in the fluid; the higher the viscosity, the slower the liquid flows; methanol and ethanol has a low viscosity while waste vegetable oil has a high viscosity.

Vitrification (magmavication): The result of the interaction between plasma and inorganic materials, in presence of a coke bed or coke-like products in the cupola or reactor, a vitrified material is produced that can be used in the manufacture of architectural tiles and construction materials.

VOCs: Volatile organic compounds – the name given to low-boiling organic hydrocarbons which escape as vapor from fuel tanks or other sources, and during the filling of tanks; VOCs contribute to smog; see volatile organic compounds.

Volatile matter: Hydrogen, carbon monoxide, methane, tar, other hydrocarbons, carbon dioxide, and water obtained on pyrolysis of a carbonaceous feedstock; those products, exclusive of moisture, given off as gas and vapor, determined by definite prescribed methods (ASTM D2361, ASTM D 3761, ASTM D 3175, ASTM D 3176, ASTM D3178, and ASTM D3179).

Volatility: Propensity of a fuel to evaporate.

Vortex flow: The whirling motion of a gas stream in a round vessel that causes separation by downward flow of solid or liquid particulates contained in the gas.

Waste streams: Unused solid or liquid byproducts of a process.

Water-cooled vibrating grate: A boiler grate made up of a tuyere grate surface mounted on a grid of water tubes interconnected with the boiler circulation system for positive cooling; the structure is supported by flexing plates allowing the grid and grate to move in a vibrating action; ash is automatically discharged.

Water gas (carbureted blue gas): A mixture of carbon monoxide and hydrogen formed by the action of air and then steam on hot coal or coke and enriched with hydrocarbon gases from the pyrolysis of oils.

Watt: The common base unit of power in the metric system; 1 w = 1 J/s, or the power developed in a circuit by a current of 1 amp flowing through a potential difference of 1 v (1 w = 3.412 Btu/h).

Wet scrubber: A device that is used to remove pollutants (contaminants) from a gas stream. In a wet scrubber, the polluted gas stream is brought into contact with the scrubbing liquid, by spraying it with the liquid, by forcing it through a pool of liquid, or by some other contact method, so as to remove the pollutants (contaminants).

Wood: A solid lignocellulosic material naturally produced in trees and some shrubs, made of up to 40 to 50% cellulose, 20 to 30% hemicellulose, and 20 to 30% lignin.

Wood alcohol: See methanol.

Xylan: A polymer of xylose with a repeating unit of $C_5H_8O_4$, found in the hemicellulose fraction of biomass – can be hydrolyzed to xylose.

Xylenes: The term that refers to all three types of xylene isomers (meta-xylene, ortho-xylene, and para-xylene); produced from crude oil; used as a solvent and in the printing, rubber, and leather industries as well as a cleaning agent and a thinner for paint and varnishes; see BTEX, BTX.

Xylose: A five-carbon sugar $C_6H_{10}O_5$; a product of hydrolysis of xylan found in the hemicellulose fraction of biomass.

About the Author

Dr. James G. Speight has doctorate degrees in Chemistry, Geological Sciences, and Petroleum Engineering. He is the author of more than 85 books in petroleum science, fossil fuel science, petroleum engineering, environmental sciences, and ethics.

Dr. Speight has more than fifty years of experience in areas associated with (i) the properties, recovery, and refining of reservoir fluids, conventional petroleum, heavy oil, extra heavy oil, tar sand bitumen, and oil shale, (ii) the properties and refining of natural gas, gaseous fuels, (iii) the production and properties of chemicals from crude oil, coal, and other sources, (iv) the properties and refining of biomass, biofuels, biogas, and the generation of bioenergy, and (v) the environmental and toxicological effects of energy production and fuels use. His work has also focused on safety issues, environmental effects, environmental remediation, and safety issues as well as reactors associated with the production and use of fuels and biofuels.

Although he has always worked in private industry with emphasis on contract-based work leading to commercialization of concepts, Dr. Speight was (among other appointments) a Visiting Professor in the College of Science, University of Mosul (Iraq) and has also been a Visiting Professor in Chemical Engineering at the Technical University of Denmark (Lyngby, Denmark) and the University of Trinidad and Tobago (Point Lisas, Trinidad).

In 1996, Dr. Speight was elected to the Russian Academy of Sciences and awarded the Academy's Gold Medal of Honor that same year for outstanding contributions to the field of petroleum sciences. In 2001, he received the Scientists without Borders Medal of Honor of the Russian Academy of Sciences and was also awarded the Einstein Medal for outstanding contributions and service in the field of Geological Sciences. In 2005, the Academy awarded Dr. Speight the Gold Medal - Scientists without Frontiers, Russian Academy of Sciences, in recognition of Continuous Encouragement of Scientists to Work Together Across International Borders. In 2007, Dr. Speight received the Methanex Distinguished Professor award at the University of Trinidad and Tobago in recognition of excellence in research. In 2018, he received the American Excellence Award for Excellence in Client Solutions from the United States Institute of Trade and Commerce, Washington, D.C.

Index

Absorption, 275, 276, 280, 292
Acid gas removal, 278
   absorption, 280
   adsorption, 276, 279
   chemisorption, 281
Adsorption, 276, 279
Alkalis, 233
Ammonia
   production, 424
   removal, 298
Autothermal reforming, 249
Biofiltration, 300
Bio-oxidation, 303
Bioscrubbing, 302
Carbon dioxide gasification, 77, 185
Catalytic gasification, 80, 134
Catalytic partial oxidation, 163
Char gasification, 131
Chemicals production, 42, 140, 311, 399, 410, 428
Claus process, 304
Coal, 16, 125
   types and properties, 128
Coal devolatilization, 131
Coal tar chemicals, 140
Combined reforming, 249
Condensable hydrocarbon removal, 289
   absorption, 292
   enrichment, 293
   extraction, 291
   fractionation, 292
Crude oil family, 10
Crude oil residua, 153
Cryogenics process, 278
Devolatilization, 118
Dimethyl ether production, 368, 429
Dry reforming, 250
Energy independence, 31
Energy sources, 1, 8, 15

Energy supply, 3, 28
   economic factors, 28
   geopolitical factors, 29
   physical factors, 29
   technological factors, 30
Enrichment process, 293
Entrained-bed gasifier, 108
Environmental benefits
   air emissions, 450
   carbon dioxide, 450
   solids generation, 450
   water use, 450
Extraction process, 291
Extra heavy crude oil, 12, 155
Feedstock properties, 224
Feedstocks, 131, 157, 158, 193, 224, 245
Fischer-Tropsch catalysts, 143
Fischer-Tropsch chemicals, 143
Fischer-Tropsch chemistry, 339, 340
Fischer-Tropsch liquids, 360, 428
Fischer-Tropsch process
   catalysts, 143
   chemicals, 143
   chemistry, 339, 340
   history and development, 313
   liquids, 360, 428
Fischer-Tropsch synthesis, 192, 359
Fixed-bed gasifier, 102
Fluid-bed gasifier, 105
Fractionation process, 292
Fuel production, 224, 311
Fuels from synthesis gas, 311, 358
Gas cleaning, 263
   absorption, 275, 276, 292
   adsorption, 276
   ammonia removal, 298
   biofiltration, 300
   bio-oxidation, 303
   bioscrubbing, 302

condensable hydrocarbons removal, 289
cryogenics, 278
enrichment, 293
extraction, 291
fractionation, 292
nitrogen removal, 296
particulate matter removal, 298
process types, 272
water removal, 274
Gaseous fuels and chemicals, 311, 425
Gaseous products, 201
   high Btu gas, 86, 138, 204
   hydrogen, 139, 140, 377, 426
   low Btu gas, 84, 136, 203
   medium Btu gas, 85, 138, 203
   synthesis gas, 86
Gasification chemistry, 68, 95, 115, 132, 177, 178, 181
Gasification feedstocks, 46, 323
   biomass, 21, 173, 193
   biomass with coal, 194
   biomass with other feedstocks, 198
   biomedical waste, 223
   bio-solids, 222
   char, 131
   coal, 16, 125, 128
   crude oil coke, 157
   effect of heat release, 121
   effect of process parameters, 120
   feedstock devolatilization, 67
   feedstock pretreatment, 66, 225
   heavy feedstocks, 149, 152, 155
   influence of feedstock quality, 63
   mixed feedstocks, 194, 198
   solvent deasphalter bottoms, 158
   waste, 217
Gasification in the refinery, 167, 349
   gasification of residua, 353
   gasification of residua with biomass, 354
   gasification of residua with coal, 354
   gasification of residua with waste, 356
   processes and feedstocks, 350
   synthetic fuel production, 358
Gasification process, 65, 187
   carbon dioxide gasification, 77, 185
   catalytic gasification, 80, 134
   catalytic partial oxidation, 163
   char gasification, 68

   commercial processes, 65, 187, 248
   feedstock devolatilization, 67
   feedstock pretreatment, 66, 225
   gaseous products, 82, 83, 84, 85, 86, 136, 138, 203, 204
   general chemistry, 68, 72, 74, 95, 115, 132, 177, 178, 181, 246
   halogens/acid gases, 231
   liquid products, 87, 205
   physical effects, 80
   primary gasification, 72
   reactions, 68, 72, 95, 115, 132, 177, 178, 181
   secondary gasification, 74
   stage-by-stage chemistry, 72, 74
Gasification products, 130, 201, 228
   high Btu gas, 86, 138, 204
   hydrogen, 139, 140, 377, 426
   low Btu gas, 84, 136, 203
   medium Btu gas, 85, 138, 203
   synthesis gas, 86
   tar, 88, 229
Gasification reactions, 68, 95, 115, 132, 177, 178, 181
Gasifier selection, 113
Gasifier types, 95, 96, 188
   entrained-bed gasifier, 108
   fixed-bed gasifier, 102
   fluid-bed gasifier, 105
   molten salt gasifier, 109
   plasma gasifier, 111, 134
   selection, 113
Gas processing
   see gas cleaning, 263
Gas production, 199
Gas purification and quality, 165
Gas streams
   cleaning, 263
   composition, 270
Halogens/acid gases, 231
High Btu gas, 86, 138, 204
Hydrocracking, 244
Hydrogasification, 78, 133, 186
Hydrogen, 139, 140, 377, 427
Hydrogen-based refinery processes, 242
Hydrogen management, 259, 394
Hydrogen production, 377, 426
   commercial processes, 381, 383

feedstocks, 382
   hydrocarbon gasification, 384
Hypro process, 385
   process chemistry, 388
   refinery gas, 387
Hydrogen purification, 257, 390
   cryogenic separation, 258, 393
   membrane systems, 258, 392
   pressure-swing adsorption, 257, 391
   wet scrubbing, 257, 391
Hydrotreating, 243
Industrial solid waste, 221
Liquid fuels production, 428
Liquid products, 87, 205
Low Btu gas, 84, 136, 203
Medium Btu gas, 85, 138, 203
Methanation, 79, 186
Methanation catalysts, 256
Methane, 139
Methanol production, 367, 428
Methanol-to-gasoline and olefins, 429
Molten salt gasifier, 109
Municipal solid waste, 221
Natural gas, 9
Natural gas hydrates, 9
Nitrogen removal process, 296
Oil shale, 19
Partial oxidation technology, 160, 163
Particulate matter, 231, 298
Particulate matter removal, 298
Petrochemicals, 412, 416, 417, 418, 419
Phillips process, 163
Plasma gasifier, 111, 134
Preprocessing, 66, 225
Process chemistry, 68, 72, 74, 95, 115, 132, 177, 178, 181, 246
Process design, 119, 133, 227
Process optimization, 135, 166
Process options, 119, 133
Process products, 201, 228
   high Btu gas, 86, 138, 204
   hydrogen, 139, 140, 377, 427
   low Btu gas, 84, 136, 203
   medium Btu gas, 85, 138, 203
   synthesis gas, 86
   tar, 88, 229
Processes requiring hydrogen, 242
Product distribution, 144

Product quality, 136
Products, 118, 164, 336
Process products, 84, 85, 86, 88, 138, 201, 203, 204, 228, 229
Refinery gasification, 167, 349
Refinery of the future, 453
Refinery processes requiring hydrogen, 242
Refining Fischer-Tropsch products, 344
Reforming processes, 239
   autothermal, 249
   catalysts, 254
   combined, 249
   dry, 250
   steam-methane reforming, 251
   steam-naphtha reforming, 253
Removal of condensable hydrocarbons, 289
   absorption, 292
   enrichment, 293
   extraction, 291
   fractionation, 292
Sabatier-Senderens process, 366
SCOT process, 305
Sewage sludge, 223
Shell gasification process, 162
Shift conversion catalysts, 256
Slag, 233
Solid products, 205
Solid waste, 25, 220, 221
Steam-methane reforming, 251
Steam-naphtha reforming, 253
Synthesis gas, 86, 201, 228, 320
Synthesis gas cleaning, 270
Synthesis gas in the refinery, 349
   gasification of residua, 353
   gasification of residua with biomass, 354
   gasification of residua with coal, 354
   gasification of residua with waste, 356
   processes and feedstocks, 350
   synthetic fuel production, 311, 358
Synthesis gas production, 41, 44, 159, 323
   catalysts, 33
   feedstocks, 323
   biomass, 52
   biomass and coal, 62
   black liquor, 59
   coal, 50
   crude oil resid, 47
   extra heavy crude oil, 47

heavy crude oil, 47
mixed feedstocks, 61
natural gas, 46
refinery coke, 50
solid waste, 56, 62
tar sand bitumen, 47
process parameters, 327
product distribution, 326
product quality, 337
products, 336
high Btu gas, 86, 138, 204
hydrogen, 139, 140, 377, 427
low Btu gas, 84, 136, 203
medium Btu gas, 85, 138, 203
synthesis gas, 86
reactors, 330
entrained-bed gasifier, 108
fixed-bed gasifier, 102
fluid-bed gasifier, 105
molten salt gasifier, 109
plasma gasifier, 111, 134
Synthetic natural gas production, 427
Synthol reactor, 363
Tail gas, 165, 259, 260
Tail gas cleaning, 303, 304, 305
Tar, 88, 229

Tar removal
   physical methods, 294
   thermal methods, 296
Tar sand bitumen, 12, 155
Technology integration, 439
   applications and products, 440
   biomass gasification, 447
   environmental benefits, 449
   hydrogen for crude oil refining, 442
   power generation, 445
   production of chemicals and fertilizers, 441
   substitute natural gas, 441
   transportation fuels, 443, 445
   waste-to-energy gasification, 446
Texaco process, 162
Upgrading Fischer-Tropsch products, 344, 362
   diesel production, 365
   gasoline production, 363
Waste types, 217, 219
   biomedical waste, 223
   industrial solid waste, 221
   municipal solid waste, 221
   solid waste, 25, 220, 221
Water gas shift reaction, 76, 184
Water removal process, 274

Also of Interest

Check out these other related titles from Scrivener Publishing

From the Same Author

Handbook of Gasification Technology: Science, Processes, and Applications, by James G. Speight, ISBN 9781118773536. Written by one of the world's foremost petroleum engineers, this is the most comprehensive and up-to-date handbook on gasification technology, covering every aspect of the subject, including energy sources, equipment, processes, applications, and the science of gasifying all types of feedstocks in an effort to reduce the world's carbon footprint. *COMING IN April 2020*

Global Climate Change Demystified, by James G. Speight, ISBN 9781119653851. Tackling one of the most controversial subjects of our time, one of the world's foremost environmental and petroleum engineers explores the potential causes and ramifications of global climate change. *NOW AVAILABLE!*

Formulas and Calculations for Drilling Operations, by James G. Speight, ISBN 9780470625996. Newly revised, this is still the "must have" guide for any drilling, production, or petroleum engineer, with thousands of handy formulas and calculations that the engineer needs on a daily basis. *NOW AVAILABLE!*

Rules of Thumb for Petroleum Engineers, by James G. Speight, ISBN 9781118595268. The most comprehensive and thorough reference work available for petroleum engineers of all levels. *NOW AVAILABLE!*

Ethics in the University, by James G. Speight, ISBN 9781118872130. Examining the potential for unethical behavior by all academic staff, both professionals and non-professionals, this groundbreaking new study uses documented examples to show where the matter could have been halted before it became an ethics issue and how to navigate the maze of today's sometimes confusing ethical academic arena. *NOW AVAILABLE!*

Ethics in Engineering, by James G. Speight and Russell Foote, ISBN 9780470626023. Covers the most thought-provoking ethical questions in engineering. *NOW AVAILABLE!*

Peak Energy: Myth or Reality? by James G. Speight and M. R. Islam, ISBN 9781118549421. This groundbreaking study, written by two of the world's foremost authorities in the energy industry, examines our planet's energy future from the perspective of the "peak oil" philosophy. *NOW AVAILABLE!*

Coal-Fired Power Generation Handbook, by James G. Speight, ISBN 9781118208465. The most complete and up-to-date handbook on power generation from coal, this book covers all of today's new, cleaner methods for creating electricity from coal, the environmental challenges and concerns involved in its production, and developing technologies. *NOW AVAILABLE!*

Bioremediation of Petroleum and Petroleum Products, by James Speight and Karuna Arjoon, ISBN 9780470938492. With petroleum-related spills, explosions, and health issues in the headlines almost every day, the issue of remediation of petroleum and petroleum products is taking on increasing importance, for the survival of our environment, our planet, and our future. This book is the first of its kind to explore this difficult issue from an engineering and scientific point of view and offer solutions and reasonable courses of action. *NOW AVAILABLE!*

An Introduction to Petroleum Technology, Economics, and Politics, by James Speight, ISBN 9781118012994. The perfect primer for anyone wishing to learn about the petroleum industry, for the layperson or the engineer. *NOW AVAILABLE!*

Advances in Natural Gas Engineering Series:

The Three Sisters: Acid Gas Injection, Carbon Capture and Sequestration, and Enhanced Oil Recovery, edited by Ying (Alice) Wu, John J. Carroll, and Yongle Hu, ISBN 9781119510062. This seventh volume in the series, Advances in Natural Gas Engineering, presents the "three sisters," the three hottest and most important topics in the industry today, both independently and how they're interrelated: acid gas injection, carbon capture and sequestration, and enhanced oil recovery. *NOW AVAILABLE!*

Carbon Dioxide Capture and Acid Gas Injection, edited by Ying Wu and John J. Carroll, ISBN 9781118938669. This sixth volume in the series, Advances in Natural Gas Engineering, offers the most in-depth and up-to-date treatment of CO2 capture and acid gas injection, two of the hottest topics in natural gas engineering. *NOW AVAILABLE!*

Acid Gas Extraction for Disposal and Related Topics, edited by Ying Wu, John Carroll, and Weiyao Zhu, ISBN 9781118938614. This fifth volume in the series, Advances in Natural Gas Engineering, offers the most in-depth and up-to-date treatment of acid gas extraction for disposal, an important innovation in natural gas engineering. *NOW AVAILABLE!*

Gas Injection for Disposal and Enhanced Recovery, edited by Ying Wu, John Carroll, and Qi Li, ISBN 9781118938560. This fourth volume in the series, Advances in Natural Gas Engineering, offers the most in-depth and up-to-date treatment of the disposal and enhanced recovery of natural gas. *NOW AVAILABLE!*

Sour Gas and Related Technologies, edited by Ying Wu, John J. Carroll, and Weiyao Zhu, ISBN 9780470948149. Written by a group of the most well-known and knowledgeable authors on the subject in the world, volume three focuses on one of the hottest topics in natural gas today, sour gas. This is a must for any engineer working in natural gas, the energy field, or process engineering. *NOW AVAILABLE!*

Acid Gas Injection and Related Technologies, edited by Ying Wu and John J. Carroll, ISBN 9781118016640. Focusing on the engineering of natural gas and its advancement as an increasingly important energy resource, this volume is a must-have for any engineer working in this field. *NOW AVAILABLE!*

Carbon Dioxide Sequestration and Related Technologies, edited by Ying Wu and John J. Carroll, ISBN 9780470938768. volume two focuses on one of the hottest topics in any field of engineering, carbon dioxide sequestration. *NOW AVAILABLE!*

Printed and bound by CPI Group (UK) Ltd, Croydon, CR0 4YY